# SCIENCE

## A CLOSER LOOK

Mc Graw Hill
Macmillan
McGraw-Hill

# Partnerships

The American Museum of Natural History in New York City is one of the world's preeminent scientific, educational, and cultural institutions, with a global mission to explore and interpret human cultures and the natural world through scientific research, education, and exhibitions. Each year the Museum welcomes around 4 million visitors, including 500,000 schoolchildren in organized field trips. It provides professional development activities for thousands of teachers; hundreds of public programs that serve audiences ranging from preschoolers to seniors; and an array of learning and teaching resources for use in homes, schools, and community-based settings. Visit www.amnh.org for online resources.

As the education arm of the **National Science Foundation**-funded Joint Oceanographic Institutions, JOI Learning brings the excitement of discovery and the scientific process to your classroom through expedition-based science, math, reading, and social studies activities. JOI Learning and Macmillan/McGraw-Hill have formed a partnership to give teachers and students access to JOI's resources through www.macmillanmh.com.

The National Science Digital Library (NSDL) is funded by the **National Science Foundation** as an online library of resources for science, technology, engineering, and mathematics education. NSDL has partnered with Macmillan/McGraw-Hill to provide teaching and learning resources that will help develop teachers' content knowledge in the topic and skill areas addressed at each grade level.

Students with print disabilities may be eligible to obtain an accessible, audio version of the pupil edition of this textbook. Please call Recording for the Blind & Dyslexic at 1-800-221-4792 for complete information.

A

*The McGraw·Hill Companies*

**Macmillan McGraw-Hill**



Send all inquiries to:
Macmillan/McGraw-Hill
8787 Orion Place
Columbus, OH 43240-4027

ISBN: 978-0-02-287982-2 *(Teacher Edition)*
MHID: 0-02-287982-X *(Teacher Edition)*
ISBN: 978-0-02-288006-4 *(Student Edition)*
MHID: 0-02-288006-2 *(Student Edition)*

 is a registered trademark of The McGraw-Hill Companies, Inc.

Printed in the United States of America.

1 2 3 4 5 6 7 8 9 10 073/055 14 13 12 11 10 09

# Program Authors

**Dr. Jay K. Hackett**
**Professor Emeritus of Earth Sciences**
University of Northern Colorado
Greeley, CO

**Dr. Richard H. Moyer**
**Professor of Science Education and Natural Sciences**
University of Michigan–Dearborn
Dearborn, MI

**Dr. JoAnne Vasquez**
**Elementary Science Education Consultant**
**NSTA Past President**
**Member, National Science Board and NASA Education Board**

**Mulugheta Teferi, M.A.**
**Principal, Gateway Middle School**
Center of Math, Science, and Technology
St. Louis Public Schools
St. Louis, MO

**Dinah Zike, M.Ed.**
Dinah Might Adventures LP
San Antonio, TX

**Kathryn LeRoy, M.S.**
**Chief Officer Curriculum Services**
Duval County Schools, FL

**Dr. Dorothy J. T. Terman**
**Science Curriculum Development Consultant**
**Former K–12 Science and Mathematics Coordinator**
Irvine Unified School District
Irvine, CA

**Dr. Gerald F. Wheeler**
**Executive Director National Science Teachers Association**

**Bank Street College of Education**
New York, NY

# Contributors and Reviewers

## Contributing Authors

**Dr. Sally Ride**
Sally Ride Science
San Diego, CA

**Lucille Villegas Barrera, M.Ed.**
Elementary Science Supervisor
Houston Independent School District
Houston, TX

**American Museum of Natural History**
New York, NY

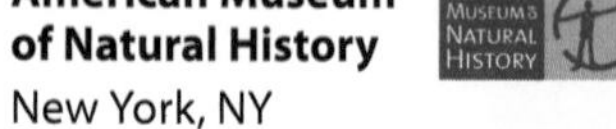

## Contributing Writer

**Ellen C. Grace, M.S.**
Consultant
Albuquerque, NM

## Content Consultants

**Paul R. Haberstroh, Ph.D.**
Mohave Community College
Lake Havasu City, AZ

**Timothy Long**
School of Earth and Atmospheric Sciences
Georgia Institute of Technology
Atlanta, GA

**Rick MacPherson, Ph.D.**
Program Director
The Coral Reef Alliance
San Francisco, CA

**Hector Córdova Mireles, Ph.D.**
Physics Department
California State Polytechnic University
Pomona, CA

**Charlotte A. Otto, Ph.D.**
Department of Natural Sciences
University of Michigan–Dearborn
Dearborn, MI

**Paul Zitzewitz, Ph.D.**
Department of Natural Sciences
University of Michigan–Dearborn
Dearborn, MI

## Editorial Advisory Board

**Deborah T. Boros, M.A.**
President, Society of Elementary Presidential Awardees
Second Grade Teacher
Mississippi Elementary
Coon Rapids, MN

**Lorraine Conrad**
K–12 Coordinator of Science
Richland County School District #2
Columbia, SC

**Kitty Farnell**
Science/Health/PE Coordinator
School District 5 of Lexington and Richland Counties
Ballentine, SC

**Kathy Grimes, Ph.D.**
Science Specialist
Las Vegas, NV

**Richard Hogen**
Fourth Grade Teacher
Rudy Bologna Elementary School
Chandler, AZ

**Kathy Horstmeyer**
Educational Consultant
Past President, Society of Presidential Awardees
Past Preschool/Elementary NSTA Director
Carefree, AZ and Chester, CT

**Jean Kugler**
Gaywood Elementary School
Prince Georges County Public Schools
Lanham, MD

**Bill Metz, Ph.D.**
Science Education Consultant
Fort Washington, PA

**Karen Stratton**
Science Coordinator K–12
Lexington District One
Lexington, SC

**Emma Walton, Ph.D.**
Science Education Consultant
NSTA Past President
Anchorage, AK

**Debbie Wickerham**
Teacher
Findlay City Schools
Findlay, OH

## Teacher Reviewers

**Barbara Adcock**
Pocahontas Elementary
Powhatan, VA

**Erma Anderson**
Educational Consultant
Needmore, PA

**Cathryn Beck-Potter**
Chestatee Elementary
Gainesville, GA

**Teri Warden Bickmore, M.Ed.**
Science Consultant
Midland, MI

**Jaime Breedlove**
Jane D. Hull Elementary
Chandler, AZ

**Jacqueline M. Brown**
Cascade Elementary
Atlanta, GA

**April M. Bruce**
Supervisor for Instruction
Lynchburg City Schools
Lynchburg, VA

**Patricia A. Cavanagh**
Merrimac Elementary
Holbrook, NY

**Meghan Ciacchella**
L'Anse Creuse Public Schools
Chesterfield, MI

**Gary L. Cooper**
Science Department Chair, Biology Teacher
MSD of Pike Township Schools
Indianapolis, IN

**Sarah M. D'Agostini**
Joseph M. Carkenord Elementary
Chesterfield, MI

**Dr. Kelly A. Decker**
University of Richmond
Richmond, VA

**Frances Pistone DeLuca**
Our Lady of Perpetual Help School
South Ozone Park, NY

**Wendy DeMers**
Hynes Charter School
New Orleans, LA

**Kellie DeRango**
Washington Elementary
Wauwatosa, WI

**Sheri Dudzinski**
Marie C. Graham Elementary
Harrison Township, MI

**Delores Dalton Dunn**
Curriculum Specialist (retired)
Virginia Department of Education
Hanover, VA

**Laura A. Edwards**
Vickery Creek Elementary
Cumming, GA

**Mary Vella Ernat**
District Elementary Science Content Leader
Wayne-Westland Community Schools
Westland, MI

**Jenny Sue Flannagan**
Elementary Science Coordinator
Virginia Beach City Public Schools
Virginia Beach, VA

**Marjorie Froberger, M.A.**
Anchor Bay Schools
New Baltimore, MI

**Clara Mackin Fulkerson**
Curriculum Resource Consultant
Nelson County Schools
Bardstown, KY

**Lou Gatto**
Hunterdon Central District School
Flemington, NJ

**Lori Gehrman**
Jane D. Hull Elementary
Chandler, AZ

**Angela Geibel**
Francis A. Higgins Elementary
Chesterfield Township, MI

**Lori Gilchrist**
Sharon Elementary
Suwanee, GA

**Connie Grubbs**
Varner Elementary
Powder Springs, GA

**Tasha Hamil**
Cumming Elementary
Cumming, GA

**Nancy Hayes**
Educational Consultant
Lemont, IL

**Carol Johnson**
Jane D. Hull Elementary
Chandler, AZ

**Jerry D. Kelley, Ed.S.**
Chestatee Elementary
Forsyth, GA

**Andrew C. Kemp**
Jefferson County Public Schools
Louisville, KY

**Heather W. Kemp**
Middletown Elementary
Louisville, KY

**Tricia Reda Kerr**
Science Specialist, EXCEL Program
The Ohio State University
Columbus, OH

**Barbara Kingston**
Blessed Sacrament School
Jackson Heights, NY

**Gina Koger**
Carroll County Public School
Westminster, MD

**Bonnie Kohler**
L'Anse Creuse Public Schools
Harrison Township, MI

**Heather LeBlanc**
Chestatee Elementary
Gainesville, GA

**Larry Lebofsky**
Senior Research Scientist
Lunar and Planetary Laboratory
University of Arizona
Tucson, AZ

**Richard MacDonald**
Science Curriculum Leader
Hampton City Schools
Hampton, VA

**Brenda S. Martin**
Coal Mountain Elementary
Cumming, GA

**Rebecca Martin**
Westridge Elementary
Frankfort, KY

**Corinne Masters**
Natoma Elementary
Natoma, KS

**Tiah E. McKinney**
Albert Einstein Fellow
National Science Foundation
Arlington, VA

**Sharon Meyer**
Barnesville Elementary
Barnesville, OH

**Janiece Mistich**
Tchefuncte Middle School
Mandeville, LA

**Anthony Molock**
Cascade Elementary
Atlanta, GA

**Sandy Morris**
Department of Learning Services
Wichita, KS

**Terri Oatis-Wilson**
Peyton Forest Elementary
Atlanta, GA

**Brenda A. Oulsnam**
Clayton County Schools (retired)
Jonesboro, GA

**Jim Peters**
Science Resource Teacher
Carroll County Board of Education
Westminster, MD

**Sharon Pinion**
Sawnee Elementary
Cumming, GA

**Amy Quick**
Joseph M. Carkenord Elementary
Chesterfield, MI

**Stacey Race**
Sharon Elementary
Suwanee, GA

**Gloria R. Ramsey**
Mathematics/Science Specialist
Memphis City Schools
Memphis, TN

**Anna Reitz**
Forsyth County Schools
Cumming, GA

**Steve A. Rich**
Science Coordinator
Georgia Youth Science & Technology Center
Carrollton, GA

**Maureen Riordan**
Fairway Elementary
Wildwood, MO

**Richard Ruiz**
Jane D. Hull Elementary
Chandler, AZ

**Ruth M. Ruud**
Millcreek Township School District
Erie, PA

**Sarah Rybarczyk**
Joseph M. Carkenord Elementary
Chesterfield, MI

**Laura W. Schaefer**
Coordinator, School Partnerships
Missouri Botanical Garden
St. Louis, MO

**Rhonda Segraves**
Settles Bridge Elementary
Suwanee, GA

**Ursula M. Sexton**
Senior Research Associate/Educational Consultant
WestEd
San Ramon, CA

**Rita Jane Shelton**
Louisa Middle School
Louisa, KY

**Matt Silberglitt**
Science Assessment Specialist
Minnesota Department of Education
Roseville, MN

**William L. Siletti**
Packer Collegiate Institute
Brooklyn, NY

**Georgia Ann Smith**
Sunflower Elementary
Lenexa, KS

**Victoria L. Thom**
Baker Elementary
Acworth, GA

**Shannon Tribble**
Daves Creek Elementary
Cumming, GA

**Shirley Whorley**
Science Coordinator, K-12
Roanoke City Public Schools
Roanoke, VA

**Laura Wilkowski**
Science Consultant
Midland, MI

**Dr. Sharon Wynstra**
Science Coordinator
Rockford Public Schools
Rockford, IL

**Brad Yohe**
Science Supervisor
Carroll County Public Schools
Westminster, MD

# Table of Contents

## Overview



## Teaching with the Student Edition



## Life Science

### UNIT A Plants and Animals

#### CHAPTER 1 Plants



#### CHAPTER 2 Animals



### UNIT B Habitats

#### CHAPTER 3 Looking at Habitats



#### CHAPTER 4 Kinds of Habitats



## Earth and Space Science

### UNIT C Our Earth

#### CHAPTER 5 Land and Water



#### CHAPTER 6 Earth's Resources



### UNIT D Weather and Sky

#### CHAPTER 7 Observing Weather

## CHAPTER 8 Earth and Space



# Physical Science

## UNIT E Matter

### CHAPTER 9 Looking at Matter



### CHAPTER 10 Changes in Matter



## UNIT F Motion and Energy

### CHAPTER 11 How Things Move



### CHAPTER 12 Using Energy



## Student Edition Reference



# Teacher Resources

# SCIENCE
# A CLOSER LOOK

*A wealth of resources that brings science to life*

## Student and Teacher Editions ▼

Grade 1

Grade 2

Grade 3

Kindergarten

***Pre-K also available***

Grade 4

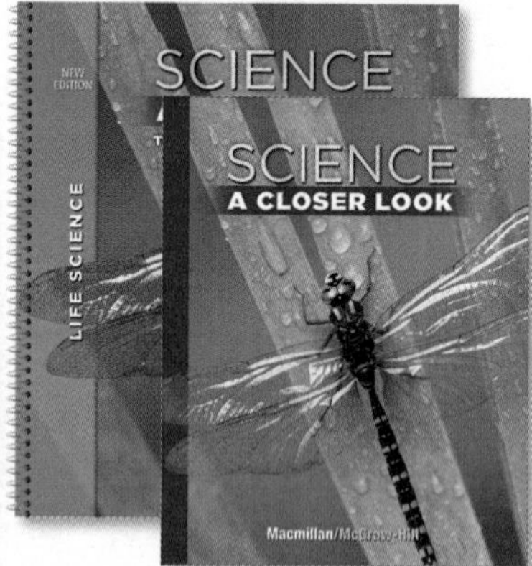
Grade 5

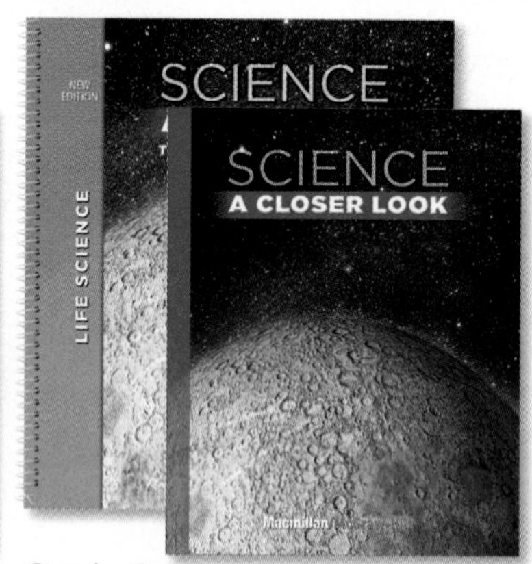
Grade 6

## Activity Resources ▶

**Materials Kits** support Science Activities.

**Grab 'n Go Activity Bags** make preparing for activities quick and easy.

**Activity Flipcharts** provide flexibility for your Science Center.

The program for today's teacher delivers...

**activities & content**

Standards-based instruction with activities that work.

**access for all**

Support for all learners.

**ease of use**

"Fast Track" to fit a busy schedule.

# Instructional Resources ▼

**Reading Resources** make science accessible and fun to read.

**Key Resources** support instruction, build skills, and promote comprehension.

**Supporting Resources** provide for a wide variety of student needs.

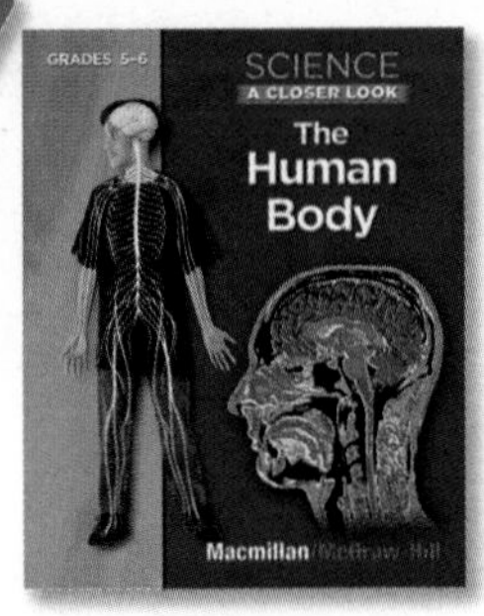

INTERACTIVE WHITEBOARD READY

**also with full technology support**

# Technology for the Student

## Practice and Activities ▼

**Science in Motion ▶**
Animated key concepts with assessment

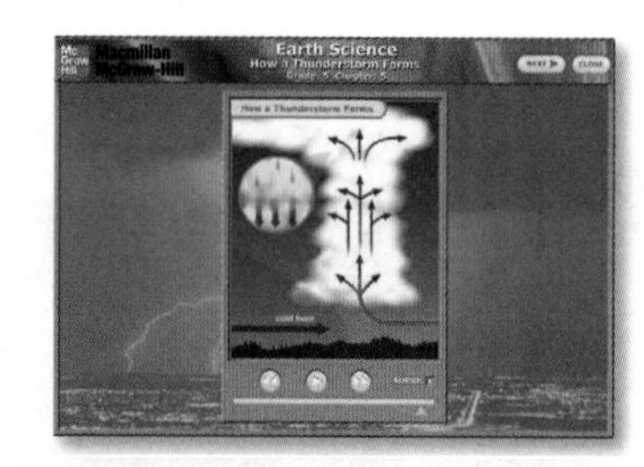

**Operation: Science Quest CD-ROM ▶**
Interactive learning simulations and problem-solving modules

**StudentWorks™ Plus CD-ROM ▶**
Student Edition on CD

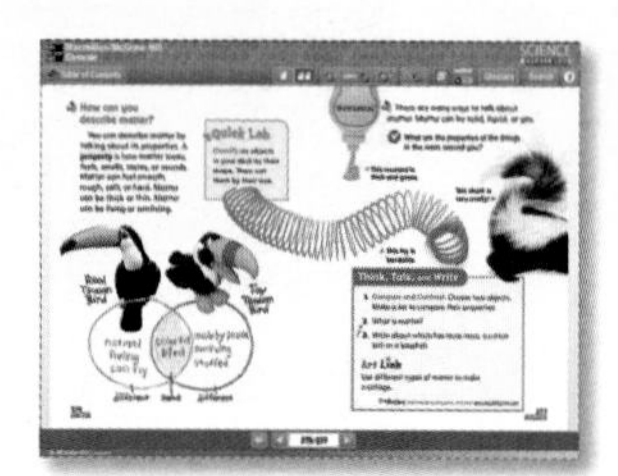

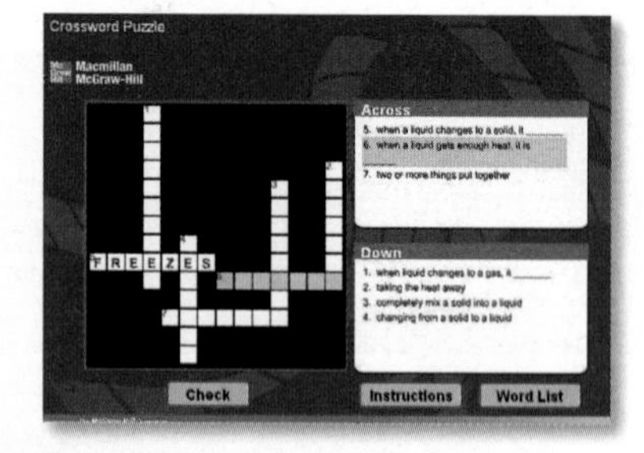

**◀ PuzzleMaker CD-ROM**
Creates puzzles to reinforce science vocabulary

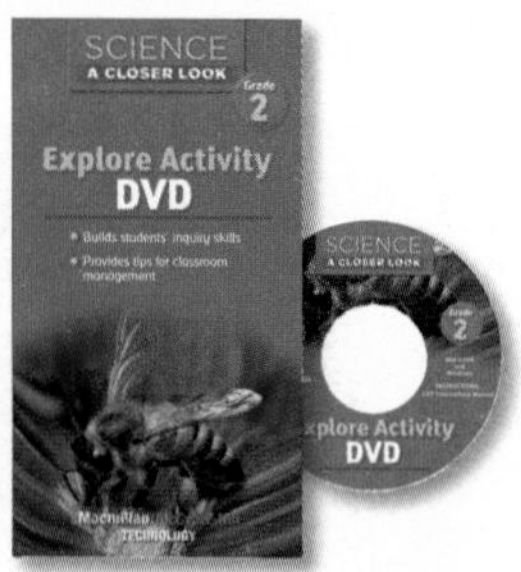

**◀ Science Songs Audio CD**
Reinforce science content with familiar tunes

**◀ Science Activity DVD**
Video modeling of the explore activities

**visit us at www.macmillanmh.com**

Provides supportive activities and enriching content for the entire program

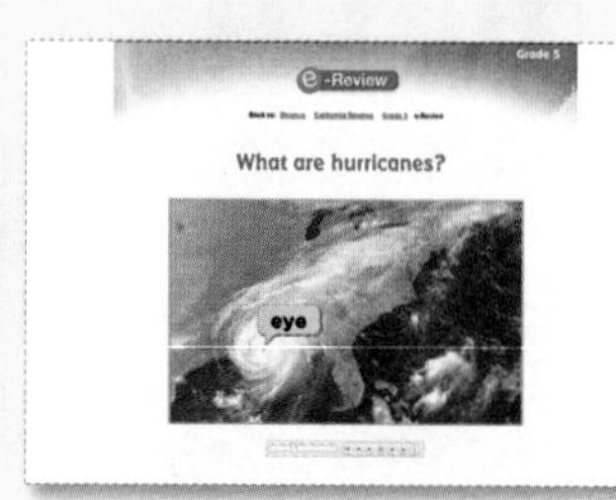

**e-Review**
Narrated, animated version, with interactive quizzes

**e-Journal**

**e-Careers**

**e-Glossary**

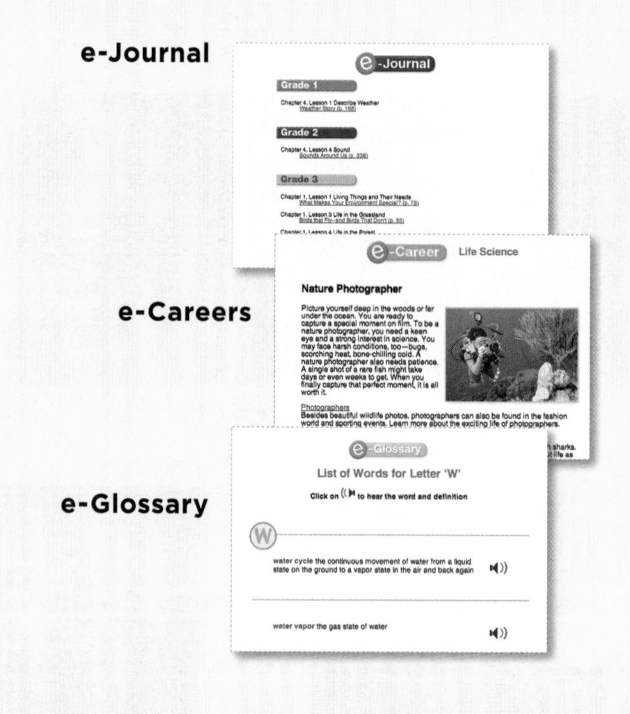

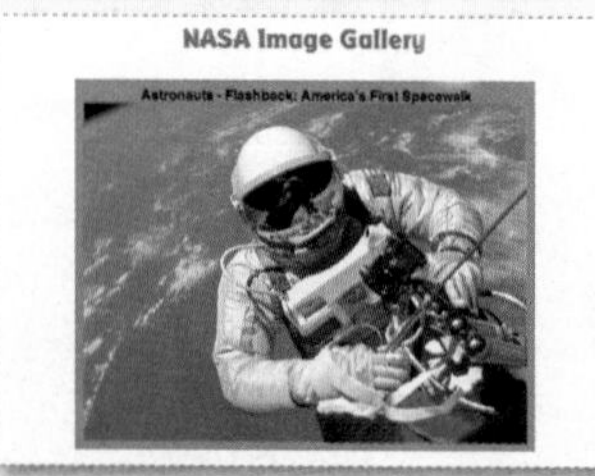

**NASA Image Gallery**

# Technology
## for the Teacher

## Planning and Instruction ▼

**TeacherWorks™ Plus CD-ROM ▶**
Electronic Teacher's Edition with lesson planning and printable resources

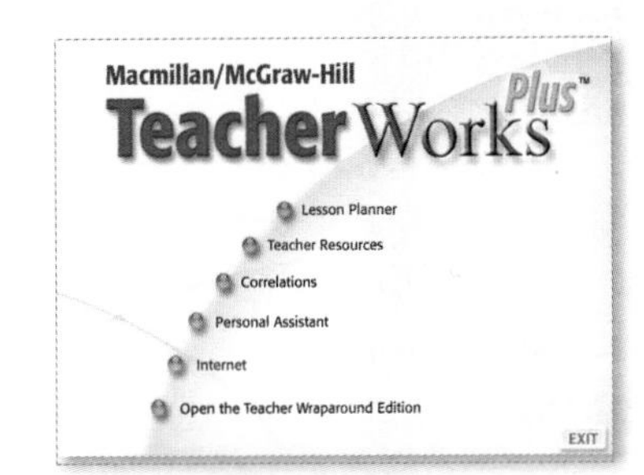


**Professional Development DVD ▶**

**Classroom Presentation Toolkit CD-ROM ▶**
PowerPoint presentations for every lesson

**ExamView® Assessment Suite CD-ROM ▶**
Student assessments via online testing

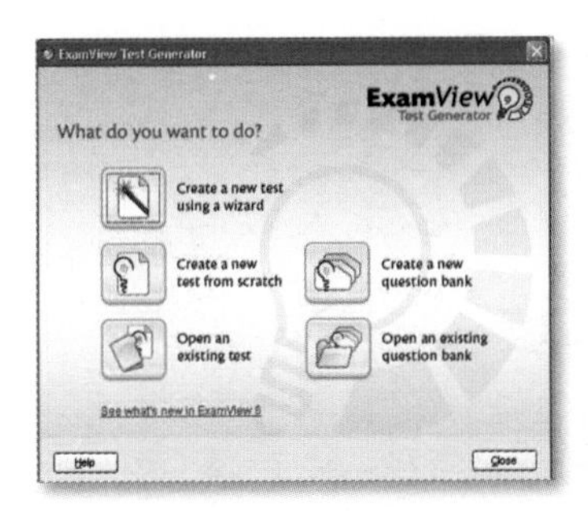


**Visit us at www.macmillanmh.com**

Offers a rich collection of professional development resources, leveled book selections and links to topic-related sites

**Professional Development**

**National Science Digital Library**

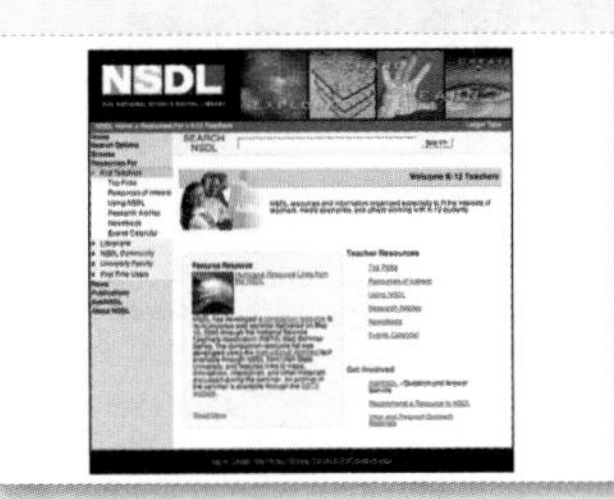


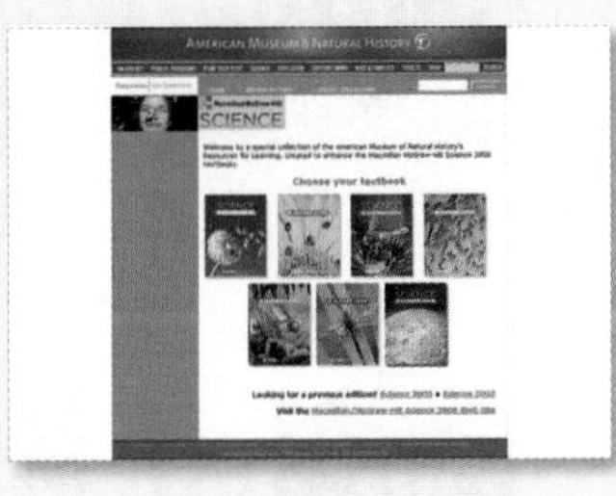


**American Museum of National History**

**Other Partnerships**

# Inquiry-based Activities

- Provide a variety of inquiry experiences
- Support structured, guided, and open inquiry
- Foster conceptual understanding

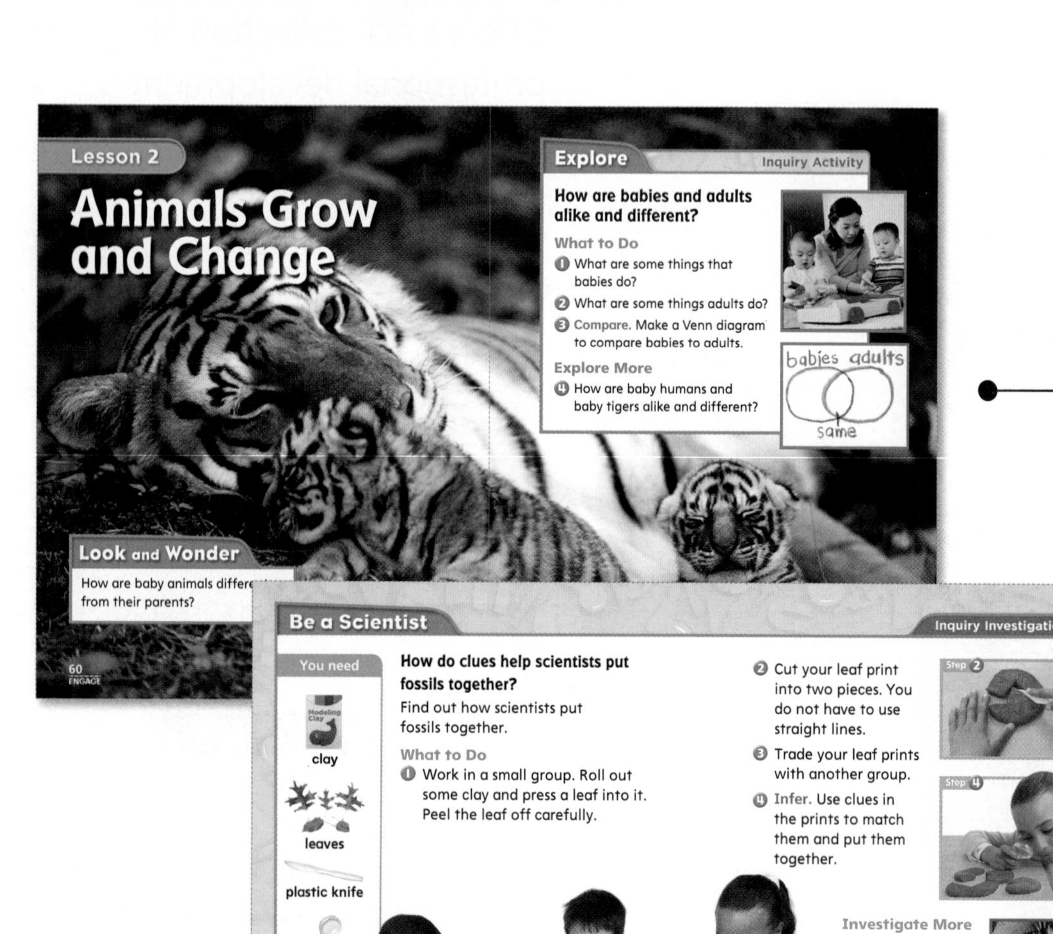


**Explore Activities** begin every lesson

**Inquiry Investigations** extend science learning

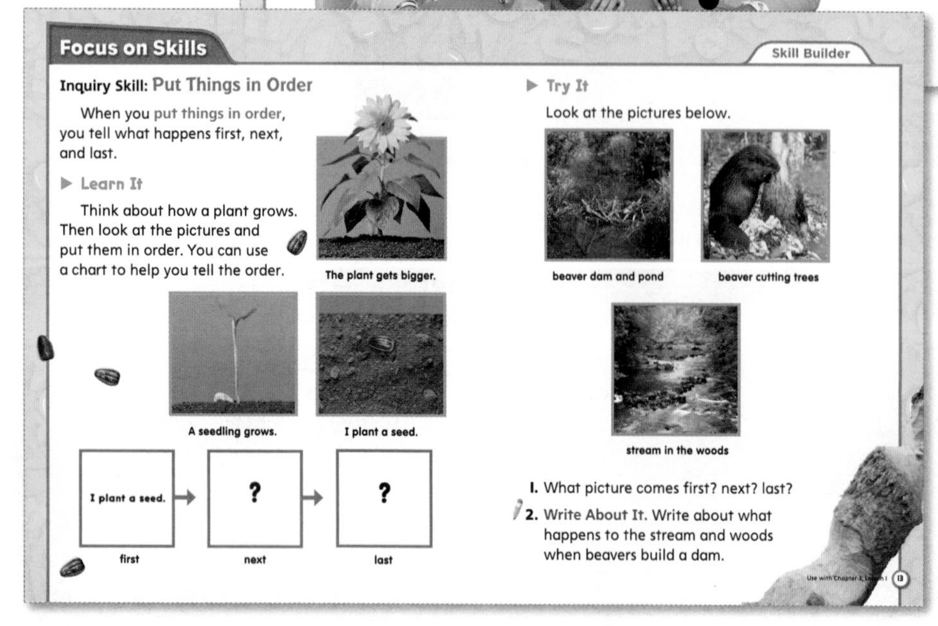


**Skill Builder Activities** develop inquiry skills

*full technology support*

- Science Activity DVD

# Standards-based Content

- Develops the big ideas of science
- Provides depth of understanding
- Supports reading skills

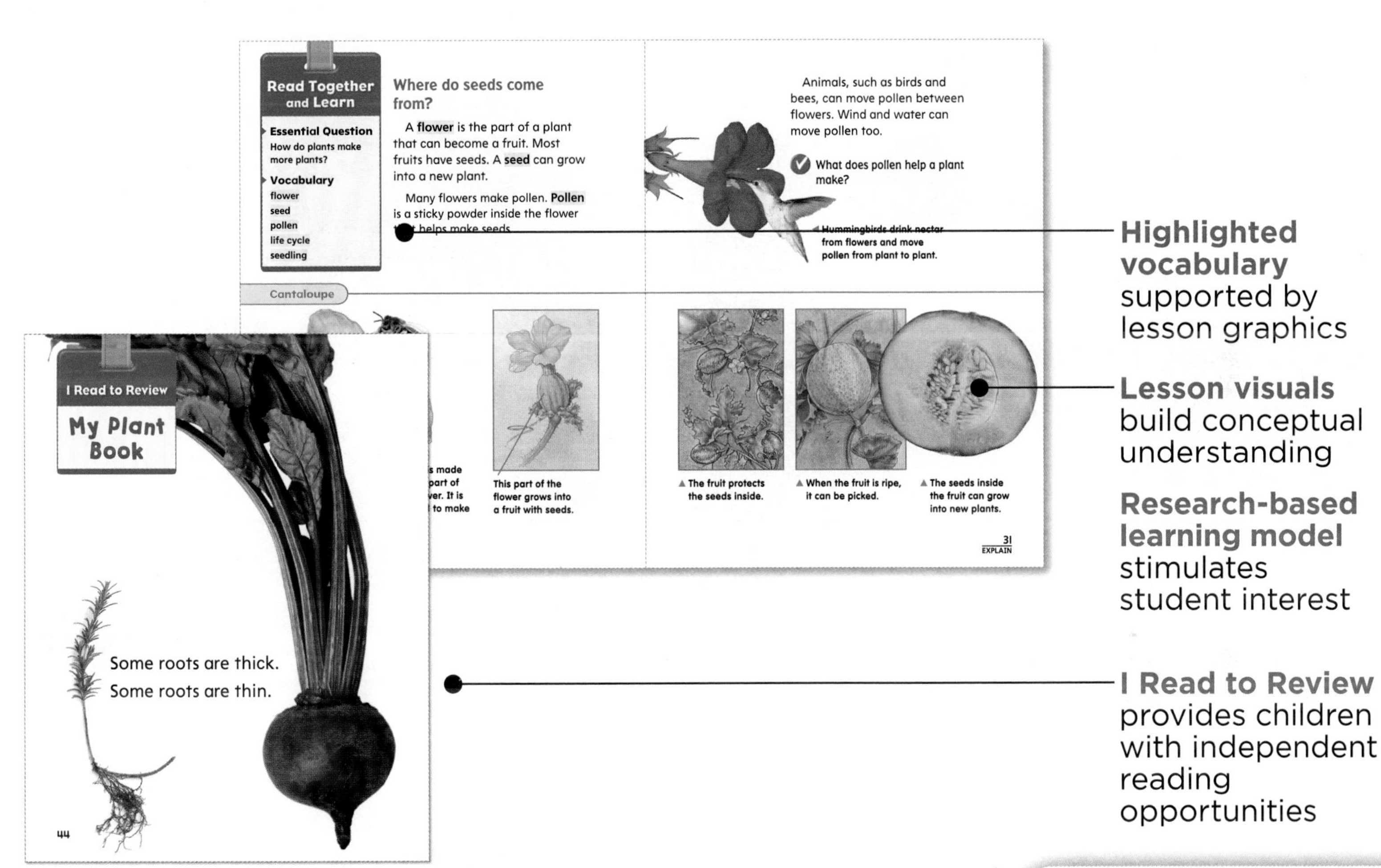

Read Together and Learn
Essential Question
How do plants make more plants?
Vocabulary
flower
seed
pollen
life cycle
seedling
Where do seeds come from?
A flower is the part of a plant that can become a fruit. Most fruits have seeds. A seed can grow into a new plant.
Many flowers make pollen. Pollen is a sticky powder inside the flower that helps make seeds.
Animals, such as birds and bees, can move pollen between flowers. Wind and water can move pollen too.
What does pollen help a plant make?
Hummingbirds drink nectar from flowers and move pollen from plant to plant.
Cantaloupe
This part of the flower grows into a fruit with seeds.
The fruit protects the seeds inside.
When the fruit is ripe, it can be picked.
The seeds inside the fruit can grow into new plants.
31
EXPLAIN
I Read to Review
My Plant Book
Some roots are thick.
Some roots are thin.
44

**Highlighted vocabulary** supported by lesson graphics

**Lesson visuals** build conceptual understanding

**Research-based learning model** stimulates student interest

**I Read to Review** provides children with independent reading opportunities

***full technology support***

*Science in Motion* See the parts of a strawberry plant at www.macmillanmh.com

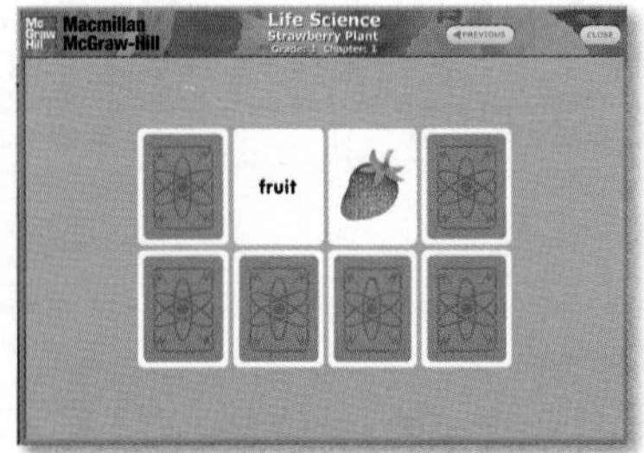


◀ **Science in Motion** animated key concepts with assessment **plus...**

- **Operation: Science Quest CD-ROM**
- **StudentWorks™ Plus CD-ROM**

activities & content

# The Learning Cycle

## Engage

Stimulates student interest and prepares them for the lesson.

## Explore

Provides a hands-on experience around which the lesson concept is developed.

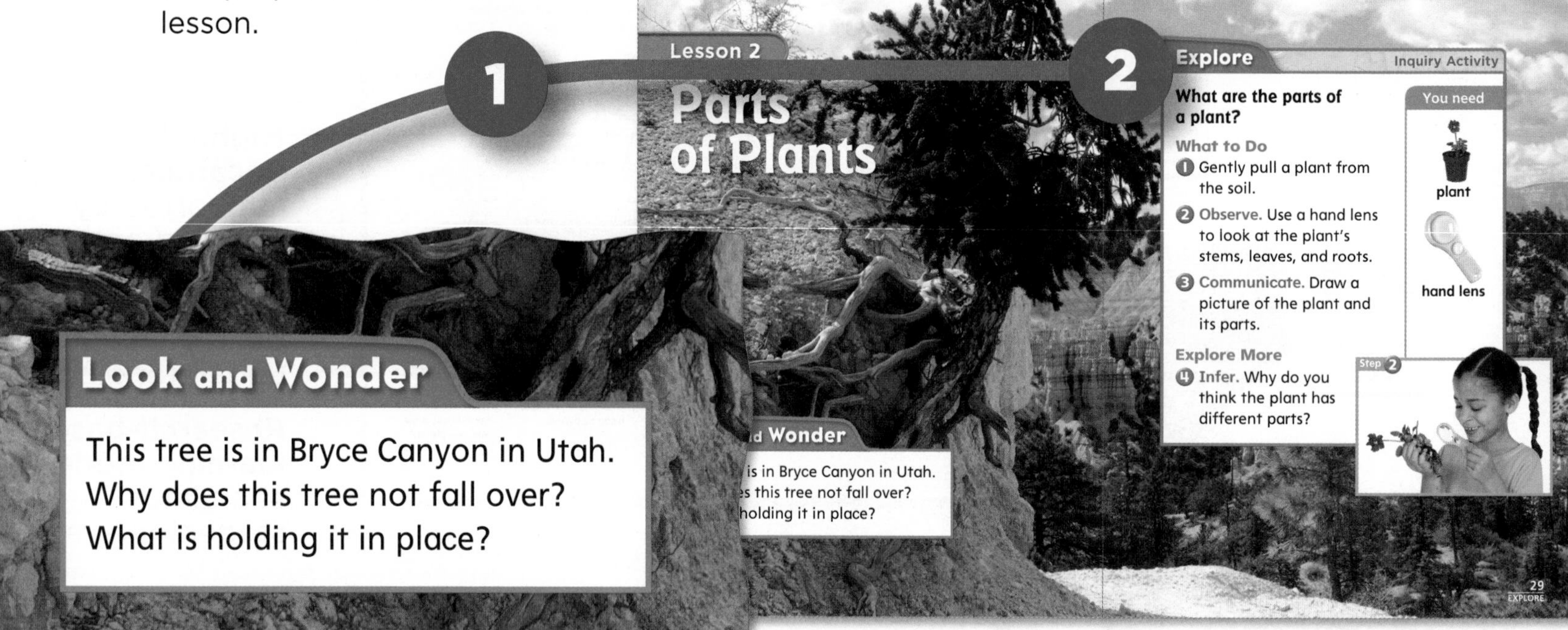

1
Lesson 2
Parts of Plants
Look and Wonder
This tree is in Bryce Canyon in Utah. Why does this tree not fall over? What is holding it in place?
2
Explore
Inquiry Activity
What are the parts of a plant?
What to Do
1 Gently pull a plant from the soil.
2 Observe. Use a hand lens to look at the plant's stems, leaves, and roots.
3 Communicate. Draw a picture of the plant and its parts.
Explore More
4 Infer. Why do you think the plant has different parts?
You need
plant
hand lens
Step 2
29
EXPLORE

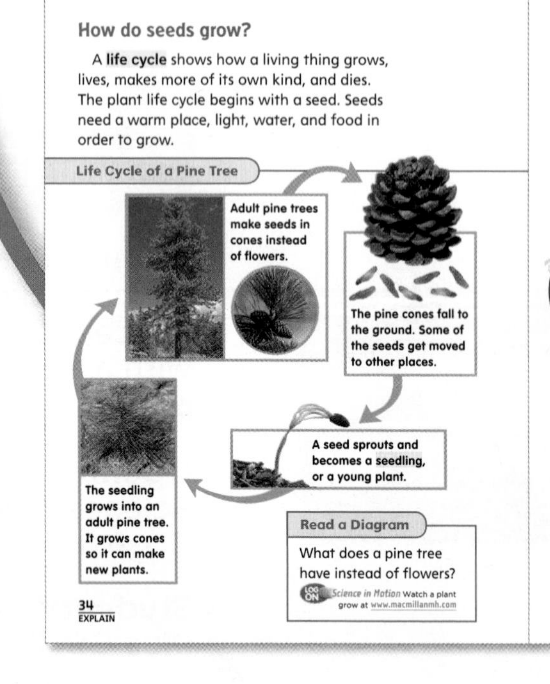

How do seeds grow?
A life cycle shows how a living thing grows, lives, makes more of its own kind, and dies. The plant life cycle begins with a seed. Seeds need a warm place, light, water, and food in order to grow.
Life Cycle of a Pine Tree
Adult pine trees make seeds in cones instead of flowers.
The pine cones fall to the ground. Some of the seeds get moved to other places.
A seed sprouts and becomes a seedling, or a young plant.
The seedling grows into an adult pine tree. It grows cones so it can make new plants.
Read a Diagram
What does a pine tree have instead of flowers?
Science in Motion Watch a plant grow at www.macmillanmh.com
34
EXPLAIN

Most plants follow the same life cycles as their parent plants. Different kinds of plants have different life cycles. Some plants live for just a few weeks. Other plants live for many years.
What will a pine seed grow into?
This flower goes through its whole life cycle in just a few months.
Redwood trees take more than two years to make cones.
Think, Talk, and Write
1. Sequence. How do flowers make new plants?
2. How would you take care of seeds to help them grow?
3. Essential Question. How do plants make more plants?
Health Link
We eat the fruit and seeds of many plants. How many can you think of? What other plant parts do we eat?
-Review Summaries and quizzes online at www.macmillanmh.com
35
EVALUATE
5

## Extend

Links the development of science big ideas to other curriculum areas.

# Explain

Introduces vocabulary and makes the science content understandable through words and visuals.

3

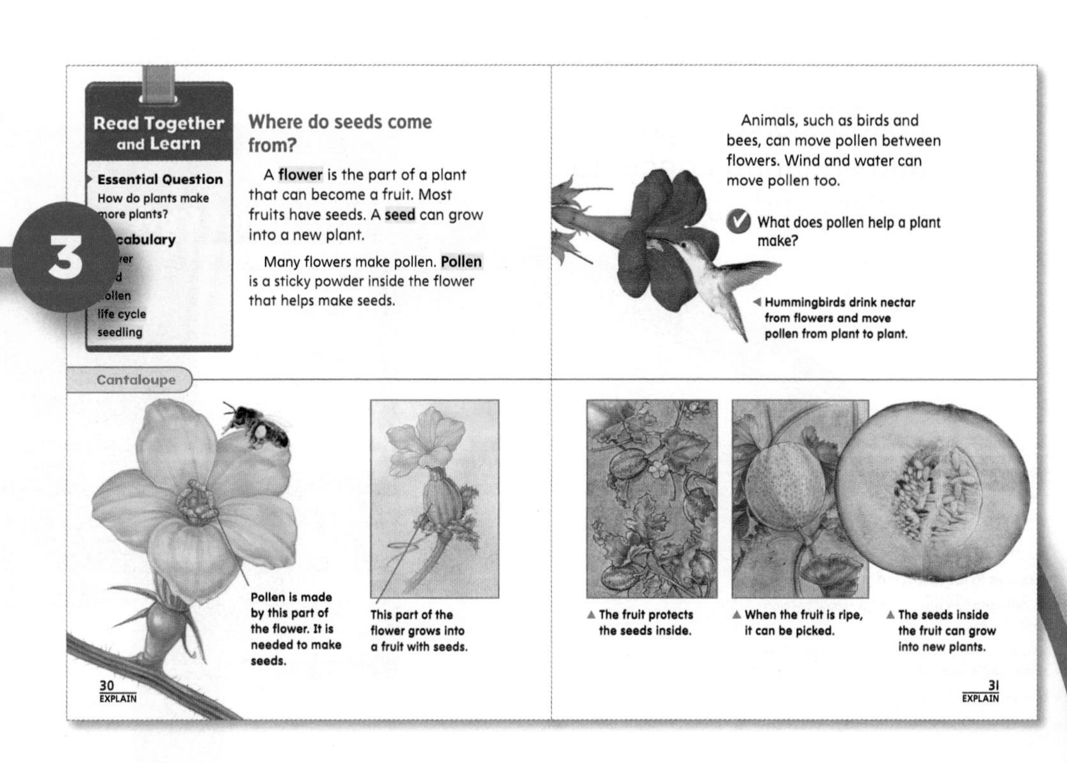

Read Together and Learn
Essential Question
How do plants make more plants?
Where do seeds come from?
A flower is the part of a plant that can become a fruit. Most fruits have seeds. A seed can grow into a new plant.
Many flowers make pollen. Pollen is a sticky powder inside the flower that helps make seeds.
Animals, such as birds and bees, can move pollen between flowers. Wind and water can move pollen too.
What does pollen help a plant make?
Hummingbirds drink nectar from flowers and move pollen from plant to plant.
Cantaloupe
Pollen is made by this part of the flower. It is needed to make seeds.
This part of the flower grows into a fruit with seeds.
The fruit protects the seeds inside.
When the fruit is ripe, it can be picked.
The seeds inside the fruit can grow into new plants.
30 EXPLAIN
31 EXPLAIN

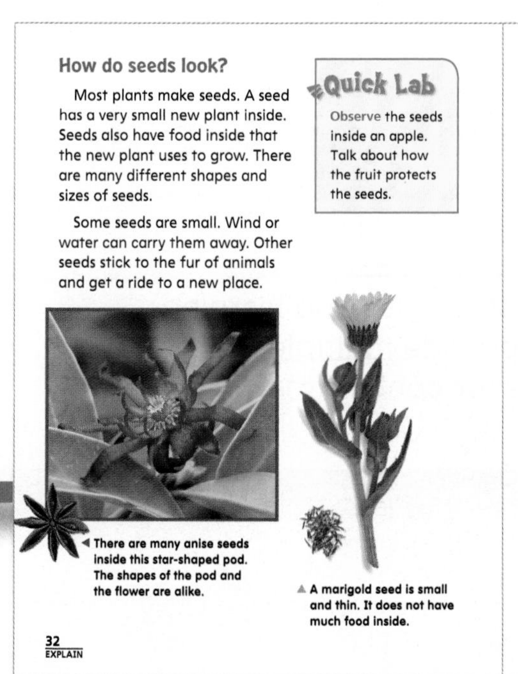

How do seeds look?
Most plants make seeds. A seed has a very small new plant inside. Seeds also have food inside that the new plant uses to grow. There are many different shapes and sizes of seeds.
Some seeds are small. Wind or water can carry them away. Other seeds stick to the fur of animals and get a ride to a new place.
Quick Lab
Observe the seeds inside an apple. Talk about how the fruit protects the seeds.
There are many anise seeds inside this star-shaped pod. The shapes of the pod and the flower are alike.
A marigold seed is small and thin. It does not have much food inside.
32 EXPLAIN

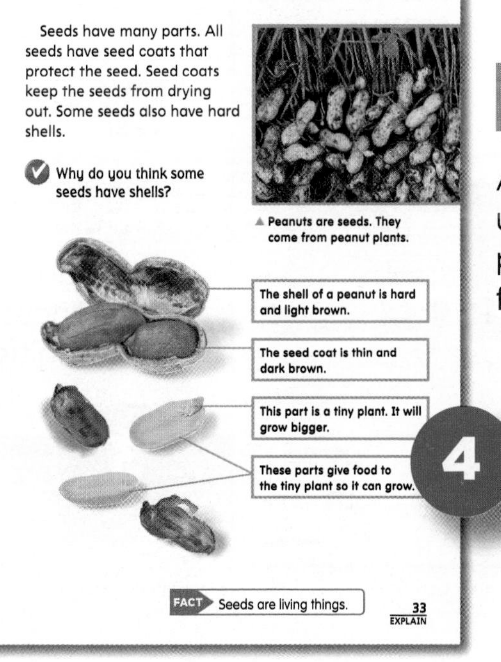

Seeds have many parts. All seeds have seed coats that protect the seed. Seed coats keep the seeds from drying out. Some seeds also have hard shells.
Why do you think some seeds have shells?
Peanuts are seeds. They come from peanut plants.
The shell of a peanut is hard and light brown.
The seed coat is thin and dark brown.
This part is a tiny plant. It will grow bigger.
These parts give food to the tiny plant so it can grow.
FACT Seeds are living things.
33 EXPLAIN

4

# Evaluate

Assesses student understanding and provides opportunities for reteaching.

# Teach for Understanding

Concept-focused instruction based on ASCD's Understanding by Design® framework.

## Identify Learning Outcomes

Focuses on overarching ideas and enduring conceptual understandings.

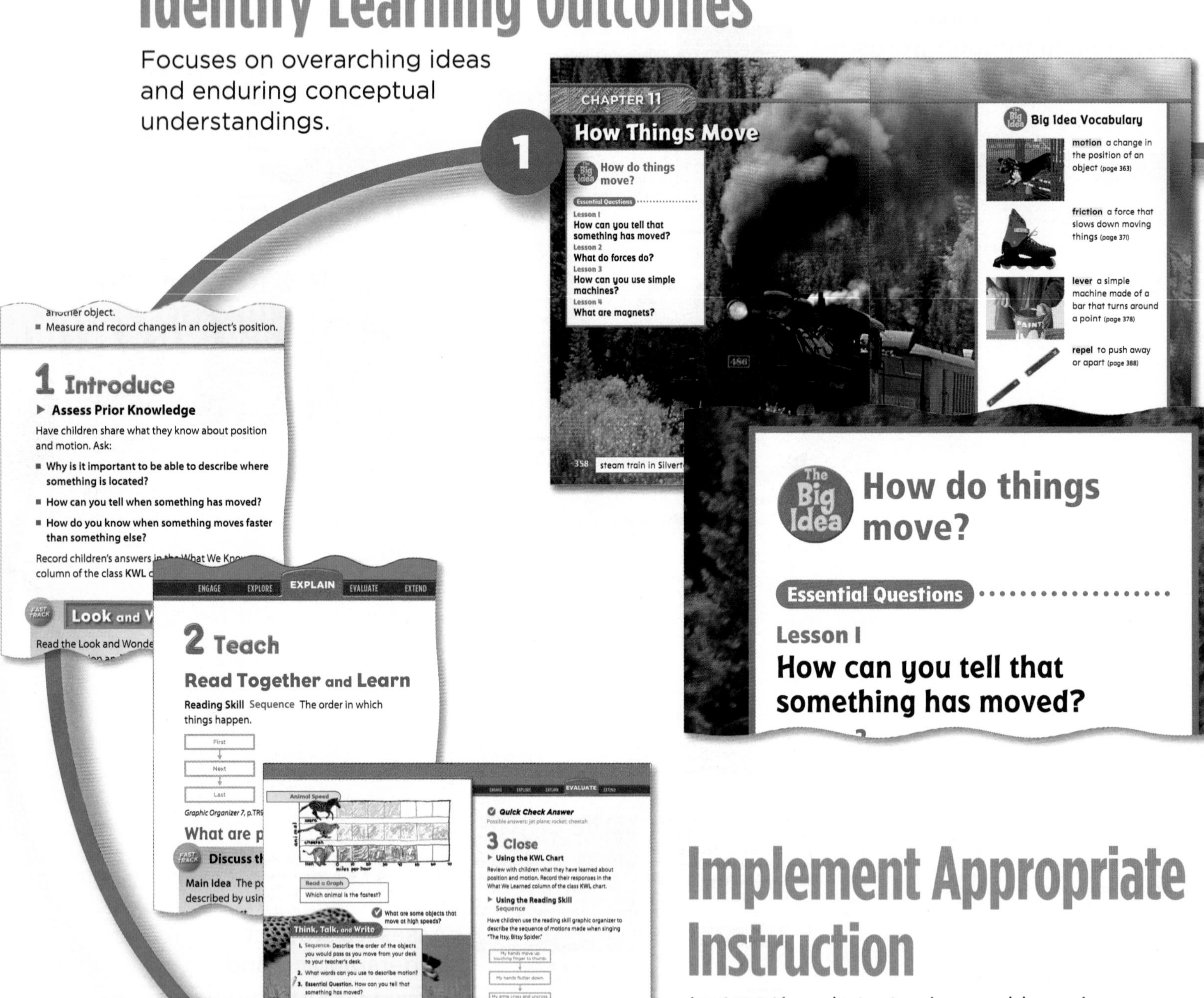

1
CHAPTER 11
How Things Move
How do things move?
Essential Questions
Lesson 1
How can you tell that something has moved?
Lesson 2
What do forces do?
Lesson 3
How can you use simple machines?
Lesson 4
What are magnets?
Big Idea Vocabulary
motion a change in the position of an object (page 363)
friction a force that slows down moving things (page 371)
lever a simple machine made of a bar that turns around a point (page 378)
repel to push away or apart (page 388)
The Big Idea
How do things move?
Essential Questions
Lesson I
How can you tell that something has moved?
Measure and record changes in an object's position.
1 Introduce
Assess Prior Knowledge
Have children share what they know about position and motion. Ask:
Why is it important to be able to describe where something is located?
How can you tell when something has moved?
How do you know when something moves faster than something else?
2 Teach
Read Together and Learn
Reading Skill Sequence The order in which things happen.
3 Close
Using the KWL Chart
Think, Talk, and Write
Social Studies Link
Formative Assessment
3


## Implement Appropriate Instruction

Instructional strategies and learning experiences provide multiple opportunities for content mastery.

# Define Evidence of Understanding

Assessments, including performance tasks and traditional assessments, target a variety of modalities and multiple depths of knowledge.

2

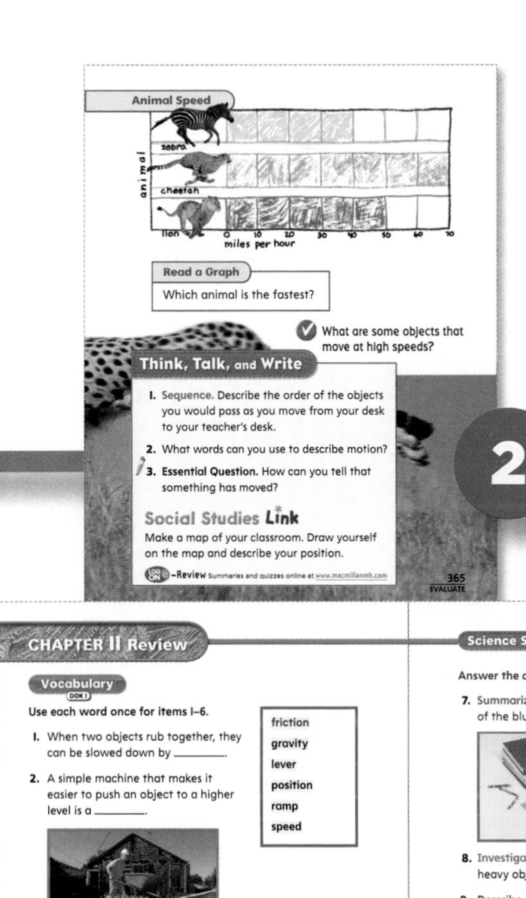
Animal Speed



Read a Graph

Which animal is the fastest?

What are some objects that move at high speeds?

**Think, Talk, and Write**

1. Sequence. Describe the order of the objects you would pass as you move from your desk to your teacher's desk.
2. What words can you use to describe motion?
3. Essential Question. How can you tell that something has moved?

**Social Studies Link**

Make a map of your classroom. Draw yourself on the map and describe your position.

-Review Summaries and quizzes online at www.macmillanmh.com

365 EVALUATE

**CHAPTER 11 Review**

**Vocabulary** DOK 1

Use each word once for items 1–6.

| friction |
|---|
| gravity |
| lever |
| position |
| ramp |
| speed |

1. When two objects rub together, they can be slowed down by ______.
2. A simple machine that makes it easier to push an object to a higher level is a ______.
3. We can tell where an object is by its ______.
4. Objects fall to the floor because of a force called ______.
5. How far an object moves in a period of time is called ______.
6. A simple machine that moves against a fulcrum is called a ______.

394 -Glossary Words and definitions online at www.macmillanmh.com

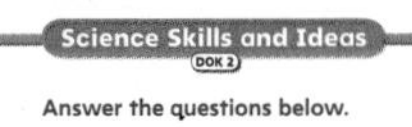

**Science Skills and Ideas** DOK 2

Answer the questions below.

7. Summarize. Describe the position of the blue paper.

8. Investigate. What can help you move a heavy object?
9. Describe some of the simple machines in this picture and how they work.

Big Idea 10. How do things move?

-Review Summaries and quizzes online at www.macmillanmh.com 395

**Test Preparation**

1. This graph shows how fast a train, a car, a plane, and a boat can move.

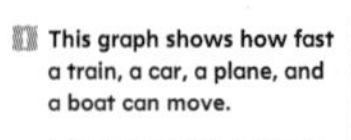

Which moves the fastest?

A a train
B a car
C a plane
D a boat

2. Look at the picture.

What force is the boy using to close the door?

A gravity
B friction
C pulling
D pushing

3. Which item will not be attracted by a magnet?

A a paper clip
B scissors
C a rubber band
D a safety pin

395B

# Differentiated Instruction

- Provides access to tested science concepts
- Includes diverse instructional tools to reach all students
- Offers strategies for English Language Learners

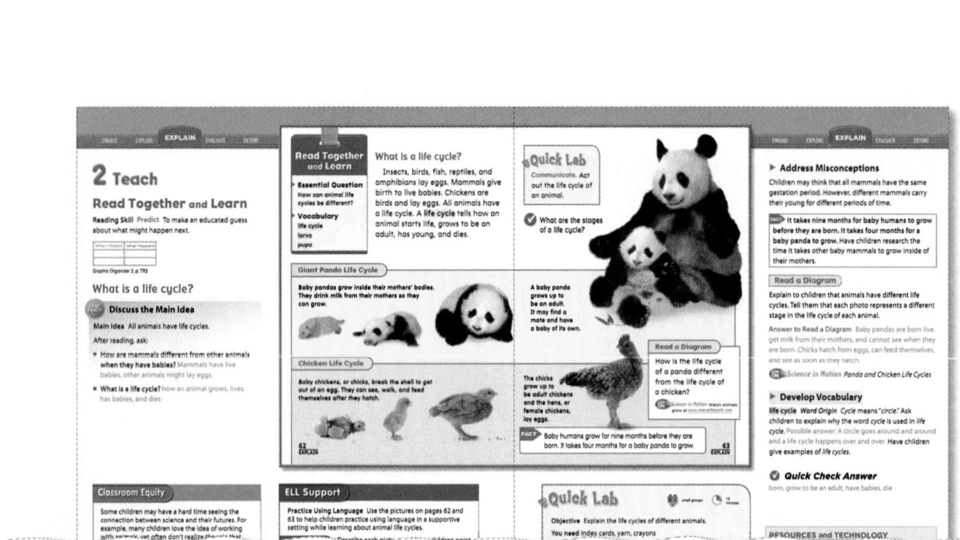

**Teacher's Edition** provides complete support for the teacher

### ELL Support

**Practice Using Language** Use the pictures on pages 62 and 63 to help children practice using language in a supportive setting while learning about animal life cycles.

BEGINNING Describe each picture and have children point to the correct picture. Ask them to name each animal.

INTERMEDIATE Read the captions on pages 62 and 63 with children. Have children look at the pictures again and describe in their own words what happens in each stage.

ADVANCED Have children explain how a baby chick is different from a baby panda. Talk together about the life cycle of a cat or dog.

### Differentiated Instruction

**Leveled Activities**

EXTRA SUPPORT Give children nature magazines and ask them to clip out pictures of animals. Once they have enough clippings, ask them to identify parts that help the animals stay alive. They should put the pictures in groups, such as *camouflage* or *claws*. Have children glue the pictures to poster board and write about how the parts of the animal help keep it alive. Encourage children to present their posters to the class.

ENRICHMENT Have children choose an animal and research its adaptations. Encourage children to group the adaptations by animal needs (shelter, food, water, air). Ask children to make a diorama of the animal in its environment. Give them sticky notes on which they may write descriptions of the animal's adaptations and place them in the appropriate areas of the diorama.

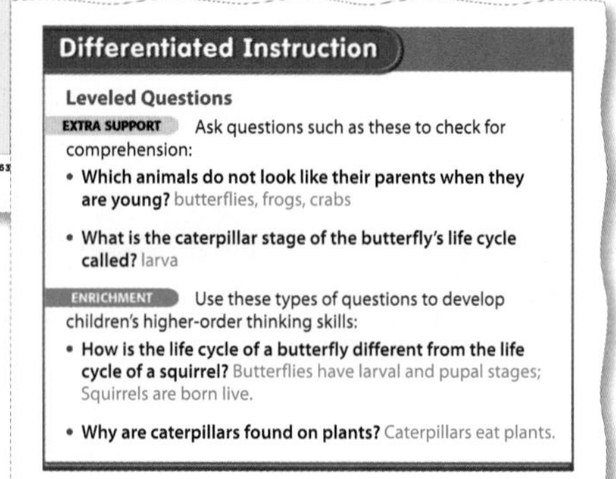

### Differentiated Instruction

**Leveled Questions**

EXTRA SUPPORT Ask questions such as these to check for comprehension:

- **Which animals do not look like their parents when they are young?** butterflies, frogs, crabs
- **What is the caterpillar stage of the butterfly's life cycle called?** larva

ENRICHMENT Use these types of questions to develop children's higher-order thinking skills:

- **How is the life cycle of a butterfly different from the life cycle of a squirrel?** Butterflies have larval and pupal stages; Squirrels are born live.
- **Why are caterpillars found on plants?** Caterpillars eat plants.

**ELL Teacher's Guide** supports language acquisition

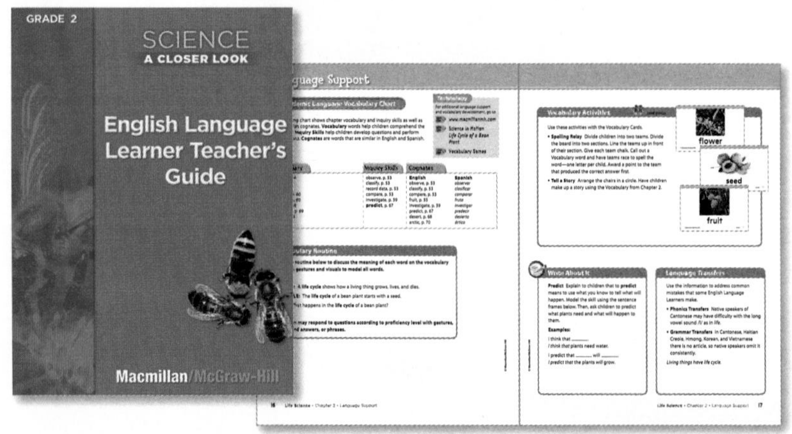


**Reading Essentials** makes content more accessible

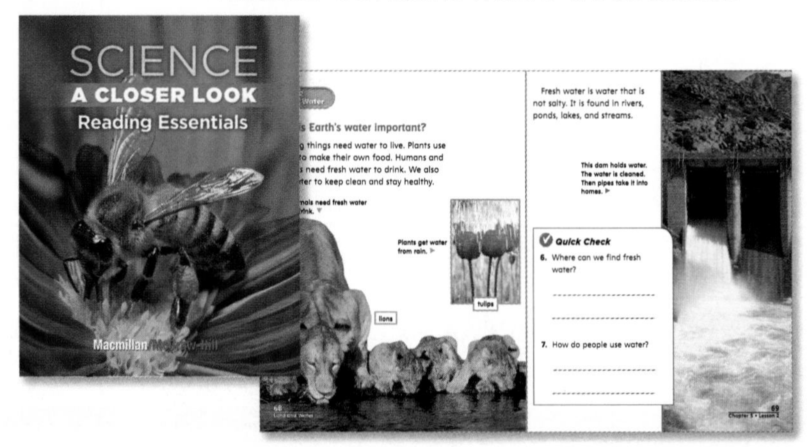

**Leveled Readers** deliver multilevel science content in trade book format

**Leveled Reader Teacher's Guides** provide additional reading strategies for using the Leveled Readers

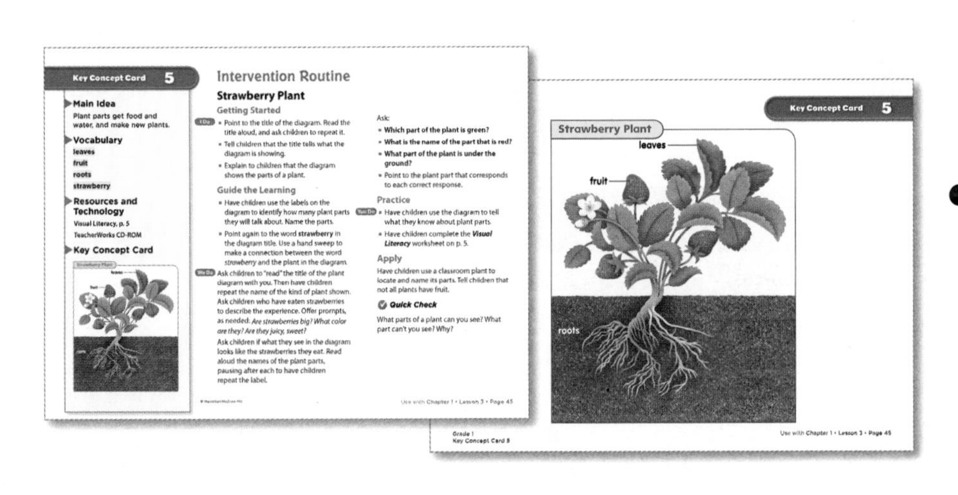

**Key Concept Cards** allow for small group or one-on-one instruction for essential concepts

full technology support

e-Review Summaries and quizzes online at www.macmillanmh.com

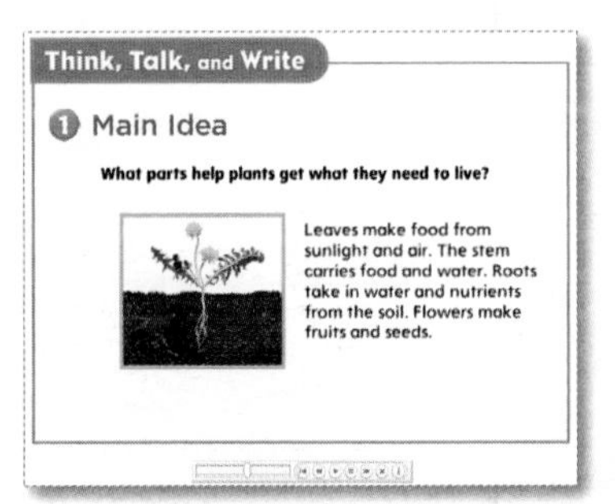


**e-Review** contains animated summaries and assessment

**plus...**

- **Leveled Readers Audio Selections**

# Manageable Organization

- Makes planning easy
- Delivers lesson resources quickly and effectively
- Maximizes the use of instructional time

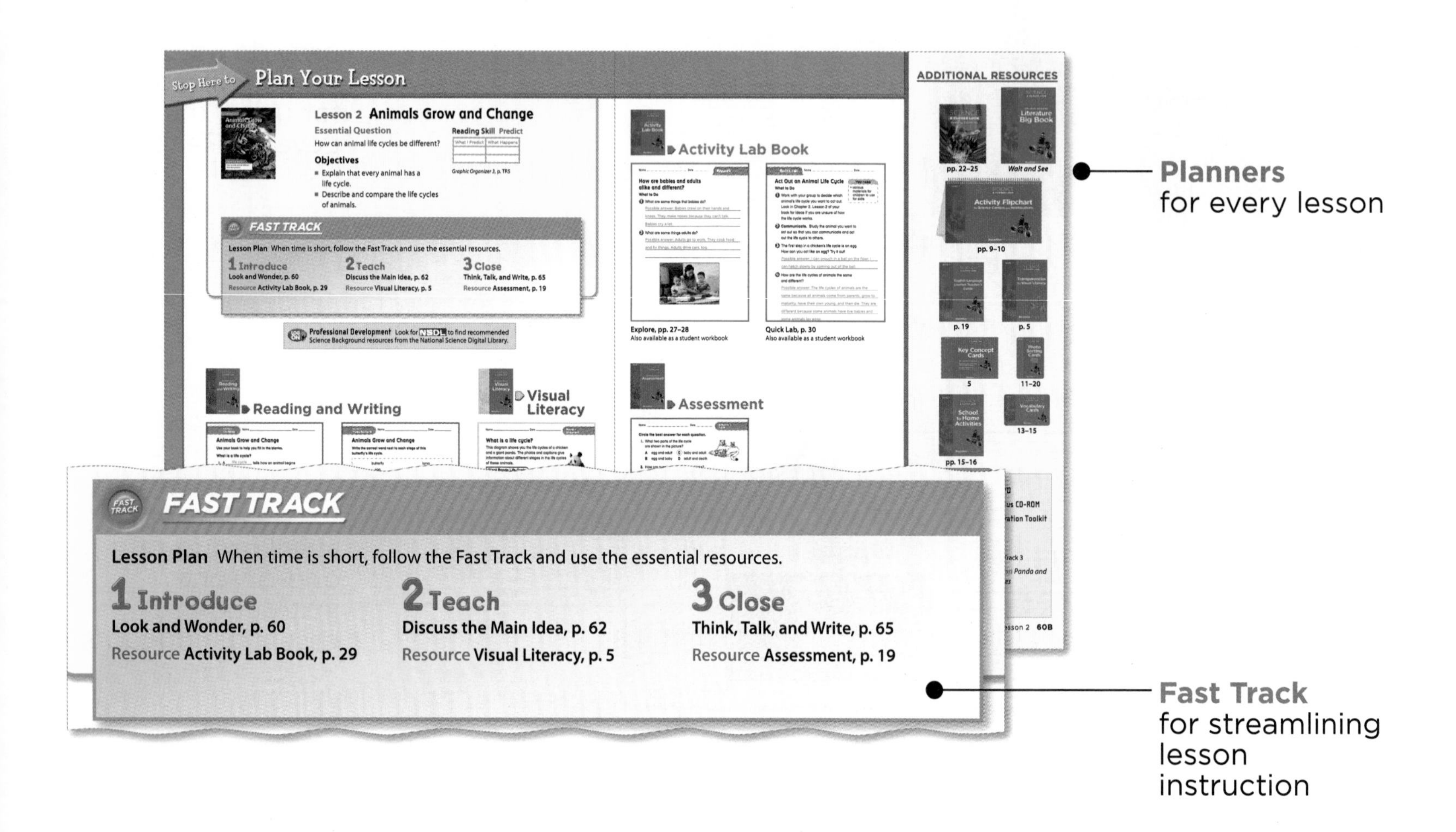

Plan Your Lesson
Lesson 2 Animals Grow and Change
Activity Lab Book
ADDITIONAL RESOURCES
Reading and Writing
Visual Literacy
Assessment
FAST TRACK
Lesson Plan When time is short, follow the Fast Track and use the essential resources.
1 Introduce
Look and Wonder, p. 60
Resource Activity Lab Book, p. 29
2 Teach
Discuss the Main Idea, p. 62
Resource Visual Literacy, p. 5
3 Close
Think, Talk, and Write, p. 65
Resource Assessment, p. 19


**Planners** for every lesson

**Fast Track** for streamlining lesson instruction

***full technology support***

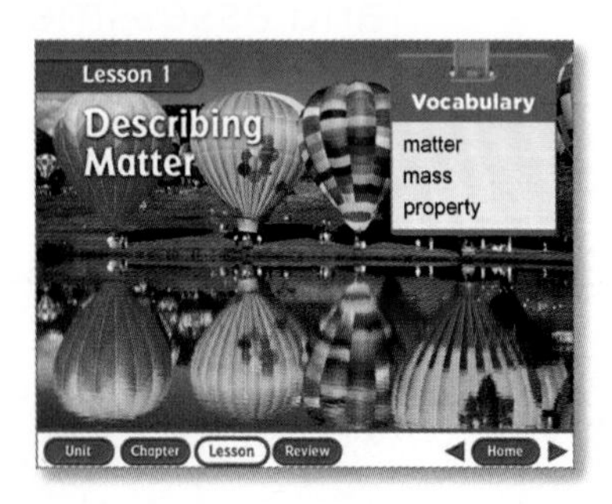


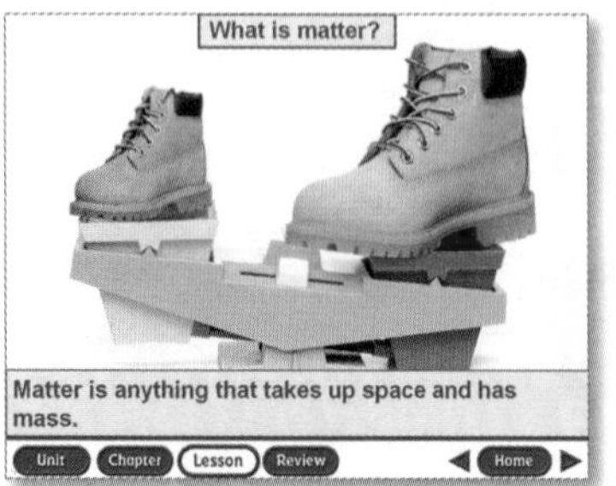


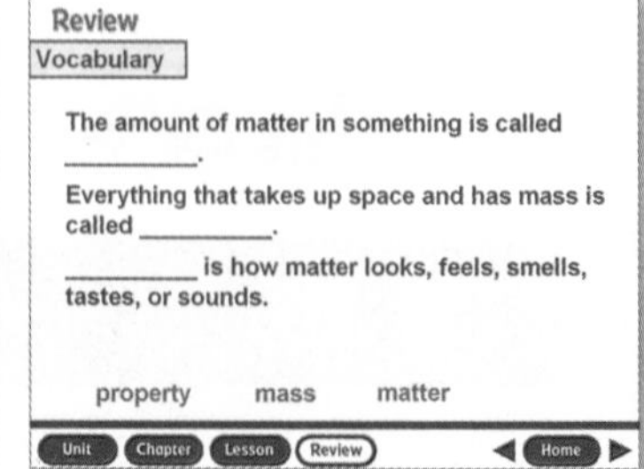

◂ **Classroom Presentation Toolkit CD-ROM** PowerPoint® delivery system for each chapter

**plus...**

- **TeacherWorks™ Plus CD-ROM**

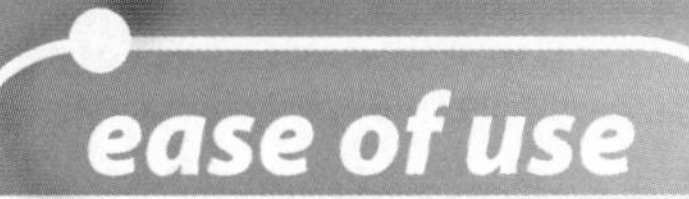


# Assessment

- Includes a variety of assessment options
- Contains tools to assess student understanding
- Helps inform instruction and tracks student progress

## 1 Introduce

▶ **Assess Prior Knowledge**

Have children share what they know about the parts of plants. Write the parts they name in a word bank for them to use throughout the lesson. Ask:

- **What parts do plants have?**
- **What do these parts do for the plant?**
- **How do plants make new plants?**
- **Why do plants make flowers?**

Record children's answers in the What We Know column of the class **KWL** chart.

**Entry-level Assessments** to help determine student readiness

### Formative Assessment

**Draw a Plant**

Have children draw a plant and label its parts. Have them explain verbally, pictorially, or in writing what the parts do and how they work together to help the plant get the things it needs.

Key Concept Cards For student intervention, see the prescribed routine on **Key Concept Card 2.**

**Formative Assessments** to check for understanding during the lesson

Plants have roots that take in water and nutrients from the soil.

Some plants have roots close to the surface of the ground. Others have long and deep roots.

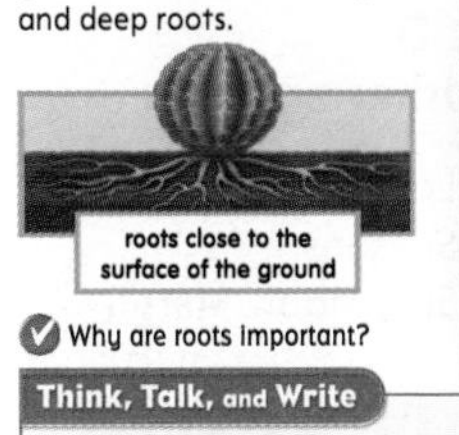


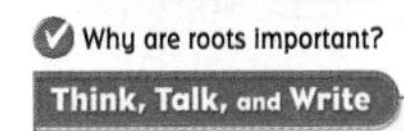


Why are roots important?

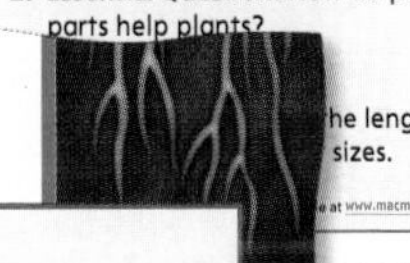

33 EVALUATE

**Summative Assessments** to determine extent of student learning

**Chapter Review** with test preparation

**Chapter Tests**

✓ Why are roots important?

### Think, Talk, and Write

1. **Summarize.** Describe plant parts.
2. **Essential Question.** How do plant parts help plants?

### Math Link

Find two plants. Measure the lengths of their stems. Compare their sizes.

**-Review** Summaries and quizzes online at www.macmillanmh.com

***full technology support***

- **Exam*View*® *Assessment Suite* CD-ROM**
- **Progress Reporter**

## program overview

| | Pre-K | Kindergarten | Grade 1 | Grade 2 |
|---|---|---|---|---|
| Life Science | **Unit A**<br>**Be a Scientist**<br><br>**Unit B**<br>**Plants** | **Unit A**<br>**Plants**<br>• Parts of Plants<br>• What Plants Need and How They Grow<br>• Leaves and Flowers and How We Use Plants | **Unit A**<br>**Plants**<br>Chapter 1<br>**Plants Are Living Things**<br>Chapter 2<br>**Plants Grow and Change** | **Unit A**<br>**Plants and Animals**<br>Chapter 1<br>**Plants**<br>Chapter 2<br>**Animals** |
| Life Science | **Unit C**<br>**Animals** | **Unit B**<br>**Animals**<br>• Animals Are Everywhere<br>• Animal Needs<br>• How Animals Grow and Change | **Unit B**<br>**Animals and Their Homes**<br>Chapter 3<br>**All About Animals**<br>Chapter 4<br>**Places to Live** | **Unit B**<br>**Habitats**<br>Chapter 3<br>**Looking at Habitats**<br>Chapter 4<br>**Kinds of Habitats** |
| Earth and Space Science | **Unit D**<br>**Our Earth** | **Unit C**<br>**Our Earth, Our Home**<br>• Soil and Rocks<br>• Land and Water<br>• Resources and Recycling | **Unit C**<br>**Our Earth**<br>Chapter 5<br>**Looking at Earth**<br>Chapter 6<br>**Caring for Earth** | **Unit C**<br>**Our Earth**<br>Chapter 5<br>**Land and Water**<br>Chapter 6<br>**Earth's Resources** |
| Earth and Space Science | **Unit E**<br>**Sky and Weather** | **Unit D**<br>**Weather and Sky**<br>• Look at Weather<br>• Seasons<br>• Sun, Moon, Stars | **Unit D**<br>**Weather and Sky**<br>Chapter 7<br>**Weather and Seasons**<br>Chapter 8<br>**The Sky** | **Unit D**<br>**Weather and Sky**<br>Chapter 7<br>**Observing Weather**<br>Chapter 8<br>**Earth and Space** |
| Physical Science | **Unit F**<br>**Matter and Motion** | **Unit E**<br>**Exploring Matter**<br>• Paper and Cloth<br>• Wood, Metal, and Clay<br>• Water | **Unit E**<br>**Matter**<br>Chapter 9<br>**Matter Everywhere**<br>Chapter 10<br>**Changes in Matter** | **Unit E**<br>**Matter**<br>Chapter 9<br>**Looking at Matter**<br>Chapter 10<br>**Changes in Matter** |
| Physical Science | | **Unit F**<br>**Moving Right Along**<br>• Wheels and Motion<br>• Gravity and Sounds<br>• Magnets | **Unit F**<br>**Motion and Energy**<br>Chapter 11<br>**On the Move**<br>Chapter 12<br>**Energy Everywhere** | **Unit F**<br>**Motion and Energy**<br>Chapter 11<br>**How Things Move**<br>Chapter 12<br>**Using Energy** |

| | Grade 3 | Grade 4 | Grade 5 | Grade 6 |
|---|---|---|---|---|
| Life Science | **Unit A Living Things**<br>Chapter 1 **A Look at Living Things**<br>Chapter 2 **Living Things Grow and Change** | **Unit A Living Things**<br>Chapter 1 **Kingdoms of Life**<br>Chapter 2 **The Animal Kingdom** | **Unit A Diversity of Life**<br>Chapter 1 **Cells and Kingdoms**<br>Chapter 2 **Parents and Offspring** | **Unit A Diversity of Life**<br>Chapter 1 **Classifying Living Things**<br>Chapter 2 **Cells** |
| | **Unit B Ecosystems**<br>Chapter 3 **Living Things in Ecosystems**<br>Chapter 4 **Changes in Ecosystems** | **Unit B Ecosystems**<br>Chapter 3 **Exploring Ecosystems**<br>Chapter 4 **Surviving in Ecosystems** | **Unit B Ecosystems**<br>Chapter 3 **Interactions in Ecosystems**<br>Chapter 4 **Ecosystems and Biomes** | **Unit B Patterns of Life**<br>Chapter 3 **Genetics**<br>Chapter 4 **Ecosystems** |
| Earth and Space Science | **Unit C Earth and Its Resources**<br>Chapter 5 **Earth Changes**<br>Chapter 6 **Using Earth's Resources** | **Unit C Earth and Its Resources**<br>Chapter 5 **Shaping Earth**<br>Chapter 6 **Saving Earth's Resources** | **Unit C Earth and Its Resources**<br>Chapter 5 **Our Dynamic Earth**<br>Chapter 6 **Protecting Earth's Resources** | **Unit C Earth and Its Resources**<br>Chapter 5 **Changes over Time**<br>Chapter 6 **Conserving Our Resources** |
| | **Unit D Weather and Space**<br>Chapter 7 **Changes in Weather**<br>Chapter 8 **Planets, Moons, and Stars** | **Unit D Weather and Space**<br>Chapter 7 **Weather and Climate**<br>Chapter 8 **The Solar System and Beyond** | **Unit D Weather and Space**<br>Chapter 7 **Weather Patterns**<br>Chapter 8 **The Universe** | **Unit D Weather and Space**<br>Chapter 7 **Weather and Climate**<br>Chapter 8 **Astronomy** |
| Physical Science | **Unit E Matter**<br>Chapter 9 **Observing Matter**<br>Chapter 10 **Changes in Matter** | **Unit E Matter**<br>Chapter 9 **Properties of Matter**<br>Chapter 10 **Matter and Its Changes** | **Unit E Matter**<br>Chapter 9 **Comparing Kinds of Matter**<br>Chapter 10 **Physical and Chemical Changes** | **Unit E Matter**<br>Chapter 9 **Classifying Matter**<br>Chapter 10 **Chemistry** |
| | **Unit F Forces and Energy**<br>Chapter 11 **Forces and Motion**<br>Chapter 12 **Forms of Energy** | **Unit F Forces and Energy**<br>Chapter 11 **Forces**<br>Chapter 12 **Energy** | **Unit F Forces and Energy**<br>Chapter 11 **Using Forces**<br>Chapter 12 **Using Energy** | **Unit F Forces and Energy**<br>Chapter 11 **Exploring Forces**<br>Chapter 12 **Exploring Energy** |

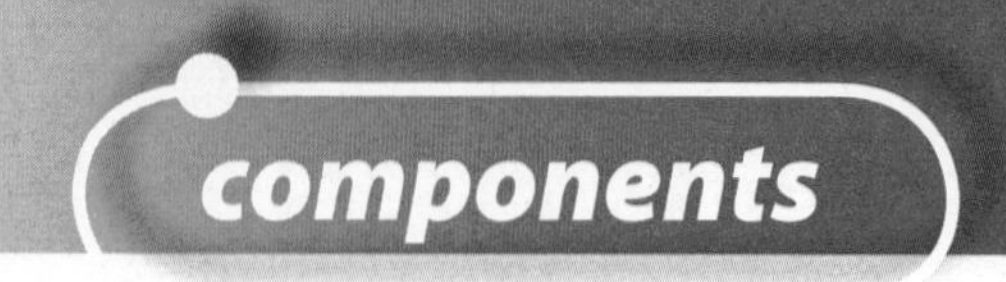


# SCIENCE
## A CLOSER LOOK

| Components | K | 1 | 2 | 3 | 4 | 5 | 6 |
|---|---|---|---|---|---|---|---|
| Student Edition* | | • | • | • | • | • | • |
| Teacher's Edition | • | • | • | • | • | • | • |
| Student Edition Unit Big Books | | • | • | | | | |
| Reading Essentials | | • | • | • | • | • | • |
| Reading and Writing* | | • | • | • | • | • | • |
| Math | | • | • | • | • | • | • |
| Activity Lab Book | | • | • | • | • | • | • |
| Visual Literacy* | | • | • | • | • | • | • |
| Assessment | | • | • | • | • | • | • |
| The Human Body and Teacher's Guide | • | • | • | • | • | • | • |
| Technology A Closer Look and Teacher's Guide | • | • | • | • | • | • | • |
| Transparencies for Visual Literacy | | • | • | • | • | • | • |
| English Language Learner Teacher's Guide | | • | • | • | • | • | • |
| Vocabulary Cards | • | • | • | • | • | • | • |
| Key Concept Cards | | • | • | • | • | • | • |
| Leveled Readers and Leveled Reader Teacher's Guide | • | • | • | • | • | • | • |
| Kindergarten Flipbook | • | | | | | | |
| Literature Big Books | • | • | • | | | | |
| A to Z Activity Book | • | | | | | | |
| Science Projects in a Pocket | | • | • | | | | |
| Science Resource Book | • | | | | | | |
| Floor Puzzles | • | | | | | | |
| Science on the Go | • | • | | | | | |
| Photo Sorting Cards | • | • | • | | | | |
| Equipment Kits | • | • | • | • | • | • | • |
| Activity Flipchart | | • | • | • | • | • | • |
| Science Fair Handbook | | • | • | • | • | • | • |
| Online Student Edition | | • | • | • | • | • | • |
| StudentWorks™ Plus CD-ROM | | • | • | • | • | • | • |
| Online Teacher's Edition | • | • | • | • | • | • | • |
| TeacherWorks™ Plus CD-ROM | • | • | • | • | • | • | • |
| Exam*View*® *Assessment Suite* CD-ROM | | • | • | • | • | • | • |
| Classroom Presentation Toolkit CD-ROM | | • | • | • | • | • | • |
| Science Activity DVD | | • | • | • | • | • | • |
| Science Songs Audio CD | • | • | • | | | | |
| PuzzleMaker CD-ROM | | • | • | • | • | • | • |
| Operation: Science Quest CD-ROM | | • | • | • | • | • | • |
| Professional Development for Science, The Master Teacher Series | • | • | • | • | • | • | • |
| Teacher's Desk Reference | • | • | • | • | • | • | • |
| Companion Web site | • | • | • | • | • | • | • |

Pre-K Components: Teacher's Edition, Flipbook, Big Science Readers, Photo Cards, Science Song Posters, and Science Songs CD

* also available in Spanish

# Be a Scientist

## Life Science

# UNIT A Plants and Animals





vi

# UNIT B Habitats

# Earth and Space Science

## UNIT C Our Earth





viii

## UNIT D Weather and Sky





ix

Physical Science

## UNIT E Matter





x

## UNIT F Motion and Energy

# Online Resources

## Animations

*Science in Motion* Animations online at www.macmillanmh.com

### LIFE SCIENCE



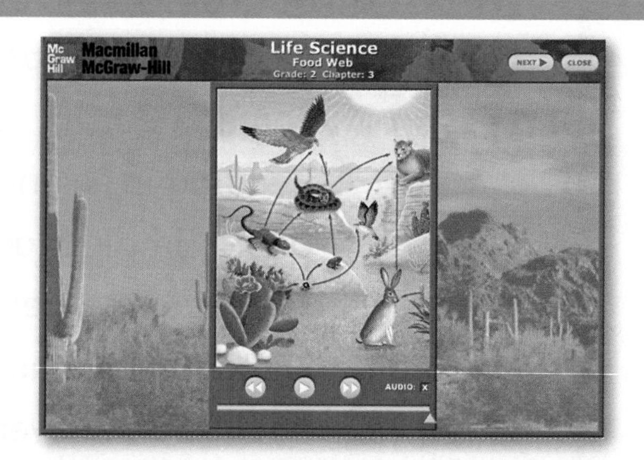


### EARTH AND SPACE SCIENCE



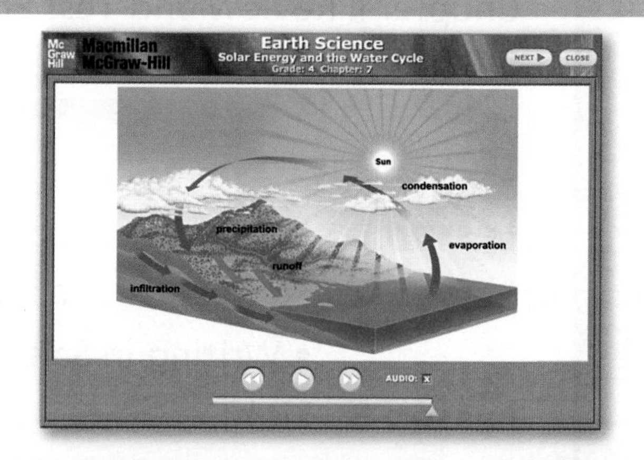


### PHYSICAL SCIENCE



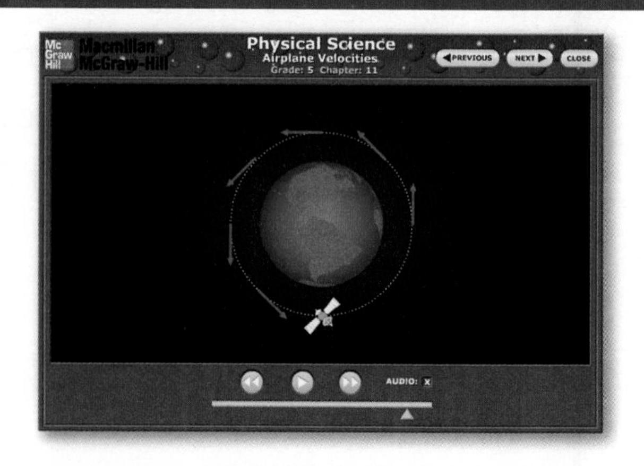




xii

# Additional Student Resources

Visit www.macmillanmh.com for additional student resources.

## Online Student Edition

See the book online.

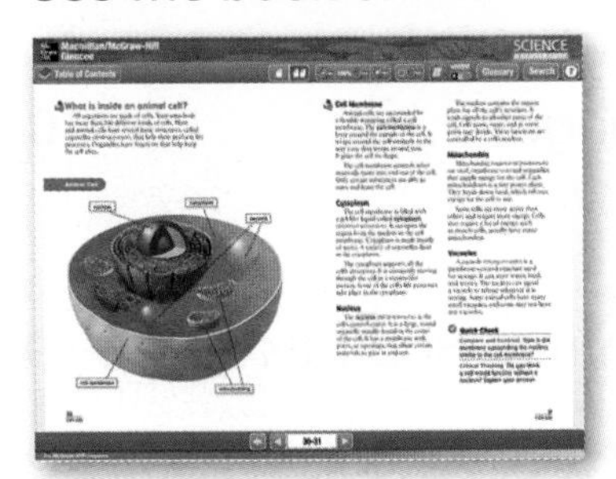

## e-Journal

Discover and write about science.

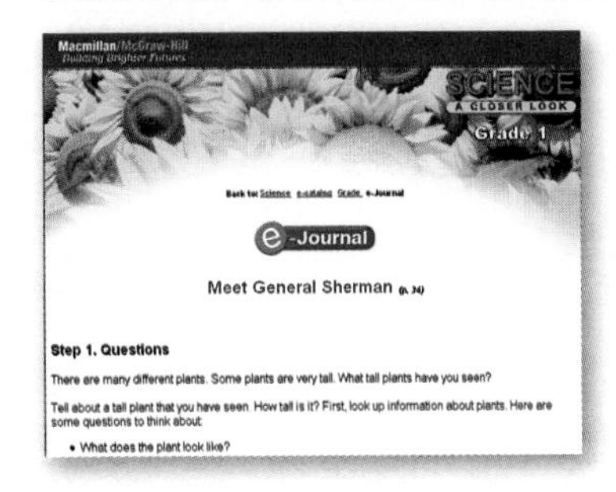

## Vocabulary Games

Review the vocabulary words.

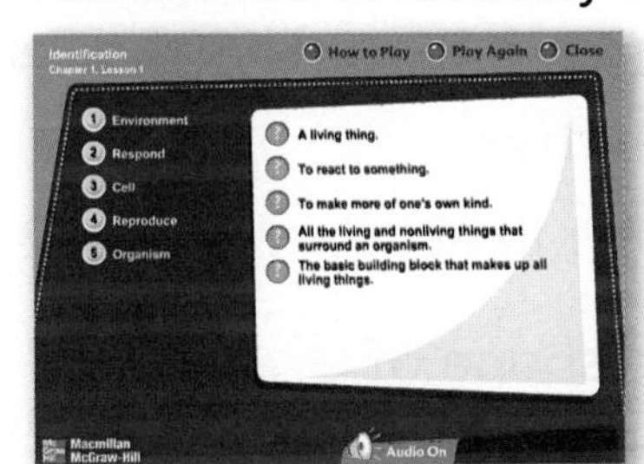

## e-Glossary

See and hear vocabulary.

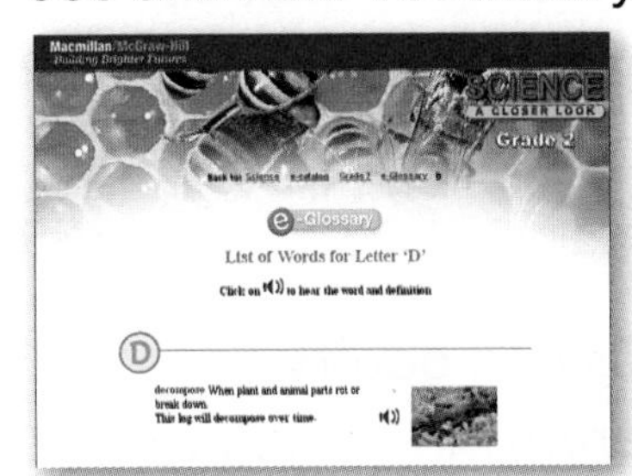

## e-Career

Learn about science careers.

## e-Review

Watch a summary and take a quiz for each lesson.

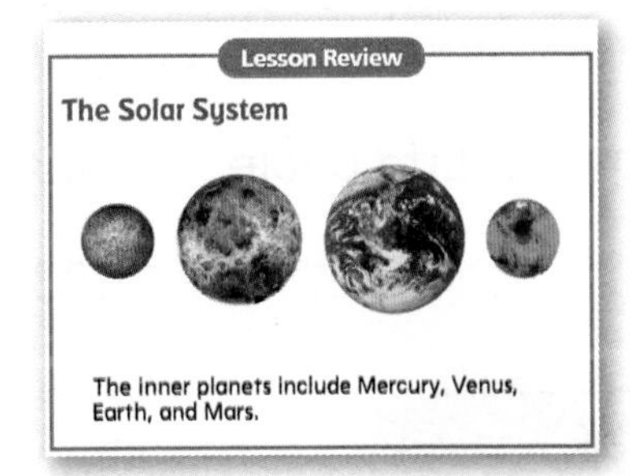


## Other Online Resources

See NASA images or how science is a part of your life every day at the American Museum of Natural History online.

# Activities and Investigations

## Life Science





xiv

Be a Scientist investigations and Focus on Skills skill builders are found in the Activities Flipchart.

# Earth and Space Science





xv

# Activities and Investigations

## Physical Science





xvi

# Be a Scientist

**Some tree frogs lay their eggs on leaves floating on water.**

Connect to

at www.macmillanmh.com

## Science Skills

### Objectives

- Identify skills scientists use to investigate questions.
- Explain how science skills are used to learn about pond animals.

Science Skills
Look and Wonder
Do you see the frog? How does it stay on the lily pad?
2
SCIENCE SKILLS


# 1 Introduce

### ▶ Assess Prior Knowledge

Create a **KWL** chart with children to determine what they already know about scientists and what they want to know. Ask:

- **What do scientists do?**
- **How do scientists work?**
- **How do scientists learn?**

Record children's answers in the What We Know column of the class **KWL** chart and note any misconceptions that they may have.

## Look and Wonder

Read the Look and Wonder questions with children.

Invite children to share their responses about the frog on the lily pad. Ask:

- **How would a scientist investigate how a frog stays on a lily pad?**

Write children's responses on the class **KWL** chart.

## Warm Up

### Start with a Discussion

Open the discussion by explaining to children that doctors, nurses, veterinarians, and astronomers all study science. Ask:

- **What kind of things might a doctor have to study?** Possible answers: how blood flows through the body; what makes people cough; how bones are connected to muscles
- **What kind of science does an astronomer study?** Possible answers: the planets and stars

Have children draw a picture of a scientist at work. Encourage children to write a caption for their drawings. Have children share their work with classmates.

### RESOURCES and TECHNOLOGY

- **Activity Lab Book,** pp. 1–3
- **Activity Flipchart,** p. 1

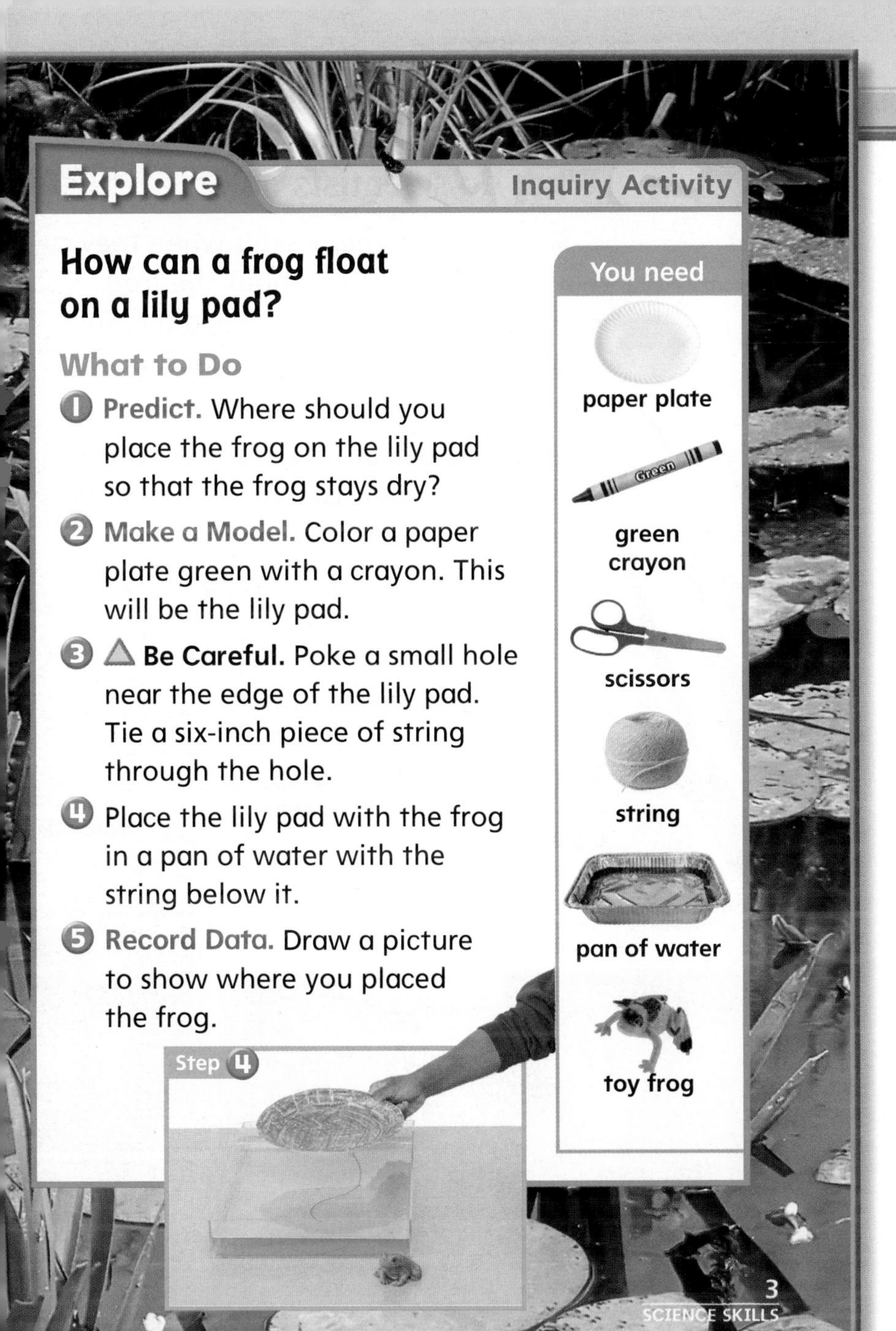
## Explore — Inquiry Activity

### How can a frog float on a lily pad?

**What to Do**

1. **Predict.** Where should you place the frog on the lily pad so that the frog stays dry?
2. **Make a Model.** Color a paper plate green with a crayon. This will be the lily pad.
3. ⚠ **Be Careful.** Poke a small hole near the edge of the lily pad. Tie a six-inch piece of string through the hole.
4. Place the lily pad with the frog in a pan of water with the string below it.
5. **Record Data.** Draw a picture to show where you placed the frog.

**You need**

- paper plate
- green crayon
- scissors
- string
- pan of water
- toy frog



3
SCIENCE SKILLS

pairs

20 minutes

**Plan Ahead** Fill the pans with water and cut 6-inch strings for all of the pairs beforehand. Children may need to take turns using the toy frog.

**Purpose** Support children's understanding about how scientists use models to investigate questions.

**Structured Inquiry** What to Do

Ask children whether they have seen lily pads, and, if so, to describe them. Ask: **How does a lily pad stay in its place in the water?** It is rooted in the soil.

1. **Predict** Encourage children to discuss the different locations to place the frog with their partner. Ask the children how they decided where to place the frog.
2. **Make a Model** Ask children to color the plates completely. Have them look closely at the picture of lily pads in their books and describe how their plates are similar to and different from the pictured lily pads.
3. Show children how to poke a hole at the edge of the paper plate with a pencil point. **Be Careful!** The pencil will be sharp. Have children make a knot at the end of the string and then pull the string through the plate hole. The knot should be at the top of the plate so the string falls below the plate to simulate roots hanging down. Children may need assistance doing this part of the activity.
4. Remind children to keep the water inside of the pans. If the pans are clear, encourage them to look through the side of the pan and describe the positions of the plate and string.
5. **Record Data** Have children share their work with classmates and compare where they placed the frog.

## Alternative Explore

### What can carry clay on water?

Provide children with a small ball of clay and materials that will float and materials that will not float, such as corks, paper, coins, craft sticks, and pattern blocks.

Have children make an object that will float while carrying the clay.

Ask children to record what they used and how they made their floating object.

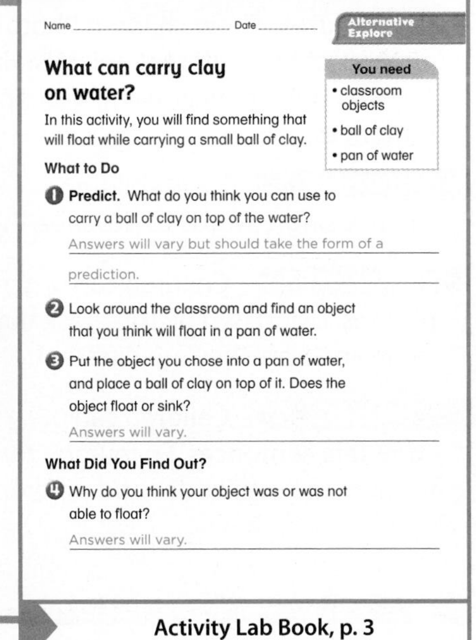
Name ______ Date ______ Alternative Explore

**What can carry clay on water?**

In this activity, you will find something that will float while carrying a small ball of clay.

**You need**
- classroom objects
- ball of clay
- pan of water

**What to Do**

1. **Predict.** What do you think you can use to carry a ball of clay on top of the water?
   Answers will vary but should take the form of a prediction.
2. Look around the classroom and find an object that you think will float in a pan of water.
3. Put the object you chose into a pan of water, and place a ball of clay on top of it. Does the object float or sink?
   Answers will vary.

**What Did You Find Out?**

4. Why do you think your object was or was not able to float?
   Answers will vary.

Activity Lab Book, p. 3

# 2 Teach

## Read Together and Learn

### What do scientists do?

#### ▶ Discuss the Main Idea

**Main Idea** Scientists make models and observe, compare, and classify their subjects of study to learn more about them.

Before reading, ask:

- **What kinds of things do you do to find an answer to a question?** Possible answer: Look up the answer in the encyclopedia.

Read the text with the children. Ask:

- **Why is making a model helpful?** Possible answer: It can show how something works.
- **How does observing something help a scientist to learn more about it?** Possible answer: A scientist can see what happens, notice details, and compare changes.
- **Why do scientists compare and classify things?** Possible answers: They can find out a lot about things by knowing how they are the same and different. Putting things in order makes it easier to study them.

## What do scientists do?

Scientists use many skills when they work. You wondered about the frog on a lily pad. Just as you did, a scientist might **make a model**. A model shows how something in real life looks.

Scientists use other skills that you can use too. Scientists **observe**, or look carefully. A scientist who observes a pond can find many amazing things.

Scientists observe the height, color, and shape of plants near a pond.

4
SCIENCE SKILLS

## ELL Support

**Use Pictures** Show children Photo Sorting Cards 11 and 16, or pictures of two different animals. Have children describe and compare the animals, using descriptive words and complete sentences.

**BEGINNING** Have children observe each animal. Children can use single words to describe them.

**INTERMEDIATE** Children can describe each animal in short phrases. Using two pictures, ask the children how they are the same and how they are different.

**ADVANCED** Children can compare the animals using complete sentences. Encourage them to make a Venn diagram to show the similarities and differences.

Scientists compare things by telling how they are alike or different. Look at the two pond animals on this page. How might a scientist compare them?

Look closely. Both animals have wings. They both live near ponds. They are also different in many other ways. Scientists find ways to classify things, or put them in groups. Insects and birds are in different animal groups.

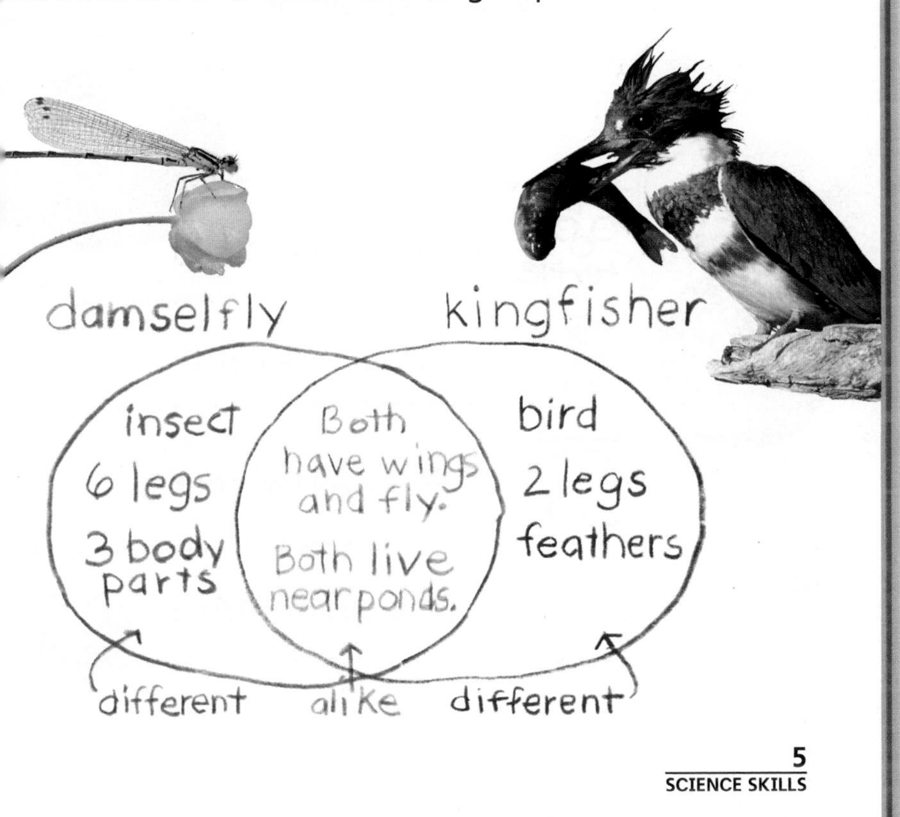


5
SCIENCE SKILLS

## Differentiated Instruction

### Leveled Activities

**EXTRA SUPPORT** Display two objects that are the same and different in various ways, such as two different plants or writing utensils. Ask children to observe the objects and describe them. Have children compare how the objects are the same and different. Repeat this activity with other objects.

**ENRICHMENT** Ask children to find two objects in the classroom. Have them create a Venn diagram comparing how the objects are the same and different. Encourage children to share their Venn diagrams with a partner.

## ▶ Use the Visuals

Have children look at the pictures on page 4 and ask them to describe what they see. Ask:

- **How are the plants the same and different?** Possible answer: Some plants have flowers, others do not.
- **How can observing the plants over a long time help scientists?** Possible answers: Scientists can observe changes in the plants. They can observe which things help plants grow.
- **If a scientist observed changes in the plants, what might the scientist try to find out?** Possible answer: If there were changes in the water of the pond.

Have children look at the pictures and Venn diagram on page 5. Explain to children that a Venn diagram is a way to compare things and show how things are different and the same.

Tell children that in a Venn diagram, the items that are different are written in the circles under the titles, and the items that are the same are written in the space where the two circles intersect. Ask:

- **How is the damselfly different from the kingfisher?** Possible answer: The damselfly has six legs and the kingfisher has two legs.
- **How are they the same?** Possible answer: They have wings, can fly, and live near ponds.
- **How does it help a scientist to compare animals?** Possible answers: They can see many different ways that animals do the same things. They can group the animals.

## ▶ Explore the Main Idea

**ACTIVITY** Have children work in pairs. Ask them to observe and describe what their partner is wearing. Have partners make a Venn diagram to compare how their clothes are the same and different.

# How do scientists work?

## ▶ Discuss the Main Idea

**Main Idea** Scientists measure, record data, put things in order, and infer to find out more about the things they are studying.

Read the text on page 6 with children. Ask:

- **What are some things people measure?** Possible answers: ingredients, temperature, lengths of wood
- **Why is it important for scientists to record data?** Possible answer: so they do not forget information; they can compare data later on; to use the data again

Read the text on page 7 together. Invite children to answer the questions. Ask:

- **Why do scientists put things in order?** Possible answers: It is easier to read the information. It helps them compare information. It organizes the data.
- **What things do you use that are put in order?** Possible answers: The telephone book is in alphabetical order. The pages of a story are in the order in which the story takes place. Nesting dolls are arranged from smallest to largest.
- **How can you infer what the weather might be like tomorrow?** Possible answers: by observing the weather today; finding out the temperature

## How do scientists work?

Look at all the eggs a scientist found near a pond! Scientists can **measure** how large or how heavy the eggs are. When you measure, you find out how long or how heavy something is. You can also find out how hot or how cold something is.

The facts scientists find are called data. When scientists **record data**, they write down what they observe.

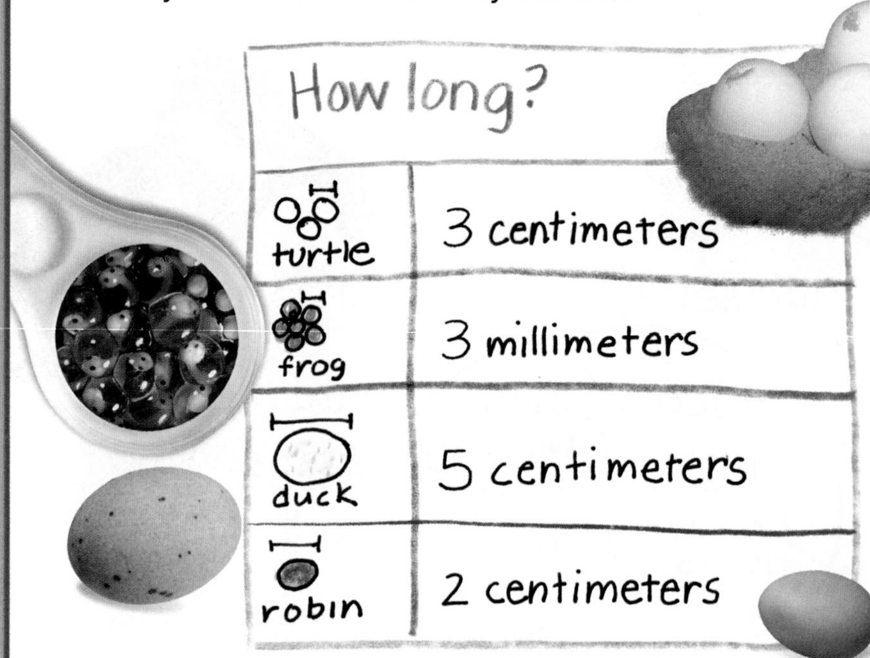


6
SCIENCE SKILLS

## ELL Support

**Demonstrate** Provide different-sized objects to children. Label each one. Measure each object with children. Write the name of each object and its length on the board.

**BEGINNING** Children can point to the longest object and shortest object. Have children put them in order from longest to shortest.

**INTERMEDIATE** Children can use short sentences to answer questions about the objects. Explain that they are providing data about the objects.

**ADVANCED** Children can explain the chart on the board. Children can explain in what order the blocks should be placed if they were to go from shortest to longest.

After scientists collect data, they can put their data in order. To **put things in order** means to arrange them in some way. For example, you can order the eggs by their size. Which egg is smallest? Which is largest?

Another skill scientists use is to **infer**. When you infer, you use what you know to figure something out. Can you infer which eggs belong to the animals on this page?

## Differentiated Instruction

### Leveled Questions

**EXTRA SUPPORT** Ask questions such as these to check children's understanding of the material.

- What do scientists do? Possible answer: They observe, measure, record data, put things in order, and infer.
- What is data? facts

**ENRICHMENT** Use these types of questions to develop children's higher-order thinking skills.

- How does using what you know help you to figure out a problem? Possible answer: Using what I know helps me identify the part of the problem I need to figure out.
- Why is it important for scientists to record measurements? Possible answer: They will remember what they measured.

## ▶ Use the Visuals

Review the chart on page 6 with children. Ask children to read the title and labels. Ask:

- **How is a chart helpful?** People can see the information clearly.
- **What information does the chart give?** Possible answer: the length of the eggs of different animals
- **Which animal has the shortest eggs?** frog
- **Which animal has the longest eggs?** duck

Have children look at the photographs of the different animals' eggs. Ask children to describe the eggs and explain how they are different. Ask:

- **Which eggs would be easiest for another animal to eat? Why?** Possible answer: The frog eggs would be easiest to eat because they do not have a hard shell

## ▶ Explore the Main Idea

**ACTIVITY** Provide children with different size boxes, and objects that fit in each box. Have children match each object to a box. Ask:

- **How did you match the boxes with the correct objects?** Possible answer: I put the boxes and the objects in size order.
- **What did you infer?** Possible answer: The largest object matches the largest box.

Have children use a ruler to measure each box. Encourage them to make a chart to record their data.

# How do scientists learn new things?

## ▶ Discuss the Main Idea

**Main Idea** Scientists investigate, predict, draw conclusions, and communicate their ideas about the results of their investigations.

Read the text together. Ask:

- **What investigations have you done at school?** Possible answers: grown plants; made games
- **How do scientists predict what the answer might be to a question?** They use what they know to guess what will happen next.
- **How does a wrong prediction help a scientist?** Possible answer: They learn from mistakes and can rule out a possible answer.

Have children predict what the young frog will look like next. Ask:

- **What helped you to predict what the frog will look like next?** Possible answer: looking at the other frog pictures

## ▶ Use the Visuals

Look at the illustrations on page 9. Explain to children that scientists record information in science journals similar to the one shown. Point out the date and labels. Then have children read the sequence of events in the notes. Ask:

- **Which words in the frog notes tell you about the order in which the frog grew?** *first, then, now*
- **What do the illustrations show?** how a frog grows
- **Why is it important for scientists to communicate their investigations?** Possible answers: Other people can learn from the investigation. Other scientists can redo the investigation to see if they get the same results.

## How do scientists learn new things?

Scientists learn new things by investigating. When you **investigate**, you make a plan and try it out.

Scientists start by asking a question. They predict what the answer might be. When you **predict**, you use what you know to tell what you think will happen.

Look at the pictures of the tadpole and the young frog. What do you predict the young frog will look like next?

8
SCIENCE SKILLS

## Differentiated Instruction

### Leveled Activities

**EXTRA SUPPORT** Provide each child with a cup of water and red and yellow food coloring. Tell children they will investigate what happens when yellow and red colors are mixed together. Make a plan together. For example, decide how much water and drops of each food coloring to use. Ask children to predict what will happen. Have children draw conclusions and communicate what happens.

**ENRICHMENT** Ask children to write a plan to investigate what happens when oil is mixed with water. Have children make a prediction and then supply them with cups of water and teaspoons of cooking oil for the experiment. Encourage children to record their observations, check their predictions, draw conclusions, and communicate their results.

When you **draw conclusions**, you use what you observe to explain what happens. Scientists draw conclusions. They conclude tadpoles live in the water, grow legs, and climb onto land.

Scientists communicate their ideas to other people. When you **communicate**, you write, draw, or tell your ideas.

September 17

long legs
First, it was a tadpole.
Then the tadpole grew legs. It still has a tail.
Now it has long back legs and no tail.

Frogs grow legs and can walk on land.

## Think, Talk, and Write

1. Which skill helps scientists put things into groups?
2. Write about what new things you might want to learn if you were a scientist.



# 3 Close

## ▶ Using the KWL Chart

Review with children what they have learned about science skills and how they are used. Record their responses in the What We Learned column of the class **KWL** chart.

## ▶ Think, Talk and Write

1. classify
2. Answers will vary. Encourage children to think about different areas of science, such as astronomy, biology, medicine, ecology, anatomy, and botany.

## Formative Assessment

**Frog Science**

Have children fold a piece of paper into thirds and number each rectangle. Ask children to look at the frog pictures on pages 8 and 9. Ask them to predict what the leaping frog will do next and write it in the first box. Have them explain the reasons for their prediction.

1. I predict the frog will eat the fly.
2.
Lily pads grow in ponds.
3. I could use a ruler to measure a frog.

In the second box, have children draw a picture showing where frogs live. Ask them to explain how they inferred that frogs live in that place.

In the last box, have children write different frog measurements they could make and the tools they would use to measure frogs.

## Scientific Method

### Objective

- Explain the steps scientists use to investigate questions.

# 1 Introduce

### Assess Prior Knowledge

Have children share what they know about how scientists investigate questions. Ask:

- **What are some things a scientist needs to do to find out answers to a question?**
- **What steps do you take when you have a problem or question to solve?**

Record children's answers in the What We Know column of the class **KWL** chart.

### Look and Wonder

Read the Look and Wonder section. Invite children to share their responses to the question about how frogs move. Ask:

- **How would a scientist investigate other ways frogs move?** Possible answer: Carefully watch frogs.

Write children's responses on the **KWL** chart and note any misconceptions they may have.

Scientific Method
Look and Wonder
This frog can swim! How else can a frog move? Scientists ask questions like this. They follow certain steps to find the answers.
10
SCIENTIFIC METHOD


### Warm Up

**Start with a Demonstration**

Ask children to think about the different ways they can move their bodies. Have volunteers demonstrate specific movements, such as walking, hopping, jumping, crawling, and rolling.

List the movements on the board. Ask:

- **How can you find out which movements will take you across the classroom faster than others?**

Help children develop a plan to test which movements will get them across the classroom most quickly. For example, identify which movements to test, determine which route children will use to move from one part of the classroom to another, find a method to measure how quickly children move, and determine a way to record the results.

**RESOURCES and TECHNOLOGY**

- **Activity Lab Book,** pp. 4–6
- **Activity Flipchart,** p. 2

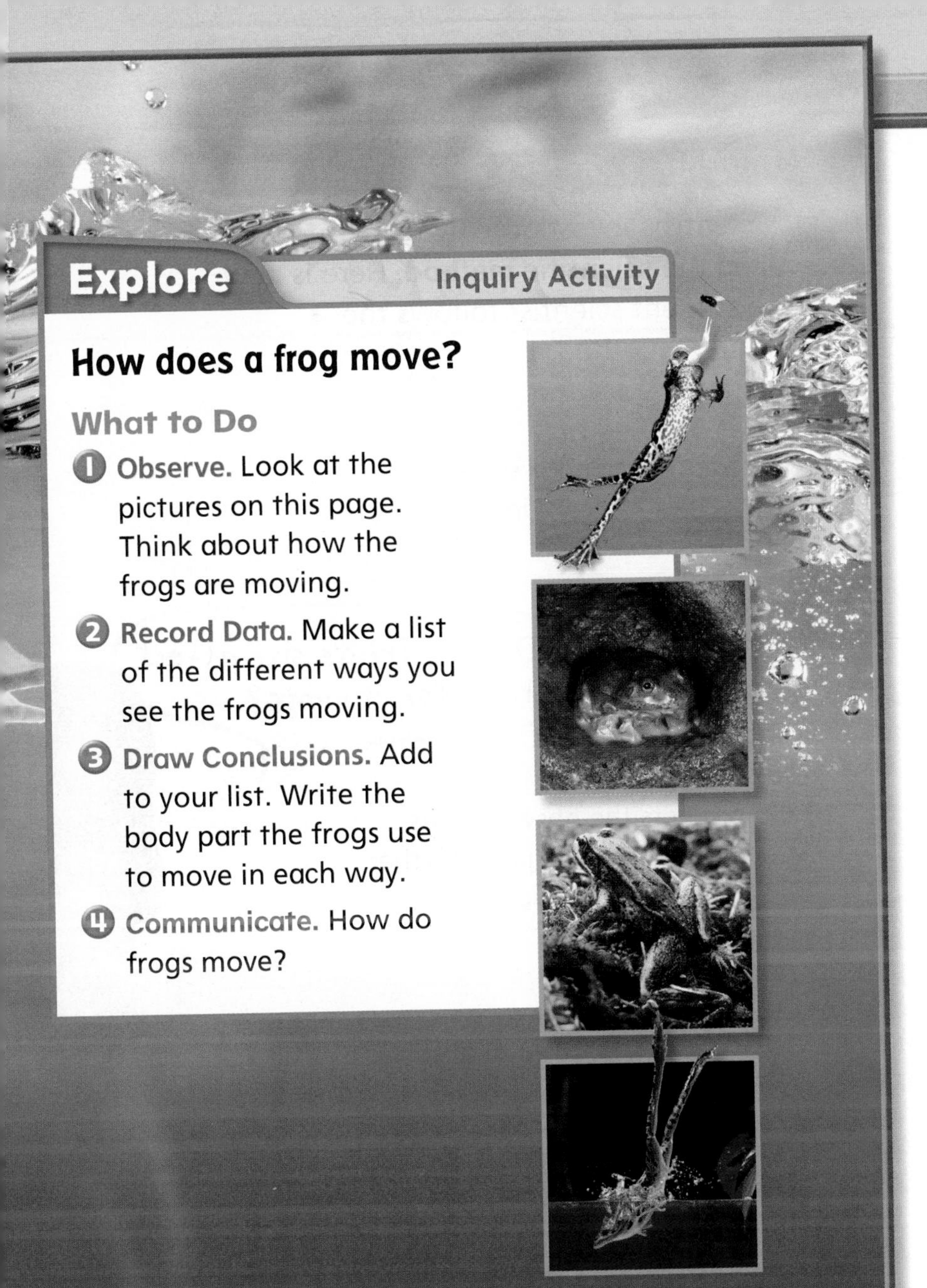

## Explore Inquiry Activity

### How does a frog move?

**What to Do**

1. **Observe.** Look at the pictures on this page. Think about how the frogs are moving.
2. **Record Data.** Make a list of the different ways you see the frogs moving.
3. **Draw Conclusions.** Add to your list. Write the body part the frogs use to move in each way.
4. **Communicate.** How do frogs move?

11
SCIENTIFIC METHOD

## Explore

individual 15 minutes

**Plan Ahead** Provide children with paper and pencils to record their observations.

**Purpose** Support children's understanding of the steps of the scientific method.

**Structured Inquiry** What to Do

1. **Observe** Remind children that observing means to look carefully.
2. **Record Data** Have children make a chart for their lists. Ask them to label the columns: *Picture 1, 2, 3,* and *4*. Encourage them to give details about the movements.
3. **Draw Conclusions** Have children describe the action in the body part, such as *legs are bent* or *legs are stretched*. Encourage children to act out the movements that they see, so they can better identify the body parts that are used.
4. **Communicate** Ask children to share their conclusions with their classmates. They should identify that frogs jump, sit, climb, and dive.

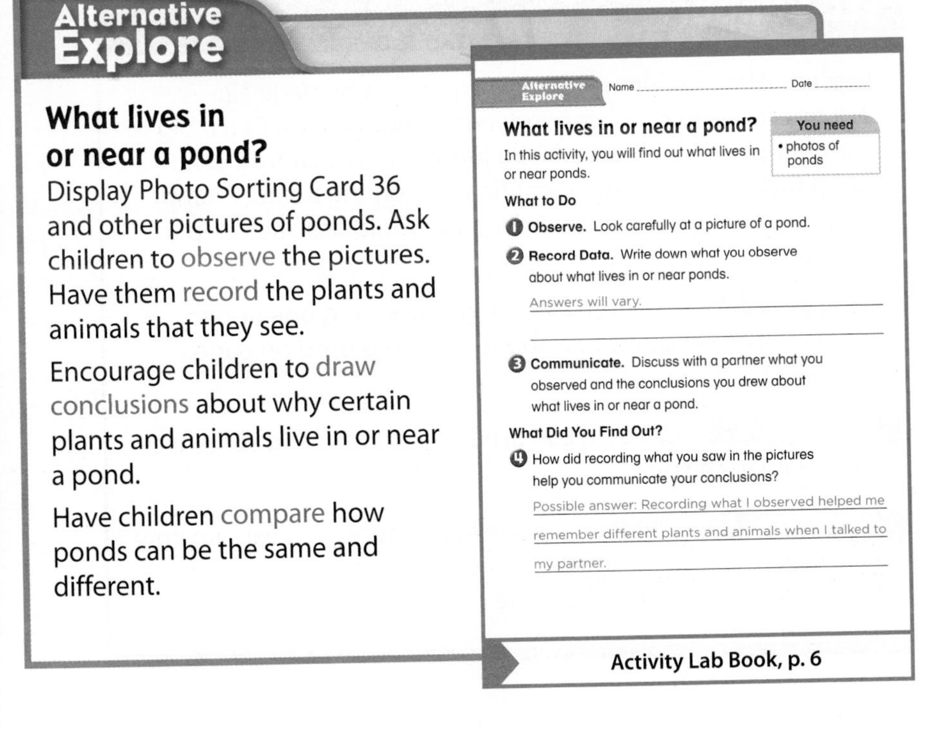

## Alternative Explore

### What lives in or near a pond?

Display Photo Sorting Card 36 and other pictures of ponds. Ask children to observe the pictures. Have them record the plants and animals that they see.

Encourage children to draw conclusions about why certain plants and animals live in or near a pond.

Have children compare how ponds can be the same and different.

Alternative Explore Name ____________ Date ________

**What lives in or near a pond?**

In this activity, you will find out what lives in or near ponds.

You need
- photos of ponds

**What to Do**

1. **Observe.** Look carefully at a picture of a pond.
2. **Record Data.** Write down what you observe about what lives in or near ponds.
   Answers will vary.
3. **Communicate.** Discuss with a partner what you observed and the conclusions you drew about what lives in or near a pond.

**What Did You Find Out?**

4. How did recording what you saw in the pictures help you communicate your conclusions?
   Possible answer: Recording what I observed helped me remember different plants and animals when I talked to my partner.

Activity Lab Book, p. 6

# 2 Teach

## Read Together and Learn

### How high can a frog jump?

#### ▶ Discuss the Main Idea

**Main Idea** The scientific method includes observing, asking questions, predicting, and making plans to learn more about something.

Before reading, ask children how they would investigate the question: *How high can a frog jump?*

After reading together, ask:

- **Why is observing important?** Possible answer: you can learn new things
- **How did Lola come up with her prediction?** Possible answer: she observed the frog's leg lengths.
- **What do you notice about Lola's plan?** Possible answers: It has numbered steps; it is written clearly so other people can understand it; she used writing and illustrations to communicate the plan.
- **What can Lola do if her plan doesn't work?** Possible answer: She can change the plan.

### How high can a frog jump?

Scientists investigate by following steps called the **scientific method**. Here is how one student scientist follows the scientific method.

**Observe**
Lola uses her science skills to observe the frogs in her classroom.

**Ask a Question**
Lola's question is: Does a frog's size affect how far it jumps?

**Make a Prediction**
Lola predicts the answer is yes. She thinks Andy will jump farther because his legs are longer.

Andy

Molly



## ELL Support

**Explain** Have children read Lola's question on page 12. Ask questions to check for understanding, such as: **Can a big frog jump far? Can a small frog jump far? Which frog can jump farther?** Write the scientific steps on the board. Prompt children to show or explain how they would use each step to answer Lola's question.

**BEGINNING** Children can use gestures and one or two words to answer questions about each step used on pages 12 and 13.

**INTERMEDIATE** Children can use short phrases and sentences to explain each step.

**ADVANCED** Children can use full sentences to explain how Lola used the steps to answer her question, and what they would do differently or the same.

### Make a Plan

Lola writes down a plan to test her idea. When she writes the plan, other people can follow it too.

My Frog Jumping Plan
① Make a starting line on the floor.
② Place one frog behind the starting line.
Clap to make it jump.
Measure how far the frog jumped.
③ Repeat step ② with the other frog.
clap!
start line
measure this distance
frog lands here

### Follow the Plan

Lola follows her plan. She changes the plan if parts of it do not work.



## ▶ Use the Visuals

Have children read Lola's plan on page 13. Explain to children that it is easier to read steps when they are numbered. Ask:

- **What is Lola measuring in her plan?** how far the frogs jumped
- **How does Lola make the frog jump?** She claps.
- **Why does Lola repeat step 2?** She needs to test the other frog.
- **Why is the drawing helpful?** It shows the idea of the plan.

## ▶ Explore the Main Idea

ACTIVITY Have children work in small groups. Distribute Photo Sorting Cards 11, 13–18, 20, or pictures of animals to each group. Ask each group to choose a picture and think of a question about the animal that they would like to research.

Have children predict an answer to their question, and make a plan about how they can find out the answer. Remind children to develop a plan that has clear steps that can be easily followed.

## Differentiated Instruction

### Leveled Questions

**EXTRA SUPPORT** Ask questions such as these to check children's understanding of the material.

- What do you call the steps that scientists use to investigate questions? the Scientific Method
- What are some of the steps of the scientific method? observe, ask a question, make a prediction, make a plan

**ENRICHMENT** Use these types of questions to develop children's higher-order thinking skills.

- Why do scientists change their plans? to make a plan work
- Why do scientists want others to follow their plans? Possible answer: to see whether other scientists get the same results

## What did you find out?

### ▶ Discuss the Main Idea

**Main Idea** The Scientific Method also includes recording data, trying plans again, drawing conclusions, and communicating ideas.

Before reading, ask:

- **What do you think needs to happen after you make a plan for an investigation?** You need to follow the plan.

Read the text together. Ask:

- **Where did Lola record her results?** on a chart
- **How else can results be recorded?** Possible answers: diagram, bar graph
- **How was Lola able to draw a conclusion?** Possible answer: from the results of her plan

### ▶ Use the Visuals

Have children look at the chart on page 14. Ask:

- **How did Lola organize her chart?** Possible answer: frogs listed in rows, each jump attempt listed in a different column
- **Which frog jumped farther?** Molly **How much further?** 5 cm
- **What does the first try tell us about Lola's prediction?** Possible answer: It might be wrong.
- **Why will it help Lola to test the frogs two more times?** Possible answer: If she gets the same results, she can make a stronger conclusion.

## What did you find out?

**Record the Results**
Lola makes a chart to show how far each frog jumps.

How far can each frog jump?

| frog | 1st try | 2nd try | 3rd try |
|---|---|---|---|
| Andy | 20 cm | | |
| Molly | 25cm | | |

**Try the Plan Again**
Lola tests each frog three times. This helps her know if her results are correct.

**Draw a Conclusion**
Lola explains what her results mean.

14
SCIENTIFIC METHOD

## Differentiated Instruction

### Leveled Activities

**EXTRA SUPPORT** Ask children to predict whether people with longer legs can jump farther. Have two children of different heights volunteer to do a broad jump. Record the data in a chart on the board. Ask children to jump two more times. Have children draw a conclusion from the data and explain it.

**ENRICHMENT** Have children do an activity to gather data, such as tossing a coin, to find out how many times it lands on heads or tails. Ask children to record the data, draw a conclusion, and explain what the results mean.

a talks to her classmates
out what her results mean.
is can lead to new questions
d new investigations.

You can follow the
Scientific Method
when you investigate too!


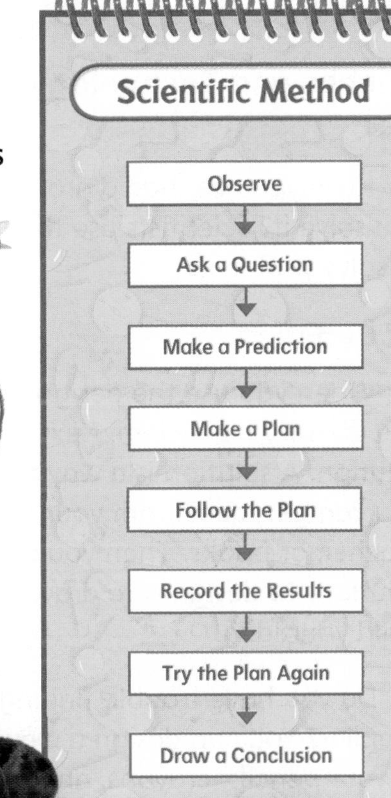

Scientific Method
Observe
Ask a Question
Make a Prediction
Make a Plan
Follow the Plan
Record the Results
Try the Plan Again
Draw a Conclusion


## Think, Talk, and Write

1. Why do you think it is important for scientists to make a plan?
2. Write about why scientists write down their plans.



# 3 Close

### ▶ Using the KWL Chart

Review with children what they have learned about the steps of the Scientific Method. Record their responses in the What We Learned column of the class **KWL** chart.

### ▶ Think, Talk and Write

1. Possible answers: The plan is important for other scientists to follow and see if they get the same results; for other people to know how the scientist got the results; for scientists to know how to change the plan if it doesn't work.
2. Possible answers: Scientists write down their plans to so they won't forget the steps they are going to take; so others can follow the plan.

## Formative Assessment

### Dissecting the Method

Show children a chart like the one on page 14, but at the top write: *How far can each frog move its tongue to catch a fly?* and in the *First Try* column write 4 cm for Andy and 2 cm for Molly.

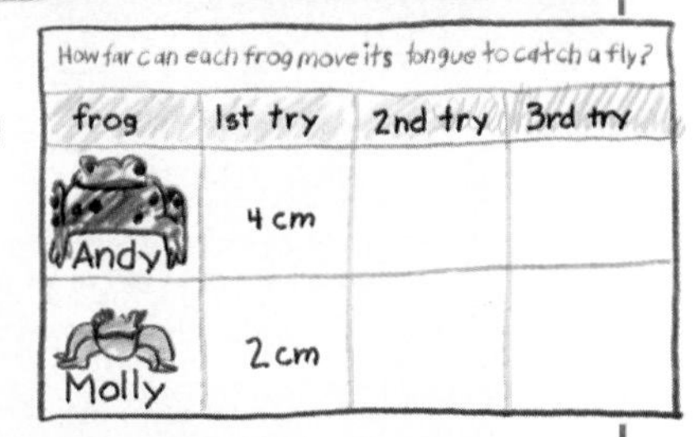

How far can each frog move its tongue to catch a fly?

| frog | 1st try | 2nd try | 3rd try |
|---|---|---|---|
| Andy | 4 cm | | |
| Molly | 2 cm | | |

Ask children to fold a piece of paper in fourths and number the boxes. In the first box, have them write an observation the scientist may have made based on the question. In the second box, ask them to write the question the scientist is researching. In the third box, have children describe the plan the scientist may follow to investigate the question. In the fourth box, have them identify the step in the Scientific Method that the chart represents.

# Focus on Skills

**Objective**

- Complete the design process by coming up with a way to organize school supplies so that they can easily be found.

**Materials:** paper, pencil

**Plan Ahead:** Have some school supplies out for children to view.

EXTEND This activity will teach children the steps involved in the design process. They will design a model of something to assist in organizing school supplies, and compare that model to those of others. They will also consider ways to improve their original designs.

## Science and Technology:

## The Design Process

### ▶ Learn It

Explain to children that the design process involves a series of steps that scientists use to develop ideas for solving a problem or meeting a need.

- **Why is it important to ask questions during the design phase of the design process?** to find ways to improve the product design so that it solves the problem
- **Why do scientists sometimes ask for feedback from other people?** Other people might have different opinions about whether the product works; other people may make suggestions for improvements.

Tell children that scientists often test their designs many times before they are satisfied with the final product.

**RESOURCES and TECHNOLOGY**

▶ **Technology: A Closer Look,** Lesson 5

## Focus on Skills

**Science and Technology:**
**The Design Process**

Have you ever had a problem? How did you solve it? Scientists use the **design process** to solve problems.

### ▶ Learn It

When you use the design process, first you identify a problem. Next you think of a solution. A solution is a way to fix a problem. You can get ideas from your friends, a teacher, or books. Then you design your solution. To design is to draw, plan, and build your idea.

Do you have trouble finding your school supplies? You can design a way to keep track of your pencils, crayons, and other supplies.

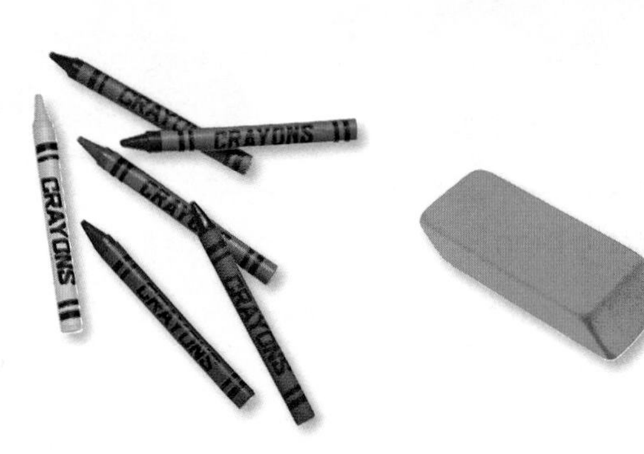

Skill Builder

## Try It

Michael designed a box to hold all of school supplies. Michael's box had a ce for his pencils, crayons, glue, and sers.

Design a way to store your school plies. Make a sketch of your idea. are your idea with your cher. Gather the terials that you need your design. Build ur invention and test ur design.

How did your design compare to Michael's?

Did your design solve your problem?

Write about it. How could you change your design to make it better?



## ▶ Try It

1. Help children brainstorm a list of school supplies they might need to consider while designing an organizer.
2. Encourage children to consider details like convenience of use and ease of building in their designs.

## ▶ Apply It

1. Have children present or discuss their designs with the class. Allow them to make constructive suggestions as to how other children's designs could be improved.
2. Ask children to compare and contrast the designs of different children in the class. Lead the class in identifying any common characteristics among the designs.
3. Allow children to revise or improve upon their designs by drawing new models following the class discussion.

## Science and Technology Background

The design process is a creative, methodical approach to solving a problem that incorporates aspects of the scientific method. This sequential process can involve repetition of certain steps based on feedback and testing. After identifying a problem, the scientist brainstorms possible solutions and sketches or jots them down. Rather than deliberating a single hypothesis, the scientist may devise multiple possible solutions to the problem, more than one of which could be suitable. Testing and assessing the validity of possible solutions is crucial. If none of the possible solutions are feasible, more possibilities need to be generated. The scientist ultimately selects a final design to the problem based on the most useful and practical possibilities. The design process is most effectively implemented by exchanging information with one's peers, revising designs based on peer review, and using creativity and logic to find solutions to a problem.

# Safety Tips

**Objective**

- Identify important safety procedures.

## ▶ Talk About It

Encourage children to share their experiences with rules and to discuss as a class why rules are made. Ask:

- **What kinds of safety rules do you have at home?**
- **How are rules for the kitchen and playing outside the same?**

Write children's responses on chart paper. Ask:

- **Why do people make rules?**

Children should understand that rules are created to keep them safe.

## ▶ Learn About It

Have a volunteer read the first sentence on page 16. Ask children to list other safety symbols they know, such as stop signs. Invite them to look through their books and find **Be Careful!** notations. Ask:

- **Why do you need to be careful when doing the activity on that page?**

Discuss the types of science activities children may do in class, and encourage them to propose safety procedures. Have a volunteer read the rest of page 16. For each safety tip, ask children to explain the rationale behind the rule. Ask:

- **How does this rule help us stay safe?**

## ▶ Try It

Divide the class into five groups and assign one safety tip from page 16 to each group. Have each group create a poster to explain and illustrate their safety tip, and encourage them to present their posters to the class.

**RESOURCES and TECHNOLOGY**

- **Activity Lab Book,** pp. v–vi
- **TeacherWorks™ Plus CD-ROM**

## Safety Tips

**When you see "△ Be Careful", follow the safety rule**

Tell your teacher about accidents and spills right away.

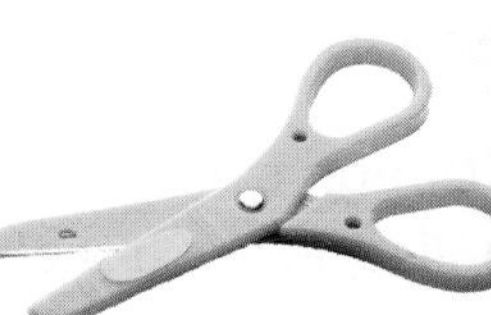

Be careful with sha objects and glass.

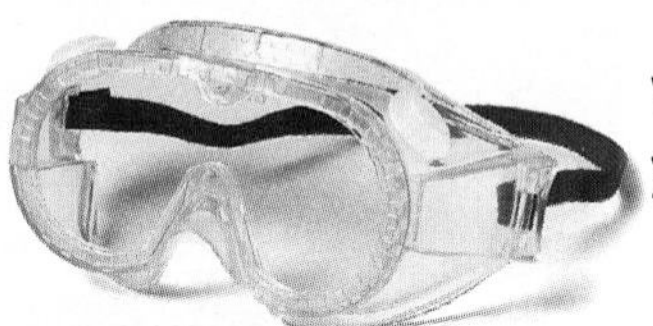

Wear goggles when you are told to.

Wash your hands after each activity.

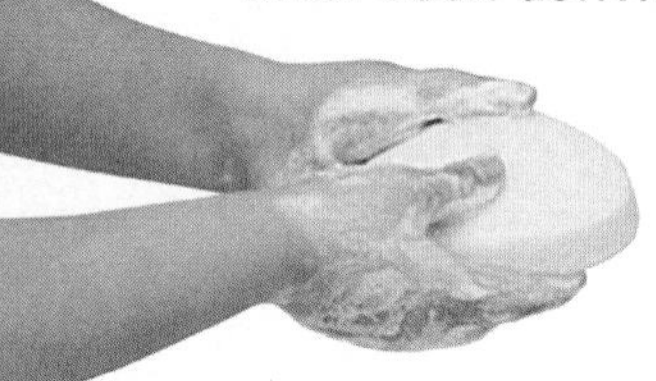

Keep your workplace neat. Clean up when you are dor

16
SAFETY

## Integrate Writing

**Introduction to the Science Kit**

Distribute Science Kit items to small groups of children. Choose items that are likely to be unfamiliar, such as goggles, funnels, hand lenses, or droppers. Have children discuss what each item is and how it may be used by scientists.

Review all the objects by displaying each item. Ask:

- **How could this be used during a science activity?**

If children have difficulty identifying an object, name the item and explain how it is used.

Once all of the unfamiliar items have been introduced, have each child choose one, draw it, label it, and write a sentence to describe how it is used.

UNIT E

# Matter

Some paints get their color from plants and minerals.

# Materials

**Materials required to complete activities in the Student Edition are listed below. These charts also list materials found in the Deluxe Activity Kit.**

## Non-Consumable Materials

Based on 6 groups

| MATERIALS | QUANTITY NEEDED PER GROUP | KIT QUANTITY | CHAPTER/ LESSON |
|---|---|---|---|
| Balance | 1 | 1 | 9/2; 10/1 |
| Classroom objects | 6 | | 9/1, 2 |
| Cup, measuring | 1 | 12 | 9/3; 10/1, 3 |
| Cup, plastic, 9 oz | 2 | 50 | 9/3; 10/3 |
| Film canister | 6 | 18 | 9/3 |
| Jar, plastic 16 oz, with lid | 1 | 12 | 10/1 |
| Metric weight set | 1 | 1 | 10/1 |
| Pan, aluminum foil (8″ x 8″) | 1 | 6 | 9/2, 3 |
| Scissors | 1 | | 10/1, 2 |
| Spoon, metal | 1 | 3 | 9/2 |
| Spoon, wood | 1 | 3 | 9/2 |

## Consumable Materials

Based on 6 groups

| MATERIALS | QUANTITY NEEDED PER GROUP | KIT QUANTITY | CHAPTER/ LESSON |
|---|---|---|---|
| Apples | ¼ | | 10/1 |
| Butter | 1 tbsp | | 10/2 |
| Chocolate pieces | 1 piece | | 10/2 |
| Clay, modeling | 1 box | 2 (1-lb) boxes | 10/1 |
| Cracker, assorted | 6 | | 9/1 |
| Cracker, fish shape | 6 | | 9/1 |
| Cracker, square shape | 6 | | 9/1; 10/1 |
| Crayons | several | | 9/1; 10/1 |
| Cream, chilled | 1/4 c | | 10/1 |
| Glue | small amount | | 10/2 |
| Juice, lemon | 3 tbsp | | 10/1 |
| Knife, plastic | 1 | 24 | 10/1, 2 |
| Magazines, nature | 3 | | 10/2 |
| Markers, colored | several | | 10/2 |
| Paper, construction | 1 sheet | | 10/2 |
| Plate, paper | 2 | 50 | 10/1, 2 |
| Salt | ¼ c | | 10/3 |
| Sand | ¼ c | 1 (5.5-kg) bag | 10/3 |
| Spoon, plastic | 1 or 2 | 24 | 9/2; 10/3 |
| Wrap, plastic | several in. | | 10/1 |

The Big Idea **What are different types of matter?**

The Big Idea **How can matter change?**

## CHAPTER 9
## Looking at Matter

 CD-ROM

**TeacherWorks™ Plus CD-ROM**
Interactive Lesson Planner, Teacher's Edition, Worksheets, and Links to Online Resources

**Classroom Presentation Toolkit CD-ROM**

**ExamView® *Assessment Suite* CD-ROM**

**PuzzleMaker CD-ROM**

DVD

**Science: Master Teacher DVD Set**

**Science Activity DVD**

*Science in Motion* *Measuring Solids*

**Online Teacher's Edition**

**Leveled Reader Database**

**Progress Reporter Assessments**

**Professional Development**

**e-Glossary**

**e-Journal**

**e-Review**

NSDL National Science Digital Library

## CHAPTER 10
## Changes in Matter

 CD-ROM

**TeacherWorks™ Plus CD-ROM**
Interactive Lesson Planner, Teacher's Edition, Worksheets, and Links to Online Resources

**Classroom Presentation Toolkit CD-ROM**

**ExamView® *Assessment Suite* CD-ROM**

**PuzzleMaker CD-ROM**

DVD

**Science: Master Teacher DVD Set**

**Science Activity DVD**

*Science in Motion* *Adding Heat to Ice*

**Online Teacher's Edition**

**Leveled Reader Database**

**Progress Reporter Assessments**

**Professional Development**

**e-Glossary**

**e-Journal**

**e-Review**

NSDL National Science Digital Library

◀ Operation: Science Quest

## CHAPTER 9
# Looking at Matter

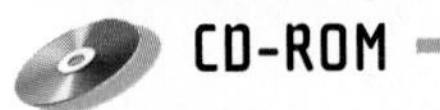

StudentWorks™ Plus CD-ROM

PuzzleMaker CD-ROM

Science Activity DVD

 www.macmillanmh.com

 *Science in Motion* *Measuring Solids*

Online Student Edition

Online Vocabulary Games

e-Careers

e-Glossary

e-Journal

e-Review

## CHAPTER 10
# Changes in Matter

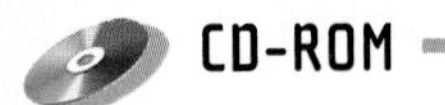

StudentWorks™ Plus CD-ROM

PuzzleMaker CD-ROM

Science Activity DVD

 www.macmillanmh.com

 *Science in Motion* *Adding Heat to Ice*

Online Student Edition

Online Vocabulary Games

e-Careers

e-Glossary

e-Journal

e-Review

Science in Motion *From Wheat to Bread* ▶

# Literature
## Poem

### Objective
- Identify how heat changes popcorn, which makes a sound.

Literature
Poem
Popcorn Hop
by Stephanie Calmenson
Everybody do the popcorn dance!
290

## Popcorn Hop

**Genre: Poetry** Poems have words placed in lines. Sometimes words at the end of the lines rhyme.

Sometimes poets use words because of the way they sound. Have children look at the words on page 291, Ask:

- **What things make a popping sound?** Possible answers: a bottle opening; games; toys

## Before Reading

Explain to children that poems are often quite musical and that poets pick the words for their poems very carefully. Have children look at the illustration and ask:

- **What does the illustration show?** Possible answer: popcorn popping
- **What does the sound of popcorn popping sound like?** Possible answer: There is a lot of popping sounds; some pops are soft and some are loud.

## During Reading

Read the poem together. Tell children that the words in capital letters should be read louder than the other words. Emphasize the popping words to create a strong rhythmic pattern. Ask:

- **What happens when the popcorn gets hot?** It pops.

Discuss movements people make when they feel warm and when they walk on hot pavement barefooted. Invite children to recite the poem again. This time have them use body movements to show what is happening to the popcorn. In the last stanza have children move to the beat of the *Popcorn Hop*.

### RESOURCES and TECHNOLOGY
- **Reading and Writing,** p. 161
- **ELL Teacher's Guide,** p. 85

### ELL Support

**Share Information** Write *popcorn* on the board and have children break down the compound word. Explain that *pop* is the sound that a kernel makes when it heats and turns into popcorn. Tell them that this is an example of heat energy changing something.

**BEGINNING** Have children use one or two words to explain how corn becomes popcorn, and give their languages' counterparts to the word *pop*.

**INTERMEDIATE** Ask children to use short sentences to explain how to make popcorn, and give their languages' counterparts to the word *pop*.

**ADVANCED** Have children use their own words to explain how a corn kernel becomes popcorn, and give their languages' counterparts to the word *pop*.

Put your popcorn
in a pot.
Wait till it gets
really hot.
When you start to
feel the heat,
Listen for the
popcorn beat:
Pop-pop-POP-pop
pop-pop-POP!
Come and do the
popcorn hop!
Talk About It
What made the popcorn
change?
291

## After Reading

Have children focus on how the poem puts in order the steps for making popcorn. Ask:

- **What does the author say to do first?** Put popcorn in a pot.
- **What does she say to do next?** Heat the popcorn.
- **What does she say to do last?** Listen for the pop, and then dance.

Point out to children that heat energy changes things. Ask:

- **If you listened for the beat before putting the pot on the stove, would you hear anything? Why not?** No; popcorn needs heat to pop.

## Talk About It

Possible answers: The hard popcorn kernels popped and became fluffy and soft. The heat made the popcorn change.

If children have difficulty understanding how heat changes things, ask:

- **How does heat change a cheese sandwich when you grill it?** It makes the cheese soft by melting it.

### More to Read

***The Popcorn Book,*** by Tomie de Paola (Scholastic, 1978).

## Differentiated Instruction

### Leveled Activities

**EXTRA SUPPORT** Give children unpopped corn kernels and ask them to describe how they look and feel. If possible, use a popper or other heat source to pop some corn, or bring in popped corn. Have children carefully observe the popcorn to locate the remains of the hard shell of the kernel. Ask them to compare unpopped corn and popcorn and identify what caused the kernels to pop.

**ENRICHMENT** Explain that things react to heat in different ways. For example, not all seeds pop when heat is added. Have children explore how other foods react to heat. Ask them to identify foods they want to test. Together, plan a safe way to investigate what happens when the foods are heated, such as microwaving. Encourage them to make predictions, observe what happens, record the results, and draw conclusions.

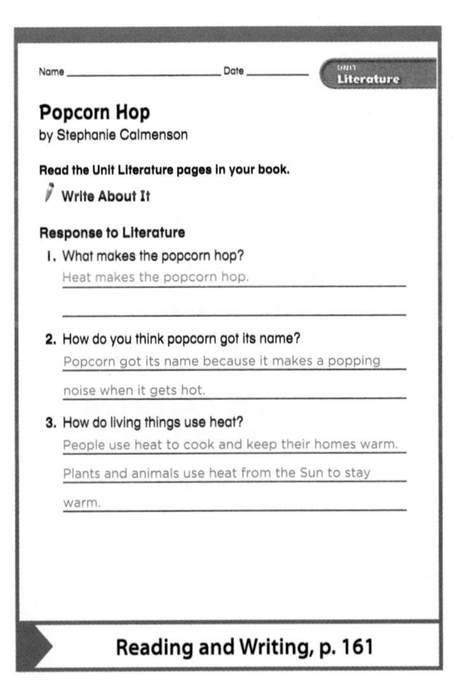
Name ____________ Date ________ Unit Literature

Popcorn Hop
by Stephanie Calmenson

Read the Unit Literature pages in your book.

Write About It

Response to Literature

1. What makes the popcorn hop?
Heat makes the popcorn hop.

2. How do you think popcorn got its name?
Popcorn got its name because it makes a popping noise when it gets hot.

3. How do living things use heat?
People use heat to cook and keep their homes warm. Plants and animals use heat from the Sun to stay warm.

Reading and Writing, p. 161

# CHAPTER 9 Planner

**Classroom Presentation Toolkit CD-ROM** Lesson Presentations

**TeacherWorks™ Plus CD-ROM** Interactive Lesson Planner, Teacher's Edition, Worksheets, and Online Resources.

| Lesson | OBJECTIVES AND READING SKILLS | VOCABULARY | RESOURCES AND TECHNOLOGY |
|---|---|---|---|
| **1 Describing Matter**<br>PAGES **294–299**<br>PACING: 3 days<br>*FAST TRACK:* 1 day | ▪ Identify matter as anything that has mass and takes up space.<br>▪ Compare and contrast different properties of matter.<br>Different Alike Different<br>**Reading Skill**<br>Compare and Contrast *Graphic Organizer 10* | **matter**<br>**mass**<br>**property** | Reading and Writing, pp. 163–166<br>Activity Lab Book, pp. 131–136<br>Visual Literacy, p. 26<br>Transparencies, p. 26 |
| **2 Solids**<br>PAGES **300–307**<br>PACING: 2 days<br>*FAST TRACK:* 1 day | ▪ Compare and contrast the properties of solids.<br>▪ Use different ways to measure solids.<br>Summary<br>**Reading Skill**<br>Summarize *Graphic Organizer 5* | **solid** | Reading and Writing, pp. 167–172<br>Activity Lab Book, pp. 137–140<br>Visual Literacy, p. 27<br>Transparencies, p. 27<br>**Science in Motion** *Measuring Solids* |
| **3 Liquids and Gases**<br>PAGES **308–315**<br>PACING: 3 days<br>*FAST TRACK:* 1 day | ▪ Describe the properties of liquids and gases.<br>▪ Compare and contrast liquids and gases.<br>**Reading Skill**<br>Classify *Graphic Organizer 11* | **liquid**<br>**volume**<br>**gas** | Reading and Writing, pp. 173–178<br>Math, pp. 17–18<br>Activity Lab Book, pp. 141–144<br>Visual Literacy, p. 28<br>Transparencies, p. 28 |
| **I Read to Review**<br>PAGES **316–319** | **Matter All Around**<br>▪ Selection for independent reading<br>**Performance Assessment** | | **Resources**<br>School to Home Activities, pp. 91–92<br>Assessment, pp. 117–118 |
| **CHAPTER 9 Review**<br>PAGES **320–321, 321A–321B** | ▪ Review chapter concepts.<br>**Resources**<br>Assessment, pp. 106–109, 113–116<br>Reading and Writing, pp. 179–180 | | **Technology**<br>**ExamView® *Assessment Suite* CD-ROM**<br>**-Review** |

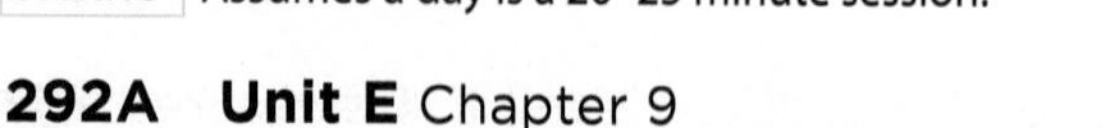

PACING Assumes a day is a 20–25 minute session.

www.macmillanmh.com for more planning resources and www.nsdl.org/refreshers/science for science resources from NSDL

**Science Activity DVD** Explore Activity demos

Materials included in the Deluxe Activity Kit are listed in *italics.*

## EXPLORE Activities

### Explore p. 295 PACING: 20 minutes 

**Objective** Classify objects based on their characteristics.

**Skills** observe, record data, classify

**Materials** crackers, crayons, paper

| | texture | shape | size | color |
|---|---|---|---|---|
| | | | | |
| | | | | |
| | | | | |
| | | | | |
| | | | | |

PLAN AHEAD Select different kinds of crackers and have enough for groups to share. **Be careful!** Be aware of children's food allergies.

### Explore p. 301 PACING: 20 minutes 

**Objective** Identify the properties of solids.

**Skills** observe, predict, record data

**Materials** *a variety of spoons, aluminum pan,* water, pencils, paper

PLAN AHEAD Collect a variety of spoons made from different materials.

### Explore p. 309 PACING: 30 minutes 

**Objective** Recognize that the volume of a liquid is determined by the size and shape of its container.

**Skills** predict, draw conclusions, infer

**Materials** *measuring cups, containers, aluminum pans,* paper, pencils

PLAN AHEAD Select containers of various sizes and shapes. Line tables with newspapers beforehand in case of any spills.

## QUICK LAB Activities

### Quick Lab p. 298 PACING: 15 minutes 

**Objective** Describe and classify objects based on their shapes and sizes.

**Skills** observe, classify

**Materials** classroom objects, crayons, paper

PLAN AHEAD Display a variety of objects around the classroom for children to classify.

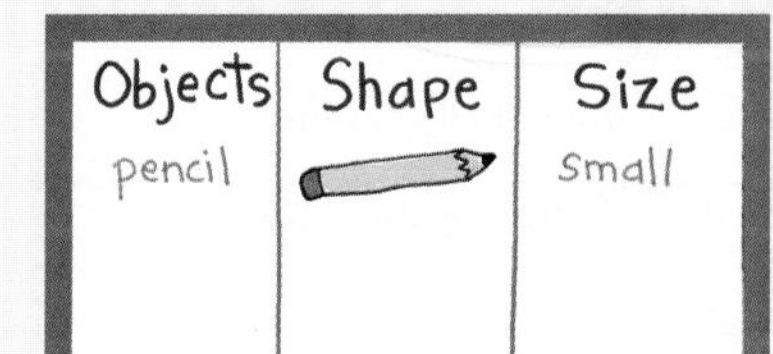


### Quick Lab p. 304 PACING: 15 minutes 

**Objective** Use a balance to determine the mass of two solids.

**Skills** measure, compare, put in order

**Materials** *balances,* classroom objects

PLAN AHEAD Use objects small enough to fit in the buckets of the balance.

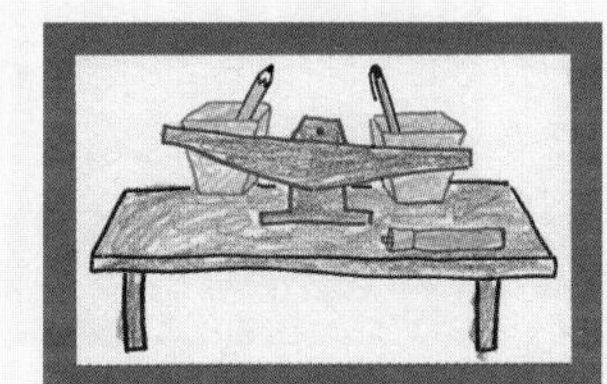

### Quick Lab p. 313 PACING: 15 minutes 

**Objective** Classify matter by different properties.

**Skills** infer

**Materials** *six cans with lids* for the entire class, small solid objects (toys) that will fit in the cans, liquids

PLAN AHEAD Collect cans with lids of equal size, shape, and material. Fill the cans prior to the lesson.

## FOR MORE ACTIVITIES

**Focus on Skills**

Teach the inquiry skill: record data, p. 41

Use the Activities in your work station. See **Activity Lab Book** for more support.

For a comprehensive list of consumable and non-consumable materials, see the back of the unit tab.

# English Language Learner Support

## Technology

For additional language support and vocabulary development, go to www.macmillanmh.com

Science in Motion

*Measuring Solids*

Vocabulary Games

## ADDITIONAL RESOURCES

pp. 86–95

## Academic Language

English language learners need help in building their understanding of the academic language used in daily instruction and science activities. The following strategies will help to increase children's language proficiency and comprehension of content and instruction words.

### Strategies to Reinforce Academic Language

- **Use Context** Academic language should be explained in the context of the task. Use gestures, expressions, and visuals to support meaning.
- **Use Visuals** Use charts, transparencies, and graphic organizers to explain key labels to help children understand classroom language.
- **Model** Use academic language as you demonstrate the task to help children understand instruction.

### Academic Language Vocabulary Chart

The following chart shows chapter vocabulary and inquiry skills as well as some Spanish cognates. **Vocabulary** words help children comprehend the main ideas. **Inquiry Skills** help children develop questions and perform investigations. **Cognates** are words that are similar in English and Spanish.

| Vocabulary | Inquiry Skills | Cognates | |
|---|---|---|---|
| | | **English** | **Spanish** |
| matter, p. 296 | observe, p. 295 | observe, p. 295 | *observar* |
| mass, p. 296 | record data, p. 295 | classify, p. 295 | *clasificar* |
| property, p. 298 | classify, p. 295 | matter, p. 296 | *materia* |
| solid, p. 302 | predict, p. 301 | mass, p. 296 | *masa* |
| liquid, p. 310 | draw conclusions, p. 309 | predict, p. 301 | *predecir* |
| volume, p. 311 | infer, p. 309 | solid, p. 302 | *sólido* |
| gas, p. 312 | | conclusion, p. 309 | *conclusión* |
| | | infer, p. 309 | *inferir* |
| | | liquid, p. 310 | *líquido* |
| | | volume, p. 311 | *volumen* |
| | | gas, p. 312 | *gas* |

## Vocabulary Routine

**Use the routine below to discuss the meaning of each word on the vocabulary list. Use gestures and visuals to model all words.**

**Define** *Property* is the look, feel, smell, sound, or taste of an object.

**Example** Small, soft, and furry are *properties* of a toy raccoon.

**Ask** What is a *property* of a pencil?

**Children may respond to questions according to proficiency level with gestures, one-word answers, or phrases.**

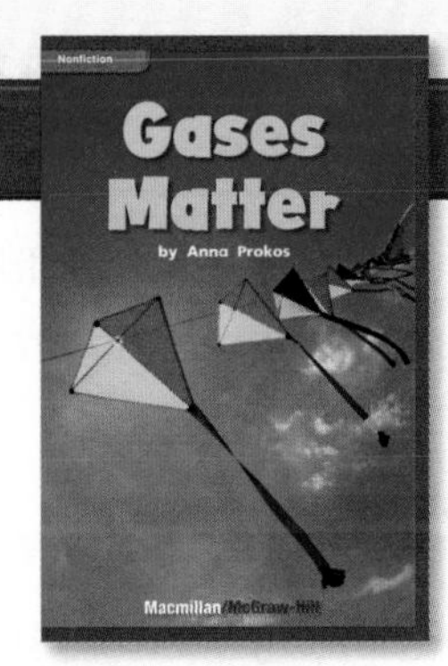


**ELL Leveled Reader**

**Gases Matter**
by Anna Prokos

**Summary** Read about gases and how they give us everything we need to live.

**Reading Skill**
Classify

## Vocabulary Activities

Help children compare and contrast objects according to their properties.

**BEGINNING** Pass around classroom objects, such as a glue stick, a board eraser, and a sheet of paper. Label a three-column chart with the names of the objects. Encourage children to brainstorm properties: *How do the objects feel? What colors are the objects?* List responses in the chart and go over the properties listed.

**INTERMEDIATE** Pass around a furry toy, a ball of clay, and a piece of sandpaper. Ask: *What special properties do these objects have?* Demonstrate squeezing, bending, and folding. Have students complete the sentence frames: *The ______ is squeezable. The ________ is bendable. The ________ is foldable.*

**ADVANCED** Working in pairs, have partners list the properties of a personal or classroom object of their choice. Have groups exchange objects and try to add other properties to the lists.

## Language Transfers

**Grammar Transfer**
Cantonese, Haitian Creole, Hmong, Korean, and Vietnamese do not use a plural marker.

*Round and smooth are property of a soccer ball.*

**Phonics Transfer**
Haitian Creole, Hmong, Korean, and Vietnamese do not have the r-controlled vowel sound /ûr/ as in *property*.

CHAPTER 9

# Looking at Matter

 **THE BIG IDEA** **What are different types of matter?**

**Chapter Preview** Have children take a picture walk through the chapter and predict what the lessons will be about.

## ▶ Assess Prior Knowledge

Before beginning the chapter, create a **KWL** chart with children. Ask the Big Idea question, and then ask:

- **How can you use your senses to describe different objects?**
- **What are some of the differences among solids, liquids, and gases?**

**Matter**

| What We Know | What We Want to Know | What We Learned |
|---|---|---|
| Things can be described by color. | How do we describe gases? | |
| Solids are hard. | How do we know gases are around us? | |
| You can not see gases. | | |

Answers shown represent sample student responses.

Follow the **Instructional Plan** at right after assessing children's prior knowledge of chapter content.

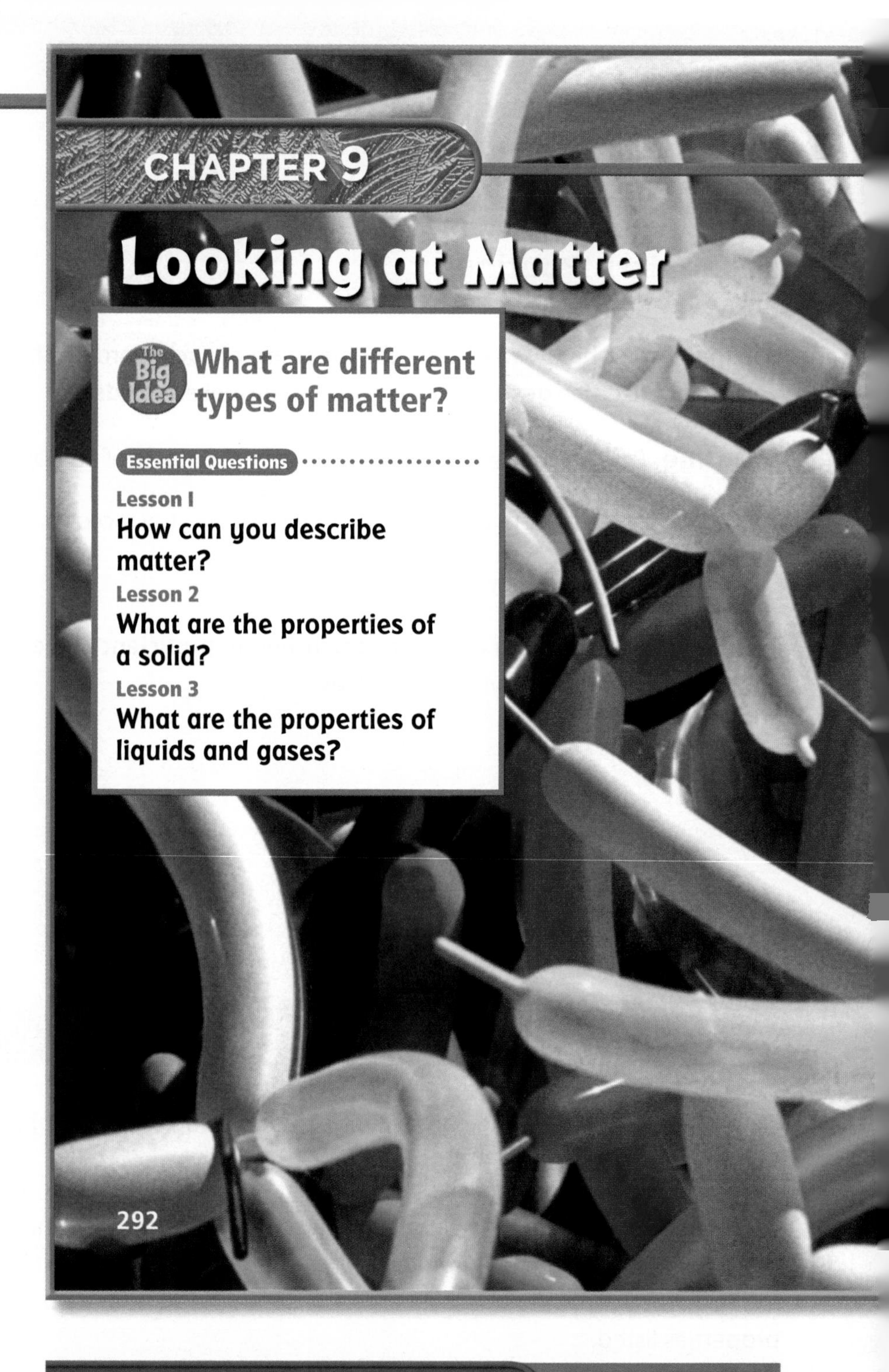
CHAPTER 9

Looking at Matter

What are different types of matter?

Essential Questions

Lesson 1
How can you describe matter?

Lesson 2
What are the properties of a solid?

Lesson 3
What are the properties of liquids and gases?

292

## Differentiated Instruction

**Instructional Plan**

**Chapter Concept** Matter exists in different states.

**EXTRA SUPPORT** Children who need to know the basic properties of objects should cover all of **Lesson 1,** pages 294–299, before continuing with the rest of the chapter.

**ON LEVEL** Children who know basic properties of objects can review the states of matter, **Lesson 1,** pages 298–299, and then go to **Lesson 2,** pages 300–307, and **Lesson 3,** pages 308–311, as well as pages 314–315, to compare properties of solids and liquids, and explore how they are measured.

**ENRICHMENT** Children who are ready to go further may explore how gases are measured, **Lesson 3,** pages 312–313.

### RESOURCES and TECHNOLOGY

- **School to Home Activities,** pp. 83–92
- **Reading and Writing,** pp. 163–180
- **Assessment,** pp. 106–118
- Classroom Presentation Toolkit CD-ROM
- PuzzleMaker CD-ROM
-  www.macmillanmh.com
- 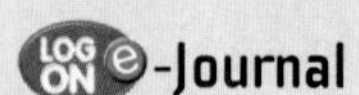 -Journal

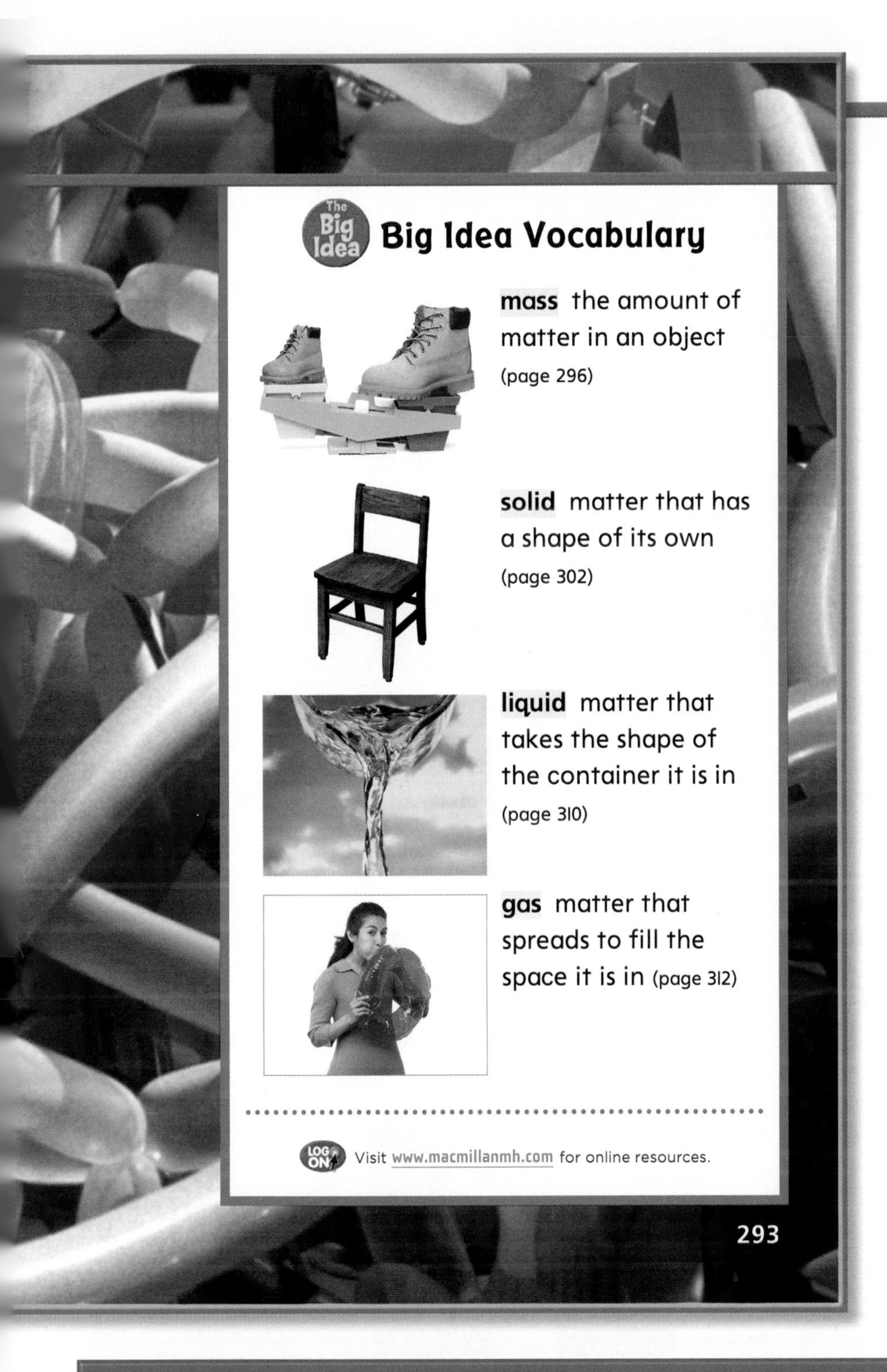
## Big Idea Vocabulary

**mass** the amount of matter in an object (page 296)

**solid** matter that has a shape of its own (page 302)

**liquid** matter that takes the shape of the container it is in (page 310)

**gas** matter that spreads to fill the space it is in (page 312)

Visit www.macmillanmh.com for online resources.

# Big Idea Vocabulary

- Have a volunteer read the **Big Idea Vocabulary** words aloud to the class. Ask children to find one or two of the words in the chapter by using the given page references. Add these words and their definitions to a class Word Wall.

- Encourage children to use the illustrated glossary in the Student Edition's reference section. Guide children to explore the **e-Glossary**, which offers audio pronunciations, definitions, and sentences using the vocabulary words.

## Science Leveled Readers

ALSO ON AUDIO CD · Leveled Reader Library

APPROACHING

**Matter and Change** Find out about the properties of solids, liquids, and gases.

**ISBN:** 978-0-02-285857-5

ON LEVEL

**Gases Matter** Read about gases and how they give us everything we need to live.

**ISBN:** 978-0-02-285865-0

BEYOND

**Hot Air Balloons** Up, up, and away! Go on a journey in a hot air balloon.

**ISBN:** 978-0-02-285872-8

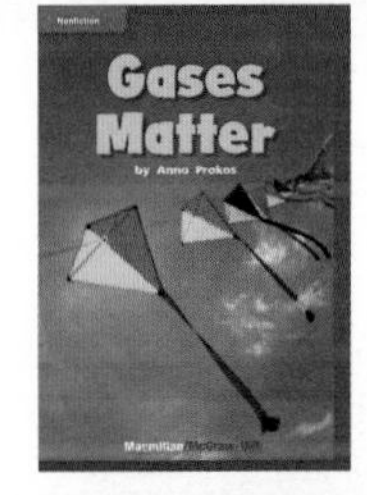


ELL

**Gases Matter** Uses sheltered language of On-Level Reader

**ISBN:** 978-0-02-283461-6

See teaching strategies in the Leveled Reader Teacher's Guide. To order, call 1-800-442-9685.

**Leveled Reader Database** Online Readers, searchable by topic, reading level, and keywords

Stop Here to

# Plan Your Lesson

## Lesson 1 Describing Matter

### Essential Question

How can you describe matter?

### Objectives

- Identify matter as anything that has mass and takes up space.
- Compare and contrast different properties of matter.

**Reading Skill** Compare and Contrast

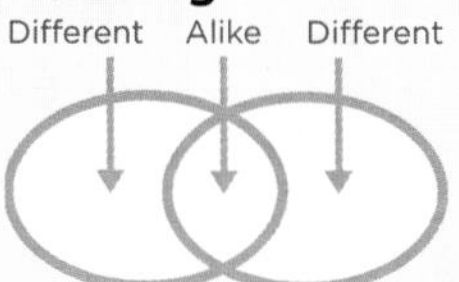


*Graphic Organizer 10,* p. TR12

## FAST TRACK

**Lesson Plan** When time is short, follow the Fast Track and use the essential resources.

**1 Introduce**
**Look and Wonder, p. 294**
Resource **Activity Lab Book, p. 133**

**2 Teach**
**Discuss the Main Idea, p. 296**
Resource **Visual Literacy, p. 26**

**3 Close**
**Think, Talk, and Write, p. 299**
Resource **Assessment, p. 110**

**Professional Development** Look for NSDL to find recommended Science Background resources from the National Science Digital Library.

## Reading and Writing

Name ______________ Date ________ Lesson Outline

**Describing Matter**

**Use your book to help you fill in the blanks.**

**What is matter?**

1. Matter is anything that takes up ___space___ and has mass.
2. Some matter can be ___made___ by people.
3. An object's mass is the amount of ___matter___ it has.
4. Objects can be made of ___different___ amounts of matter.
5. A ___balance___ is used to measure and compare mass.

**How can you describe matter?**

6. Matter can be described by talking about its ___properties___.
7. A ___property___ is how matter looks, feels, smells, tastes, or sounds.

**Outline, pp. 163–164**
Also available as a student workbook

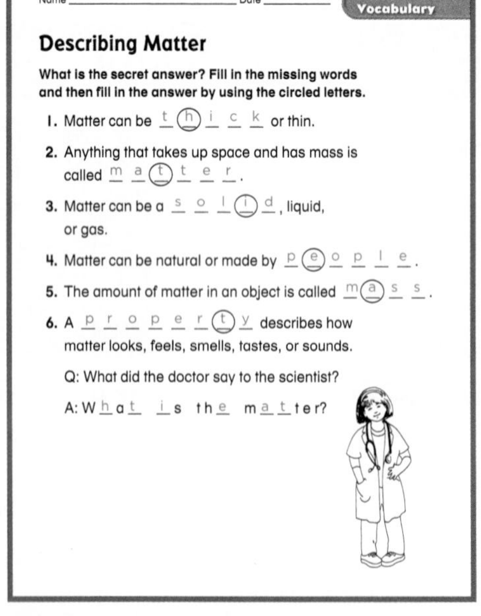
Name ______________ Date ________ Lesson Vocabulary

**Describing Matter**

**What is the secret answer? Fill in the missing words and then fill in the answer by using the circled letters.**

1. Matter can be t (h) i c k or thin.
2. Anything that takes up space and has mass is called m a (t) t e r.
3. Matter can be a s o l (i) d, liquid, or gas.
4. Matter can be natural or made by p (e) o p l e.
5. The amount of matter in an object is called m (a) s s.
6. A p r o p e r (t) y describes how matter looks, feels, smells, tastes, or sounds.

Q: What did the doctor say to the scientist?

A: What is the matter?

**Vocabulary, p. 165**
Also available as a student workbook

Read a Photo Name ______________ Date ________

**What is matter?**

Photographs can help you understand ideas. The photograph below helps to explain what matter is.

Using Matter

1. Yes. Water takes up space so it is matter.
2. Answers may vary but should include the boy, the dog, the tub, the towel, the water in the tub, the soap suds, the boy's clothes, and the air surrounding the boy and dog.

**Answer the questions.**

1. The boy is using water to wash the dog. Is water matter? Why or why not?
2. List all of the matter you see in the photograph.

**Read a Photo, p. 26**
Also available as a transparency

# ADDITIONAL RESOURCES

## Activity Lab Book

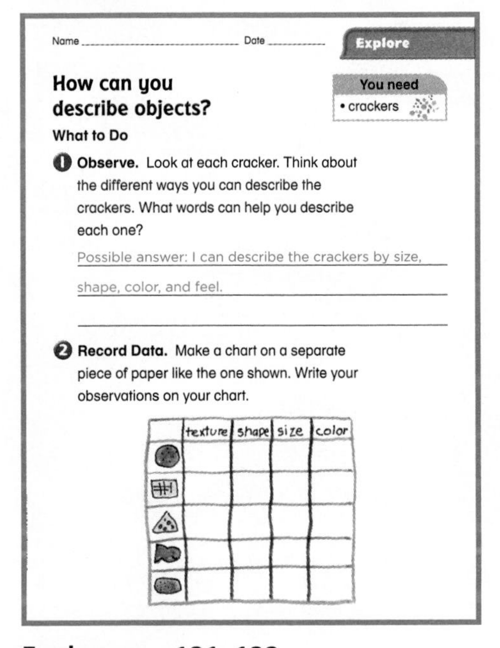

Name ________ Date ________ **Explore**

**How can you describe objects?**

**You need**
- crackers

**What to Do**

1. **Observe.** Look at each cracker. Think about the different ways you can describe the crackers. What words can help you describe each one?

   Possible answer: I can describe the crackers by size, shape, color, and feel.

2. **Record Data.** Make a chart on a separate piece of paper like the one shown. Write your observations on your chart.

| | texture | shape | size | color |
|---|---|---|---|---|
| | | | | |
| | | | | |
| | | | | |
| | | | | |
| | | | | |

**Explore, pp. 131–132**
Also available as a student workbook

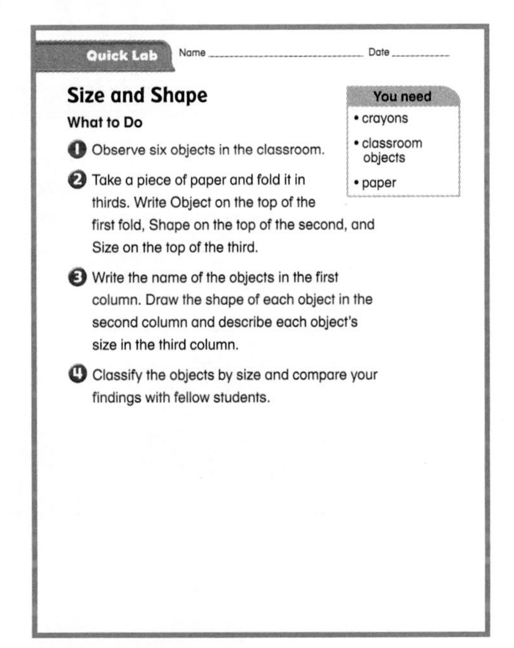

**Quick Lab** Name ________ Date ________

**Size and Shape**

**You need**
- crayons
- classroom objects
- paper

**What to Do**

1. Observe six objects in the classroom.
2. Take a piece of paper and fold it in thirds. Write Object on the top of the first fold, Shape on the top of the second, and Size on the top of the third.
3. Write the name of the objects in the first column. Draw the shape of each object in the second column and describe each object's size in the third column.
4. Classify the objects by size and compare your findings with fellow students.

**Quick Lab, p. 134**
Also available as a student workbook

## Assessment

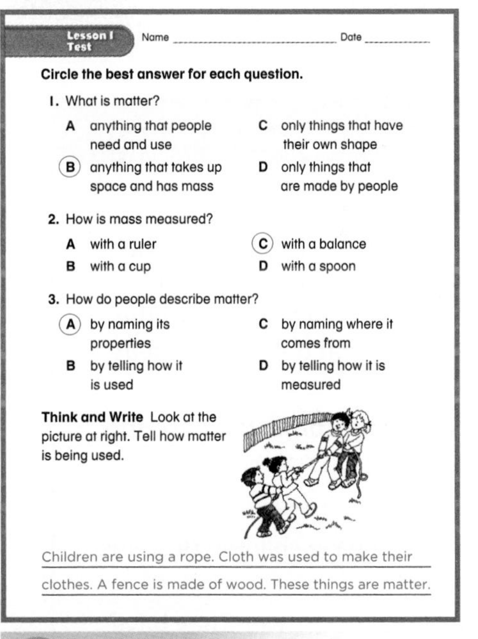

**Lesson 1 Test** Name ________ Date ________

**Circle the best answer for each question.**

1. What is matter?
   - A anything that people need and use
   - (B) anything that takes up space and has mass
   - C only things that have their own shape
   - D only things that are made by people
2. How is mass measured?
   - A with a ruler
   - B with a cup
   - (C) with a balance
   - D with a spoon
3. How do people describe matter?
   - (A) by naming its properties
   - B by telling how it is used
   - C by naming where it comes from
   - D by telling how it is measured

**Think and Write** Look at the picture at right. Tell how matter is being used.

Children are using a rope. Cloth was used to make their clothes. A fence is made of wood. These things are matter.

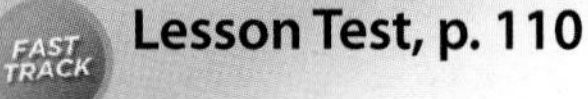

**Lesson Test, p. 110**

pp. 128–131

*Gases Matter*

pp. 40–41

p. 90

p. 26

26

81–90

pp. 85–86

67–69

### Technology

- Science Activity DVD
- TeacherWorks™ Plus CD-ROM
- Classroom Presentation Toolkit CD-ROM
-  e-Review
-  NSDL

# Lesson 1 Describing Matter

**Objectives**

- Identify matter as anything that has mass and takes up space.
- Compare and contrast different properties of matter.

## 1 Introduce

### Assess Prior Knowledge

Have children share what they know about matter. Ask:

- **What things are made up of matter?**
- **How can you describe objects that have matter?**

Record children's answers in the What We Know column of the class **KWL** chart.

### Look and Wonder

Read the Look and Wonder question, and list children's responses as they name what they see. Have children use as many descriptive words as possible to describe each object. Remind children to discuss the object's size, shape, color(s), and what it might be made of. Ask:

- **With what are the balloons filled?** air
- **What one thing does everything in the picture have in common?** Everything is made of matter.

Write children's responses on the class **KWL** chart and note any misconceptions that they may have.

**RESOURCES and TECHNOLOGY**

- **Activity Lab Book,** pp. 131–133
- **Activity Flipchart,** p. 40
- Science Activity DVD

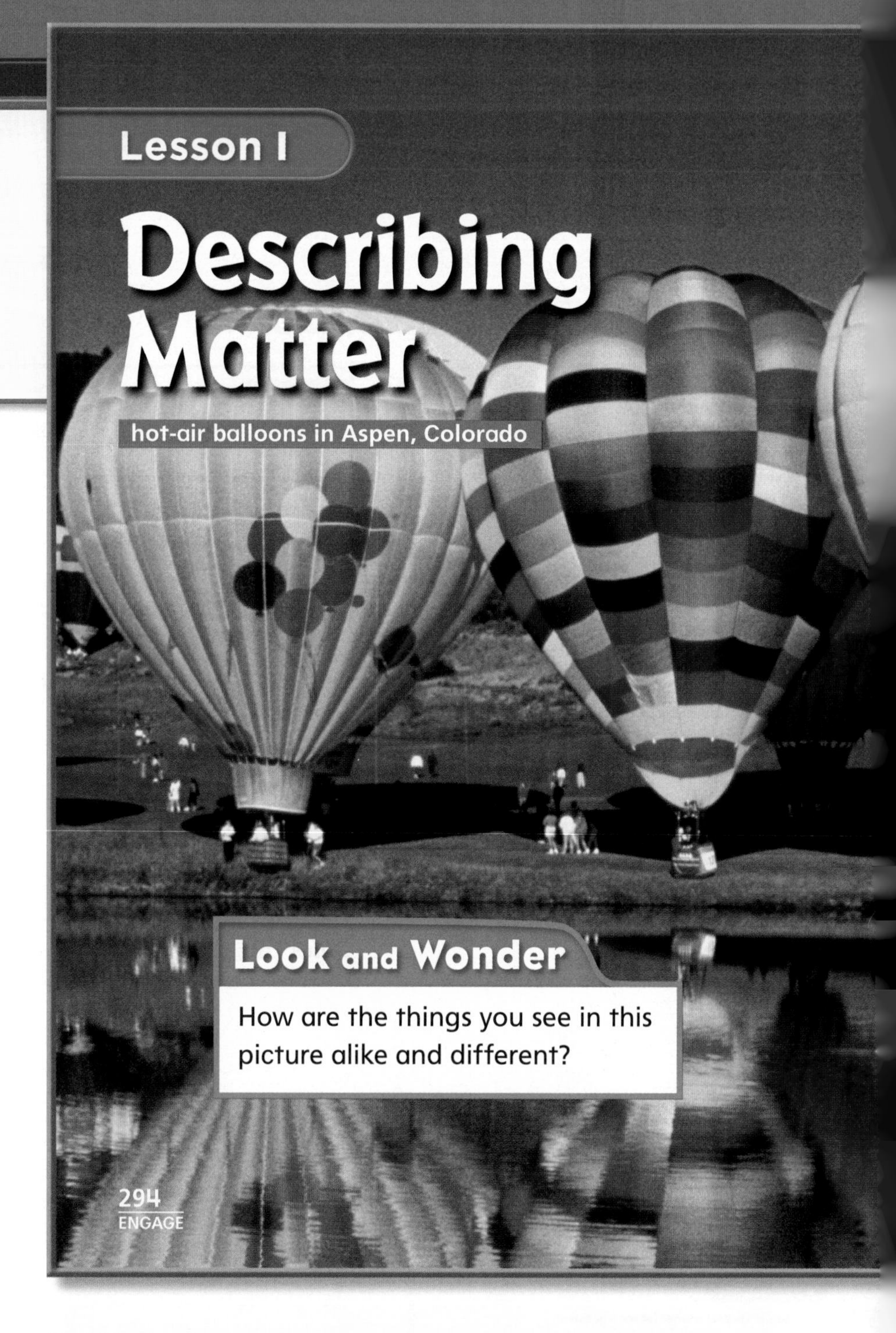

Lesson I
Describing Matter
hot-air balloons in Aspen, Colorado
Look and Wonder
How are the things you see in this picture alike and different?
294
ENGAGE


### Warm Up

**Start with a Demonstration**

Put a stapler inside a paper bag and tape it shut. Put another object that weighs about the same as the stapler, but is a different shape, in another bag. Pass the first bag around the classroom and have children describe what they feel. Repeat with the second bag. Ask:

- **If you could see the objects, what else could you describe about them?** Possible answers: color; what they're made from; what they're used for

Remove the objects from the bags and have children describe the objects. Ask:

- **How are the objects different?** Possible answer: They have different shapes.
- **How are the objects alike?** Possible answer: similar weight

## Explore

Inquiry Activity

### How can you describe objects?

**You need**

crackers

**What to Do**

1. **Observe.** Look at each cracker. Think about the different ways you can describe the crackers. What words can help you describe each one?
2. **Record Data.** Make a chart like the one shown. Write your observations on your chart.
3. **Classify.** Use your chart to help you sort the crackers.

**Explore More**

4. How else can you sort the crackers?

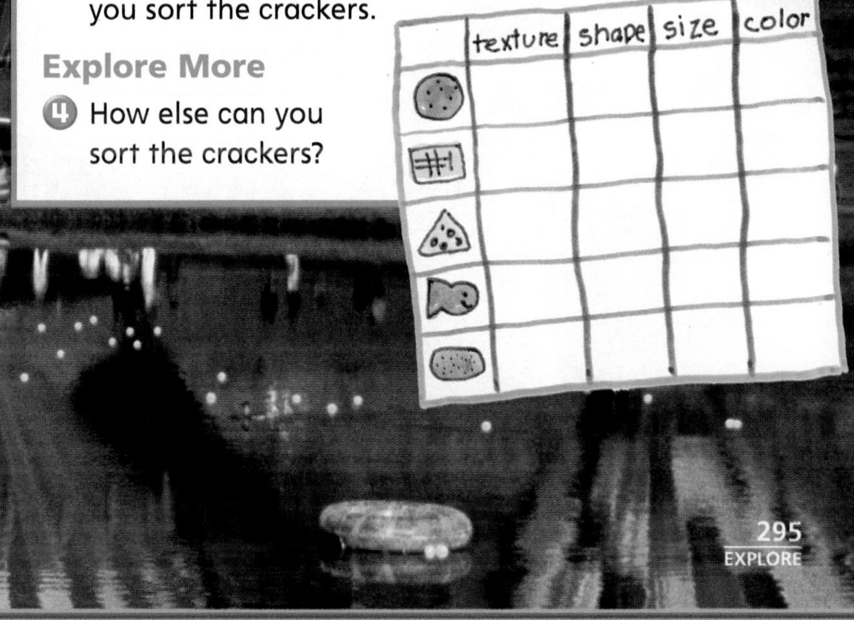


295
EXPLORE

## Explore

pairs or small groups

20 minutes

**Plan Ahead** Select different kinds of crackers and have enough for groups to share. **Be Careful!** When selecting crackers, be aware of children's food allergies!

**Purpose** This activity will help children describe and classify matter based on their observations. Using the chart to record their findings will help children visually organize their observations.

**Structured Inquiry** What to Do

Have children identify the senses they use to describe things. Ask: **Which sense do you use to describe color?** sight Explain to children that they are going to use all of their senses to carefully observe, record, and classify their findings.

1. **Observe** Encourage children to work together to create a list of words to describe the crackers' appearance.
2. **Record Data** Review with children how to read a chart. Point out that each row will represent a cracker. The column headings will help describe each cracker.
3. **Classify** Have children use their charts to group the crackers that are alike in a specific color. For example, they can record crackers that are small in red crayon so when they look at the chart they can immediately tell which crackers are small.

**Guided Inquiry** Explore More

4. Have groups share how they sorted the crackers. Allow children to eat one of each type of cracker and add taste as a category to their charts.

**Open Inquiry**

Encourage children to think about other things they might want to learn about the crackers. Discuss which tools might help, such as a hand lens, a ruler, or the ingredients listed on the boxes.

Provide children with the cracker boxes so they can research additional information about the crackers.

FAST TRACK

## Alternative Explore

### How are objects alike and different?

Have children choose two objects in the classroom that are the same color. Have them observe each object, and record their observations by completing a Venn diagram.

Have children compare their work with a partner. Ask children to list any additional properties of the objects.

Name ______________ Date ______________

Alternative Explore

**How are objects alike and different?**

**You need**
- pencil
- paper

In this activity, you will use a Venn diagram to record how two objects are alike and different.

**What to Do**

1. Look around the classroom and select two different objects that are the same color.
2. On a separate piece of paper, draw a Venn diagram. Label the Venn diagram with the name of each object. In each oval, list some properties of one of the objects. In the middle section, list the properties that both objects have in common.
3. Ask a partner to check your work. If they can think of more ways to describe each object, add their ideas to your diagram.

**What Did You Find Out?**

4. How did you determine which objects were alike and which ones were different?
Possible answer: I used my senses to observe the objects. I used a Venn diagram to record my results.

Activity Lab Book, p. 133

# 2 Teach

## Read Together and Learn

**Reading Skill** Compare and Contrast To compare is to decide how things are alike. To contrast is to decide how things are different.

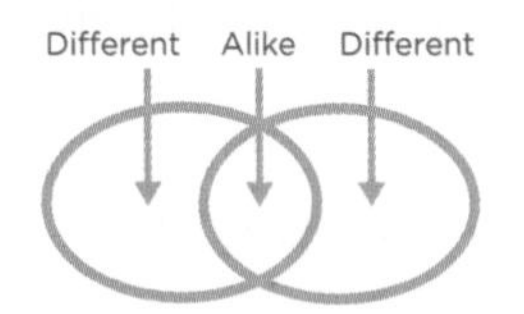


*Graphic Organizer 10,* p. TR12

## What is matter?

### FAST TRACK Discuss the Main Idea

**Main Idea** Matter is anything that has mass and takes up space.

Before reading the lesson, review how a balance can measure the mass of two objects.

After reading together, ask:

- **What happens when two things on a balance have the same mass?** The two sides will be level.

Have children look around the classroom for two objects that look as if they have similar mass. Ask:

- **Why did you pick those two objects?** Possible answer: They are similar in size.

### Read Together and Learn

**Essential Question**
How can you describe matter?

**Vocabulary**
matter
mass
property

### What is matter?

**Matter** is anything that takes up space and has mass. **Mass** is the amount of matter in an object. The water you drink is matter. The air you breathe is matter. Matter can be natural or made by people. We use matter every day.

**Using Matter**

**Read a Photo**
How is this boy using matter?

296
EXPLAIN

### Science Background

**Matter** Matter occupies space and cannot share that space with other matter. Matter is composed of tiny particles called atoms and molecules. The amount of energy the particles have determines whether a substance is a solid, a liquid, or a gas. One measurement of matter is known as mass. Mass is the amount of matter in an object.

See **Science Yellow Pages**, in the Teacher Resources section, for background information.

**Professional Development** For more Science Background and resources from NSDL visit http://nsdl.org/refreshers/science

### ELL Support

**Compare and Contrast** Choose two classroom objects that share some common properties. Have children use complete sentences and Venn diagrams to describe how the objects are similar and different.

BEGINNING Show children two objects. Have them work in pairs to describe the properties of each. Help children use those words in simple, complete sentences.

INTERMEDIATE Give each child two objects and have them describe each object and tell one way that they are alike.

ADVANCED Give each child two objects and have them describe at least two ways the objects are alike, and two ways they are different. Encourage children to make a Venn diagram.

Different objects have different amounts of mass. A truck has a lot of mass. A pencil has a little mass. Does a book have more mass than a flower? Yes! A book feels heavier if you try to pick it up. We can use a balance to measure and compare mass.

he larger shoe has nore mass than the maller shoe. ▶

ometimes a smaller bject can have more nass than a larger bject. ▶

✓ What are some examples of matter found in your desk?



**Read a Photo**

Before answering the question, have children identify all of the things in the photograph that are matter.

**Answer to Read a Photo** Possible answers: He is using water and soap to wash the dog; a metal pail to hold the water and the dog; and a cotton towel to dry the dog.

## ▶ Develop Vocabulary

**matter** Remind children that *matter* is anything that has mass and takes up space. Have children make a two-column chart with the labels: *Living Things* and *Nonliving Things*. Ask children to identify matter for each category.

**mass** Explain that *mass* is the amount of matter in an object. The more matter in an object, the heavier that object will be. Show children two objects and ask them to describe each object. Have a volunteer hold the objects to determine which has the most mass, and make sure they use the word *mass* in their descriptions. Repeat using other objects made of different types of matter.

## ✓ *Quick Check Answer*

Possible answers: pencils, erasers, folders, books

## Differentiated Instruction

### Leveled Activities

**EXTRA SUPPORT** Have children use a balance to determine the amount of mass in three objects. First, ask them to test two objects to determine which is heavier. Then, have them test the third object with the other two to determine the amount of mass from least to most.

**ENRICHMENT** Have children locate objects of similar size in the classroom. Ask them to compare the objects and to decide which has more mass. Encourage them to use a balance to check whether the item they chose has the most mass. Challenge children to find the smallest object in the class that has the most mass.

### RESOURCES and TECHNOLOGY

- **Reading and Writing,** pp. 163–165
- **Visual Literacy,** p. 26
- PuzzleMaker CD-ROM
- Classroom Presentation Toolkit CD-ROM
- LOG ON e-Glossary

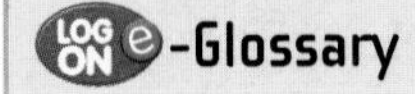

## What are properties?

### ▶ Discuss the Main Idea

**Main Idea** Matter has properties that can be described in many ways.

Read the blue question at the top of page 298 and ask children to discuss how they can use their senses to describe matter.

### ▶ Develop Vocabulary

**property** ***Scientific vs. Common Use*** *Property* often refers to something that belongs to someone. In science, *property* refers to the characteristics of an object, such as the way it tastes, sounds, looks, and feels.

Identify an object, such as a red rubber ball, and ask children to write a sentence in which they list the object's different properties.

### ▶ Use the Visuals

Look at the pictures on page 299 and read the captions. Ask:

- **How would you describe the properties of the mustard in the picture?** Possible answers: The mustard is yellow, thick, and probably has a sour taste.
- **How would you describe the properties of the toy in the picture?** Possible answers: It is green, it can bend, and it has a long shape that twirls.
- **Which object is a liquid?** the mustard

## What are properties?

You can describe matter by talking about its properties. A **property** is how matter looks, feels, smells, tastes, or sounds. Matter can feel smooth, rough, soft, or hard. Matter can be thick or thin. Matter can be living or nonliving.

**Quick Lab**

**Classify** six objects in your desk by their shape. Then sort them by their size.

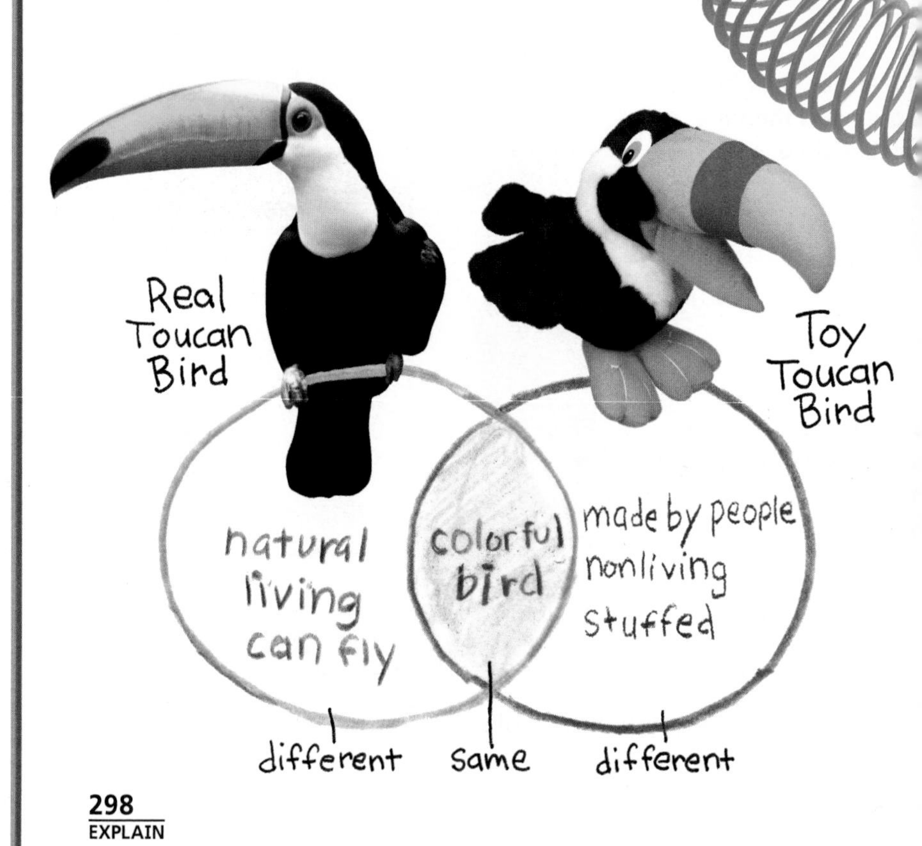


298
EXPLAIN

**Quick Lab**

**Objective** Describe and classify objects based on their shapes and sizes.

**You need** classroom objects, crayons, paper

1. Have children observe six objects in the classroom. Ask them to fold their paper in thirds and write the labels: *Object, Shape,* and *Size.* Have them write the object's name in the first column, draw the shape of each object in the second column, and describe its size in the third column.
2. Ask children to classify the objects by size.

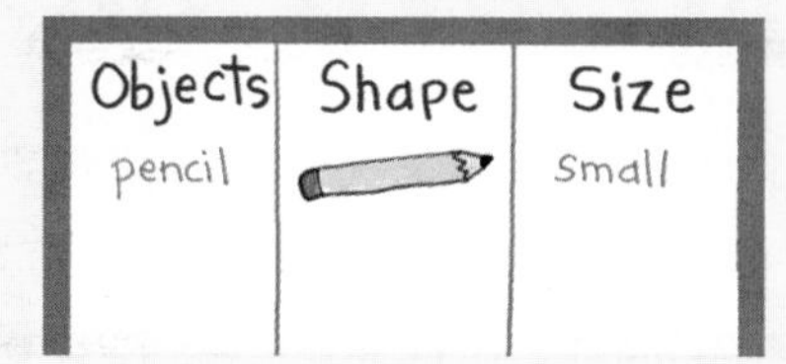


**RESOURCES and TECHNOLOGY**

- **School to Home Activities,** pp. 85–86
- **Reading and Writing,** p. 166
- **Activity Lab Book,** p. 134
- LOG ON **e-Review** Narrated Summary and Quiz
- ***ExamView® Assessment Suite* CD-ROM**

There are many ways to talk about matter. Matter can be solid, liquid, or gas.

What are the properties of the things in the room around you?

◀ This mustard is thick and gooey.

This skunk is very smelly! ▶

▲ This toy is bendable.

## Think, Talk, and Write

1. Compare and Contrast. Choose two objects. Make a list to compare their properties.
2. What is matter?

3. **Essential Question.** How can you describe matter?

### Art Link

Use different types of matter to make a collage.

e-Review Summaries and quizzes online at www.macmillanmh.com



## Formative Assessment

**Compare Matter**

Give pairs of children two objects, such as a few sheets of paper and a book. Ask them to list the properties of each, and use a balance to determine which has more mass.

Have children record their observations and draw a picture to show what the balance looked like when they measured the masses. Have them write a sentence to describe the mass of each object.

The paper has less mass than the book.

Key Concept Cards For student intervention, see the prescribed routine on **Key Concept Card 26.**

### Quick Check Answer

Possible answers: The desks are hard and made of metal and plastic; the erasers are soft and pink; the books are made of paper; the chairs are made of metal and wood.

# 3 Close

### ▶ Using the KWL chart

Review with children what they have learned about matter and the properties of matter. Record their responses in the What We Learned column of the class **KWL** chart.

### ▶ Using the Reading Skill

**Compare and Contrast**

Use the reading skill graphic organizer to compare and contrast matter. Show children a pencil and a marker. Ask: **How can you compare and contrast these two objects?**

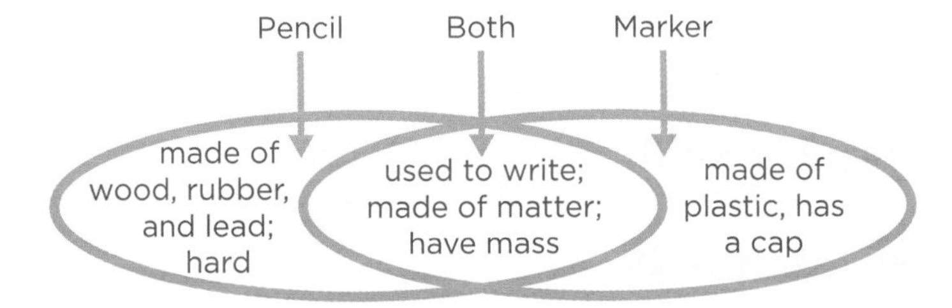


*Graphic Organizer 10,* p. TR12

## FAST TRACK Think, Talk and Write

1. Compare and Contrast Encourage children to use a two-column chart or Venn diagram to describe and compare the two objects they select.
2. Matter is anything that has a mass and takes up space. Matter can be either a solid, a liquid, or a gas.
3. **Essential Question** Possible answer: You can measure the mass of matter using a balance. You can describe how matter looks, feels, smells, tastes, or sounds. You can describe matter as a solid, a liquid, or a gas.

### Art Link

Give children a piece of colored poster board to use as a background for their collage. Supply children with materials of various textures, colors, shapes, and sizes for their collage.

## Focus on Skills

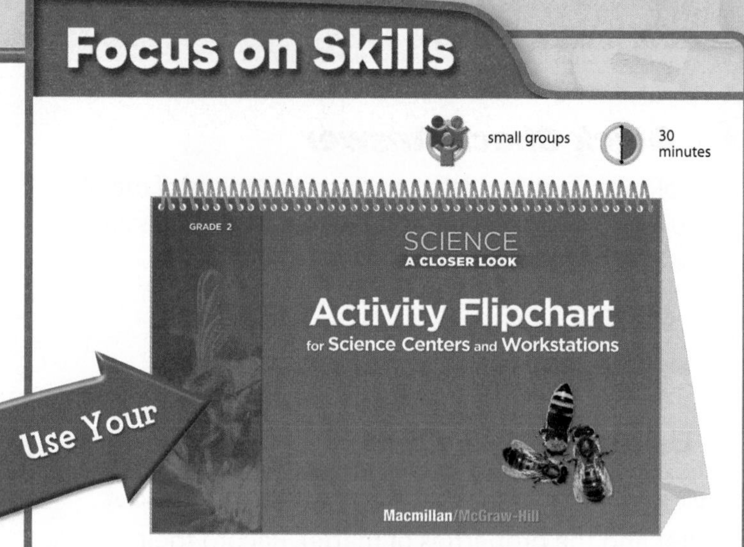


Activity Flipchart, p. 41

### Objective

- Describe and sort matter into two groups: *solid* and *liquid*.

**You need** paper, pencil, crayons

**EXTEND** This inquiry skill activity will help children sort pictured items into two groups: *human-made* and *natural.*

## Inquiry Skill: Record Data

### ▶ Learn It

Read the left side of page 41 together. Explain that when people record data they organize and write down the results of their work. Have children discuss the results of Joanie's work. Ask:

- **Were there more solids or liquids in the lunches?** solids
- **Why were more solids in the lunches?** Possible answer: People usually have one drink, but eat more than one solid food for lunch.

Discuss how the graph makes it easy to interpret and analyze the data that Joanie collected. Ask:

- **What are some other ways to record data?** Possible answers: make a list; draw a diagram; make a chart

## Focus on Skills

**Inquiry Skill: Record Data**

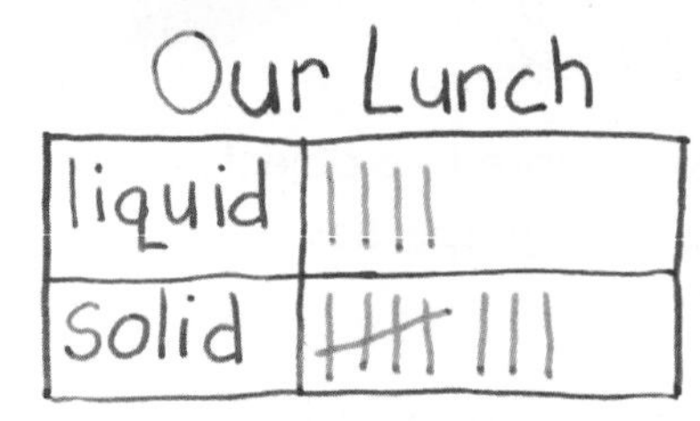

When you **record data**, you write down what you observe.

**▶ Learn It**

Joanie talked to each of her classmates about what they had for lunch. She made a tally chart to help her count the kinds of foods they ate. She recorded what was a liquid and what was a solid.

Then she made a bar graph from her results. A bar graph is a good way to compare data in different groups.

Our Lunch

| liquid | \|\|\|\| |
|---|---|
| Solid | ~~\|\|\|\|~~ \|\|\| |

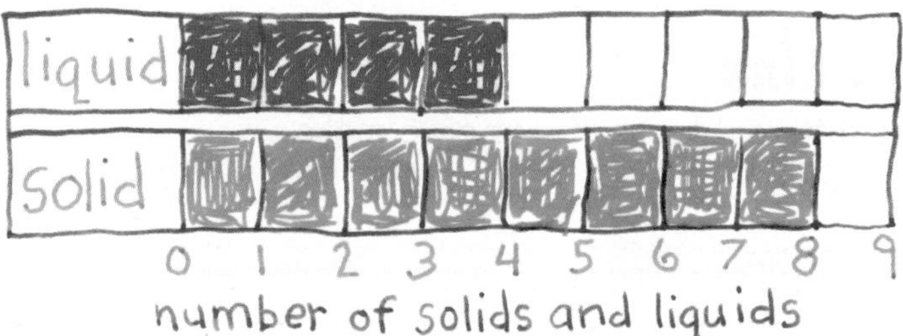


### Integrate Math

**Graphing**

Each day for a week, have children record what they ate for breakfast on a class chart. Have them add a tally mark for each type of food, such as cereal, milk, juice, bread, fruit, eggs, or pancakes.

At the end of the week, have the class work together to make a bar graph of the results.

Help children discuss their findings. Ask:

- **What does our graph tell us about the kinds of food we eat for breakfast?** Possible answer: Many children eat cereal in the morning.
- **Who might be interested in knowing this information?** Possible answers: cooks who prepare breakfast in the cafeteria; restaurants that serve breakfast

## Skill Builder

### Try It

Look at this picture. Some things are
atural and some are made by people.
Make a tally chart to show how many of
ach thing you see. Then display your data
n a bar graph.

1. How many things in the picture were made by people?
2. What kind of chart can help you record your data?
3. **Write About It.** How can a bar graph help you compare data?



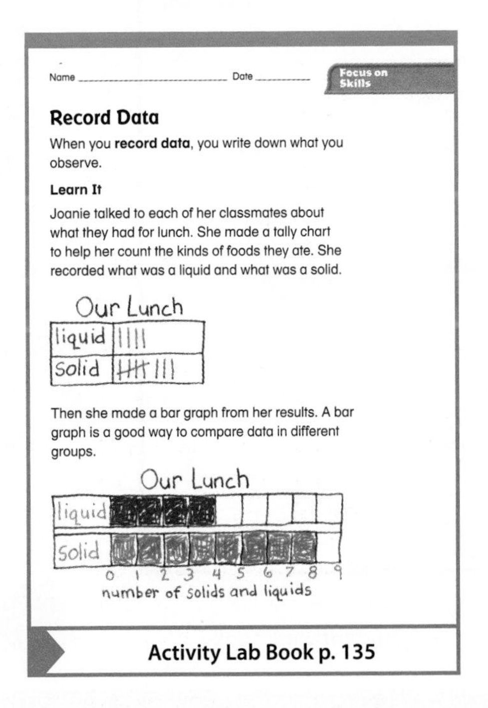

Name ______________ Date ________ Focus on Skills

**Record Data**

When you **record data**, you write down what you observe.

**Learn It**

Joanie talked to each of her classmates about what they had for lunch. She made a tally chart to help her count the kinds of foods they ate. She recorded what was a liquid and what was a solid.

Our Lunch

| liquid | \|\|\|\| |
|---|---|
| Solid | 卌 \|\|\| |

Then she made a bar graph from her results. A bar graph is a good way to compare data in different groups.



Activity Lab Book p. 135

## ▶ Try It

Read the right side of page 41 together and discuss the difference between human-made things and natural things.

1. Possible answer: these two things are made by people: the bridge and buildings Discuss which items in the picture are natural. Ask children to explain how they determined which were natural.
2. Possible answers: a bar graph; a two-column chart Explain that when children made their bar graph they filled in a box instead of making a tally mark. Ask: **How do you transfer the information from your tally chart to a bar graph?** count the tally marks in each column
3. Children should describe in their own words why the visual representation of numbers on the bar graph makes it easy to see which category has more and which has fewer, and exactly how many more or fewer.

## ▶ Apply It

Create a sorting jar by filling a large, clear container with a variety of small materials that have multiple attributes. For example: multi-colored buttons of various shapes and sizes that have two, four, or no holes.

Have children work in small groups to decide which properties of the solids in the jar they will use to sort the contents into three or more groups. Have them record the results of their work in bar graphs and share them with the class.

Challenge each group to see how many different ways they can sort the contents of the jar.

**RESOURCES and TECHNOLOGY**

- **Activity Lab Book,** pp. 135–136
- **TeacherWorks™ Plus CD-ROM**

Stop Here to

# Plan Your Lesson

## Lesson 2 Solids

### Essential Question

What are the properties of a solid?

### Objectives

- Compare and contrast the properties of solids.
- Use different ways to measure solids.

**Reading Skill** Summarize

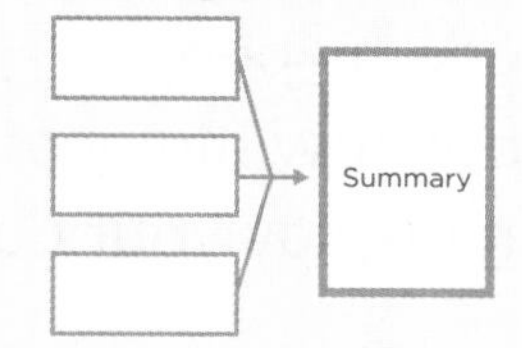


*Graphic Organizer 5*, p. TR7

## FAST TRACK

**Lesson Plan** When time is short, follow the Fast Track and use the essential resources.

**1 Introduce**
**Look and Wonder, p. 300**
Resource **Activity Lab Book, p. 139**

**2 Teach**
**Discuss the Main Idea, p. 302**
Resource **Visual Literacy, p. 27**

**3 Close**
**Think, Talk, and Write, p. 305**
Resource **Assessment, p. 111**

**Professional Development** Look for NSDL to find recommended Science Background resources from the National Science Digital Library.

## Reading and Writing

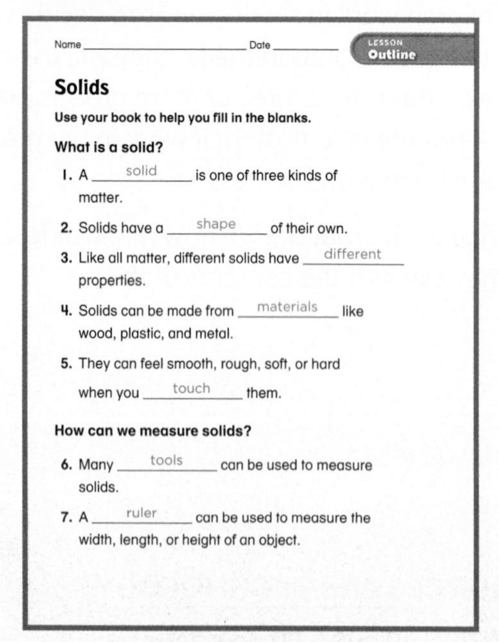

Name ______ Date ______ Lesson Outline

**Solids**

**Use your book to help you fill in the blanks.**

**What is a solid?**

1. A ___solid___ is one of three kinds of matter.
2. Solids have a ___shape___ of their own.
3. Like all matter, different solids have ___different___ properties.
4. Solids can be made from ___materials___ like wood, plastic, and metal.
5. They can feel smooth, rough, soft, or hard when you ___touch___ them.

**How can we measure solids?**

6. Many ___tools___ can be used to measure solids.
7. A ___ruler___ can be used to measure the width, length, or height of an object.

**Outline, pp. 167–168**
Also available as a student workbook

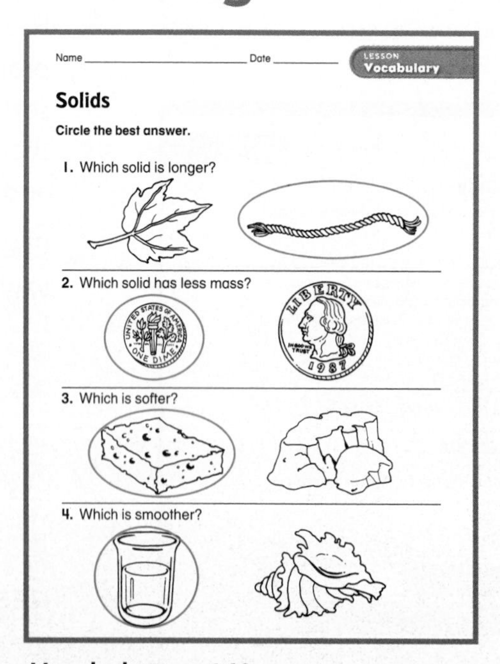

Name ______ Date ______ Lesson Vocabulary

**Solids**

**Circle the best answer.**

1. Which solid is longer?
2. Which solid has less mass?
3. Which is softer?
4. Which is smoother?

**Vocabulary, p. 169**
Also available as a student workbook

## Visual Literacy

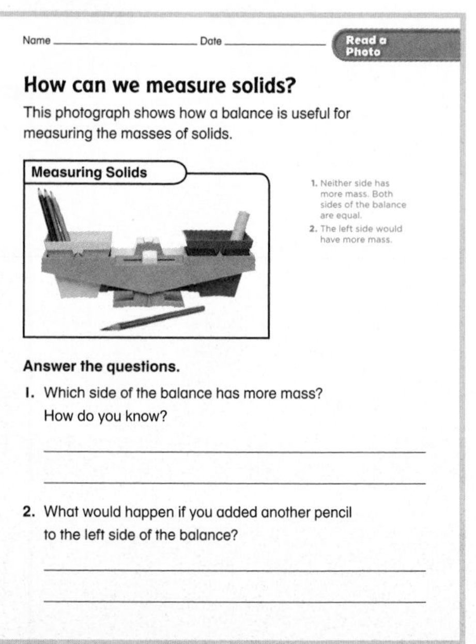

Name ______ Date ______ Read a Photo

**How can we measure solids?**

This photograph shows how a balance is useful for measuring the masses of solids.

Measuring Solids

1. Neither side has more mass. Both sides of the balance are equal.
2. The left side would have more mass.

**Answer the questions.**

1. Which side of the balance has more mass? How do you know?
2. What would happen if you added another pencil to the left side of the balance?

**Read a Photo, p. 27**
Also available as a transparency

## Activity Lab Book

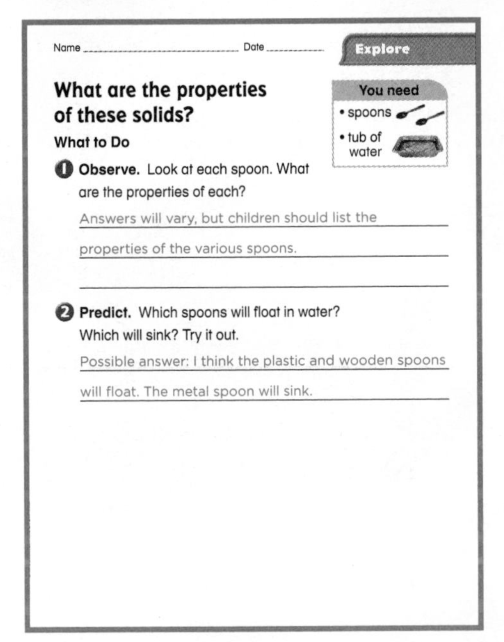
Name ______ Date ______ Explore

**What are the properties of these solids?**

You need
- spoons
- tub of water

**What to Do**

1. **Observe.** Look at each spoon. What are the properties of each?
Answers will vary, but children should list the properties of the various spoons.

2. **Predict.** Which spoons will float in water? Which will sink? Try it out.
Possible answer: I think the plastic and wooden spoons will float. The metal spoon will sink.

**Explore, pp. 137–138**
Also available as a student workbook

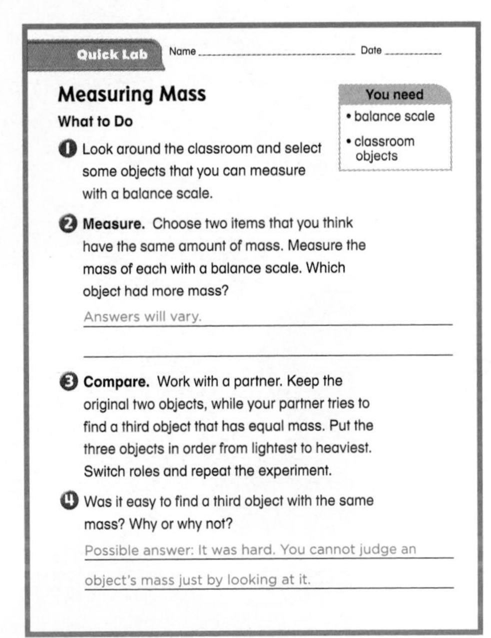
Quick Lab Name ______ Date ______

**Measuring Mass**

You need
- balance scale
- classroom objects

**What to Do**

1. Look around the classroom and select some objects that you can measure with a balance scale.
2. **Measure.** Choose two items that you think have the same amount of mass. Measure the mass of each with a balance scale. Which object had more mass?
Answers will vary.
3. **Compare.** Work with a partner. Keep the original two objects, while your partner tries to find a third object that has equal mass. Put the three objects in order from lightest to heaviest. Switch roles and repeat the experiment.
4. Was it easy to find a third object with the same mass? Why or why not?
Possible answer: It was hard. You cannot judge an object's mass just by looking at it.

**Quick Lab, p. 140**
Also available as a student workbook

## Assessment

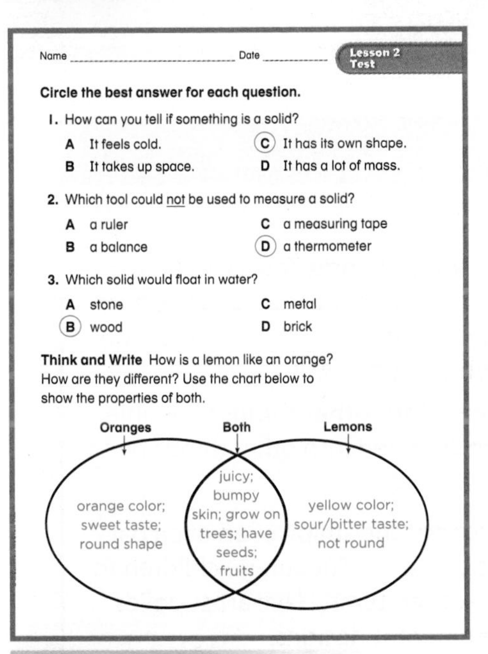
Name ______ Date ______ Lesson 2 Test

**Circle the best answer for each question.**

1. How can you tell if something is a solid?
A It feels cold. (C) It has its own shape.
B It takes up space. D It has a lot of mass.
2. Which tool could not be used to measure a solid?
A a ruler C a measuring tape
B a balance (D) a thermometer
3. Which solid would float in water?
A stone C metal
(B) wood D brick

**Think and Write** How is a lemon like an orange? How are they different? Use the chart below to show the properties of both.

Oranges: orange color; sweet taste; round shape
Both: juicy; bumpy skin; grow on trees; have seeds; fruits
Lemons: yellow color; sour/bitter taste; not round

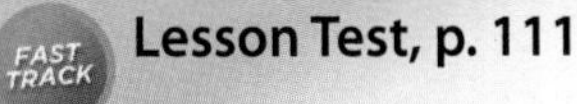
**Lesson Test, p. 111**

## ADDITIONAL RESOURCES

pp. 132–135

*Gases Matter*

p. 42

p. 91

p. 27

27

81–90

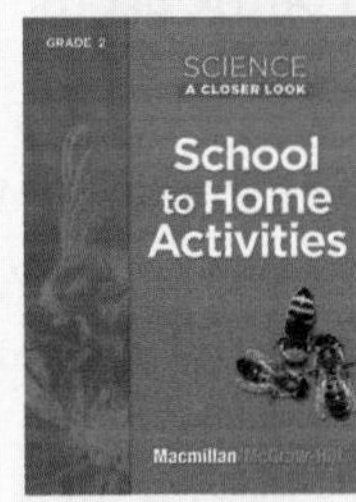
pp. 87–88

70

### Technology

- Science Activity DVD
- TeacherWorks™ Plus CD-ROM
- Classroom Presentation Toolkit CD-ROM

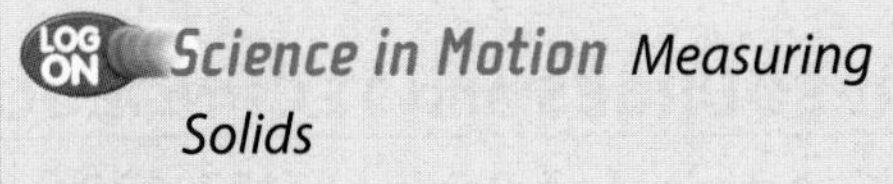
*Science in Motion* Measuring Solids

e-Review

NSDL

# Lesson 2 Solids

## Objectives

- Compare and contrast the properties of solids.
- Use different ways to measure solids.

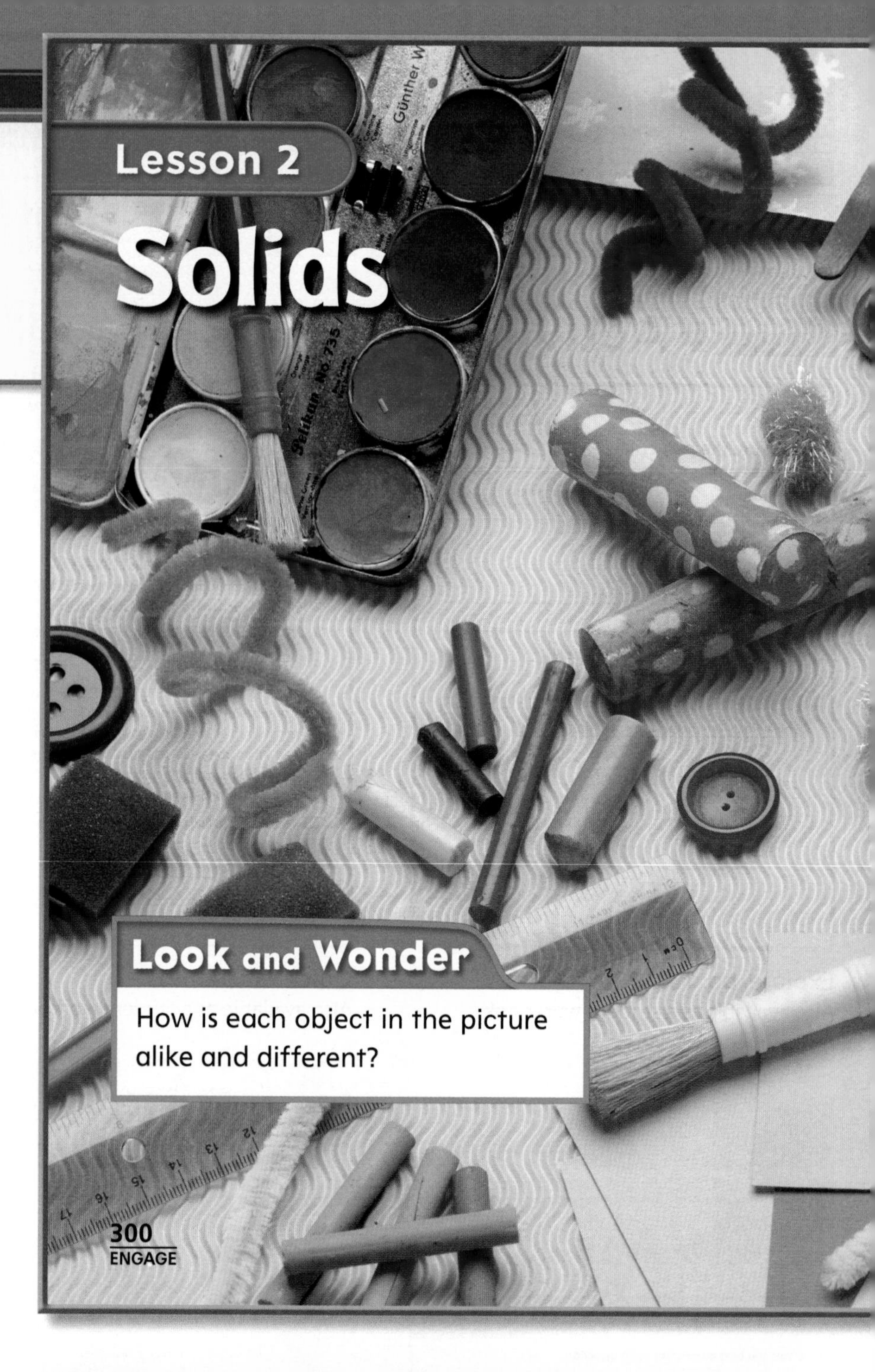

Lesson 2
Solids
Look and Wonder
How is each object in the picture alike and different?
300
ENGAGE


## 1 Introduce

### ▶ Assess Prior Knowledge

Have children share what they know about solids. Invite them to look around the classroom. Ask:

- **How can you tell that something is a solid object?**
- **What are some words you use to describe solid things?**
- **What are the properties of solids?**

Record children's answers in the What We Know column of the class **KWL** chart.

### FAST TRACK Look and Wonder

Read and discuss the Look and Wonder question. Write the words *Alike* and *Different* on the board. Record children's descriptions of the art materials under the appropriate heading. Ask:

- **What are some things that all of the objects have in common?** Possible answers: They can all be used to make art. They all have mass. They all take up space. They are all solids.

### RESOURCES and TECHNOLOGY

- **Activity Lab Book,** pp. 137–139
- **Activity Flipchart,** p. 42
- Science Activity DVD

### Warm Up

**Start with a Book**

Read ***Joseph Had a Little Overcoat,*** by Simms Taback (Viking Press, 1999). In this story, Joseph, a tailor, reuses the fabric from his beloved, plaid overcoat. Have children predict what Joseph can make from his old and worn overcoat. Ask:

- **How can you make an overcoat into other things?** Possible answers: The overcoat can be changed into other things by cutting, sewing, folding, and tearing.

After reading, ask children to cut out the shape of an overcoat from a large piece of construction paper. Encourage children to make the cut-out overcoat into other items. Have art supplies available for children to create different designs.

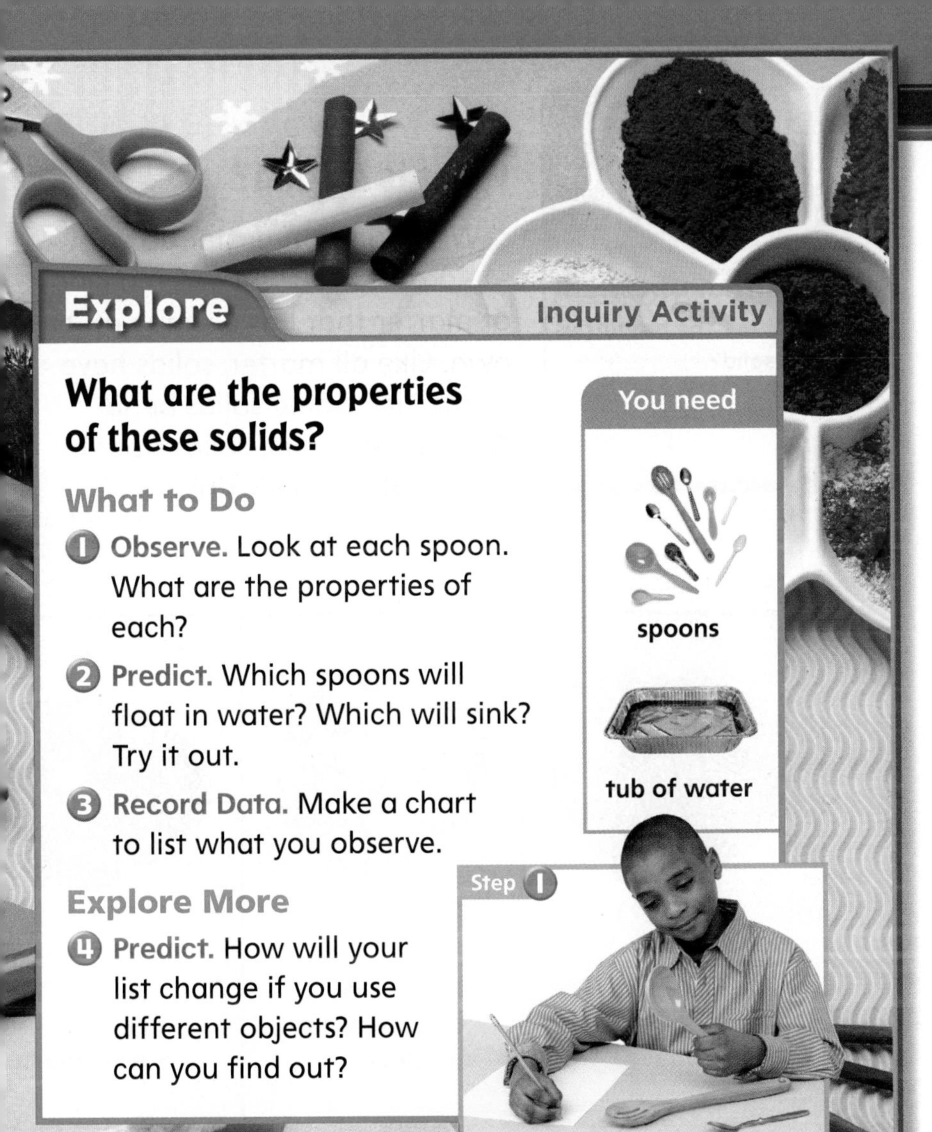

## Explore — Inquiry Activity

### What are the properties of these solids?

**What to Do**

1. **Observe.** Look at each spoon. What are the properties of each?
2. **Predict.** Which spoons will float in water? Which will sink? Try it out.
3. **Record Data.** Make a chart to list what you observe.

**Explore More**

4. **Predict.** How will your list change if you use different objects? How can you find out?

**You need**
- spoons
- tub of water

Step 1



## Explore

 pairs 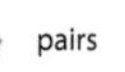  20 minutes

**Plan Ahead** Collect a variety of spoons made from different materials (wooden, metal, plastic).

**Purpose** Children will observe that some properties of solids can be identified by experimentation.

**Structured Inquiry** What to Do

Have children identify things that float in water. Ask: **What did those floating objects look like?**

1. **Observe** Remind children to feel each spoon and describe its texture. Have children hold each spoon to determine which one is the heaviest.
2. **Predict** Have children use what they know about objects that sink and float to predict which spoons will sink and which will float. Ask them to record their predictions before they put the spoons in water.
3. **Record Data** Have children make a chart with the different properties as column labels and each spoon in a row. Help them identify the number of rows and columns they will need, and the different property category labels they will need to record all of their observations.

**Guided Inquiry** Explore More

4. **Predict** Have children list category titles they would need if they were to record the properties of three different fruits. Next, ask them to think of other objects and predict what categories they would need.

**Open Inquiry**

Ask children what else they might discover by investigating other properties of solids. Ask: **What solids would you like to experiment with to learn more about them?**

If children need help generating questions to explore, model by asking: **How do some solids change over time? Are all solids made of hard materials?**

FAST TRACK

## Alternative Explore

### What happens when solids fall?

Have children drop a pencil and record what happens. Repeat using a pink eraser and a piece of paper.

Encourage children to observe the properties of the objects and to infer what caused each to behave as it did.

To explore further, ask children to predict how other solids would behave if dropped from the same height.

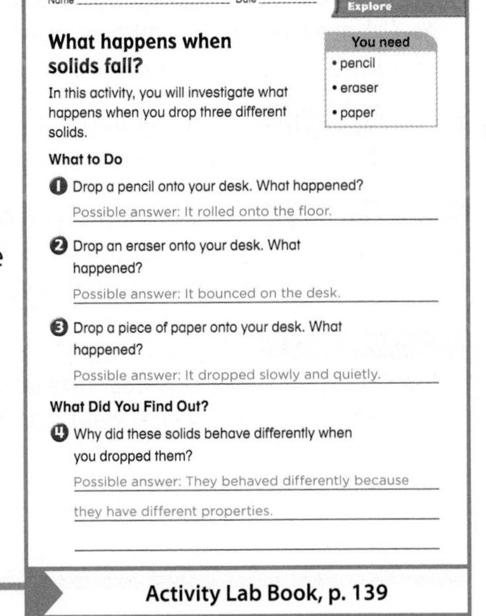

Name ______________ Date ________ Alternative Explore

**What happens when solids fall?**

In this activity, you will investigate what happens when you drop three different solids.

**You need**
- pencil
- eraser
- paper

**What to Do**

1. Drop a pencil onto your desk. What happened?
   Possible answer: It rolled onto the floor.
2. Drop an eraser onto your desk. What happened?
   Possible answer: It bounced on the desk.
3. Drop a piece of paper onto your desk. What happened?
   Possible answer: It dropped slowly and quietly.

**What Did You Find Out?**

4. Why did these solids behave differently when you dropped them?
   Possible answer: They behaved differently because they have different properties.

Activity Lab Book, p. 139

# 2 Teach

## Read Together and Learn

**Reading Skill** Summarize To retell the most important ideas from a reading selection.

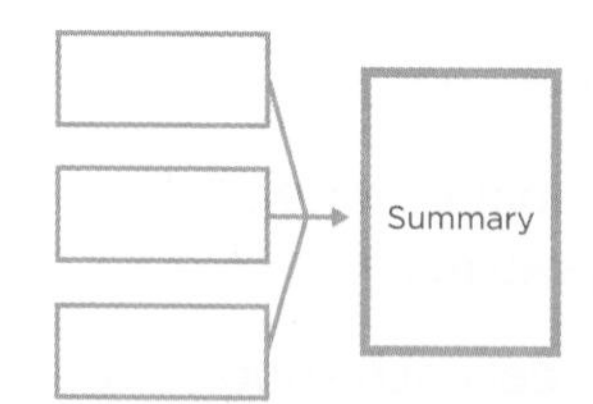


*Graphic Organizer 5*, p. TR7

## What is a solid?

### Discuss the Main Idea

**Main Idea** Solids have many different properties, but all solids have definite shapes.

Before reading, remind children that everything is made of matter, and matter can be described by its properties.

After reading together, have children identify the shapes of solids they see around the classroom. Ask:

- **What are some other properties of those solids?** Possible answers: size, color, texture
- **What are some ways to change solids?** Possible answers: cut, fold, bend, tear, twist, stretch

**Read Together and Learn**

**Essential Question**
What are the properties of a solid?

**Vocabulary**
solid

### What is a solid?

What kind of matter do you see around you? A **solid** is a kind of matter that has a shape of its own. Like all matter, solids have properties. Some solids bend. Others tear. Some solids float in water. Other solids sink.

Some Properties of Solids
rock
hard
speckled
jagged
glass
smooth
breakable
clear
yarn
soft
colorful
long and thir


**FACT** Not all solids are hard.

302
EXPLAIN

### Science Background

**Solids** A solid is matter with a definite shape and volume. Solids have different textures and can be made of different materials. The molecules in solids are held together closely and arranged in fixed positions. The arrangement and amount of molecules in a solid determines its hardness and density. Lighter solids will float in water and more dense solids will sink.

See **Science Yellow Pages**, in the Teacher Resources section, for background information.

**Professional Development** For more Science Background and resources from NSDL visit http://nsdl.org/refreshers/science

### ELL Support

**Practice Using Language** Select a variety of solid objects and share them with children. Have children practice using complete sentences to identify the various solids.

**BEGINNING** Show children an object and help them complete this sentence frame: *This* _________ *is a solid.*

**INTERMEDIATE** Show children an object and have them describe one property of the solid. For example, when shown a pencil, the child might say: *It is yellow.*

**ADVANCED** Show children an object and have them identify the object and describe at least two of its properties. For example: *The pencil is yellow and made of wood.*

Solids are made of different materials. Some metals, woods, and plastics are hard. Materials can be smooth or rough when you touch them. The chart below shows the properties of some solids.

What are some solids you use every day?

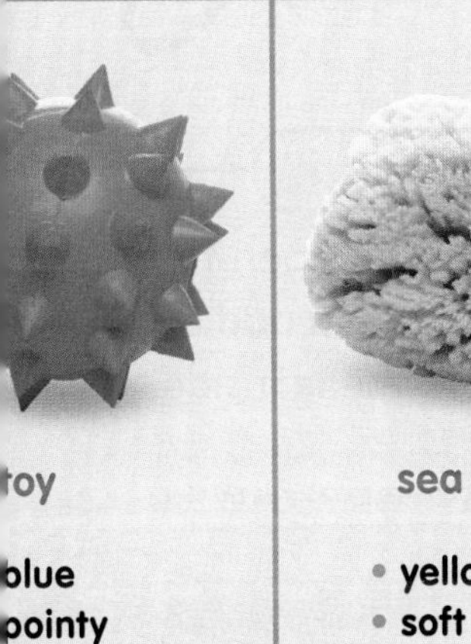

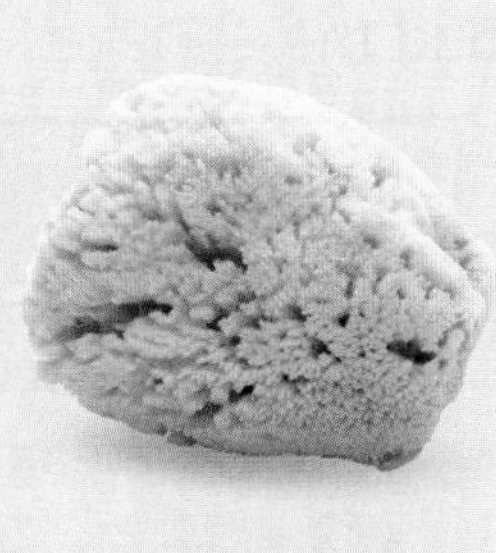

| toy | sea sponge | clay |
|---|---|---|
| • blue<br>• pointy<br>• plastic | • yellow<br>• soft<br>• scratchy | • sticky<br>• bendable<br>• firm |



## Differentiated Instruction

### Leveled Questions

**EXTRA SUPPORT** Ask questions such as these to check children's understanding of the material.

- **What is a solid?** Possible answers: Matter that has a shape of its own.
- **What kinds of materials are used to make solids?** Possible answers: wood, minerals, clay, metal, fabric, rubber

**ENRICHMENT** Use these types of questions to develop children's higher-order thinking skills.

- **What are some examples of soft solids?** Possible answers: cotton, clay, rubber, sponge
- **How can you identify a particular solid?** Possible answer: by describing its properties

## ▶ Use the Visuals

After reading the descriptions of the solids illustrated in the chart, ask children to compare two of the solids on the chart.

Encourage them to describe how the solids are alike and different. Use the following sentences as a model: *The toy and glass are alike because they are both solids. They are different because the toy is pointy and the glass is smooth.*

## ▶ Address Misconceptions

Children may think that only items that are hard can be considered a solid.

FACT **Not all solids are hard.** Identify items on the chart on pages 302 and 303 that are soft but have fixed shapes. Encourage children to identify classroom items that are soft solids.

## ▶ Develop Vocabulary

**solid** ***Scientific vs. Common Use*** Explain to children that the word *solid* has many meanings. In common use, it is often used as an adjective that means "of good quality" (a solid performance) or "pure" (solid gold).

Ask each child to use the word in a sentence. After they listen to the sentence, ask the class to give a thumbs up if they heard the scientific meaning and a thumbs down if they heard the common use of the word.

## Quick Check Answer

Possible answers: cereal bowl, pencil, desk, sponge

### RESOURCES and TECHNOLOGY

- **Reading and Writing,** pp. 167–169
- PuzzleMaker CD-ROM
- Classroom Presentation Toolkit CD-ROM
- 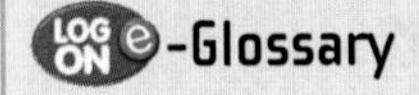e-Glossary

# How can we measure solids?

## ▶ Discuss the Main Idea

**Main Idea** Solids can be measured to determine their length, width, height, and mass.

Read the blue question on page 304 and have children identify the tools they use to measure different things, such as temperature and ingredients for experiments.

### Read a Photo

Have children put their finger on the arrow point at the base of the balance and look at the vertical line in the middle of the balance arm. Discuss how the line shifts to the right or left of the point when one bucket contains more mass than the other. Ask:

- **What can you tell about the mass of the pencils and the chalk from looking at the photograph?** The pencils and the chalk have about the same mass.
- **What would happen to the balance if another stick of chalk was added to the right bucket?** The right side of the balance would move farther down.

**Answer to Read a Photo** Adding another pencil will cause the left side to move farther down. Four pencils may have more mass than the chalk.

## ▶ Develop Vocabulary

Reinforce chapter vocabulary by having children make a word web for *properties* of *solids*. Ask children to add words to the web that describe properties of solids.

### RESOURCES and TECHNOLOGY

- **School to Home Activities,** pp. 87–88
- **Reading and Writing,** p. 170
- **Activity Lab Book,** p. 140
- **Visual Literacy,** p. 27
- LOG ON *Science in Motion* *Measuring Solids*
- LOG ON e-Review Narrated Summary and Quiz
- Exam*View*® *Assessment Suite* CD-ROM

## How can we measure solids?

We can use tools to measure solids. A ruler tells how long, wide, or high a solid is. Some rulers measure length in a unit called a centimeter. Other rulers measure in a unit called an inch. Many rulers give both measurements.

A balance tells how much mass something has. You can measure the same object in different ways. You can measure the mass and the length of a piece of chalk.

### Quick Lab

Measure the mass of things in your classroom with a balance.

### Read a Photo

What will happen to the balance if you add one more pencil to the left side?

LOG ON *Science in Motion* See how a balance measures matter at www.macmillanmh.com

**Measuring Solids**

### Quick Lab

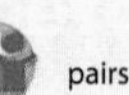
pairs

15 minutes

**Objective** Use a balance to determine the mass of two solids.

**You need** balances, classroom objects

1. Give children two objects that are similar in size, such as a pen and pencil. Have them use a balance to measure the mass of each, and draw a picture to show which classroom object has more mass.
2. Ask them to compare a third object to the first two, and put the objects in order from lightest to heaviest.

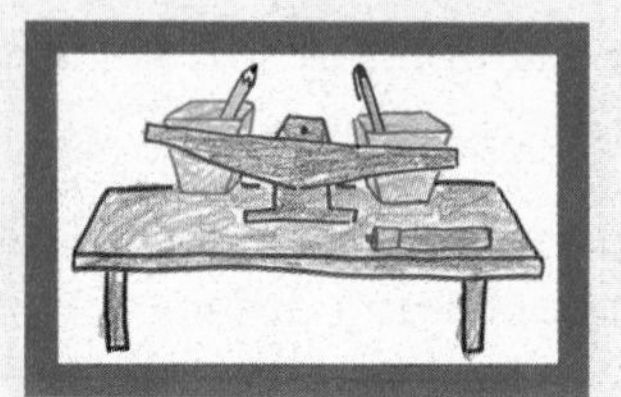

The chalk is 10 centimeters, or about 4 inches, long.

Measure the distance around the chalk with string.

Then measure the string with a ruler.

 What tools can we use to measure solids?

## Think, Talk, and Write

1. **Summarize.** What are some examples of solid matter?
2. How can you measure a solid?
3. **Essential Question.** What are the properties of a solid?

### Art Link

Find solids around the classroom. Make a piece of art showing some of their properties.

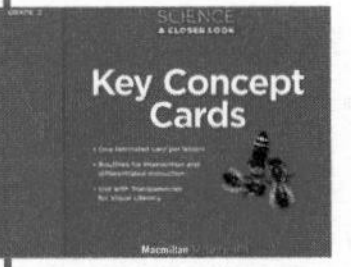

e-Review Summaries and quizzes online at www.macmillanmh.com



## Formative Assessment

### Measure and Describe Solids

Have children use a balance and ruler to measure two solid objects. Ask them to draw the objects on a piece of paper. Underneath the drawings, have children describe how the objects are alike and how they are different.

CRAYONS
The crayons and eraser are solids. The crayons have more mass than the eraser.


Key Concept Cards For student intervention, see the prescribed routine on **Key Concept Card 27.**

## Quick Check Answer

Possible answers: ruler, balance

# 3 Close

### Using the KWL chart

Review with children what they have learned about solids, their properties, and how to measure them. Record children's responses in the What We Learned column of the class **KWL** chart.

### Using the Reading Skill

**Summarize**

Use the reading skill graphic organizer to summarize the lesson.

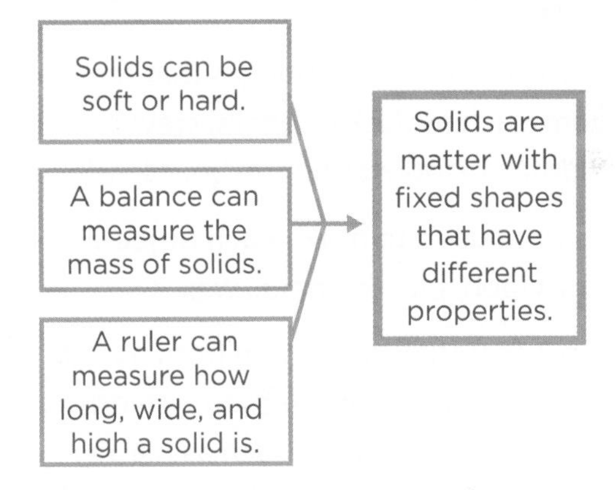

Solids can be soft or hard.
A balance can measure the mass of solids.
A ruler can measure how long, wide, and high a solid is.
Solids are matter with fixed shapes that have different properties.


*Graphic Organizer 5,* p. TR7

## FAST TRACK Think, Talk and Write

1. **Summarize** Possible answers: rocks, a wood desk, plastic toys, clay, a sponge
2. Possible answer: You can use a ruler to measure length. You can use a balance to measure mass.
3. **Essential Question** Possible answers: A solid has a shape of its own and takes up space. Solids can be soft or hard, bendable or firm, smooth or scratchy.

### Art Link

Give children glue to create a sculpture from recycled classroom materials. Have paint, fabric scraps, tissue paper, buttons, or other art materials available for children to decorate their sculptures.

# Reading in Science

**Objective**

- Summarize the difference between solids made from natural and human-made materials.

## Natural or Made by People?

**Genre: Nonfiction** Stories or books about real people and events.

Ask:

- **What information can you learn from the photos in this article?** Possible answer: what a factory looks like

### Before Reading

Ask children to describe some solid objects that they see in the classroom, such as desks, pencils, crayons, and pens. List children's responses on the board. Ask:

- **What materials were used to make each object?** Possible answers: wood, plastic, metal, wax

Remind children that materials that come from nature are called *natural resources.* Explain that some things are made from non-natural materials, such as plastic, which is made by people. Write the labels *Natural* and *Human-Made* next to the list on the board. Ask children to decide under which title each material belongs.

### During Reading

Explain to children that this article is about natural and human-made solids. Ask them to search for sentences that define *natural* and *human-made* in the article. Ask:

- **How do people get wood?** People cut down trees to get wood.
- **How do people make plastics?** People combine chemicals together to create plastics.

Ask children to think about how their lives would be different if there were no human-made solids.

**RESOURCES and TECHNOLOGY**

- **Reading and Writing,** pp. 171–172
- **Technology: A Closer Look,** Lesson 2

# Natural or Made by People?

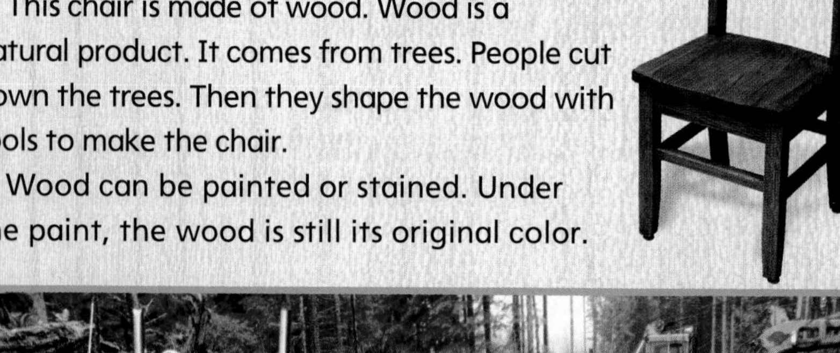

This chair is made of wood. Wood is a natural product. It comes from trees. People cut down the trees. Then they shape the wood with tools to make the chair.

Wood can be painted or stained. Under the paint, the wood is still its original color.

306
EXTEND

## ELL Support

**Use Labels** Write the labels *Natural* and *Human-Made* on sticky notes and review the meanings of each with children. Write labels for different natural and human-made solids, such as wood and plastic. Have children place the labels of solids under each category as they say the word.

**BEGINNING** Have children place the solid labels under the *Natural* or *Human-Made* labels and name each solid aloud.

**INTERMEDIATE** Ask children to place the solid labels under the *Natural* or *Human-Made* labels and then say a phrase or simple sentence to describe how each solid was made.

**ADVANCED** Have children place the solid labels under the *Natural* or *Human-Made* labels. Ask them to use complete sentences to describe what the solids can be made into.

## Science, Technology, and Society

This chair is made of plastic. Plastic made by people. People combine hemicals to create plastic. Then they hape it in molds.

There are many different kinds f plastic. Plastic can be hard or endable. People can also add a color o the chemicals in plastic. The plastic hen becomes that color.

Which solids in your classroom are atural? Which are made by people?

### Talk About It

**Summarize.** What is the difference between natural solids and solids made by people?

Connect to American Museum & Natural History at www.macmillanmh.com

307 EXTEND

## After Reading

Review children's list of natural and human-made classroom items. Discuss any items that may have been difficult for children to classify.

Have children explain how they could tell a friend how to identify materials made of wood or plastic. For example, they may look for wood grain lines in wood. Ask:

- **What are some important things that you learned about natural and human-made solids in this article?**

Record children's responses on the board. Have children read the list aloud and circle the three most important facts. Display Graphic Organizer 5 and write those three facts in the boxes on the left.

Explain to children that summarizing means retelling the most important ideas in a reading selection. Help children create a one-sentence summary from the facts on the left. Write it in the box on the right.

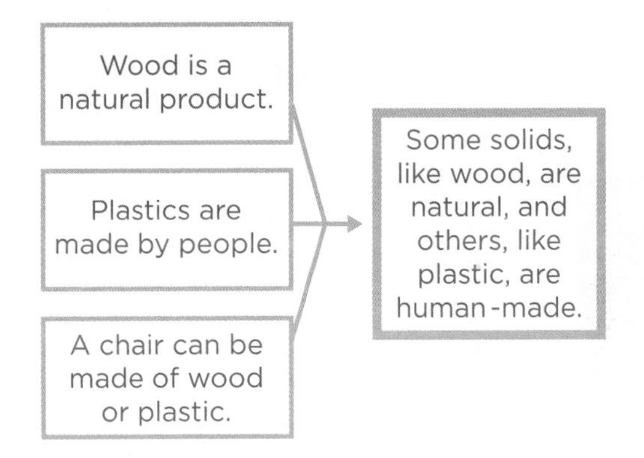


*Graphic Organizer 5,* p. TR7

## Talk About It

Natural solids contain materials that can be found in nature; human-made solids contain materials that are made by people.

If children have difficulty answering the question, provide them with examples of wood and plastic materials. For example, show children a plastic pattern block and a wood block. Encourage children to discuss which block is natural and which is human-made, and to explain how they can identify each material.

## Extended Reading

### Visit the Library

Read aloud ***Be a Friend to a Tree (Let's-Read-and-Find-Out Science),*** by Patricia Lauber (HarperCollins, 1994).

Review the difference between natural and human-made solids. Ask:

- **What natural solid is mentioned in the book?**
- **What products can be made from it?**

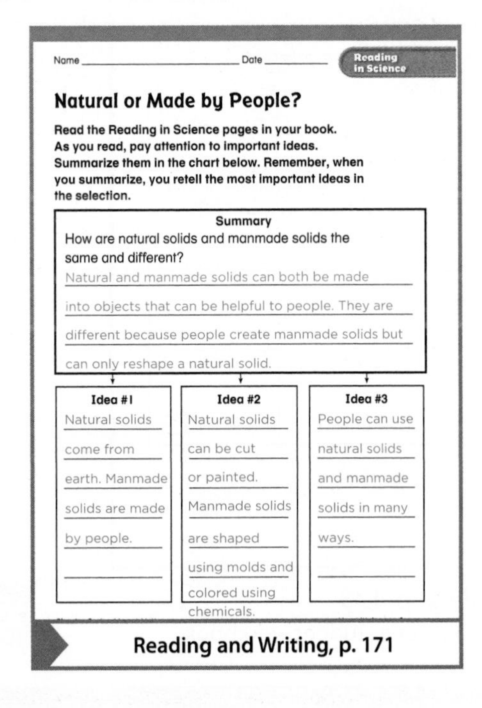

Name ______ Date ______ Reading in Science

**Natural or Made by People?**

**Read the Reading in Science pages in your book. As you read, pay attention to important ideas. Summarize them in the chart below. Remember, when you summarize, you retell the most important ideas in the selection.**

**Summary**

How are natural solids and manmade solids the same and different?

Natural and manmade solids can both be made into objects that can be helpful to people. They are different because people create manmade solids but can only reshape a natural solid.

| Idea #1 | Idea #2 | Idea #3 |
| --- | --- | --- |
| Natural solids come from earth. Manmade solids are made by people. | Natural solids can be cut or painted. Manmade solids are shaped using molds and colored using chemicals. | People can use natural solids and manmade solids in many ways. |

**Reading and Writing, p. 171**

# Plan Your Lesson

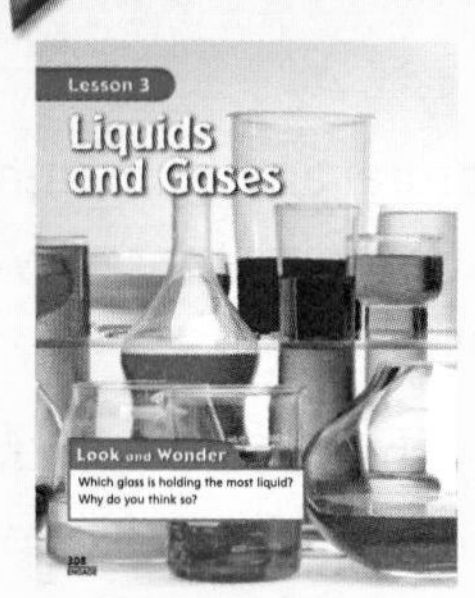


## Lesson 3 Liquids and Gases

### Essential Question

What are the properties of liquids and gases?

### Objectives

- Describe the properties of liquids and gases.
- Compare and contrast liquids and gases.

**Reading Skill** Classify

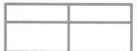

*Graphic Organizer 11*, p. TR13

**Lesson Plan** When time is short, follow the Fast Track and use the essential resources.

| 1 Introduce | 2 Teach | 3 Close |
|---|---|---|
| Look and Wonder, p. 308 | Discuss the Main Idea, p. 310 | Think, Talk, and Write, p. 313 |
| Resource **Activity Lab Book, p. 143** | Resource **Visual Literacy, p. 28** | Resource **Assessment, p. 112** |

**Professional Development** Look for **NSDL** to find recommended Science Background resources from the National Science Digital Library.

## Reading and Writing

## Visual Literacy

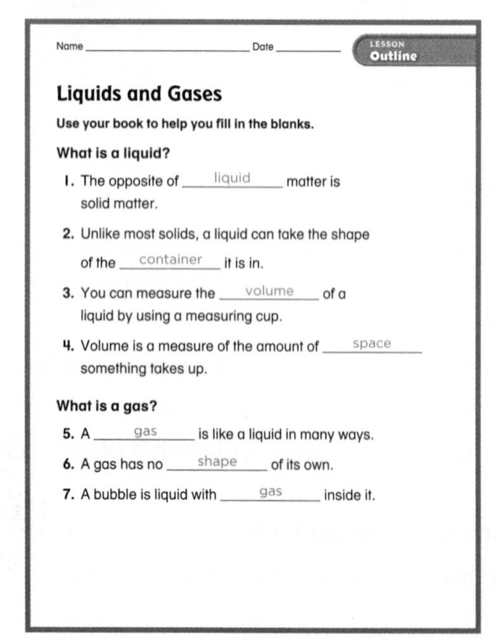

Name ______ Date ______ LESSON Outline

**Liquids and Gases**

**Use your book to help you fill in the blanks.**

**What is a liquid?**

1. The opposite of ___liquid___ matter is solid matter.
2. Unlike most solids, a liquid can take the shape of the ___container___ it is in.
3. You can measure the ___volume___ of a liquid by using a measuring cup.
4. Volume is a measure of the amount of ___space___ something takes up.

**What is a gas?**

5. A ___gas___ is like a liquid in many ways.
6. A gas has no ___shape___ of its own.
7. A bubble is liquid with ___gas___ inside it.

**Outline, pp. 173–174**
Also available as a student workbook

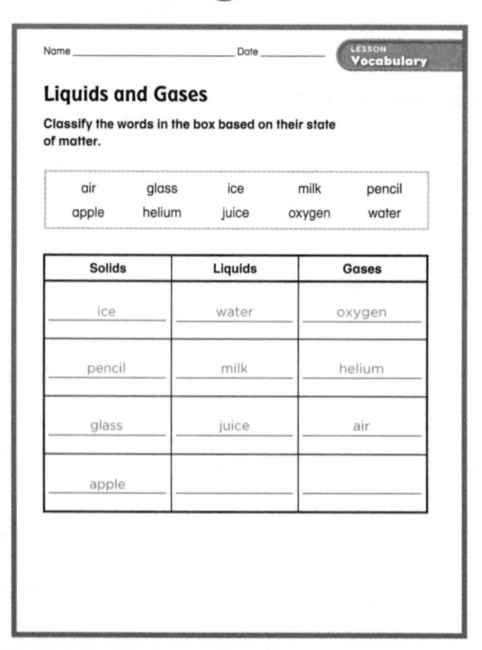

Name ______ Date ______ LESSON Vocabulary

**Liquids and Gases**

**Classify the words in the box based on their state of matter.**

air, glass, ice, milk, pencil, apple, helium, juice, oxygen, water

| Solids | Liquids | Gases |
|---|---|---|
| ice | water | oxygen |
| pencil | milk | helium |
| glass | juice | air |
| apple | | |

**Vocabulary, p. 175**
Also available as a student workbook

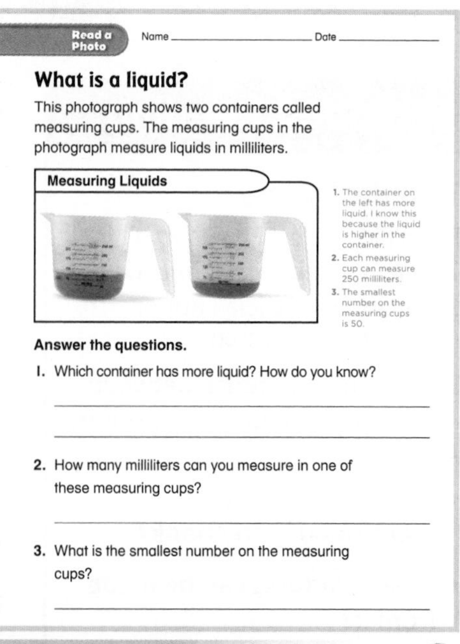

Read a Photo Name ______ Date ______

**What is a liquid?**

This photograph shows two containers called measuring cups. The measuring cups in the photograph measure liquids in milliliters.

**Measuring Liquids**

1. The container on the left has more liquid. I know this because the liquid is higher in the container.
2. Each measuring cup can measure 250 milliliters.
3. The smallest number on the measuring cups is 50.

**Answer the questions.**

1. Which container has more liquid? How do you know?
2. How many milliliters can you measure in one of these measuring cups?
3. What is the smallest number on the measuring cups?

**Read a Photo, p. 28**
Also available as a transparency

# Activity Lab Book

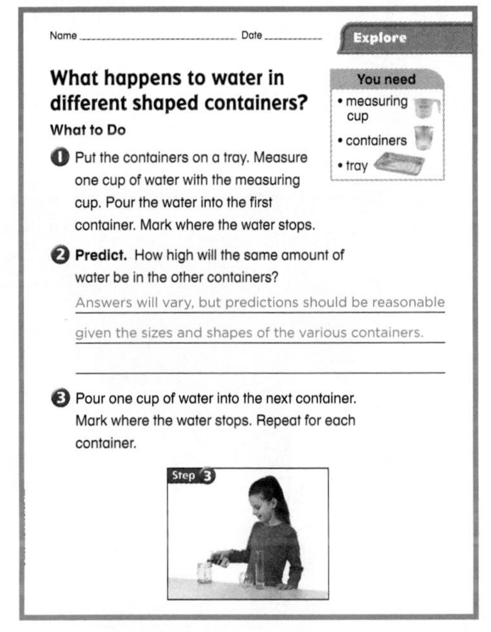

Name ______ Date ______ **Explore**

**What happens to water in different shaped containers?**

**What to Do**

**You need**
- measuring cup
- containers
- tray

1. Put the containers on a tray. Measure one cup of water with the measuring cup. Pour the water into the first container. Mark where the water stops.
2. **Predict.** How high will the same amount of water be in the other containers?
   Answers will vary, but predictions should be reasonable given the sizes and shapes of the various containers.
3. Pour one cup of water into the next container. Mark where the water stops. Repeat for each container.

Step 3

**Explore, pp. 141–142**
Also available as a student workbook

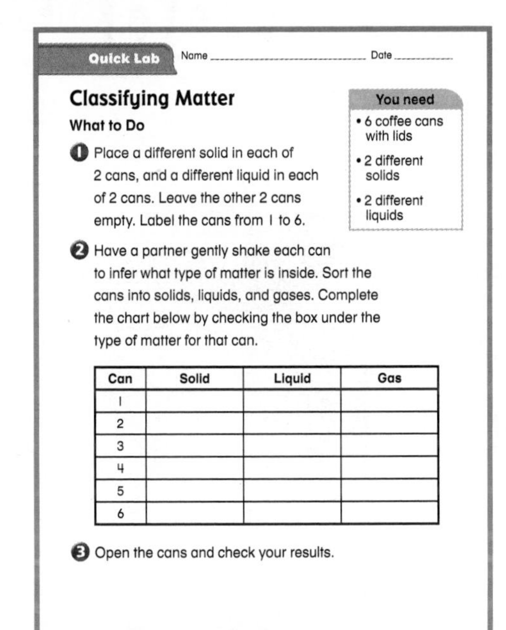

**Quick Lab** Name ______ Date ______

**Classifying Matter**

**What to Do**

**You need**
- 6 coffee cans with lids
- 2 different solids
- 2 different liquids

1. Place a different solid in each of 2 cans, and a different liquid in each of 2 cans. Leave the other 2 cans empty. Label the cans from 1 to 6.
2. Have a partner gently shake each can to infer what type of matter is inside. Sort the cans into solids, liquids, and gases. Complete the chart below by checking the box under the type of matter for that can.

| Can | Solid | Liquid | Gas |
|---|---|---|---|
| 1 | | | |
| 2 | | | |
| 3 | | | |
| 4 | | | |
| 5 | | | |
| 6 | | | |

3. Open the cans and check your results.

**Quick Lab, p. 144**
Also available as a student workbook

# Assessment

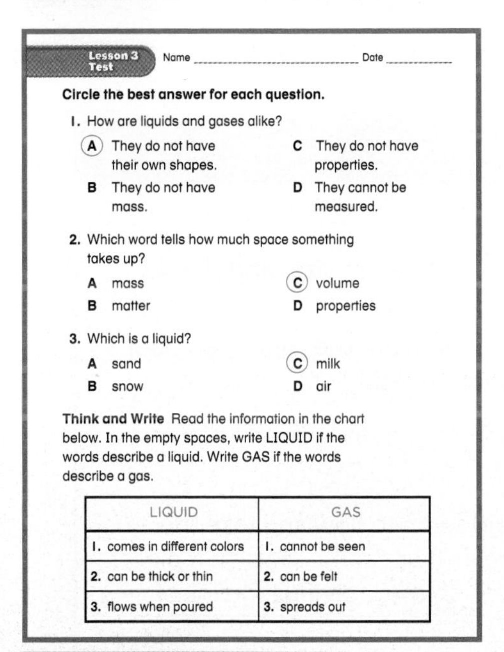

**Lesson 3 Test** Name ______ Date ______

**Circle the best answer for each question.**

1. How are liquids and gases alike?
   - (A) They do not have their own shapes.
   - B They do not have mass.
   - C They do not have properties.
   - D They cannot be measured.
2. Which word tells how much space something takes up?
   - A mass
   - B matter
   - (C) volume
   - D properties
3. Which is a liquid?
   - A sand
   - B snow
   - (C) milk
   - D air

**Think and Write** Read the information in the chart below. In the empty spaces, write LIQUID if the words describe a liquid. Write GAS if the words describe a gas.

| LIQUID | GAS |
|---|---|
| 1. comes in different colors | 1. cannot be seen |
| 2. can be thick or thin | 2. can be felt |
| 3. flows when poured | 3. spreads out |

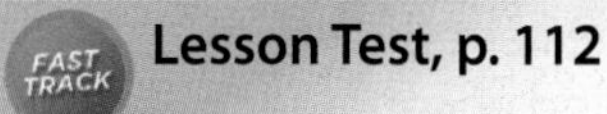

**Lesson Test, p. 112**

## ADDITIONAL RESOURCES

pp. 136–139

*Gases Matter*

p. 43

p. 92

p. 28

28

81–90

pp. 89–90

71–73

### Technology

- Science Activity DVD
- TeacherWorks™ Plus CD-ROM
- Classroom Presentation Toolkit CD-ROM

- e-Review

- NSDL

# Lesson 3 Liquids and Gases

## Objectives

- Describe the properties of liquids and gases.
- Compare and contrast liquids and gases.

## 1 Introduce

### ▶ Assess Prior Knowledge

Have children share what they know about liquids and gases. Ask:

- **How are liquids and gases like solids?**
- **How are liquids and gases different than solids?**

Record children's answers in the What We Know column of the class **KWL** chart.

### Look and Wonder

Read the Look and Wonder questions about liquids.

To help children organize their thinking about the amount of liquid, ask:

- **If all the bottles were filled up to the top, which bottle would hold the most liquid?** the largest bottle
- **Why is it difficult to tell whether the bottles hold the same amount of liquid?** Possible answer: because the bottles are different shapes and sizes

Write children's responses on the class **KWL** chart and note any misconceptions that they may have.

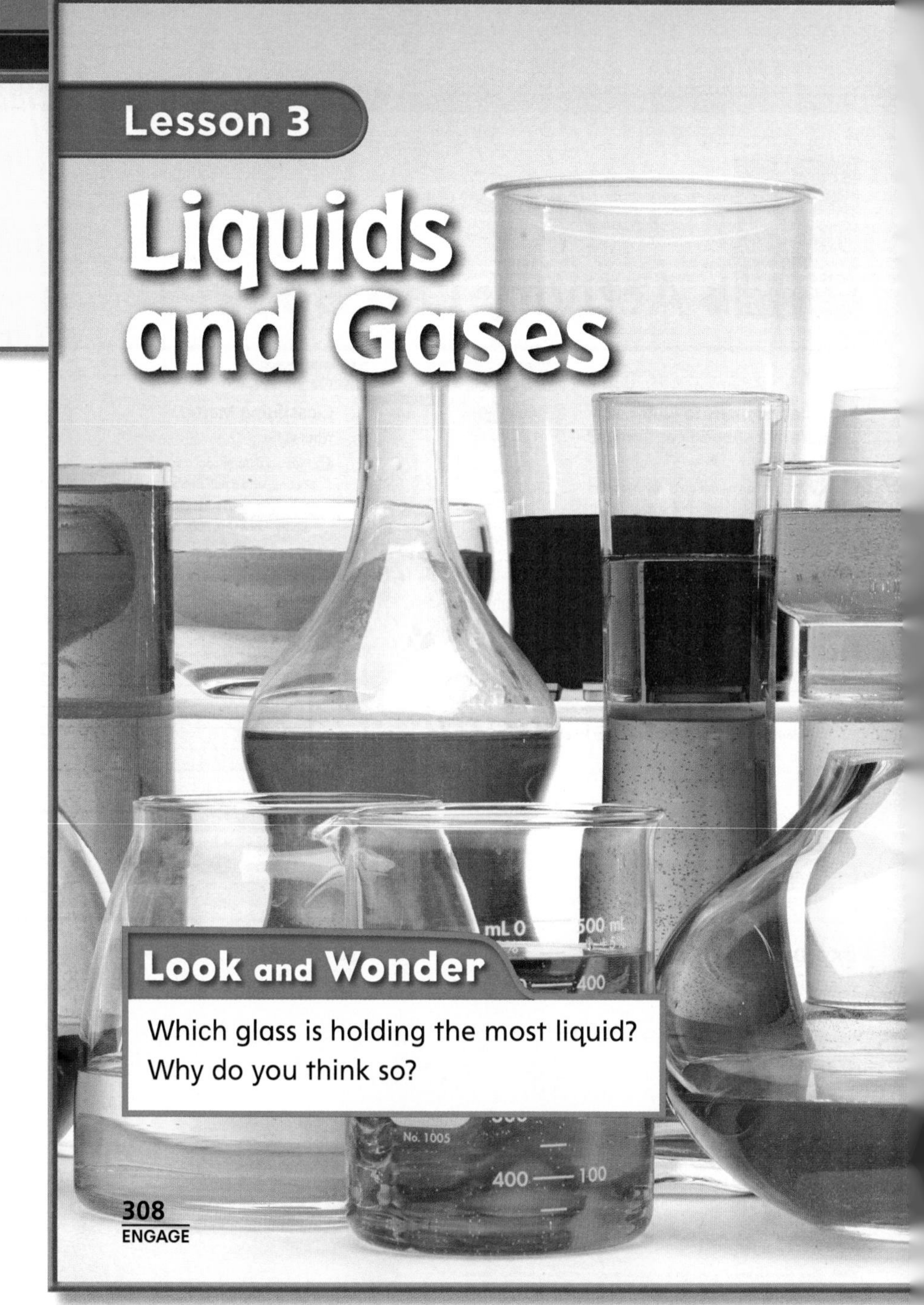
Lesson 3
Liquids and Gases
Look and Wonder
Which glass is holding the most liquid? Why do you think so?
mL 0
500 mL
400
No. 1005
400
100
308
ENGAGE

### Warm Up

**Start with a Demonstration**

Use water, sand, and containers of different sizes and shapes for this demonstration.

First, pour sand from one container to another, and then pour water from one container to another. Ask:

- **How did the sand change?**
- **How did the water change?**

Discuss how the solid and liquid seem to flow and take the shape of the container they are in. Place some sand on a piece of paper and have children observe the individual grains of sand with a hand lens.

Explain that the individual grains of sand have kept their own unique shapes. Put sand in the science area and encourage children to observe the tiny solid shapes with a hand lens.

### RESOURCES and TECHNOLOGY

- **Activity Lab Book,** pp. 141–143
- **Activity Flipchart,** p. 43
- Science Activity DVD

## Explore — Inquiry Activity

### What happens to water in different shaped containers?

#### What to Do

1. Put containers on a tray. Measure one cup of water with the measuring cup. Pour the water into the first container. Mark where the water stops.
2. **Predict.** How high will the same amount of water be in the other containers?
3. Pour one cup of water into the next container. Mark where the water stops. Repeat for each container.
4. **Draw Conclusions.** Were your predictions correct? Explain.

#### Explore More

5. **Infer.** Would the activity change if you used juice instead of water? Why or why not?

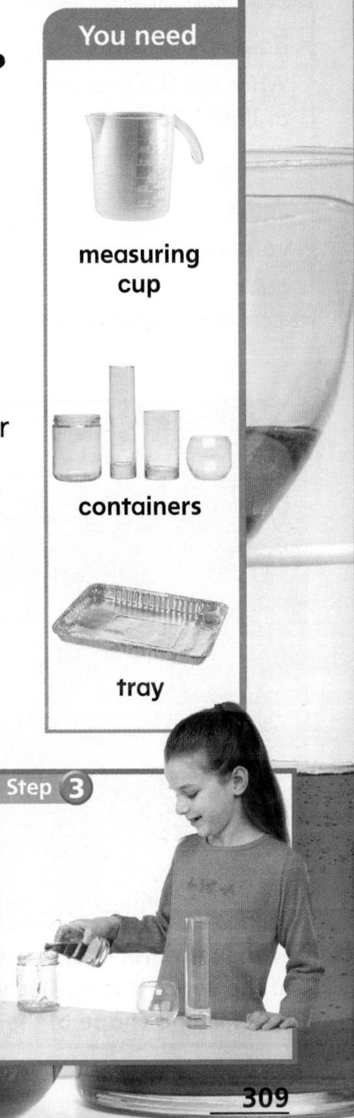




## Explore

small groups

30 minutes

**Plan Ahead** Select plastic containers of different sizes and shapes that will hold at least one cup of water. Have paper towels handy to clean up any spills.

**Purpose** The experiment will help children observe and draw conclusions about how liquids are measured and described based on how they are contained.

### Structured Inquiry What to Do

Have children identify and describe the kinds of containers they have seen that hold liquids. Ask: **What kinds of containers can you use to hold water?**

1. Review how to measure water in a measuring cup. Have children number each container.
2. **Predict** Have children observe the shapes and sizes of the containers and use that information to predict whether the water level in the other containers will be higher, lower, or about the same. Have them record their predictions.
3. Ask children to record the results for each container. Later, they may make a chart of the results.
4. **Draw Conclusions** After children have compared the results with their predictions, ask: **Which container shape had the highest water level?** Possible answer: the narrow container **Why did that happen?** Possible answer: The water could not spread out in the narrow container, so it had to go up.

### Guided Inquiry Explore More

5. **Infer** Have children use what they know about the properties of liquids to infer what would happen if another liquid was used. Provide other liquids for children to check the accuracy of their inferences.

### Open Inquiry

Point out to children that not all liquids are thin. Have them investigate the properties of thicker liquids, such as mustard, ketchup, or milk shakes. Encourage them to list the properties of each and to compare them with thinner liquids.

FAST TRACK

## Alternative Explore

### How high will the liquid go?

Have children compare the heights of different liquids in different shaped containers. Give children a selection of plastic containers (short, wide and tall, narrow), measuring cups, water, and another liquid (juice or milk).

Ask children to measure a cup of each liquid into a short container. Have them predict how high the liquid will be in the taller container. Ask them to pour the liquid into the taller container to see if the predictions were correct.

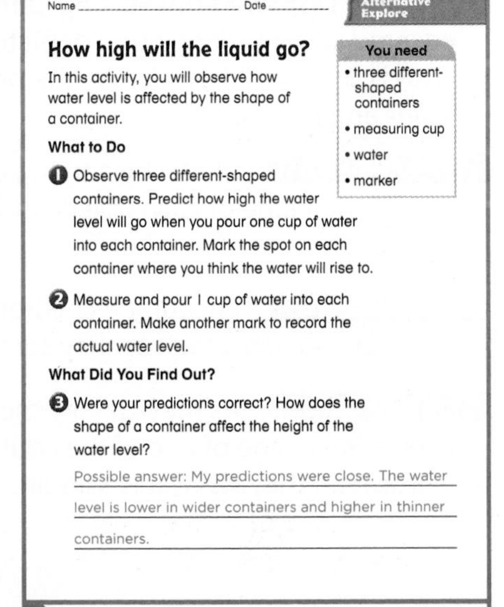
Name ____________ Date ________ Alternative Explore

**How high will the liquid go?**

In this activity, you will observe how water level is affected by the shape of a container.

You need
- three different-shaped containers
- measuring cup
- water
- marker

**What to Do**

1. Observe three different-shaped containers. Predict how high the water level will go when you pour one cup of water into each container. Mark the spot on each container where you think the water will rise to.
2. Measure and pour 1 cup of water into each container. Make another mark to record the actual water level.

**What Did You Find Out?**

3. Were your predictions correct? How does the shape of a container affect the height of the water level?
Possible answer: My predictions were close. The water level is lower in wider containers and higher in thinner containers.

Activity Lab Book, p. 143

# 2 Teach

## Read Together and Learn

**Reading Skill** Classify To put things that are alike into groups.

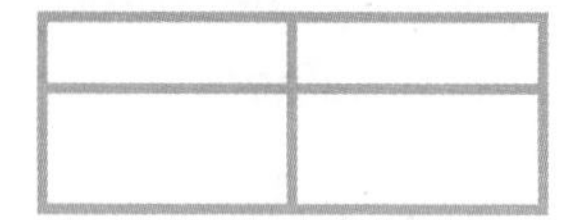

*Graphic Organizer 11*, p. TR13

## What is a liquid?

### Discuss the Main Idea

FAST TRACK

**Main Idea** Liquids have mass and their volumes can be measured.

Before reading, ask children to identify words that can be used to describe the properties of liquids.

After reading together, ask:

- **Which liquids did you use this morning?**
- **What kind of container did the liquids come in?**

Discuss how a liquid takes the shape of its container. Ask:

- **How would you describe the shape of a glass of water?** Possible answer: by the shape of the glass

**Read Together and Learn**

**Essential Question**
What are the properties of liquids and gases?

**Vocabulary**
liquid
volume
gas

## What is a liquid?

A **liquid** is a kind of matter that takes the shape of the container it is in. Without a container, liquids flow and have no shape.

All liquids have mass. Liquids can be thin like milk or thick like honey.

**Even in nature, liquid takes the shape of the space it is in. This waterfall flows and fills the shape of the lake.**

310
EXPLAIN

### Classroom Equity

Young children can be influenced by what others think of them, especially classmates, family members, and teachers. Many children dream of becoming a space scientist, but as they grow older, they may be influenced by peer pressure or stereotypes that suggest these options are not for them. Show all children that you believe they are capable of excelling in science by setting high expectations.

### ELL Support

**Classify** Display Photo Sorting Cards 81–87 (with any labels covered with masking tape) or pictures of both solids and liquids. Have children classify the photos into two groups: *solids* and *liquids*.

**BEGINNING** Model by naming each object. Have children repeat the name of each object and place the image in the correct group.

**INTERMEDIATE** Encourage children to use words and phrases to describe each image as they sort the pictures.

**ADVANCED** As they classify the pictures, have children choose an image of one of the liquids and explain in their own words how they know it is a liquid.

The amount of space something akes up is called **volume**. You can neasure the volume of a liquid with a neasuring cup. Liquids are measured n milliliters or ounces.

The measuring cups in the picture can hold the same amount. One cup s holding a greater volume of liquid han the other.

**What are some properties of liquids?**

**Measuring Volume**

**Read a Photo**

How many milliliters of liquid are in each container?

FACT Solids and gases also have volume.



## Differentiated Instruction

### Leveled Activities

**EXTRA SUPPORT** Help children use measuring cups marked with milliliters and ounces to determine volume. Pour different amounts of water into identical, labeled containers. Have children use the measuring cup to measure and record the volume of water in each container, and to compare how the volume of water differs from container to container.

**ENRICHMENT** Tell children that they are going to measure the volume of liquid they consume in a day. Give them measuring cups (marked with milliliters and ounces) and plastic drinking cups. Have children pour as much water into the drinking cup as they usually drink, and then measure the amount of water. Have them write down the number of cups of liquid they usually drink during the day and add up the totals to find out how much liquid they drink in a day.

## ▶ Addressing Misconceptions

Children may think that solids and gases do not have volume, but anything that takes up space has volume. The volume of a solid can be measured with a ruler. Gas does not have a fixed volume but can be measured by other means.

FACT **Solids and gases also have volume.** Display a book and a blown up balloon so children can see that solids and gases take up space.

### Read a Photo

Before reading the measurements of each measuring cup, have children estimate how much more liquid is in the container on the left. Ask:

- **What are the containers measuring?** volume

Point out to children that the millimeter measurements are increasing by fifty. Help children read each measurement on a similar measuring cup before answering the question.

**Answer to Read a Photo** left: 100 mL; right: 50 mL

## ▶ Develop Vocabulary

**liquid** Ask children to make a word web of *liquids*. Encourage them to include at least three liquids and words that describe the properties of liquids.

**volume** ***Scientific vs. Common Use*** *Volume* is commonly used as a noun to indicate the amount of sound heard. In scientific terms, *volume* refers to the amount of space an object uses. Ask children to write sentences that reflect both meanings of the word.

## Quick Check Answer

Possible answers: Liquids have mass, volume, and take the shape of the container they are in. They also can be thick or thin.

### RESOURCES and TECHNOLOGY

- **Reading and Writing,** pp. 173–175
- **Visual Literacy,** p. 28
- **PuzzleMaker CD-ROM**
- **Classroom Presentation Toolkit CD-ROM**
- 

## What is a gas?

### ▶ Discuss the Main idea

**Main Idea** Gases are a type of matter that cannot be seen, but have mass and take up space.

Read the text together and discuss with children other objects that are filled with gas, such as bicycle tires or hot air balloons.

### ▶ Use the Visuals

Read the caption for each photograph and discuss how the gases are changing the shape of the parachutes, the small beach tube, and the liquid soap. Ask:

- **How can you tell that the air inside a balloon has mass?** The air takes up space inside the balloon when it is blown up. The balloon filled with air is heavier than the balloon with no air.
- **How does air get into the red beach tube?** The girl blows air into it.
- **What is inside the soap bubbles?** gas, air

### ▶ Develop Vocabulary

**gas** ***Scientific vs. Common Use*** Explain that the word *gas* is often used as an abbreviation for the word *gasoline*. In scientific use, *gas* is matter that has no definite shape and expands in every direction to fill the space it is in. Have children identify an object that holds air to complete this sentence frame:
*A __________ holds gas when it is blown up.*

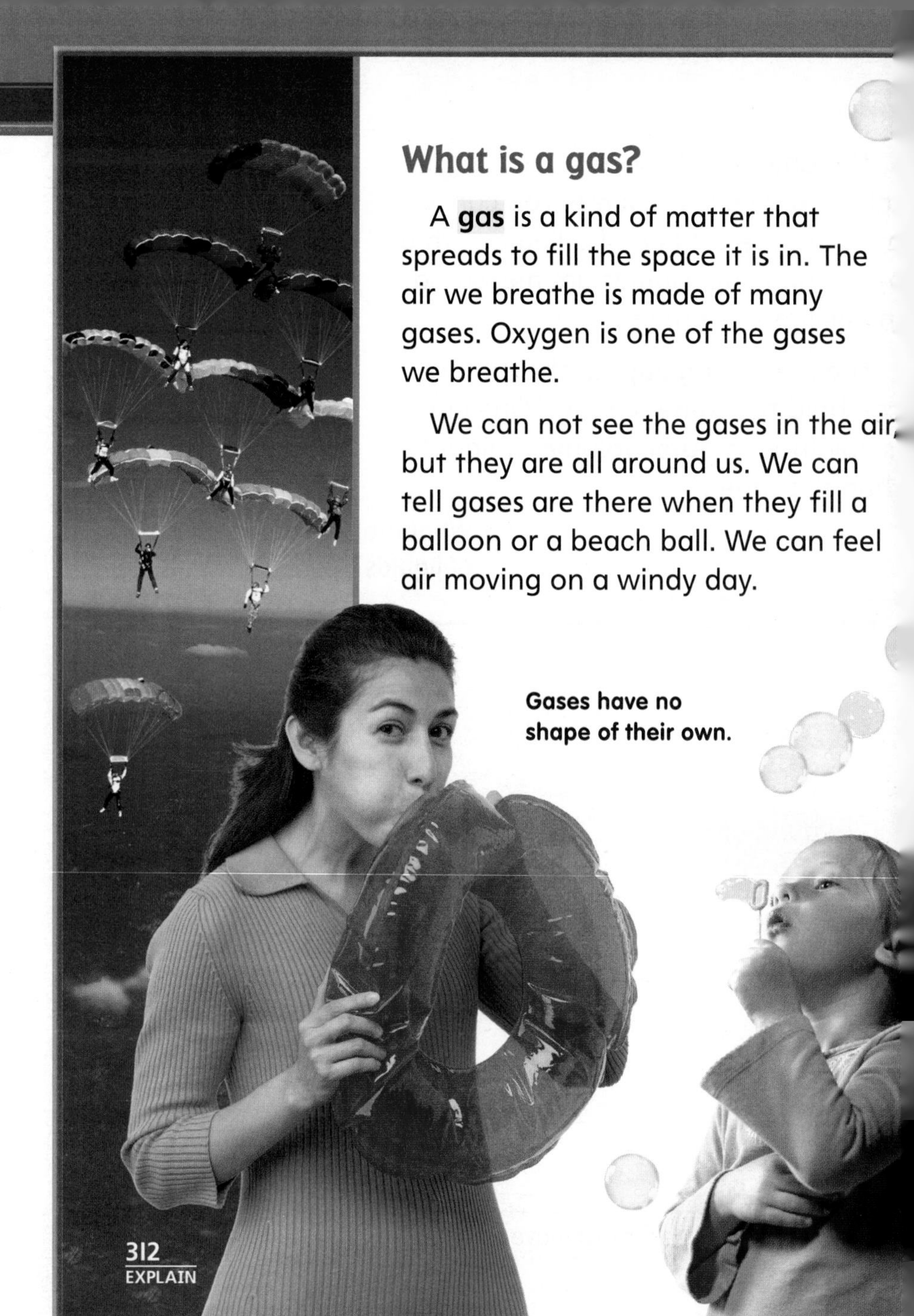

## What is a gas?

A **gas** is a kind of matter that spreads to fill the space it is in. The air we breathe is made of many gases. Oxygen is one of the gases we breathe.

We can not see the gases in the air, but they are all around us. We can tell gases are there when they fill a balloon or a beach ball. We can feel air moving on a windy day.

**Gases have no shape of their own.**

312
EXPLAIN

## Quick Lab

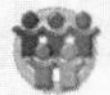 whole class  15 minutes

**Objective** Classify matter by different properties.

**You need** six cans with lids for the entire class, small solid objects, liquids

1. Place a different solid in two cans, a different liquid in two cans, and leave two cans empty. Label the cans one to six.
2. Have children gently shake the cans to infer what type of matter is inside and sort them into three groups.
3. Have children open the cans to verify their sorting predictions.

### RESOURCES and TECHNOLOGY

- **School to Home Activities,** pp. 89–90
- **Reading and Writing,** p. 176
- **Activity Lab Book,** p. 144
- **e-Review** Narrated Summary and Quiz
- ***ExamView® Assessment Suite* CD-ROM**

Remember that anything
hat takes up space is matter.
ll matter has mass. How can
ou tell that air has mass?
ook at the picture.

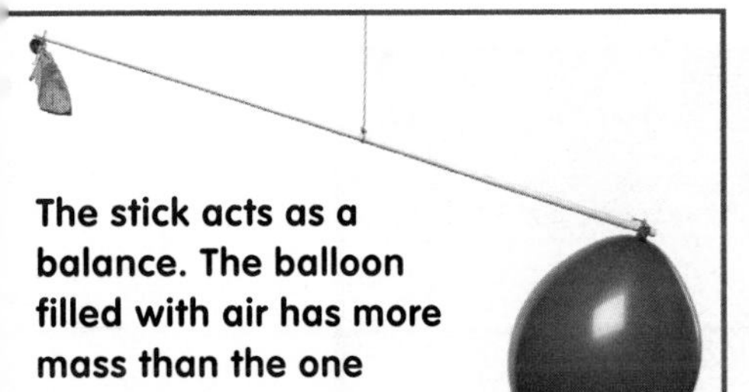

The stick acts as a balance. The balloon filled with air has more mass than the one without air.

## Quick Lab

Fill containers with different kinds of matter. Have a partner classify the matter as solid, liquid, or gas.

What are some properties of a gas?

## Think, Talk, and Write

1. Classify. List some of the items in your refrigerator. Sort them as solid, liquid, or gas.
2. How is a gas different from a liquid?
3. **Essential Question.** What are the properties of liquids and gases?

### Health Link

Make a list of liquids that are good for you.

-Review Summaries and quizzes online at www.macmillanmh.com



## Formative Assessment

### Identify Liquids and Gases

Distribute several index cards to each child. Challenge children to think of liquids and gases, and draw one on each card.

Ask them to write a sentence that describes what they have drawn. Remind children that they can illustrate a gas by showing what happens when gas fills a container, such as a balloon.

**Key Concept Cards** For student intervention, see the prescribed routine on **Key Concept Card 28.**

### Quick Check Answer

Possible answers: not visible, has no fixed shape, spreads to fill evenly whatever container it is in

## 3 Close

### ▶ Using the KWL Chart

Review with children what they have learned about liquids and gases and their properties. Record their responses in the What We Learned column of the class **KWL** chart.

### ▶ Using the Reading Skill Classify

Use the reading skill graphic organizer to help children classify solids, liquids, and gases.

| Solids | Liquids | Gases |
|---|---|---|
| Solids have a fixed shape. | Liquids flow and take up space. | Gases can not be seen but they have mass and take up space. |

*Graphic Organizer 11, p.* TR13

### FAST TRACK Think, Talk and Write

1. **Classify** Possible answers: Liquids: milk, juice, water; Solids: cheese, bread, fruit; Gases: cold air
2. Possible answer: Gas expands to fill a space evenly. Liquids flow into containers and may not take up all of the space.
3. **Essential Question** Possible answers: Both liquids and gases flow and take the shape of their container. Liquids can not spread to fill a space. Gases can spread to fill a space.

### Health Link

Explain to children that human bodies are composed mostly of water. It is important to drink lots of healthy liquids throughout the day. Discuss why sugary drinks are not as good as water or juice.

Encourage children to make a poster of the healthy liquids from their lists.

# Writing in Science

**Objective**

- Describe a fun water activity in a personal narrative.

## Fun with Water

### Talk About It

Direct children's attention to the picture.

Invite children to share their fun experiences in water. Encourage children to recall details about their experiences. Ask:

- **How did you feel before you went in the water?**
- **How did the water feel?**

### Learn About It

Read the top paragraph and Remember box together with children. Use Graphic Organizer 7 to show children how to organize ideas to tell what happened first, next, and last. Explain that this helps readers understand the order in which things happened.

### Write About It

Encourage children to discuss their story ideas quietly with a partner and choose one story to tell. To help them organize their ideas before they write, have children complete Graphic Organizer 7 to sequence the events for their stories.

After they have written their stories, have children draw a picture to illustrate them. Put children's stories together to make a class book titled *Fun with Water.*

**RESOURCES and TECHNOLOGY**

- **Reading and Writing,** pp. 177–178

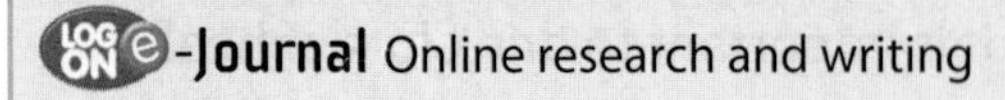

**Writing in Science**

# Fun with Water

This girl is having fun in the water! How do you enjoy water?

### Write About It

Think of times that you have had fun in water. Draw and write about what you did. Remember to add details to your story.

**Remember**

Details help your reader know what happened and how you felt.

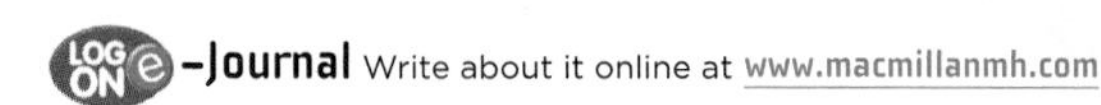

-Journal Write about it online at www.macmillanmh.com


314
EXTEND

### Integrate Writing

**Water As a Solid**

Review with children what they learned in the chapter about water as a solid. Ask:

- **What fun activities can you have with ice or snow?** Possible answers: sledding; making snow sculptures; ice skating; making ice treats

Have children draw and write about a time they had fun with water in the form of ice or snow.

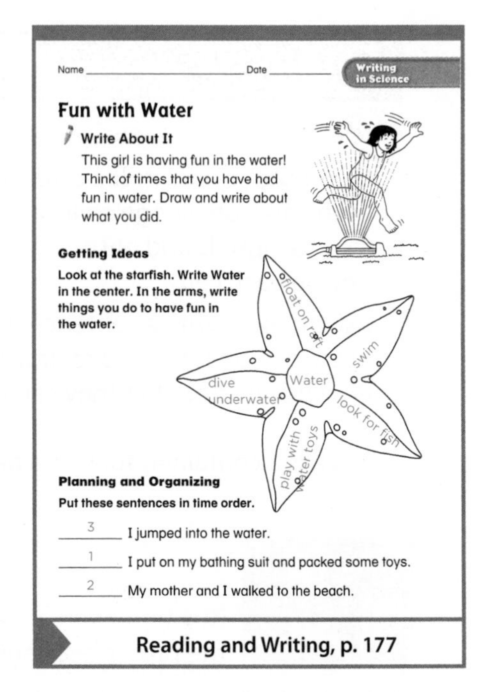

Name ______ Date ______ Writing in Science

**Fun with Water**

Write About It

This girl is having fun in the water! Think of times that you have had fun in water. Draw and write about what you did.

**Getting Ideas**

Look at the starfish. Write Water in the center. In the arms, write things you do to have fun in the water.



**Planning and Organizing**

Put these sentences in time order.

3 I jumped into the water.

1 I put on my bathing suit and packed some toys.

2 My mother and I walked to the beach.

Reading and Writing, p. 177

Math in Science

# Which Has More Volume?

Matt put juice in two measuring cups. What can you tell about the two cups of juice? Which has more volume?

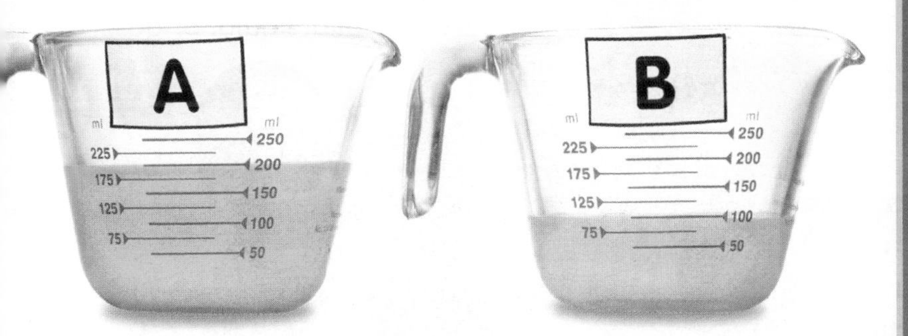


## Write a Number Sentence

Cup A has 200 milliliters (mL) of juice. Cup B has 100 mL of juice. How many more mL are in Cup A?

Write a number sentence to show how you found the answer.

**Remember**
Think about which operation to choose.

315
EXTEND

## Integrate Math

### How Much Ingredients?

Tell children that the recipe for 1 smoothie requires: $1\frac{1}{2}$ cups of juice, a $\frac{1}{4}$ cup of bananas, and a $\frac{1}{2}$ cup of berries. Ask:

- **How many cups of ingredients are used all together?** $2\frac{1}{4}$ cups
- **How much of each ingredient do you need for 2 smoothies?** 3 cups of juice, a $\frac{1}{2}$ cup of bananas, and 1 cup of berries.

Ask children to write number sentences to show how they solved the problems.

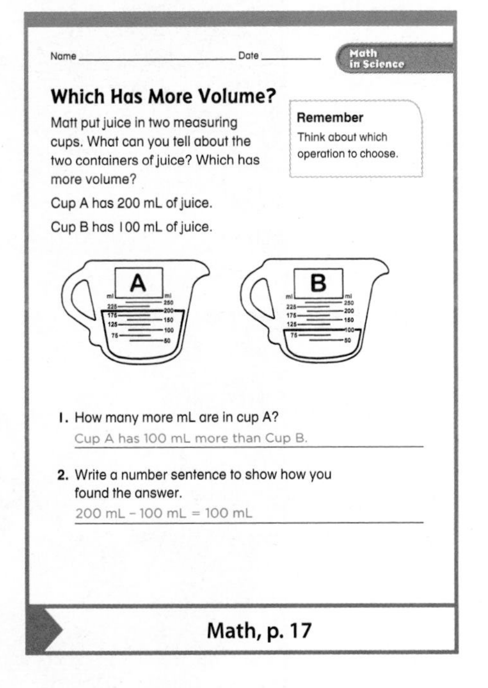
Name ______________ Date ________ Math in Science

**Which Has More Volume?**

Matt put juice in two measuring cups. What can you tell about the two containers of juice? Which has more volume?

Cup A has 200 mL of juice.
Cup B has 100 mL of juice.

**Remember**
Think about which operation to choose.

1. How many more mL are in cup A?
Cup A has 100 mL more than Cup B.

2. Write a number sentence to show how you found the answer.
200 mL − 100 mL = 100 mL

Math, p. 17



# Math in Science

**Objective**

- Compare volumes of liquids.

## Which Has More Volume?

### Talk About It

Remind children that the amount of space something takes up is volume. Have children compare the two measuring cups in the picture. Ask:

- **What is being measured in the picture?** Possible answer: the volume of the liquids
- **How can you tell which measuring cup has more volume?** Possible answer: by observing which measuring cup holds the most liquid

### Learn About It

Read the top paragraph with children. Review examples of number sentences with children. For example, have children find the total amount of the two cups together. Ask:

- **What operation would you use?** addition

Ask a volunteer to write a number sentence for the addition problem on the board. 200 + 100 = 300

### Try It

Have children read the bottom paragraph and find the words that will help them solve the problem. Ask:

- **Which words help to tell you what operation to use?** how many more

Ask a volunteer to write the number sentence that solves the problem on the board. 200 − 100 = 100

Have children write another problem using volume and exchange problems with a partner. Have children write a number sentence for their partner's problem.

**RESOURCES and TECHNOLOGY**

- Math, pp. 17–18

# I Read to Review

**Objective**

- Reinforce the properties of matter with independent reading.

### Buddy Reading

- During small-group instructional time, have children work together in pairs, taking turns to read the selection to one another.
- Pair weaker readers with stronger partners to ensure success during this work period.

### Independent Reading

- Give children copies of School to Home Activities, pages 91 and 92. Have them assemble the pages to make their own books.
- Encourage children to take the book home and read it aloud to a family member or friend.

**PAGE 316 ▶**

Ask: **How would you describe the properties of the water and the slides?** Possible answers: Water is liquid, flowing, and wet; the slides are solid, red, blue, yellow, curvy, slippery, and smooth.

**PAGES 318–319 ▶**

Ask: **How can you tell that water takes the shape of the half pipe and the pool?** Possible answers: The water flows in a curve like the pipe. Water fills the whole pool.

Liquids take the shape of the container they are in. When a liquid spills, it has no shape of its own. This girl is gliding on the flowing water.
318

**RESOURCES and TECHNOLOGY**

- **School to Home Activities**, pp. 91–92
- **Assessment**, pp. 117–118
- **TeacherWorks™ Plus CD-ROM**

A solid has a size and a shape. You can talk about its feel and color. Sometimes you can talk about its sound, smell, or taste.

317

Gases fill the spaces they are in. Air fills up these rafts. The rafts float on the water. Look at the matter you see all around you!

319

## Performance Assessment

### Matter

 individual  20 minutes

**Materials** index cards, chart paper, colored pencils, markers, or crayons

1. Make a three-column chart that can be placed on a wall either in the classroom or in the hallway. Label the three columns: *Solids*, *Liquids*, and *Gases*. Make sure that the chart is low enough so that each child can reach it independently.
2. Give each child six blank index cards. Have them draw an example of a solid, a liquid, and a gas on three of the cards. Remind them to clearly illustrate the matter so others can observe as many properties as possible. On the other three cards have them write a list of words that describe the type of matter that they drew and how they know whether it is a solid, a liquid, or a gas.
3. Once children have completed the task, have them post their cards on the class chart.

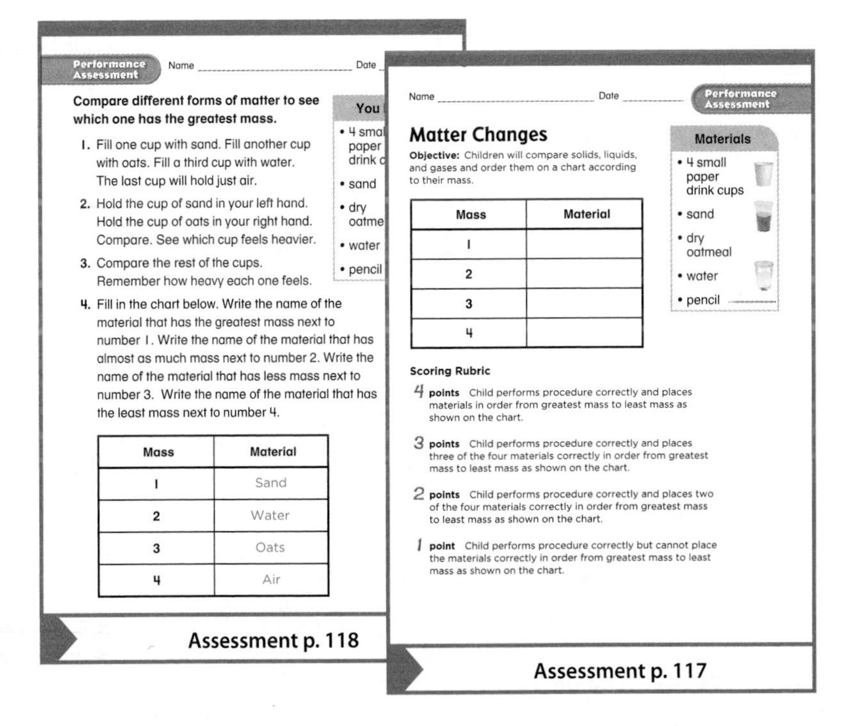

Performance Assessment Name ______ Date ______

**Compare different forms of matter to see which one has the greatest mass.**

1. Fill one cup with sand. Fill another cup with oats. Fill a third cup with water. The last cup will hold just air.
2. Hold the cup of sand in your left hand. Hold the cup of oats in your right hand. Compare. See which cup feels heavier.
3. Compare the rest of the cups. Remember how heavy each one feels.
4. Fill in the chart below. Write the name of the material that has the greatest mass next to number 1. Write the name of the material that has almost as much mass next to number 2. Write the name of the material that has less mass next to number 3. Write the name of the material that has the least mass next to number 4.

You
- 4 smal paper drink c
- sand
- dry oatme
- water
- pencil

| Mass | Material |
|---|---|
| 1 | Sand |
| 2 | Water |
| 3 | Oats |
| 4 | Air |

Assessment p. 118

Name ______ Date ______ Performance Assessment

**Matter Changes**

**Objective:** Children will compare solids, liquids, and gases and order them on a chart according to their mass.

| Mass | Material |
|---|---|
| 1 | |
| 2 | |
| 3 | |
| 4 | |

Materials
- 4 small paper drink cups
- sand
- dry oatmeal
- water
- pencil

**Scoring Rubric**

4 **points** Child performs procedure correctly and places materials in order from greatest mass to least mass as shown on the chart.

3 **points** Child performs procedure correctly and places three of the four materials correctly in order from greatest mass to least mass as shown on the chart.

2 **points** Child performs procedure correctly and places two of the four materials correctly in order from greatest mass to least mass as shown on the chart.

1 **point** Child performs procedure correctly but cannot place the materials correctly in order from greatest mass to least mass as shown on the chart.

Assessment p. 117

# CHAPTER 9 Review

 For information on Depth of Knowledge levels, see page 321B.

### ▶ Use the KWL chart

Review the **KWL** chart that the class made at the beginning of the chapter. Help children compare what they know about matter now with what they knew then. Add any additional information to the What We Learned column of the **KWL** chart.

### ▶ Make a FOLDABLES Study Guide

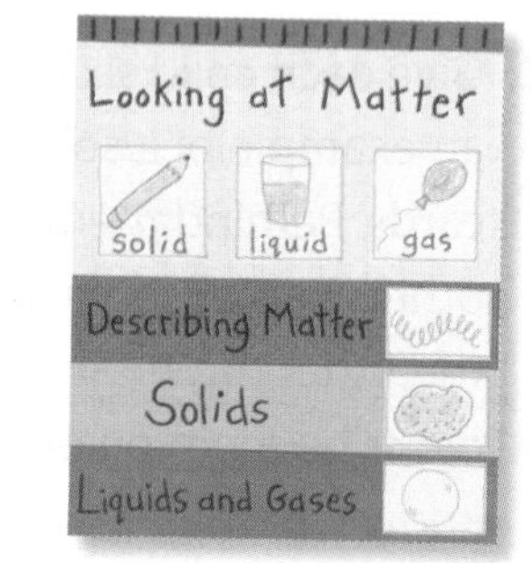


Create a layered-book project Foldables for the class. Start by dividing children into three groups. Give each child a large index card.

Ask the Lesson 1 group members to paste a picture of an object to the top of each card and list properties of the pictured object underneath. Glue the cards under the first tab.

Ask the Lesson 2 group to paste pictures of solids on the cards. In writing, have them describe the pictured objects' properties and describe how they can be measured. Glue the cards under the second tab.

Have the Lesson 3 group glue pictures of liquids and gases to cards. In writing, have them describe the properties of the pictured objects, and explain how the object can be measured. Glue the cards under the third tab.

See page TR38 in the back of the Teacher's Edition for more information on Foldables.

### Vocabulary

1. matter
2. mass
3. gas
4. solid
5. liquid
6. volume

### RESOURCES and TECHNOLOGY

- **Reading and Writing,** pp. 179–180
- **Assessment,** pp. 106–109, 113–116
- **ExamView® *Assessment Suite* CD-ROM**
- **PuzzleMaker CD-ROM**
-  **Vocabulary Games**

## CHAPTER 9 Review

### Vocabulary

DOK I

**Use each word once for items 1–6.**

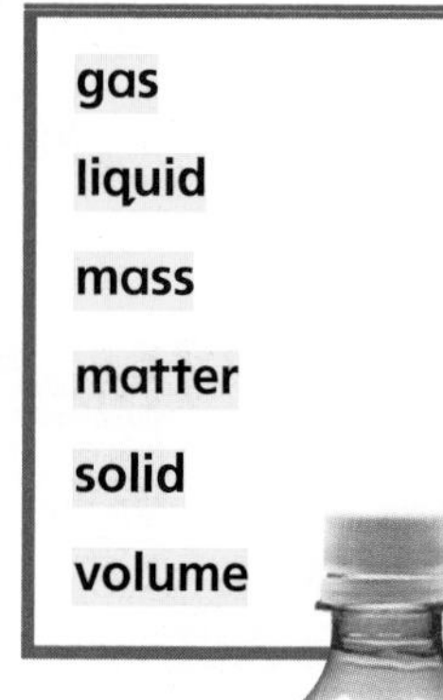

gas
liquid
mass
matter
solid
volume

1. Everything that takes up space and has mass is called ________.
2. The amount of matter in something is called ________.
3. Some matter can not always be seen. It spreads to fill the space it is in and is called a ________.
4. Matter that has a shape of its own is called a ________.
5. Matter that flows and takes the shape of the container it is in is called a ________.
6. The bottle on the right can hold a larger ________ of water than the bottle on the left.

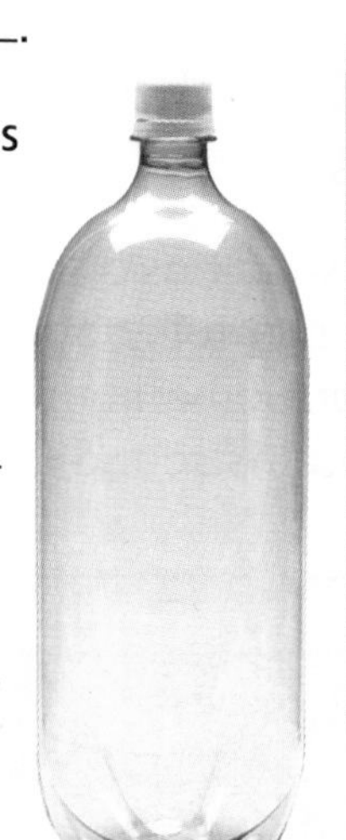

  -Glossary Words and definitions online at www.macmillanmh.com

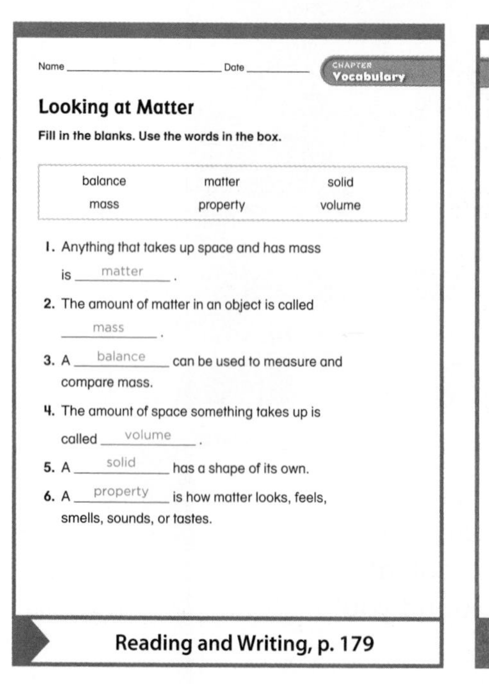

Name ________ Date ________ CHAPTER Vocabulary

**Looking at Matter**

**Fill in the blanks. Use the words in the box.**

| balance | matter | solid |
|---|---|---|
| mass | property | volume |

1. Anything that takes up space and has mass is ___matter___.
2. The amount of matter in an object is called ___mass___.
3. A ___balance___ can be used to measure and compare mass.
4. The amount of space something takes up is called ___volume___.
5. A ___solid___ has a shape of its own.
6. A ___property___ is how matter looks, feels, smells, sounds, or tastes.

**Reading and Writing, p. 179**

Chapter Test A Name ________ Date ________

**Looking at Matter**

**Write the word that best completes each sentence in the spaces below. Words may be used only once.**

| balance | mass | properties | volume |
|---|---|---|---|
| gas | matter | ruler | |
| liquids | object | solid | |

1. All things are made of ___matter___.
2. Different objects have different amounts of ___mass___.
3. Size and shape are ___properties___ of matter.
4. All ___liquids___ take the shape of the container they are in.
5. A(n) ___solid___ has a shape all its own.
6. The amount of space something takes up is called ___volume___.
7. A form of matter that can be felt but not seen is ___gas___.
8. Anything that can be seen or touched is a(n) ___object___.
9. A tool used to measure length is a(n) ___ruler___.
10. A tool used to measure mass is a(n) ___balance___.

**Assessment, p. 106**

## Science Skills and Ideas

DOK 2

**Answer the questions below.**

**7.** Record Data. How are the two balls alike and different? Which ball has more mass?

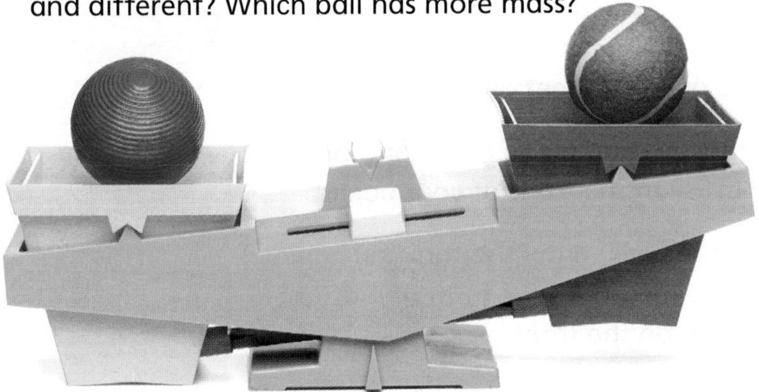

**8.** Summarize. What tools can you use to measure matter in different ways?

**9.** What type of matter has filled this balloon?

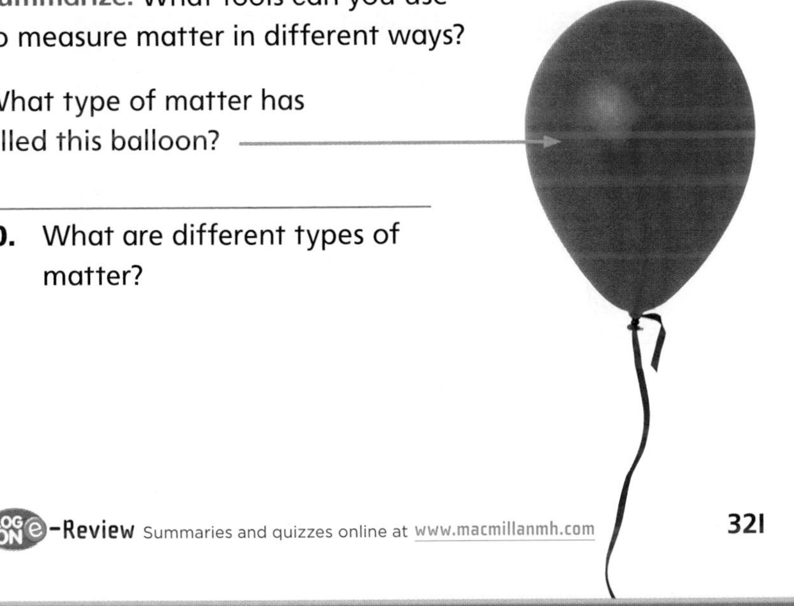

**10.** What are different types of matter?

e-Review Summaries and quizzes online at www.macmillanmh.com

321

## Science Skills and Ideas

**7. Record Data** Encourage children to complete a Venn diagram like the one below.

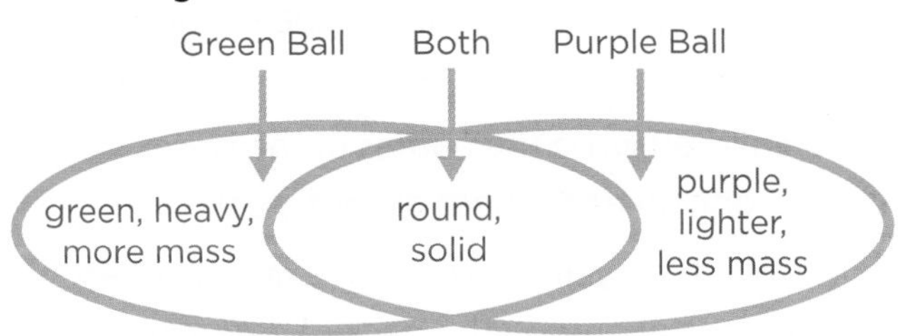


*Graphic Organizer 10,* p. TR12

**8. Summarize** Encourage children to complete a summarize graphic organizer like the one below.

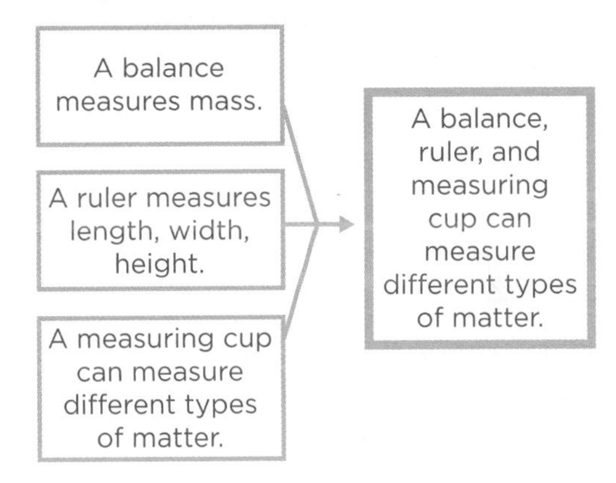

A balance measures mass.
A ruler measures length, width, height.
A measuring cup can measure different types of matter.
A balance, ruler, and measuring cup can measure different types of matter.


*Graphic Organizer 5,* p. TR7

**9.** gas

**10.** Accept all reasonable answers. Children should address concepts taught in each lesson: Matter can be a solid, liquid, or gas. Properties of matter may include color, shape, size, texture, and weight. Solids have a fixed shape of their own that does not change if the solids are moved. Liquids and gases have no fixed shape and can change shape when moved to different containers. Matter can be measured using a variety of tools.

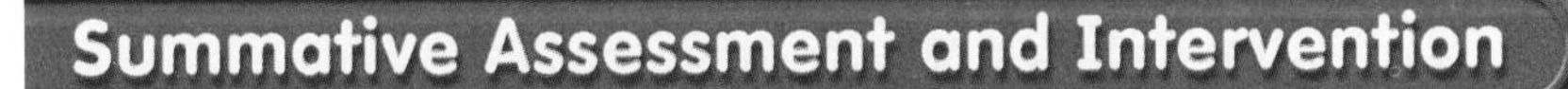

## Summative Assessment and Intervention

**Assessment** provides a summative test for Chapter 9.

**Leveled Readers** may be used to reteach lesson content in an alternative format. Leveled readers deliver chapter content at different readability levels. The back cover of each reader provides comprehension building activities specific to the book's content (see page 293).

**Key Concept Cards 26–28** contain prescribed routines for student intervention.

# Performance Assessment

## Matter of Fact Cards

**Materials** heavy card stock, file folder, pencils, crayons

### ▶ Teaching Tips

1. Explain to children that they will be making fact cards about matter.
2. Have children write the names of the states of matter, one on each sheet of card stock.
3. Tell children they are to draw an example of each state of matter and on the back of each card, they should draw a measuring tool they could use to measure that form of matter.
4. Children should then write a description of what and how they measure each form of matter.

**Performance Assessment** DOK 3

## Matter of Fact Cards

How do we measure different forms of matter?

- ▶ Write the name of each type of matter, one on each sheet of card stock paper.
- ▶ Draw an example of each form of matter on the front of your paper.
- ▶ On the back, draw the tool you could use to measure your form of matter shown on the front.
- ▶ Write a sentence telling what you are measuring and how you would do it.

321A

## Scoring Rubric

**4 Points** Child accurately draws and labels each state of matter. Child draws an appropriate measuring tool and writes a descriptive sentence.

**3 Points** Child accurately draws and labels each state of matter. Child draws an appropriate measuring tool and writes a descriptive sentence for two out of the three states of matter.

**2 Points** Child accurately draws and labels each state of matter. Child draws an appropriate measuring tool and writes a descriptive sentence for one out of the three states of matter.

**1 Point** Child accurately draws and labels each state of matter but can not identify the measuring tool or what they measure.

## Test Preparation

**1** **Look at the chart.**

**Which phrase belongs in the top circle?**



**A** Things we can not see
**B** Types of properties
**C** Things that flow
**D** Types of matter
DOK 1

**2** **Which of these describes both a solid and a gas?**

**A** It spreads to fill the container it is in.
**B** You can measure it.
**C** You can bend it, fold it, or tear it.
**D** It can be rough or smooth when you touch it.
DOK 2

**3** **Look at the pictures below.**

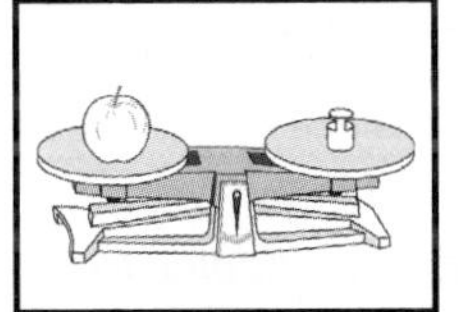
balance

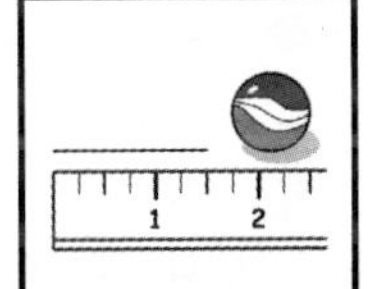
ruler

measuring cup

scale

**Which tool measures volume?**

**A** a balance
**B** a ruler
**C** a measuring cup
**D** a scale
DOK 1

321B

## Depth of Knowledge

**Level 1 Recall** Level 1 requires memory of a fact, a definition, or a procedure. At this level, there is only one correct answer.

**Level 2 Skill/Concept** Level 2 requires an explanation or the ability to apply a skill. At this level, the answer reflects a deep understanding of the topic.

**Level 3 Strategic Reasoning** Level 3 requires the use of reasoning and analysis, including the use of evidence or supporting information. At this level, there may be more than one correct answer.

**Level 4 Extended Reasoning** Level 4 requires the completion of multiple steps and requires synthesis of information from multiple sources or disciplines. At this level, the answer demonstrates careful planning and complex reasoning.

1. D: Types of matter. Solids, liquids, and gases have different properties. We can see solids and liquids. Solids do not flow.
2. B: You can measure it. Gases expand, but liquids do not. Solids have a texture and can be bent, folded, or torn. Gases and liquids do not have texture and cannot be bent, folded, or torn.
3. C: a measuring cup. A balance measures mass. A ruler measures length. A scale measures weight.

# CHAPTER 10 Planner

**Classroom Presentation Toolkit CD-ROM** Lesson Presentations

**TeacherWorks™ Plus CD-ROM** Interactive Lesson Planner, Teacher's Edition, Worksheets, and Online Resources.

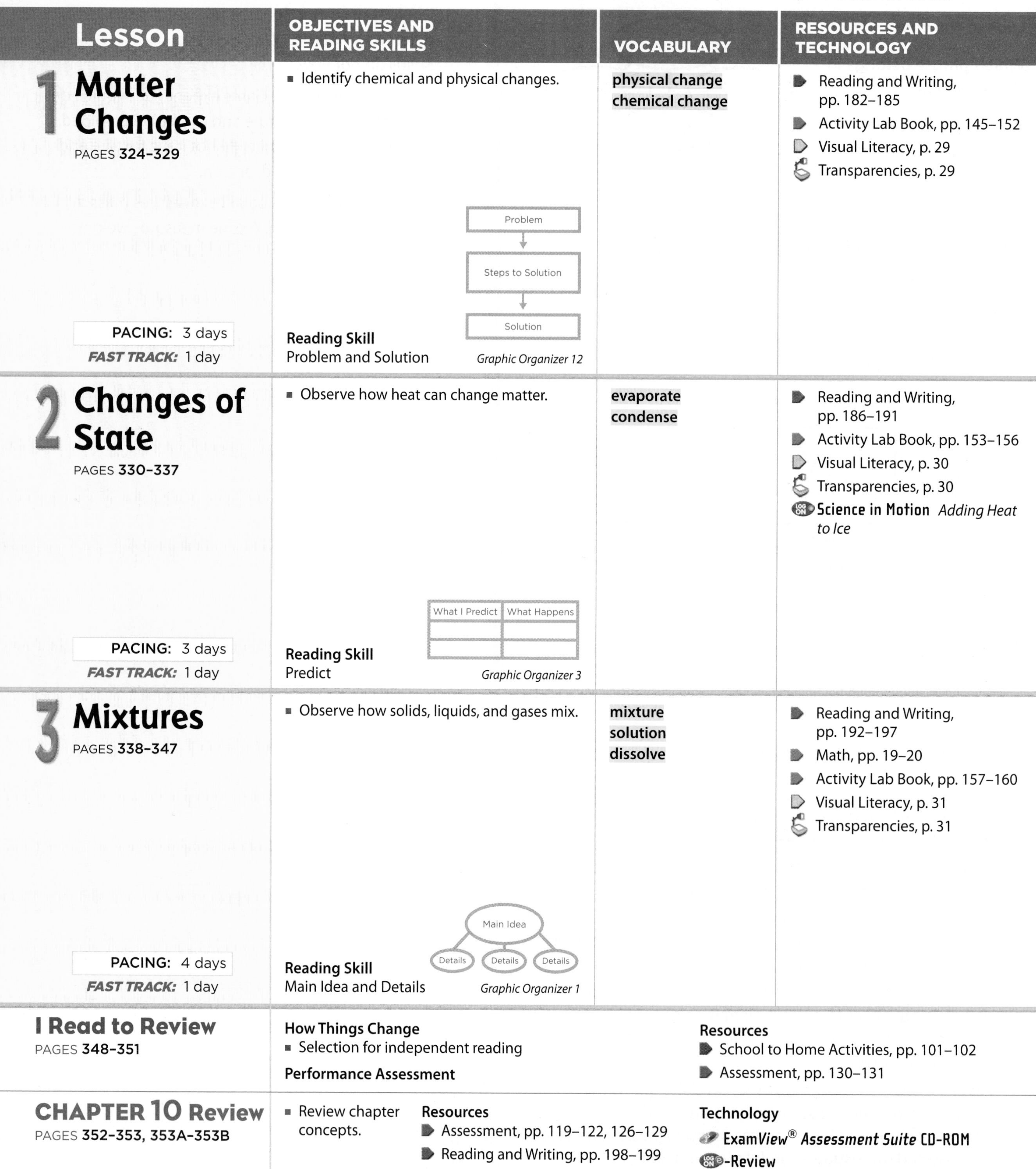

| Lesson | OBJECTIVES AND READING SKILLS | VOCABULARY | RESOURCES AND TECHNOLOGY |
|---|---|---|---|
| **1 Matter Changes** PAGES **324–329**<br>**PACING:** 3 days<br>***FAST TRACK:*** 1 day | ▪ Identify chemical and physical changes.<br>Problem → Steps to Solution → Solution<br>**Reading Skill**<br>Problem and Solution<br>*Graphic Organizer 12* | **physical change**<br>**chemical change** | Reading and Writing, pp. 182–185<br>Activity Lab Book, pp. 145–152<br>Visual Literacy, p. 29<br>Transparencies, p. 29 |
| **2 Changes of State** PAGES **330–337**<br>**PACING:** 3 days<br>***FAST TRACK:*** 1 day | ▪ Observe how heat can change matter.<br>What I Predict / What Happens<br>**Reading Skill**<br>Predict<br>*Graphic Organizer 3* | **evaporate**<br>**condense** | Reading and Writing, pp. 186–191<br>Activity Lab Book, pp. 153–156<br>Visual Literacy, p. 30<br>Transparencies, p. 30<br>**Science in Motion** *Adding Heat to Ice* |
| **3 Mixtures** PAGES **338–347**<br>**PACING:** 4 days<br>***FAST TRACK:*** 1 day | ▪ Observe how solids, liquids, and gases mix.<br>Main Idea / Details / Details / Details<br>**Reading Skill**<br>Main Idea and Details<br>*Graphic Organizer 1* | **mixture**<br>**solution**<br>**dissolve** | Reading and Writing, pp. 192–197<br>Math, pp. 19–20<br>Activity Lab Book, pp. 157–160<br>Visual Literacy, p. 31<br>Transparencies, p. 31 |
| **I Read to Review** PAGES **348–351** | **How Things Change**<br>▪ Selection for independent reading<br>**Performance Assessment** | | **Resources**<br>School to Home Activities, pp. 101–102<br>Assessment, pp. 130–131 |
| **CHAPTER 10 Review** PAGES **352–353, 353A–353B** | ▪ Review chapter concepts.<br>**Resources**<br>Assessment, pp. 119–122, 126–129<br>Reading and Writing, pp. 198–199 | | **Technology**<br>**ExamView® *Assessment Suite* CD-ROM**<br>**-Review** |

PACING Assumes a day is a 20–25 minute session.

www.macmillanmh.com for more planning resources and www.nsdl.org/refreshers/science for science resources from NSDL

**Science Activity DVD** Explore Activity demos

Materials included in the Deluxe Activity Kit are listed in *italics*.

## EXPLORE Activities

### Explore *p. 325* PACING: 30 minutes 

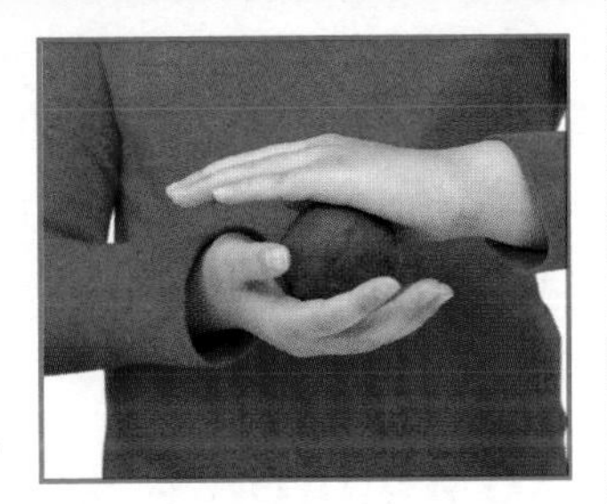

**Objective** Observe how mass may remain constant or change based on the shape of the object.

**Skills** measure, predict, draw conclusions, investigate

**Materials** *modeling clay*, *balances*, *metric weight set*, *plastic knives*, paper, pencils

PLAN AHEAD Prepare enough modeling clay for children to have two pieces of equal mass. Collect plastic knives or other utensils for children to cut the clay.

### Explore *p. 331* PACING: 15 minutes 

**Objective** Observe how butter and chocolate will change due to heat.

**Skills** predict, observe, communicate

**Materials** *paper plates*, butter, chocolate, *plastic knives*

PLAN AHEAD Provide enough chocolate and butter for small groups. **Be Careful!** Ask parents ahead of time if their children are allergic to dairy or chocolate. Supervise children when using the plastic knives, as the knives may be sharp.

### Explore *p. 339* PACING: 20 minutes 

**Objective** Compare a solution to a non-solution.

**Skills** measure, compare, investigate

**Materials** *measuring cups*, two *clear plastic cups* per grouping, two *plastic spoons* per grouping, salt, *sand*, water

PLAN AHEAD Prepare enough clear cups, measuring cups, sand, salt, and spoons for children to work in pairs.

## QUICK LAB Activities

### Quick Lab *p. 328* PACING: 15 minutes 

**Objective** Observe a chemical change in an apple.

**Skills** compare, infer

**Materials** apples, knife (teacher use only), lemon juice, *plastic wrap*, plates

PLAN AHEAD Check with parents to make sure children are not allergic to these ingredients. Do not cut the apples until just before the activity; they will begin to turn brown as soon as they are cut.

### Quick Lab *p. 335* PACING: 15 minutes 

**Objective** Classify water as solid, liquid, or gas.

**Skills** classify

**Materials** magazines, construction paper, glue, scissors, markers

PLAN AHEAD Collect enough magazines for all of the children to use.

### Quick Lab *p. 344* PACING: 15 minutes 

**Objective** Identify how evaporation helps separate a solution.

**Skills** observe, infer

**Materials** salt, water, *shallow plastic containers*

PLAN AHEAD Use warm water to help the solution dissolve more thoroughly.

## FOR MORE ACTIVITIES

| Focus on Skills | Be a Scientist |
|---|---|
| Teach the inquiry skill: communicate, p. 45 | What happens when you shake cream?, p. 46 |

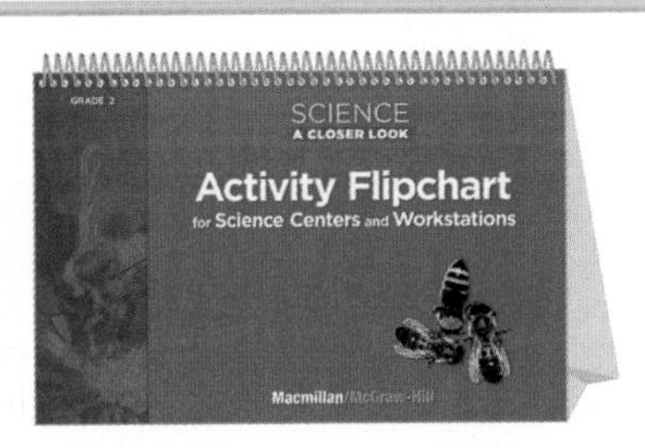


Use the Activities in your work station. See **Activity Lab Book** for more support.

For a comprehensive list of consumable and non-consumable materials, see the back of the unit tab.

# English Language Learner Support

**Technology**

For additional language support and vocabulary development, go to www.macmillanmh.com

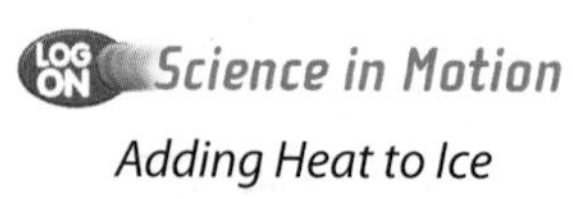
*Adding Heat to Ice*

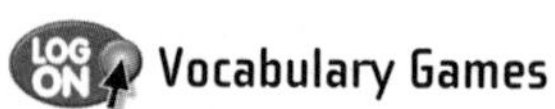

**ADDITIONAL RESOURCES**

pp. 96–104

## Academic Language

English language learners need help in building their understanding of the academic language used in daily instruction and science activities. The following strategies will help to increase children's language proficiency and comprehension of content and instruction words.

### Strategies to Reinforce Academic Language

- **Use Context** Academic language should be explained in the context of the task. Use gestures, expressions, and visuals to support meaning.
- **Use Visuals** Use charts, transparencies, and graphic organizers to explain key labels to help children understand classroom language.
- **Model** Use academic language as you demonstrate the task to help children understand instruction.

### Academic Language Vocabulary Chart

The following chart shows chapter vocabulary and inquiry skills as well as some Spanish cognates. **Vocabulary** words help children comprehend the main ideas. **Inquiry Skills** help children develop questions and perform investigations. **Cognates** are words that are similar in English and Spanish.

| Vocabulary | Inquiry Skills | Cognates | |
|---|---|---|---|
| | | **English** | **Spanish** |
| physical change, p. 326 | measure, p. 325 | predict, p. 325 | *predecir* |
| chemical change, p. 328 | predict, p. 325 | conclusion, p. 325 | *conclusión* |
| evaporate, p. 333 | draw conclusions, p. 325 | investigate, p. 325 | *investigar* |
| condense, p. 334 | investigate, p. 325 | physical, p. 326 | *físico* |
| mixture, p. 340 | observe, p. 331 | observe, p. 331 | *observar* |
| solution, p. 343 | communicate, p. 331 | communicate, p. 331 | *comunicar* |
| dissolve, p. 343 | compare, p. 339 | evaporate, p. 333 | *evaporar* |
| | | condense, p. 334 | *condensar* |
| | | compare, p. 339 | *comparar* |
| | | solution, p. 343 | *solución* |
| | | dissolve, p. 343 | *disolver* |

## Vocabulary Routine

**Use the routine below to discuss the meaning of each word on the vocabulary list. Use gestures and visuals to model all words.**

**Define** A *mixture* is two or more things put together.

**Example** When you pour salt into water, you make a *mixture*.

**Ask** What is an example of a *mixture*?

**Children may respond to questions according to proficiency level with gestures, one-word answers, or phrases.**

**ELL Leveled Reader**

**Make a Pizza**
by Joanne Mattern

**Summary** Watch how a pizza changes when it is put into a hot oven.

**Reading Skill**
Cause and Effect

## Vocabulary Activities

Help children understand that there are many kinds of mixtures.

**BEGINNING** Show the photographs on page 342. Encourage children to compare and contrast the mixtures. *How are the mixtures alike? How are the mixtures different?* Record responses in a Venn diagram. Then go over it with children.

**INTERMEDIATE** Show the photographs on page 342. Have children explain why it is easy to take apart the mixture before blending and difficult to do so after blending. Challenge them to think of ways in which they could separate the blended mixture into its parts.

**ADVANCED** Using the photographs on page 342, help children write a numbered list of steps for making a smoothie: *Put milk in blender. Cut up fruit if necessary. Put fruit in blender. Blend until mixed.* Challenge children to explain why a smoothie is a mixture.

## Language Transfers

**Grammar Transfer**
In Korean and Spanish, subject pronouns are dropped because verb endings indicate number and/or gender.

*Juanita made a mixture. Made it with water and salt.*

**Phonics Transfer**
Cantonese, Hmong, and Vietnamese do not have the /ks/ sound as in *mixture*.

# CHAPTER 10

## Changes in Matter

 **THE BIG IDEA** **How can matter change?**

**Chapter Preview** Have children take a chapter picture walk and predict what the lessons will be about.

### ▶ Assess Prior Knowledge

Before reading the chapter, create a **KWL** chart with children, ask the Big Idea question, then ask:

- **How can things change?**
- **What is a mixture?**
- **Are there different kinds of mixtures?**
- **Can heat change matter?**

**Changes in Matter**

| What We Know | What We Want to Know | What We Learned |
|---|---|---|
| Ice can melt. | How does water change to a solid? | |
| Pancake batter is a mixture. | Can a solid and liquid mix? | |

Answers shown represent sample student responses.

Follow the **Instructional Plan** at right after assessing children's prior knowledge of chapter content.

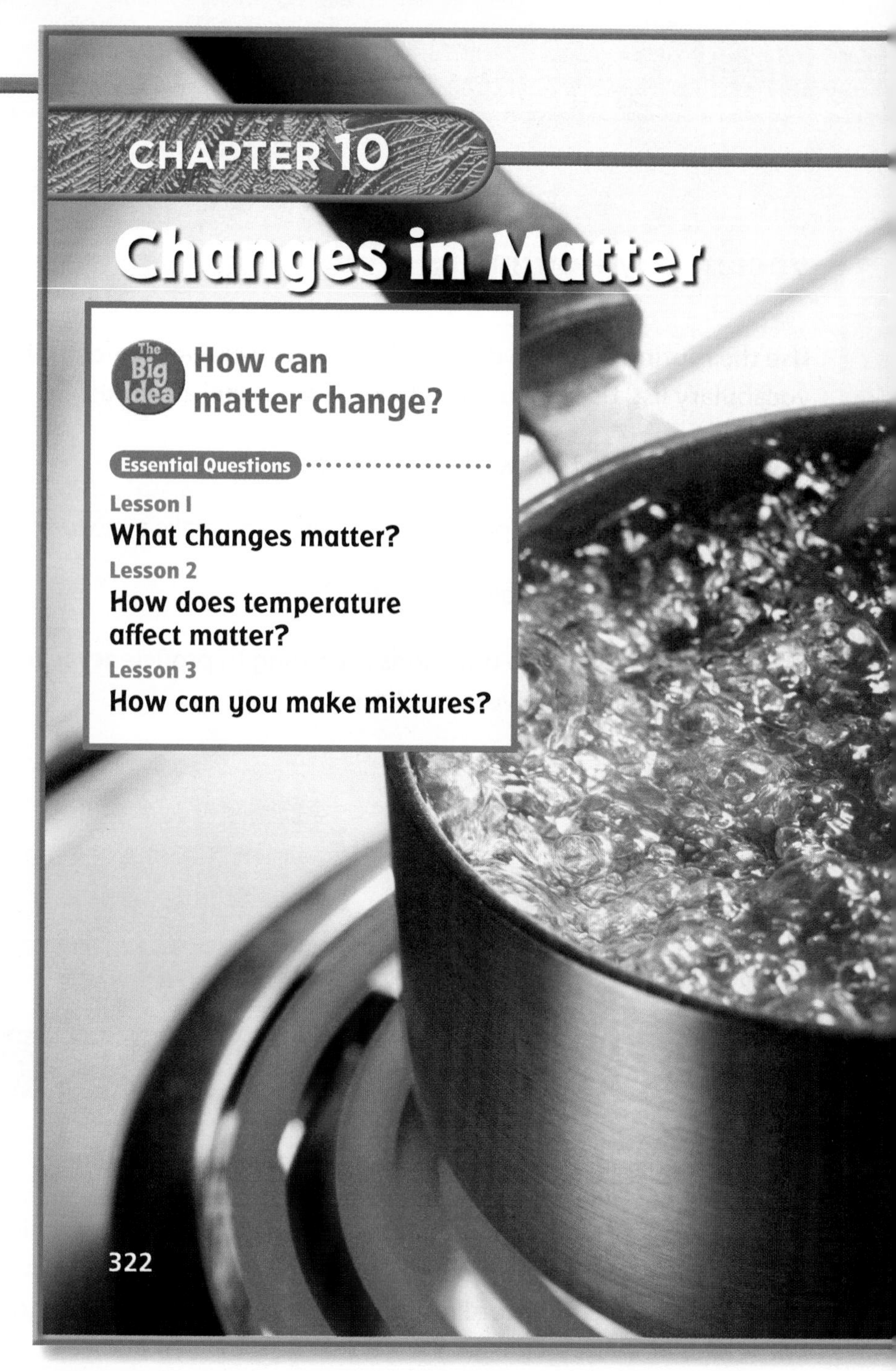


**RESOURCES and TECHNOLOGY**

- **School to Home Activities,** pp. 93–102
- **Reading and Writing,** pp. 181–199
- **Assessment,** pp. 119–131
- Classroom Presentation Toolkit CD-ROM
- PuzzleMaker CD-ROM
-  www.macmillanmh.com
-  -Journal

### Differentiated Instruction

**Instructional Plan**

**Chapter Concept** Matter can change its state.

**EXTRA SUPPORT** Children who need to understand physical changes of matter should cover all of **Lesson 1,** pages 324–329, before continuing with the rest of the chapter.

**ON LEVEL** Children who can describe physical changes, can explore chemical changes **Lesson 1,** pages 328–329, and then go directly to **Lesson 2,** pages 330–337, to compare how heating and cooling changes the state of matter.

**ENRICHMENT** For children who are ready to go further, **Lesson 3,** pages 338–347, enriches the topic of mixtures from Grade 1 by exploring how mixtures separate.

## Big Idea Vocabulary

**physical change**
a change in the size or shape of matter
(page 326)

**chemical change**
when matter changes into different matter
(page 328)

**mixture** two or more things mixed together that keep their own properties
(page 340)

**solution** a kind of mixture with parts that do not easily separate (page 343)

Visit www.macmillanmh.com for online resources.

323

# Big Idea Vocabulary

- Have a volunteer read the **Big Idea Vocabulary** words aloud to the class. Ask children to find one or two of the words in the chapter by using the given page references. Add these words and their definitions to a class Word Wall.

- Encourage children to use the illustrated glossary in the Student Edition's reference section. Guide children to explore the **e-Glossary**, which offers audio pronunciations, definitions, and sentences using the vocabulary words.

## Science Leveled Readers

ALSO ON AUDIO CD — Leveled Reader Library

APPROACHING

**Mix It Up** Shake and mix solids and liquids and see what happens.

**ISBN:** 978-0-02-285858-2

ON LEVEL

**Make a Pizza** Watch how a pizza changes when it is put into a hot oven.

**ISBN:** 978-0-02-285866-7

BEYOND

**Bicycle Metals** Let's make a bicycle and get ready to ride!

**ISBN:** 978-0-02-285873-5

ELL

**Make a Pizza** Uses sheltered language of On-Level Reader.

**ISBN:** 978-0-02-283494-4

See teaching strategies in the Leveled Reader Teacher's Guide. To order, call 1-800-442-9685.

 **Leveled Reader Database** Online Readers, searchable by topic, reading level, and keywords

Stop Here to

# Plan Your Lesson

## Lesson 1 Matter Changes

### Essential Question

What changes matter?

### Objective

- Identify chemical and physical changes.

**Reading Skill** Problem and Solution

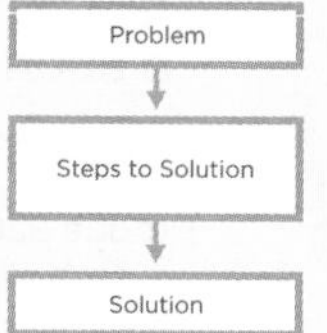


*Graphic Organizer 12,* p. TR14

## FAST TRACK

**Lesson Plan** When time is short, follow the Fast Track and use the essential resources.

**1 Introduce**
**Look and Wonder, p. 324**
Resource **Activity Lab Book, p. 147**

**2 Teach**
**Discuss the Main Idea, p. 326**
Resource **Visual Literacy, p. 29**

**3 Close**
**Think, Talk, and Write, p. 329**
Resource **Assessment, p. 123**

**Professional Development** Look for NSDL to find recommended Science Background resources from the National Science Digital Library.

## Reading and Writing

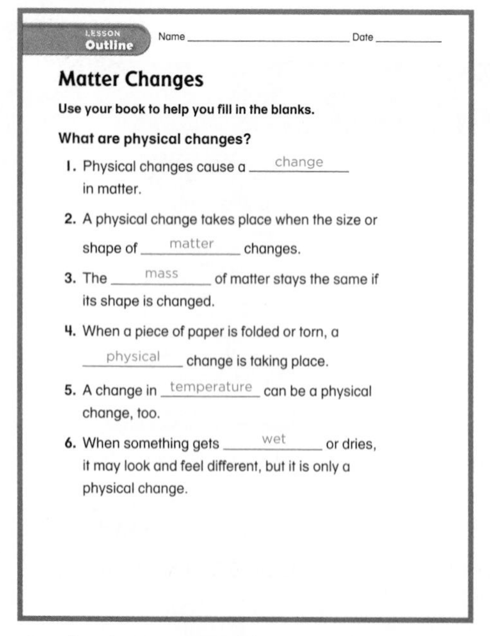
Lesson Outline Name ________ Date ________

**Matter Changes**

**Use your book to help you fill in the blanks.**

**What are physical changes?**

1. Physical changes cause a change in matter.
2. A physical change takes place when the size or shape of matter changes.
3. The mass of matter stays the same if its shape is changed.
4. When a piece of paper is folded or torn, a physical change is taking place.
5. A change in temperature can be a physical change, too.
6. When something gets wet or dries, it may look and feel different, but it is only a physical change.

**Outline, pp. 182–183**
Also available as a student workbook

Lesson Vocabulary Name ________ Date ________

**Matter Changes**

**Identify each description as a physical change or a chemical change.**

1. An iron screw rusts in the rain. chemical change
2. A piece of paper is folded. physical change
3. A rock breaks down into soil. physical change
4. Water freezes and turns into ice. physical change
5. A peach turns brown. chemical change
6. A ball gets wet. physical change
7. A slice of cheese melts. physical change
8. An egg is fried. chemical change

**Vocabulary, p. 184**
Also available as a student workbook

## Visual Literacy

Name ________ Date ________ Read a Diagram

**What are chemical changes?**

The diagram below shows three objects before and after a chemical change.

| Chemical Changes | | |
|---|---|---|
| Before | After | Cause |
| | | Heat causes the matchstick to burn. The properties of the matchstick have changed. |
| | | Water and air can cause metal to rust. Rust is a chemical change that happens slowly. |
| | | Water and air do not change the properties of plastic. |

**Use the diagram to answer the questions below.**

1. Why did the match change? How do you know?
2. Did the screw change? How do you know?

1. The match changed when light and heat caused it to burn. I know this because the chart shows that the match looks different in the "After" column. I also know this because the "Cause" column tells me so.
2. No, the plastic screw did not change. I know this because the photographs in the "Before" and "After" columns are the same. I also know this because the "Cause" column tells me so.

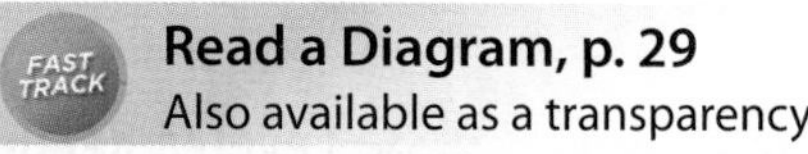

**Read a Diagram, p. 29**
Also available as a transparency

# Activity Lab Book

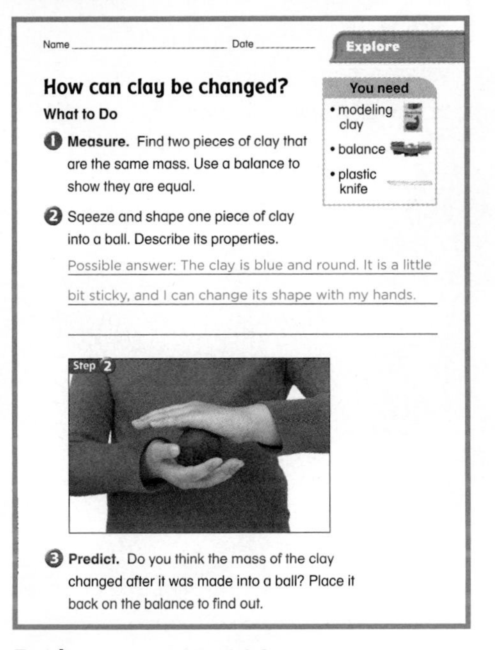

Name ______ Date ______ **Explore**

**How can clay be changed?**

**What to Do**

**You need**
- modeling clay
- balance
- plastic knife

1. **Measure.** Find two pieces of clay that are the same mass. Use a balance to show they are equal.
2. Sqeeze and shape one piece of clay into a ball. Describe its properties.
   Possible answer: The clay is blue and round. It is a little bit sticky, and I can change its shape with my hands.

Step 2

3. **Predict.** Do you think the mass of the clay changed after it was made into a ball? Place it back on the balance to find out.

**Explore, pp. 145–146**
Also available as a student workbook

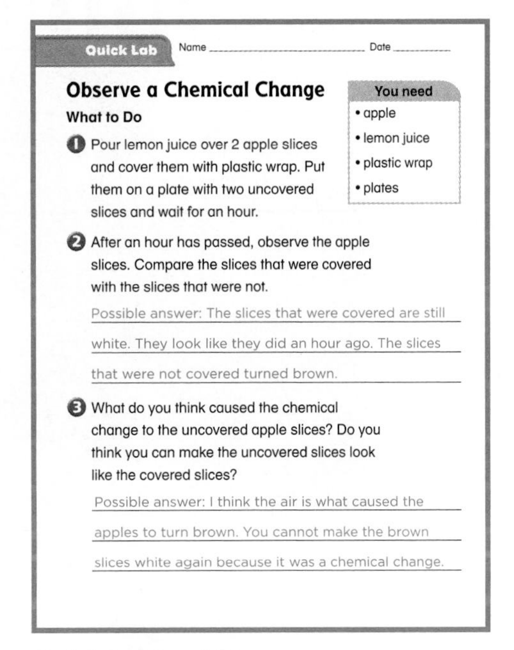

**Quick Lab** Name ______ Date ______

**Observe a Chemical Change**

**What to Do**

**You need**
- apple
- lemon juice
- plastic wrap
- plates

1. Pour lemon juice over 2 apple slices and cover them with plastic wrap. Put them on a plate with two uncovered slices and wait for an hour.
2. After an hour has passed, observe the apple slices. Compare the slices that were covered with the slices that were not.
   Possible answer: The slices that were covered are still white. They look like they did an hour ago. The slices that were not covered turned brown.
3. What do you think caused the chemical change to the uncovered apple slices? Do you think you can make the uncovered slices look like the covered slices?
   Possible answer: I think the air is what caused the apples to turn brown. You cannot make the brown slices white again because it was a chemical change.

**Quick Lab, p. 148**
Also available as a student workbook

# Assessment

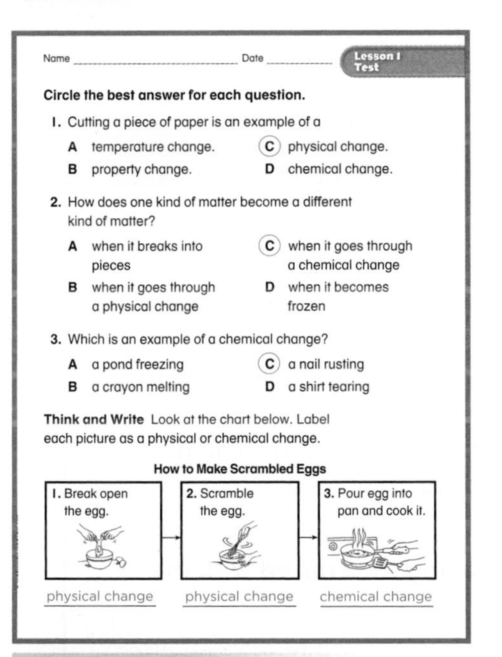

Name ______ Date ______ **Lesson 1 Test**

**Circle the best answer for each question.**

1. Cutting a piece of paper is an example of a
   - A temperature change.
   - B property change.
   - (C) physical change.
   - D chemical change.
2. How does one kind of matter become a different kind of matter?
   - A when it breaks into pieces
   - B when it goes through a physical change
   - (C) when it goes through a chemical change
   - D when it becomes frozen
3. Which is an example of a chemical change?
   - A a pond freezing
   - B a crayon melting
   - (C) a nail rusting
   - D a shirt tearing

**Think and Write** Look at the chart below. Label each picture as a physical or chemical change.

**How to Make Scrambled Eggs**

| 1. Break open the egg. | 2. Scramble the egg. | 3. Pour egg into pan and cook it. |
|---|---|---|
| physical change | physical change | chemical change |

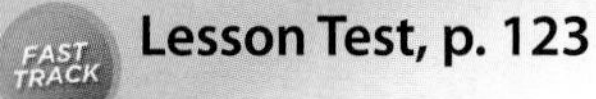
Lesson Test, p. 123

## ADDITIONAL RESOURCES

pp. 142–145

*Make a Pizza*

pp. 44–46

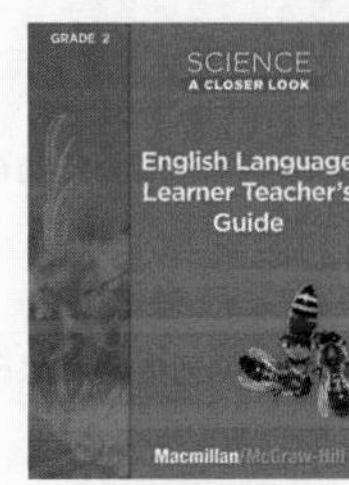
p. 100

p. 29

29

91–100

pp. 95–96

74–75

### Technology

- Science Activity DVD
- TeacherWorks™ Plus CD-ROM
- Classroom Presentation Toolkit CD-ROM

e-Review

NSDL

# Lesson 1 Matter Changes

**Objective**

- Identify chemical and physical changes.

## 1 Introduce

### ▶ Assess Prior Knowledge

Have children share what they know about how matter changes. Ask:

- **What are some ways that things can change?**
- **What are some things that can be changed and then changed back?**
- **What are some things that can be changed, but cannot be changed back?**

Record children's answers in the What We Know column of the class **KWL** chart.

### Look and Wonder

Read the Look and Wonder question. Invite the class to share their responses to the question. Ask:

- **How did the girl change the clay?** Possible answers: rolling; mixing; making new shapes; cutting
- **How could she change the clay back to how it looked before?** Possible answers: Separate the different-colored shapes and roll them back into balls. Sometimes the colors get all mixed up, which would make it hard for her to turn it back to what it was before.

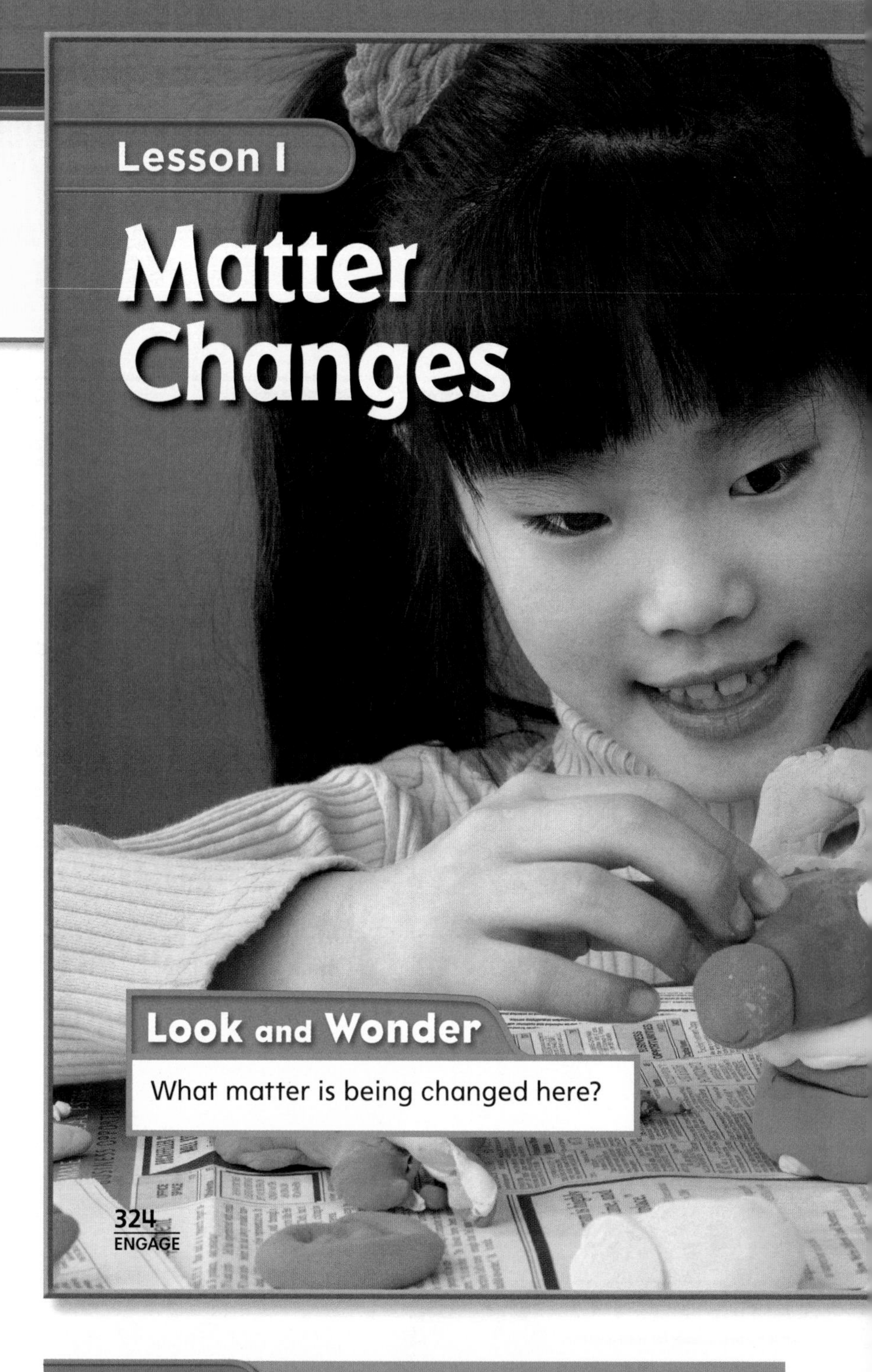

Lesson I
Matter Changes
Look and Wonder
What matter is being changed here?
324
ENGAGE


### Warm Up

**Start with an Activity**

Tell children they are going to make a volcano. Distribute vinegar, baking soda, and plastic containers. **Be Careful!** Vinegar is acidic and should be handled with care.

Have children put $\frac{1}{4}$ cup of the baking soda into the container. Ask children to carefully and slowly add vinegar. When the vinegar is poured into the baking soda, it causes a chemical change, in which gas is released and bubbles up.

Ask:

- **What happened when the baking soda mixed with the vinegar?** Gas bubbled up.
- **What was left in the container after it bubbled?** a mixture of materials

**RESOURCES and TECHNOLOGY**

- **Activity Lab Book,** pp. 145–147
- **Activity Flipchart,** p. 44
- Science Activity DVD

## Explore — Inquiry Activity

### How can clay be changed?

**What to Do**

1. **Measure.** Find two pieces of clay that are the same mass. Use a balance to show they are equal.
2. Squeeze and shape one piece of clay into a ball. Describe its properties.
3. **Predict.** Do you think the mass of the clay changed after it was made into a ball? Place it back on the balance to find out.
4. ⚠ **Be Careful!** Cut the clay ball into two halves with a plastic knife. Make the two pieces into two figures.
5. **Draw Conclusions.** How did you change the clay?

**Explore More**

6. **Investigate.** What other ways can you change clay? Will the mass change?

**You need**

modeling clay

balance

plastic knife

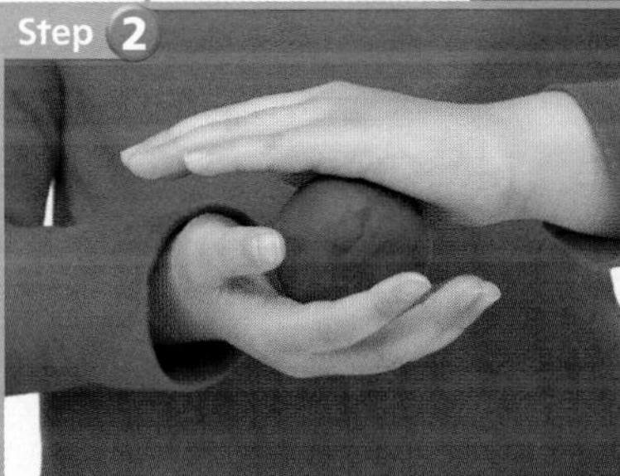

Step 2

325
EXPLORE

## Explore

**Plan Ahead** Prepare enough modeling clay for each child to have two pieces of equal mass. Collect enough plastic knives or other utensils for children to cut the clay.

**Purpose** This activity will encourage children to investigate whether mass changes as the shape of an object changes.

**Structured Inquiry** What to Do

Review how to use a balance. Explain to children that they are going to determine whether the mass of a ball of clay changes when the shape of the clay changes.

1. **Measure** Discuss with children that the two pieces of clay should be about the same size. Ask: **Can two different-sized pieces of clay have the same mass?** yes
2. Encourage children to discuss all the physical properties of the clay: color, consistency, shape, and smell.
3. **Predict** Ask children to explain the reasons for their predictions about the mass of the clay.
4. **Be Careful!** Plastic knives are sharp.
5. **Draw Conclusions** Ask: **What methods did you use to change the shape of the clay? Are the masses of the figures different?** no **How can you tell?** weigh them

**Guided Inquiry** Explore More

6. **Investigate** Have children list different ways they can change the shape of the clay. The mass of the clay will stay the same regardless of the shape.

**Open Inquiry**

Ask: **What are some other materials people use and change?** If children need a prompt, encourage them to think of materials used and changed in the kitchen. Ask: **How could you check whether those changes affect the mass of the material?**

Provide children with other materials they can use to change the shape, such as pipe cleaners or crackers.

## Alternative Explore

### What is the mass of water?

Give volunteers two differently-shaped containers, and have them place each on opposite sides of a balance. They should use small weights to balance the containers.

Ask them to take the containers off of the balance and fill each with one cup of water. Have children predict which container will weigh more, and then place containers back on the balance to check their prediction.

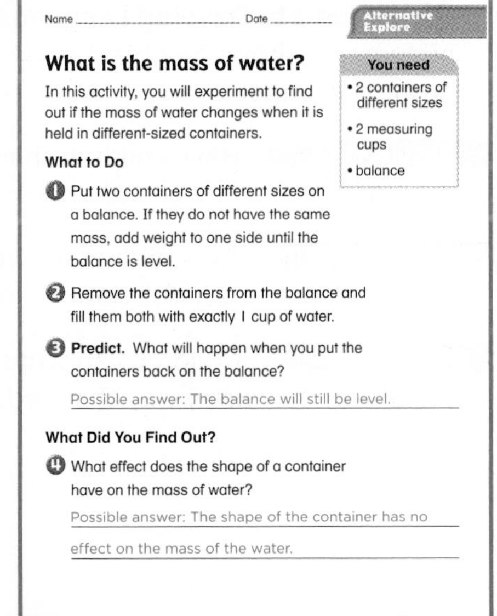

Name ______________ Date ________ Alternative Explore

**What is the mass of water?**

In this activity, you will experiment to find out if the mass of water changes when it is held in different-sized containers.

**You need**
- 2 containers of different sizes
- 2 measuring cups
- balance

**What to Do**

1. Put two containers of different sizes on a balance. If they do not have the same mass, add weight to one side until the balance is level.
2. Remove the containers from the balance and fill them both with exactly 1 cup of water.
3. **Predict.** What will happen when you put the containers back on the balance?
   Possible answer: The balance will still be level.

**What Did You Find Out?**

4. What effect does the shape of a container have on the mass of water?
   Possible answer: The shape of the container has no effect on the mass of the water.

Activity Lab Book, p. 147

# 2 Teach

## Read Together and Learn

**Reading Skill** **Problem and Solution** A problem is what needs to be done, found out, or changed. A solution fixes the problem.

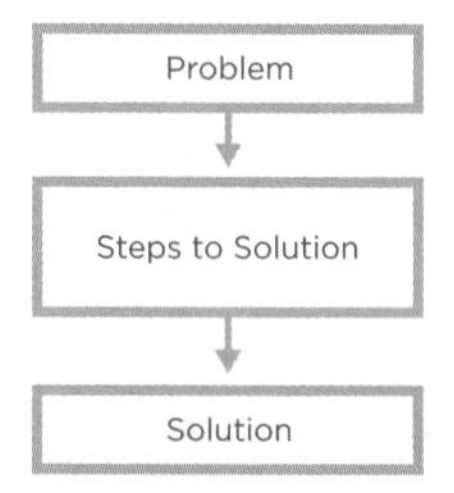


*Graphic Organizer 12*, p. TR14

## What are physical changes?

### FAST TRACK Discuss the Main Idea

**Main Idea** Physical changes affect size and shape but do not change the type of matter.

After reading together, ask:

- **Why does the paper on the left side of the balance equal the mass of the paper on the right side?** The paper on the right only had physical changes made to it.
- **What would happen to the balance if half of the sheet of paper on the left was torn off?** The balance would tip to the right.

**Read Together and Learn**

**Essential Question**
What changes matter?

**Vocabulary**
physical change
chemical change

### What are physical changes?

Matter can change in different ways. You can change the size or shape of matter. This is called a **physical change**.

When you cut, bend, fold, or tear matter, you cause a physical change. You can change the shape or size of paper by cutting or folding it. It is still paper.

◀ **Folding and writing on paper are physical changes.**

**When you only change the shape of matter, its mass stays the same.** ▼

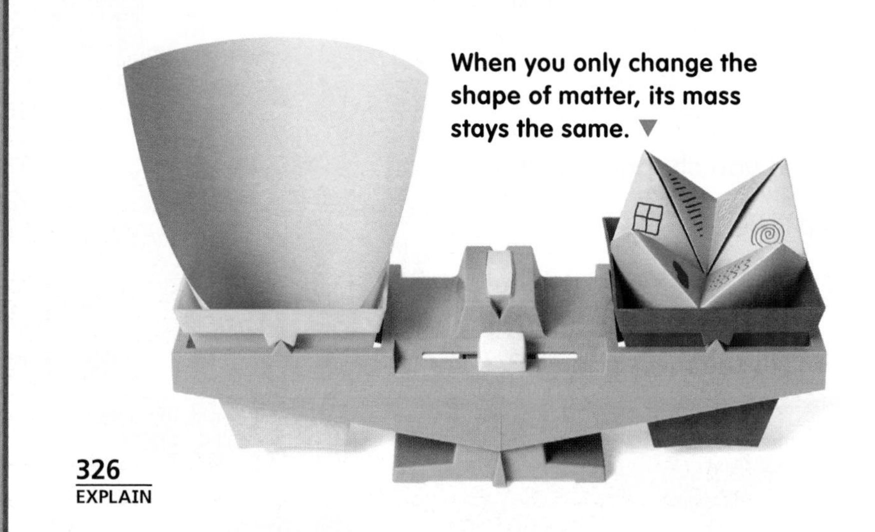



## Science Background

**Matter Changes** Matter can go through two types of changes: physical and chemical. A physical change, such as cutting or bending, is one in which the matter is still the same kind of matter after the change. Physical changes are sometimes reversible, such as ice thawing back into water. After a chemical change, such as burning, a new type of matter is formed. Chemical changes are very difficult and sometimes impossible to reverse.

See **Science Yellow Pages**, in the Teacher Resources section, for background information.

**Professional Development** For more Science Background and resources from NSDL visit http://nsdl.org/refreshers/science

## ELL Support

**Use Pictures to Develop Language** Ask children to draw a picture that shows an object before and after it undergoes a physical change.

**BEGINNING** Have children share their pictures as they say the name of the object and how it changed.

**INTERMEDIATE** Ask children to write labels on their picture, using a word to describe how the object changed.

**ADVANCED** Children can write a sentence describing how the object they drew was changed.

Sometimes, the temperature
f matter changes. On a cold
ay, water can change
ɔ ice. This is a physical
hange.

Wetting and drying can
e physical changes too.
Vet mud looks and feels
ifferent from dry mud.

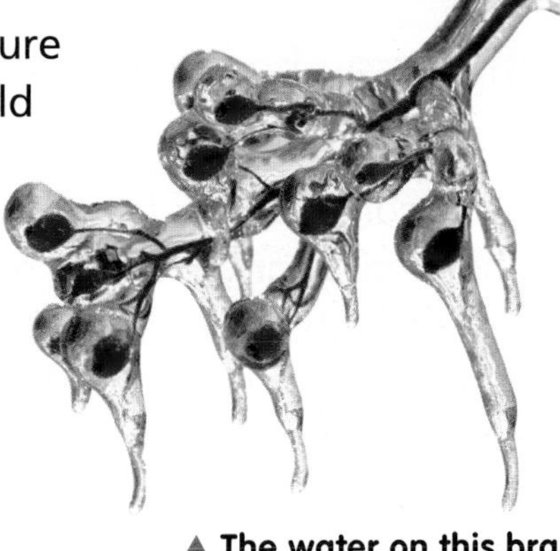

▲ The water on this branch has changed to ice.

✓ What is a physical change you could make to juice?

The color of dry mud is different from the color of wet mud. Mud feels squishy when it is wet and hard when it is dry.



## Differentiated Instruction

### Leveled Activities

**EXTRA SUPPORT** Give children a piece of paper. Ask them to find as many ways to change the appearance of the paper as possible, and record their findings in a chart labeled *Physical Changes to Paper*. Have them repeat the activity with paper clips.

**ENRICHMENT** Have children write a paragraph describing the different physical changes that can happen to water. To facilitate their writing, suggest that children revisit information they learned in Chapter 7, Lesson 2 about the water cycle in their text.

## ▶ Use the Visuals

Help children discuss what they see in the photos on page 327. Ask:

- **What will happen to the ice on the branch?** Possible answer: It will melt and turn back into water.
- **What caused the mud to change its color and texture?** Possible answers: Heat caused the water to evaporate. It turned from wet, dark mud to dry, light-colored mud.
- **How do you think it could change back to wet mud?** Possible answer: Rain would change it back to wet mud.

## ▶ Address Misconceptions

Explain to children that although the properties of ice are not the same as water, changing water to ice is a physical change because it can easily be changed back to water.

## ▶ Develop Vocabulary

**physical change** Explain to children that in this term, *physical* means "related to the body or things." Ask children to relate the meanings of the two words by showing various examples of *physical changes*, such as cut paper, a melting ice cube, and a clay ball. Have them explain which part of the example is a physical thing and how each example demonstrates a physical change. Children should use the term in their explanations.

## ✓ *Quick Check Answers*

Possible answers: Juice can be poured into different-shaped containers and the temperature of juice can be changed; Juice can be frozen into juice pops.

### RESOURCES and TECHNOLOGY

- **Reading and Writing,** pp. 182–184
- **PuzzleMaker CD-ROM**
- **Classroom Presentation Toolkit CD-ROM**
-  **e-Glossary**

# What are chemical changes?

## ▶ Discuss the Main Idea

**Main Idea** A chemical change has occurred when the properties of matter have changed.

Read the text together. Ask:

- **How are chemical changes different from physical changes?** Possible answers: Physical changes are often reversible, but chemical changes are not. Chemical changes alter the properties of matter, but physical changes do not.
- **How did the egg change?** Possible answer: Heat caused it to change from a liquid to a solid.

### Read a Chart

Explain to children that the chart shows before and after pictures of some objects that have undergone a chemical change. Ask:

- **Which object on the chart did not undergo a chemical change?** plastic screw
- **What would cause the plastic screw to undergo a physical change?** Possible answer: heating it until it melts

**Answer to Read a Chart** Water and air caused rust to form on the surface of the nail.

## ▶ Develop Vocabulary

**chemical change** Explain to children that a *chemical change* is "when matter changes into different matter." Have children write a funny comic strip about how an egg goes through a chemical change. They should use the term *chemical change* and illustrate their stories.

### RESOURCES and TECHNOLOGY

- **School to Home Activities,** pp. 95–96
- **Reading and Writing,** p. 185
- **Activity Lab Book,** p. 148
- **Visual Literacy,** p. 29
- **e-Review** Narrated Summary and Quiz
- **ExamView® *Assessment Suite* CD-ROM**



## What are chemical changes?

Sometimes the properties of matter can change. This is called a **chemical change**. When matter goes through a chemical change, it is not easy to change it back. It becomes a new kind of matter with different properties.

When you burn paper, you can not change it back. Seeing light and feeling heat are clues that a chemical change may be happening. All matter does not change in the same way.

### Quick Lab

Observe a slice o[f] apple. Infer wh[at] causes the apple to go through a chemical change

**Chemical Changes**

| Before | After | Cause |
|---|---|---|
| | | Heat causes the matchstick to burn. The properties of the matchstick have changed. |
| | | Water and air can cause metal to rust. Rusting is a chemical change that happens slowly. |
| | | Water and air do not change the properties of plastic. |

**Read a Chart**

How did the metal nail change



## Quick Lab

pairs

15 minutes

**Objective** Observe a chemical change in an apple.

**You need** apples, knife (teacher use only), lemon juice, plastic wrap, plates

1. Give each pair of children an apple cut into four pieces.
2. Have children pour lemon juice over two pieces and cover them with plastic wrap. Ask them to put all the pieces (the covered and uncovered apples) on a plate.
3. After an hour, have children compare the apple pieces and draw what they see.
4. Ask children to infer what happened to the uncovered apples. Air caused a chemical change.

Heat causes the egg to go through a chemical change you can see and smell.

How can you tell if a chemical change has happened?

## Think, Talk, and Write

1. **Problem and Solution.** Describe how you could keep a bicycle from rusting.
2. What are three examples of physical changes?
3. **Essential Question.** What changes matter?

### Math Link

Does the mass of an object change when you fold the object? How could you find out?

e-Review Summaries and quizzes online at www.macmillanmh.com

329
EVALUATE

### Quick Check Answer

Possible answer: A chemical change occurs when the properties of the matter have changed; the change cannot be easily reversed.

# 3 Close

## ▶ Using the KWL Chart

Review the **KWL** chart and ask children to describe what they now know about chemical and physical changes. Record their responses in the What We Learned column of the class **KWL** chart.

## ▶ Using the Reading Skill

### Problem and Solution

Use the reading skill graphic organizer to identify problems and solutions in the lesson. Ask: **What physical change can we make to very dry soil to make sure the seeds planted in it become plants?**

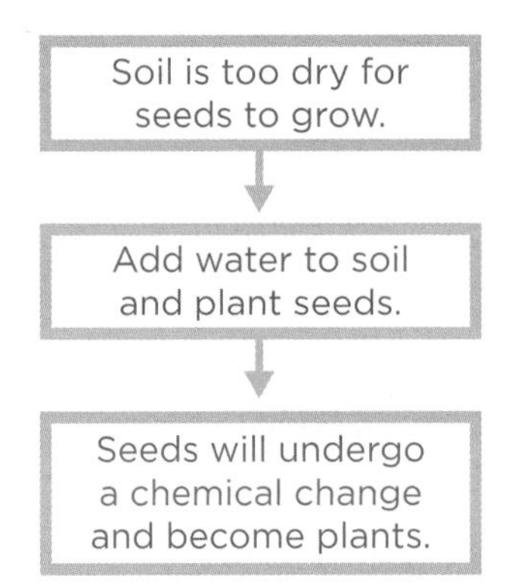

Soil is too dry for seeds to grow.
Add water to soil and plant seeds.
Seeds will undergo a chemical change and become plants.


*Graphic Organizer 12*, p. TR14

## Formative Assessment

### Draw a Sequence

Tell children to draw three pictures showing how an object undergoes either a physical or chemical change. Remind them to show the object before the change, show what was used to make the change, and show the object after the change. Children should label the picture as either *physical* or *chemical* change.

Key Concept Cards For student intervention, see the prescribed routine on **Key Concept Card 29.**

## FAST TRACK Think, Talk, and Write

1. **Problem and Solution** Possible answer: Keep it away from water by storing it in a shed or house.
2. Possible answers: tearing, folding, cutting
3. **Essential Question** Bending, folding, and cutting are physical changes that change the size and shape of matter. Rusting and burning are chemical changes that change the properties of matter.

### Math Link

Have children weigh a piece of paper, then fold it and weigh it again. Have them compare the two measurements to see if the mass of the paper changed.

## Focus on Skills

individual 25 minutes

Use Your

GRADE 2
SCIENCE
A CLOSER LOOK
Activity Flipchart
for Science Centers and Workstations
Macmillan/McGraw-Hill

Activity Flipchart, p. 45

### Objective

- List the steps taken to do something in order.

**You need** paper, scissors, pencils, crayons

**EXTEND** Children use organized lists to communicate the steps they used to change the physical characteristics of a piece of paper.

### Inquiry Skill: Communicate

### ▶ Learn It

Read the Learn It section on page 45 together. Discuss how it can be helpful to use an organized list to tell someone how something was done.

Study the photographs and list together. Ask:

- **What did Joanne do to change the clay?** She rolled, pinched, squeezed, and poked it.
- **Do you think the order of her list matters?** Possible answer: If you want to create the same shape, it does matter.
- **What other things take more than one step to do?** Possible answers: washing clothes; tying shoes; making pizza

## Focus on Skills

### Inquiry Skill: Communicate

You **communicate** when you draw, write, or share your ideas with others.

### ▶ Learn It

Joanne changed a ball of clay. She wrote a list to show others how she changed it.

Changing Clay
1. I rolled the clay.
2. I pinched the clay.
3. I squeezed the clay
4. I poked the clay.

## Integrate Math

### Mystery Numbers

Have each child select a mystery number from a hat. Remind them to keep the number hidden so no one else can see it.

Challenge children to make a list of clues that their classmates will use to discover their mystery number. For example, if they chose 12, they can say: *start with 5, add 4, and then add 3 more.*

Children can use as many steps as they would like to get to their number. Encourage students to use both subtraction and addition in their clues.

When children are finished creating their list of clues, ask them to share them with a partner and let their classmates figure out what their mystery number is.

Skill Builder

Try It

How many ways can you change a piece of paper?

1. Use a chart like Joanne's to communicate how you changed the paper.
2. Share your chart with a classmate.
3. **Write About It.** Tell how your charts are alike and how they are different.

Use with Chapter 10, Lesson 1 45

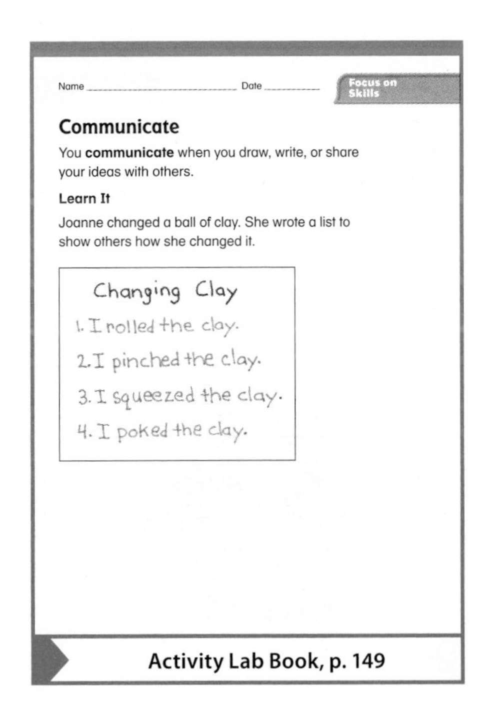

Name ______ Date ______ Focus on Skills

**Communicate**

You **communicate** when you draw, write, or share your ideas with others.

**Learn It**

Joanne changed a ball of clay. She wrote a list to show others how she changed it.

Changing Clay

1. I rolled the clay.
2. I pinched the clay.
3. I squeezed the clay.
4. I poked the clay.

Activity Lab Book, p. 149

## ▶ Try It

Read the Try It section with the class and discuss how the girl is changing the paper. Distribute paper, writing utensils, and scissors to children so they may try the activity.

Encourage children to think about ways they can change a piece of paper. Ask: **What are some other ways to change paper?** Possible answers: rolling, tearing, folding, wetting, stapling, coloring

1. Ask: **What has the girl done to the paper in the picture?** She folded it and is cutting it. Have children change their papers and record their steps for changing the paper in a numbered list.
2. Encourage children to read their partner's list and follow it in order to change paper. Ask: **Were you able to follow your partner's list to change paper the way your partner did? Why or why not?**
3. Provide children with copies of Graphic Organizer 10 to compare the lists.

## ▶ Apply It

Use the idea of making an ordered list to help record how something was done. Ask: **When is it helpful to write down the steps for a list?** Possible answers: when saving the steps for use in the future; when communicating the steps to others **Why is it important to put the steps in the proper order?** Possible answers: to make it easy to follow the steps, to make it easy for someone to repeat your results

Have children write directions for making their favorite food. Remind them to put their directions in order and to use the words *first, next,* and *last*.

Invite them to exchange their recipes with a partner and let them try to guess what the food is. Ask:

- **Was your partner able to figure out your food? Why or why not?**
- **How could you change your directions to make them more clear?**

**RESOURCES and TECHNOLOGY**

- **Activity Lab Book,** pp. 149–150
- **TeacherWorks™ Plus CD-ROM**

# Be a Scientist

small groups · 45 minutes

GRADE 2 SCIENCE A CLOSER LOOK **Activity Flipchart** for Science Centers and Workstations — Macmillan/McGraw-Hill

Use Your

Activity Flipchart, p. 46

**Skills** measure, observe, draw conclusions

## Objective

- Observe a liquid become a solid.

**You need** measuring cups, very cold cream, resealable plastic jars, crackers, plastic knives

**Plan Ahead Be Careful!** Check health records to determine whether children are lactose intolerant or allergic to dairy products.

**EXTEND** Children will put cream in a covered jar and shake it until it thickens.

**Structured Inquiry** What to Do

## What happens when you shake cream?

1. **Measure** Before measuring, encourage children to spread a little cream on the crackers with plastic knives and taste it. **Be Careful!** Plastic knives can be sharp, so remind children to handle them carefully.

Review how to use measuring cups. Ask: **How do you know when you've reached one-quarter cup?**

2. **Be Careful!** Make sure lids are secure.
3. Use a timer and have children take turns shaking the jars every minute to prevent them from getting too tired. Children should shake them until the cream thickens.

## Be a Scientist

**You need**

measuring cup

cream

plastic jar

crackers

### What happens when you shake cream?

Find out what will happen to cream when you shake it.

**What to Do**

1. **Measure.** Put $\frac{1}{4}$ cup of cream into the measuring cup.

Step 1

2. Pour the cream into the jar. Put the lid on tightly.

3. Take turns shaking the jar.

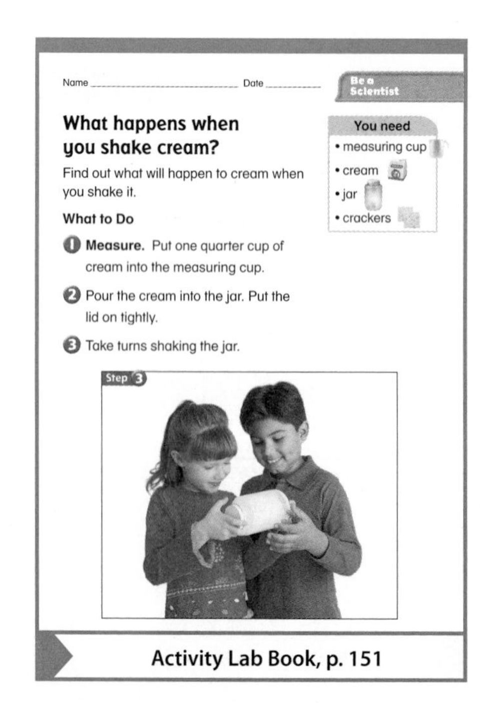

Name ______ Date ______ Be a Scientist

**What happens when you shake cream?**

Find out what will happen to cream when you shake it.

You need
- measuring cup
- cream
- jar
- crackers

**What to Do**

1. **Measure.** Put one quarter cup of cream into the measuring cup.
2. Pour the cream into the jar. Put the lid on tightly.
3. Take turns shaking the jar.

Step 3

Activity Lab Book, p. 151

**RESOURCES and TECHNOLOGY**

- **Activity Lab Book,** pp. 151–152
- **TeacherWorks™ Plus CD-ROM**

Inquiry Investigation

4 Observe. What happened to the cream? How did it change? Put it on a cracker.

5 Draw Conclusions. How do we use cream? Discuss your answers with a partner.

Investigate More

Observe. What happens if you freeze the cream while shaking it?

This old-fashioned machine stirs cream when you turn the handle. ▶

Use with Chapter 10, Lesson 1 46

## Integrate Writing

### Things Change

Distribute copies of the Sequence Graphic Organizer 7 from page TR9 of this book. Ask children to explain what happened first, next, and last in the cream experiment.

Encourage children to think of something else that changes from a liquid to a solid or from a solid to a liquid. Have them fill in the three boxes of the graphic organizer to indicate the change in the order in which it would happen.

When they are done, have volunteers share what they wrote in the graphic organizers. Before a volunteer reads what they wrote in the second and third boxes, have children try to predict how the object changes.

4 Observe Ask: **What changes do you observe in the cream?** The cream should have thickened into butter. After children have tasted the butter, discuss its consistency, color, smell, and taste. Ask: **Did some butter become thicker than others? Why?** Possible answer: Yes, because some jars were shaken faster or harder than others.

5 Draw Conclusions Ask: **Why would you shake cream?** Possible answers: to make it thick for making whipped cream; to make butter; to put it in dessert and drinks

**Guided Inquiry** **Investigate More**

Observe Invite children to look at the picture on page 46. Ask: **What is it?** It is an ice cream maker. Ask: **How do you think the ice cream maker works?** It freezes and mixes ice cream ingredients. Ask: **What materials do you need to freeze the cream?** ice cubes, salt Ask: **What clues in the picture show how the ice cream maker freezes the cream?** The ice makes the cream cold and the handle stirs the cream, so all of the cream can be mixed and cooled.

**Open Inquiry**

Ask children to think about different liquids that can be changed by cooling or shaking. If they do not come up with their own research questions, they may consider the following:

- **Does oil freeze?**
- **Will juice form a solid when it is shaken?**
- **At what temperature does water become a solid?**

Give children materials to explore. Help them investigate how cold different substances need to be in order to turn from a liquid to a solid.

# Stop Here to Plan Your Lesson

## Lesson 2 Changes of State

### Essential Question

How does temperature affect matter?

### Objective

- Observe how heat can change matter.

### Reading Skill Predict

| What I Predict | What Happens |
|---|---|
| | |
| | |

*Graphic Organizer 3*, p. TR5

## FAST TRACK

**Lesson Plan** When time is short, follow the Fast Track and use the essential resources.

**1 Introduce**
**Look and Wonder, p. 330**
Resource **Activity Lab Book, p. 155**

**2 Teach**
**Discuss the Main Idea, p. 332**
Resource **Visual Literacy, p. 30**

**3 Close**
**Think, Talk, and Write, p. 335**
Resource **Assessment, p. 124**

**Professional Development** Look for NSDL to find recommended Science Background resources from the National Science Digital Library.

## Reading and Writing

**Lesson Outline** Name ____________ Date ________

**Changes of State**

**Use your book to help you fill in the blanks.**

**How can heating change matter?**

1. Heat can change ___matter___ in different ways.
2. When a solid gets enough ___heat___, it melts.
3. When something melts, it changes from a ___solid___ to a liquid.
4. When heat is added to ice, it turns into ___liquid___ water.
5. Different solids can ___melt___ at different temperatures.
6. Some liquids ___boil___ when they get enough heat.
7. When liquid water boils, it ___evaporates___, or changes into a gas.
8. This gas is called ___water vapor___.

**Outline, pp. 186–187**
Also available as a student workbook

**Lesson Vocabulary** Name ____________ Date ________

**Changes of State**

**Solve the riddles and fill in the puzzle.**

**Down**

1. I keep my shape when I'm cool. If it gets too warm, I melt. ___ice___
2. You can add me or take me away to change matter. ___heat___
4. This happens when I get very cold. ___freeze___
6. When I start out very hot and then become cool, I turn into liquid. ___gas___

**Across**

3. This is what I do when 6 Down happens. ___condense___
5. This is how I turn solids into liquids. ___melting___
7. This is how I go into the air when I'm boiling. ___evaporate___

**Vocabulary, p. 188**
Also available as a student workbook

## Visual Literacy

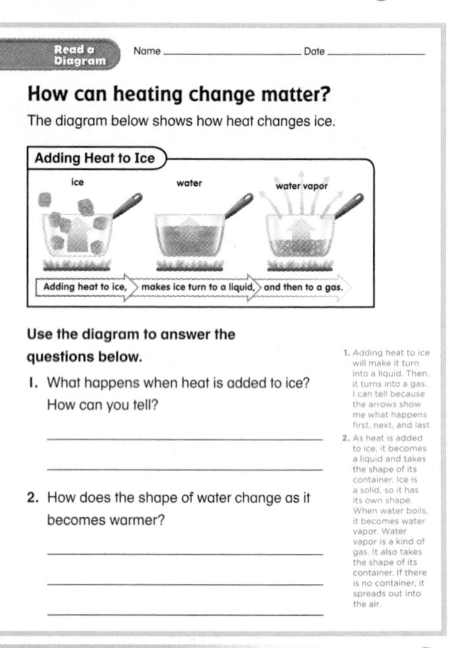

**Read a Diagram** Name ____________ Date ________

**How can heating change matter?**

The diagram below shows how heat changes ice.

Adding Heat to Ice
ice
water
water vapor
Adding heat to ice, makes ice turn to a liquid, and then to a gas.

**Use the diagram to answer the questions below.**

1. What happens when heat is added to ice? How can you tell?
2. How does the shape of water change as it becomes warmer?

1. Adding heat to ice will make it turn into a liquid. Then, it turns into a gas. I can tell because the arrows show me what happens first, next, and last.
2. As heat is added to ice, it becomes a liquid and takes the shape of its container. Ice is a solid, so it has its own shape. When water boils, it becomes water vapor. Water vapor is a kind of gas. It also takes the shape of its container. If there is no container, it spreads out into the air.

**Read a Diagram, p. 30**
Also available as a transparency

# Activity Lab Book

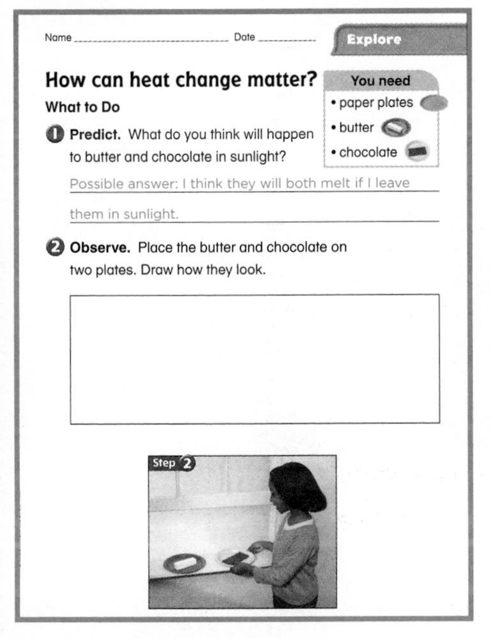
Name ________ Date ________ **Explore**

**How can heat change matter?**

**What to Do**

1. **Predict.** What do you think will happen to butter and chocolate in sunlight?
   Possible answer: I think they will both melt if I leave them in sunlight.
2. **Observe.** Place the butter and chocolate on two plates. Draw how they look.

**You need**
- paper plates
- butter
- chocolate

Step 2

**Explore, pp. 153–154**
Also available as a student workbook

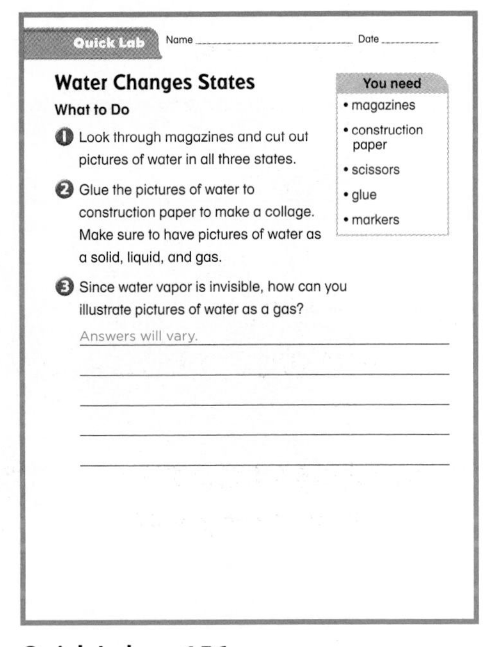
**Quick Lab** Name ________ Date ________

**Water Changes States**

**What to Do**

1. Look through magazines and cut out pictures of water in all three states.
2. Glue the pictures of water to construction paper to make a collage. Make sure to have pictures of water as a solid, liquid, and gas.
3. Since water vapor is invisible, how can you illustrate pictures of water as a gas?
   Answers will vary.

**You need**
- magazines
- construction paper
- scissors
- glue
- markers

**Quick Lab, p. 156**
Also available as a student workbook

# Assessment

Lesson 2 Test Name ________ Date ________

**Circle the best answer for each question.**

1. How does butter change when it is heated?
   - A it evaporates
   - (B) it melts
   - C it freezes
   - D it condenses
2. What do the bubbles in a pot of boiling water show?
   - A a solid that is melting
   - B a gas that is condensing
   - C a liquid that is freezing
   - (D) a liquid that is turning into a gas
3. Which is true about all liquids?
   - A They melt at low temperatures.
   - B They turn into gases when cooled.
   - (C) They freeze at different temperatures.
   - D They condense into solids when heated.

**Think and Write** What happens when heat is added to ice?
When heat is added to ice, the ice melts and becomes a liquid. If enough heat is added to the liquid, it will boil and become a gas.

**Lesson Test, p. 124**

## ADDITIONAL RESOURCES

pp. 146–149

*Make a Pizza*

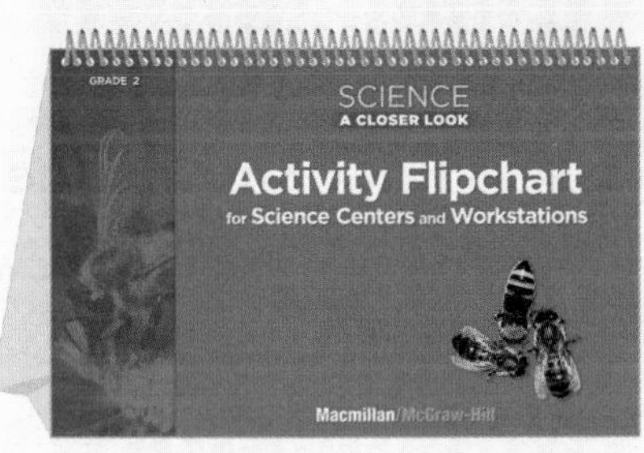

p. 47

p. 101

p. 30

30

91–100

pp. 97–98

76–77

### Technology

- Science Activity DVD
- TeacherWorks™ Plus CD-ROM
- Classroom Presentation Toolkit CD-ROM

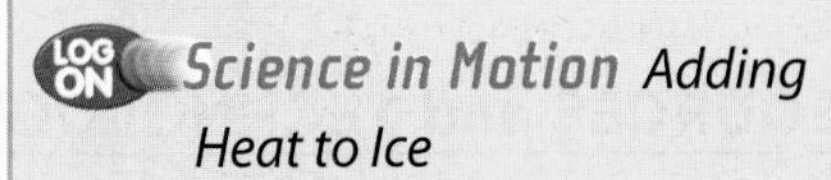
*Science in Motion* Adding Heat to Ice

e-Review

NSDL

# Lesson 2 Changes of State

**Objective**

- Observe how heat can change matter.

Lesson 2
Changes of State
Kilauea Volcano, Hawaii
Look and Wonder
Volcanoes are so hot that rocks can melt and flow like a liquid. How else can heat change things?
330
ENGAGE


## 1 Introduce

### ▶ Assess Prior Knowledge

Have children share what they know about how heat can change matter.

To facilitate the discussion, ask:

- **What happens when water is heated?**
- **What happens when a cake is baked?**
- **How can metal be changed to a liquid?**

Record children's answers in the What We Know column of the class **KWL** chart.

### Look and Wonder

Read the Look and Wonder statement and question. Discuss children's responses to the question.

Ask children to describe what is happening in the photograph, then ask:

- **How can the outside of a volcano change?** Possible answer: When a volcano erupts, it can melt the rocks that make up the walls of the volcano.
- **What are some things that are changed by heat in a kitchen?** Possible answers: water when it is boiled; food when it is cooked

### Warm Up

**Start with a Demonstration**

Show children ice cubes. Ask: **What state is the water in?** Possible answers: solid or frozen

Put the ice cubes into two plastic bags. Add hot water to one of the bags. Have children predict what will happen in each bag. After they observe the bags for a few minutes, ask:

- **Why did the ice cubes melt more quickly in the bag with the hot water?** Because there was more heat in the bag.

### RESOURCES and TECHNOLOGY

- **Activity Lab Book,** pp. 153–155
- **Activity Flipchart,** p. 47
- Science Activity DVD

## Explore — Inquiry Activity

### How can heat change matter?

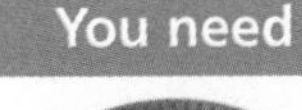

paper plates

butter

chocolate

#### What to Do

1. **Predict.** What do you think will happen to butter and chocolate in sunlight?
2. **Observe.** Place the butter and chocolate on two plates. Draw how they look.
3. **Predict.** How will the Sun's heat change each thing? Find a sunny spot. Leave the plates in the sunlight.
4. **Communicate.** What happens to each thing after one hour? Draw how they look. Compare your pictures.

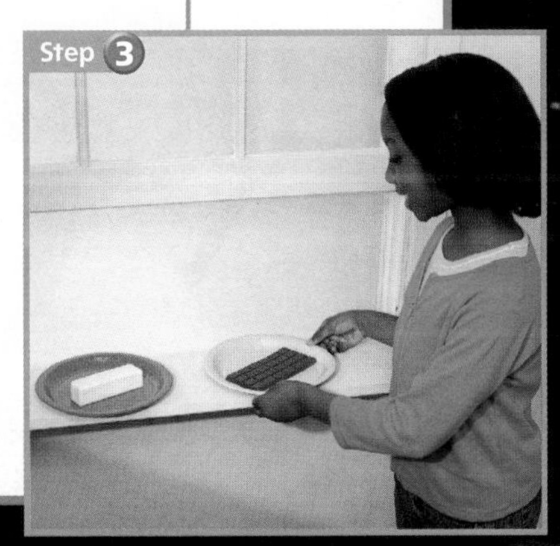


#### Explore More

5. Now try another item. How will it change?



## Explore

small groups

15 minutes

**Plan Ahead** Provide enough chocolate bars and butter sticks for small groups. **Be Careful!** Some children may be lactose intolerant or allergic to chocolate or dairy. If this is the case, use white chocolate and don't allow children who are lactose intolerant or allergic to dairy to taste the butter.

**Purpose** This activity will encourage children to investigate how the heat of the Sun can change matter.

### Structured Inquiry — What to Do

1. **Predict** Ask children to think about what happens to objects that have been in the sun. Ask: **What do you think will happen to the butter and chocolate?** The butter and chocolate will melt.
2. **Observe** Encourage children to consider the consistency of the butter and chocolate by trying to cut it with a plastic knife.
3. **Predict** Ask: **How might the characteristics of the butter and chocolate change?** Possible answers: they will get soft; they will melt; they will become gooey. **Will they change into a new substance?** No, they will go through a physical change only.
4. **Communicate** Even if the butter and chocolate are not completely melted into liquids, have children use the plastic knife to compare the consistency of the materials before and after they were in the sun.

### Guided Inquiry — Explore More

5. Encourage children to test items with low melting points, such as ice or wax. Ask: **What other sources of heat can be used to change objects?** Possible answers: stoves, campfires, grills

### Open Inquiry

Discuss with children what they have discovered about how heat changes matter. Ask: **What else can be changed by heat? Why are some things melted by sunlight while others are not? Do all things melt in heat? How would you find out?**

## Alternative Explore

### How does matter change?

Tell children to think of an ice-cream cone and a marshmallow. Have them work with a partner to write or draw answers to these questions:

- **Which melts faster?** the ice cream cone
- **What does it take to make a marshmallow melt?** more heat
- **If you add heat to an ice-cream cone and a marshmallow, what state do they become?** more liquid than solid

Name ____________ Date ________ Alternative Explore

**How does matter change?**

In this activity, you will work with a partner to explore what happens to substances in a solid state when heat is added.

You need
- pencil
- colored pencils

**What to Do**

1. Discuss with your partner the differences between an ice cream cone and a marshmallow.
2. What would happen if you left both on the table?
Possible answer: The ice cream would start to melt.
The marshmallow would not melt.
3. Predict what would happen if you held a marshmallow over a fire.
Possible answer: It would get brown and soft.

**What Did You Find Out?**

4. **Infer.** Why do some solids melt differently?
Possible answer: Some solids need a lot of heat to melt.
Other solids melt with a little bit of heat.

Activity Lab Book, p. 155

# 2 Teach

## Read Together and Learn

**Reading Skill** **Predict** To make an educated guess about what might happen next.

| What I Predict | What Happens |
|---|---|
| | |
| | |

*Graphic Organizer 3,* p. TR5

## How can heating change matter?

### Discuss the Main Idea

**Main Idea** Heat can change the state of matter.

After reading together, ask:

- **What are some other things that will melt with the addition of only a little heat?** Possible answers: candles, crayons, ice cream
- **What takes a lot of heat to melt?** Possible answers: iron, silver, plastic

**Read Together and Learn**

**Essential Question**
How does temperature affect matter?

**Vocabulary**
evaporate
condense

### How can heating change matter?

Have you ever left a bar of chocolate in your pocket in summer? When you reached in to get it, it was probably melting.

Melting is a change from a solid to a liquid. Some solids, such as gold and glass, will only melt when they are very hot. Other solids, such as ice and butter, melt at much lower temperatures.

◀ When gold melt you can pour it into molds. As t gold cools, it wi harden.



### Science Background

**Changes of State** When water is heated, it can change from one state to another, but it still remains the same substance. For example, water vapor that's released over a pot of boiling water can condense and become a drop of water again. If the drop of water is put in a freezer, it would become a solid. In all three phases, vapor, liquid, and ice, it is still water with the same chemical properties.

See **Science Yellow Pages**, in the Teacher Resources section, for background information.

**Professional Development** For more Science Background and resources from NSDL visit http://nsdl.org/refreshers/science

### ELL Support

**Practice Using Language** Bring in a variety of household products that are solids and liquids. Have children describe them and sort them into two groups: *solids* and *liquids*. Discuss what happens if you apply heat to each item.

**BEGINNING** Help children name and describe each item. Have them sort items into two groups: *liquids* and *solids*.

**INTERMEDIATE** Have children take turns describing and labeling the items. Talk together about what might happen if heat was applied to each item.

**ADVANCED** Each child should choose one solid and one liquid to describe and they should explain, in their own words, what they think will happen if heat is applied to each.

olid ice cubes
elt when
ft at room
mperature.

Water can change to a gas when it is heated. **Evaporate** means to change from liquid to gas and go into the air.

If enough heat is added to water, it will boil. When water boils, you can see bubbles. The bubbles show that the water is changing to a gas called water vapor. We can not see water vapor.

## Adding Heat to Ice

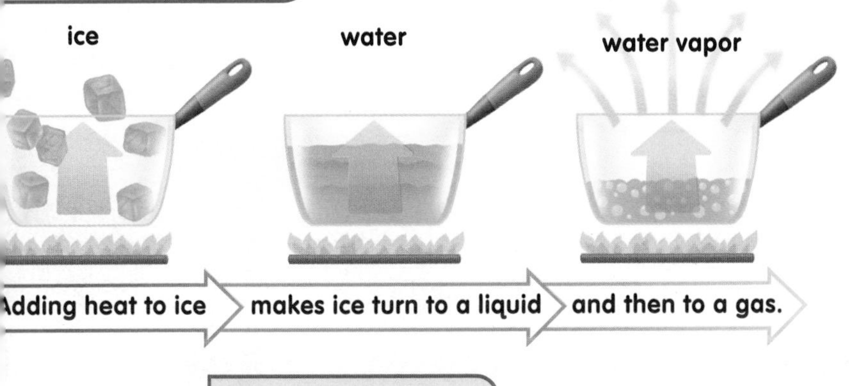


### Read a Diagram

How does ice change when it is heated?

 *Science in Motion* Watch what happens when heat melts ice at www.macmillanmh.com

 How can heat change solids?



## Differentiated Instruction

### Leveled Activities

**EXTRA SUPPORT** Have children create a short book that shows how an ice cube can be changed from a solid to a liquid, and then to a gas. Encourage children to show appropriate heat sources in their drawings.

**ENRICHMENT** Ask children to use books or the Internet to learn more about the melting point of different types of matter. Ask them to compile a list of matter that melts and arrange it from lowest to highest melting point. Encourage them to illustrate their list and share it with the class.

### Read a Diagram

Have children look carefully at the diagram and discuss with a partner what they think it shows. After their discussions, ask the group:

- **What would happen if the flame was turned off after the ice melted?** Possible answer: A smaller amount of the water would turn into gas.
- **What is the name of the gas in the diagram?** water vapor

**Answer to Read a Diagram** It changes from a solid to a liquid and then to a gas.

 *Science in Motion* *Adding Heat to Ice*

## ▶ Develop Vocabulary

**evaporate** ***Word Origin*** Write *evaporate* on the board. Challenge children to find a smaller word inside the word *evaporate*. Hint: They learned this word in the Read a Diagram on page 333. Circle the word *vapor*. Explain that *vapor* is "water that has turned to gas." Ask: **What parts of speech are these two words?** *Evaporate* is a verb; *vapor* is a noun. Have children use the words in a sentence.

## *Quick Check Answer*

Heat can change a solid into a liquid or a liquid into a gas.

### RESOURCES and TECHNOLOGY

- **Reading and Writing,** pp. 186–188
- **Visual Literacy,** p. 30
- **PuzzleMaker CD-ROM**
- **Classroom Presentation Toolkit CD-ROM**
- 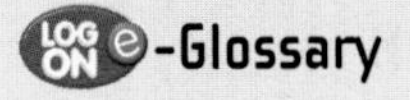 **-Glossary**

# How can cooling change matter?

## ▶ Discuss the Main Idea

**Main Idea** Cooling can change matter.

Read the text together. Then ask:

- **How does cooling affect liquids and gases?** Possible answer: Cooling can change liquids to solids, and gases to liquids.
- **Where else have you seen water vapor condense?** Possible answers: on the walls of the shower; on the bathroom mirror; on car windows
- **Will water condense on the sides of a hot glass?** No, because gases must be cooled to condense.

## ▶ Develop Vocabulary

**condense** Write *condense* on the board and circle the word *dense*. Explain to children that *dense* means "crowded closely together." Ask: **How does this relate to a gas?** When a gas condenses, it becomes a liquid because its parts move closer together.

## ▶ Address Misconceptions

Help children understand that when the warm, indoor air meets the cold glass on a wintery day, some of the water vapor in the air is cooled and condenses. It changes from a gas to a liquid. That's why drops of water will form on the inside of a window on a cold day.

FACT **Water that has condensed on a window comes from air inside the room.** Show children actual instances of water vapor condensing on a drinking glass or a window in the classroom.

**RESOURCES and TECHNOLOGY**

- **School to Home Activities,** pp. 97–98
- **Reading and Writing,** p. 189
- **Activity Lab Book,** p. 156
- LOG ON *Science in Motion* *Adding Heat to Ice*
- LOG ON **e-Review** Narrated Summary and Quiz
- ***ExamView® Assessment Suite* CD-ROM**

## How can cooling change matter?

Matter can also change by cooling, or taking away heat. Gases condense when they are cooled. **Condense** means to change from a gas to a liquid.

Water vapor in the air condenses when it touches cool objects. This is why you see small drops of water on the outside of a cold glass.

▲ **Water vapor condenses on th outside of a bot**

FACT Water that has condensed on a window comes from air inside the room.

334
EXPLAIN

**Quick Lab**

 Individual   15 minutes

**Objective** Classify water as solid, liquid, or gas.

**You need** magazines, construction paper, glue, scissors, markers

1. Invite children to find pictures of water in all three states.
2. Have children classify the pictures they found and create a collage of the images.
3. Challenge children to think of creative ways to illustrate water in its gaseous state.

When liquids cool, they can ·eeze, or become solid. Wax ·d some other liquids will ·eeze at room temperature. ·her liquids, such as water, ·ed to be much colder to ·eeze.

How does water change when it is cooled?

## Quick Lab

Classify pictures of water from magazines as solid, liquid, or gas.

**After a candle burns, the wax will cool and become solid.**

## Think, Talk, and Write

1. Predict. What will happen to a puddle of water on a sunny day?
2. What happens when water vapor condenses?
3. **Essential Question.** How does temperature affect matter?

## Math Link

Do you think the mass of ice changes when it melts? How could you find out?

**-Review** Summaries and quizzes online at www.macmillanmh.com

335
EVALUATE

### Quick Check Answer

Possible answer: Water can freeze if it is cooled enough.

# 3 Close

### Using the KWL Chart

Review the **KWL** chart and ask children to describe what they know now about how heating and cooling can affect matter. Record their responses in the What We Learned column of the class **KWL** chart.

### Using the Reading Skill
**Predict**

Use the reading skill graphic organizer to reinforce lesson content. Ask: **How does heat change water?**

| What I Predict | What Happens |
|---|---|
| When ice is heated, it turns to water. When water is heated, it turns to water vapor. | When heat is applied to a solid, it turns to a liquid. When heat is applied to a liquid, it turns to a gas. |

*Graphic Organizer 3,* p. TR5

### FAST TRACK Think, Talk, and Write

1. **Predict** Possible answer: The heat from the Sun will make the water evaporate into the air. The puddle will get smaller and eventually disappear.
2. The gas becomes a liquid.
3. **Essential Question** Possible answer: Heat can melt a solid or cause a liquid to evaporate. Cold can cause a liquid to freeze and a gas to condense.

## Math Link

Have children place ice cubes in a container and weigh it. Once the ice has completely melted, have the children weigh the container again and compare the results. The mass will stay the same.

## Formative Assessment

**Draw a Sequence**

Have children make a sequence of drawings about what happens to a candy bar that is in strong sunlight all day. Then have them do drawings of the candy left out overnight when it gets cold. Ask them to write captions below each drawing to explain what happened.

The solid chocolate bar is still cool and hard. | The heat from the Sun melted the chocolate. | Without the Sun shining on it, the chocolate is a solid again.

**Key Concept Cards** For student intervention, see the prescribed routine on **Key Concept Card 30.**

# Reading in Science

**Objective**

- Identify how wax changes at different temperatures.

## Colorful Creations

**Genre: Nonfiction** Stories or books about real people and events.

Have children look at the pictures. Ask:

- **What is this story is about?** Possible answer: how crayons are made

## Before Reading

Give children some crayons to pass around. Ask them to describe how the crayons feel. Then ask:

- **Are the crayons solid, liquid, or gas?** solid
- **From what materials are crayons made?** Possible answers: wax, dye

Display Graphic Organizer 3. Explain to children that when they make predictions, they use what they know to tell what they think will happen. Ask:

- **What will happen to wax when it is heated?** Possible answers: It will melt. It will get gooey.

Write children's predictions in the first column of the graphic organizer.

## During Reading

As children read, have them think about what happens to the wax during the crayon-making process. Read the text together. Then ask:

- **What happens to the wax when dye is added?** The wax turns into the color of the dye.
- **What happens to the wax when it is heated?** It melts and becomes a liquid.
- **What happens to the wax when it is cooled with cold water?** It hardens into a solid.

**RESOURCES and TECHNOLOGY**

- **Reading and Writing,** pp. 190–191

Real World **Reading in Science**

# Colorful Creations

There are all kinds of colors inside your crayon box. How were those crayons made?

Most crayons are made of wax. This man adds special dye to a tub of wax to give the wax color.

The colored wax is melted into a liquid. Then a worker pours this hot wax into a mold.

Inside the mold there are hundreds of holes shaped like crayons. The wax fills each hole. Then the mold is cooled with cold water.

336
EXTEND

## ELL Support

**Use Sentence Frames** Have children point to a picture of a liquid and a picture of a solid on pages 336 and 337. Next, help children describe the wax at different stages in the process by identifying words in the captions, such as *hot*. Make a list of other adjectives and say the words together. Point to the photos and ask questions, such as: *What color is the wax?*

**BEGINNING** Ask children to finish this sentence frame: *The wax is ____.* liquid

**INTERMEDIATE** Have children finish this sentence frame: *When it is ____, wax becomes a ____.* heated, liquid *When it is ____, wax becomes a ____.* cooled, solid

**ADVANCED** Use the captions under the photos to create sentence frames that children can complete.

Science, Technology, and Society

This machine packs the crayons into boxes.

is woman checks the ayons by hand to make re they are good.

**Talk About It**

**Predict.** What will happen if the mixture of wax is left out at room temperature?

Connect to American Museum of Natural History at www.macmillanmh.com

337 EXTEND

## After Reading

Display Graphic Organizer 3 and read aloud the predictions children made before reading the article. Together, look back through the article and have children find the photo and caption that tells what happens to wax when it is heated. Circle the correct prediction in the graphic organizer. Have children tell what happens to wax when it is heated. Write their responses in the second column of the graphic organizer.

| What I Predict | What Happens |
| --- | --- |
| The wax will melt when it is heated. | The wax melts when it is heated. |

*Graphic Organizer 3,* p. TR5

## Talk About It

Possible answers: If hot liquid wax is left out at room temperature, it will cool and harden into the shape of its container. If the solid wax mixture (crayon) is left out at room temperature, nothing will happen.

If children have difficulty predicting what will happen, encourage them to share their experiences with crayons. Ask whether children have ever had a crayon that melted, and discuss what happened to cause the crayon to melt.

## Extended Reading

### Visit the Library

Read aloud ***From Wax to Crayon,*** by Michael H. Forman (Children's Press, 1997).

Have children retell the steps of the crayon-making process. Ask:

- **What happens at the color mill?**
- **Why is the colored wax blended or stirred, over and over?**

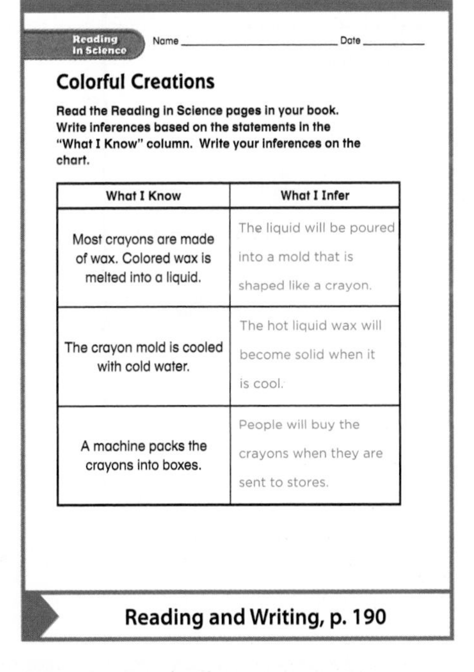
Reading in Science Name ______ Date ______

**Colorful Creations**

Read the Reading in Science pages in your book. Write inferences based on the statements in the "What I Know" column. Write your inferences on the chart.

| What I Know | What I Infer |
| --- | --- |
| Most crayons are made of wax. Colored wax is melted into a liquid. | The liquid will be poured into a mold that is shaped like a crayon. |
| The crayon mold is cooled with cold water. | The hot liquid wax will become solid when it is cool. |
| A machine packs the crayons into boxes. | People will buy the crayons when they are sent to stores. |

Reading and Writing, p. 190

# Plan Your Lesson

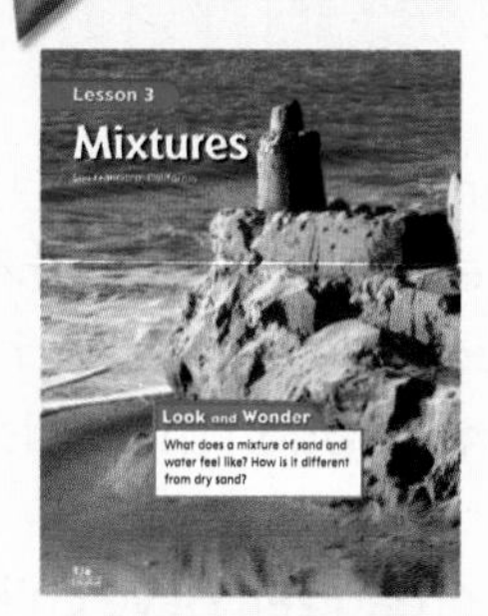


## Lesson 3 Mixtures

### Essential Question

How can you make mixtures?

### Objective

- Observe how solids, liquids, and gases mix.

**Reading Skill** Main Ideas and Details

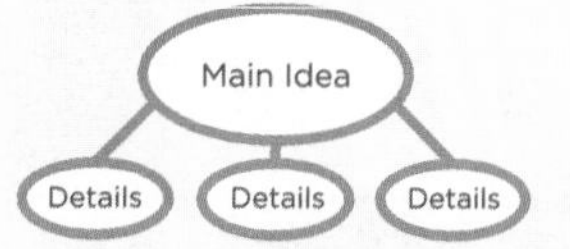


*Graphic Organizer 1*, p. TR3

## FAST TRACK

**Lesson Plan** When time is short, follow the Fast Track and use the essential resources.

**1 Introduce**

**Look and Wonder, p. 338**

Resource **Activity Lab Book, p. 159**

**2 Teach**

**Discuss the Main Idea, p. 340**

**Develop Vocabulary, p. 343**

Resource **Visual Literacy, p. 31**

**3 Close**

**Think, Talk, and Write, p. 345**

Resource **Assessment, p. 125**

**LOG ON** **Professional Development** Look for NSDL to find recommended Science Background resources from the National Science Digital Library.

## Reading and Writing

LESSON Outline Name ________ Date ________

**Mixtures**

**Use your book to help you fill in the blanks.**

**What are mixtures?**

1. When two or more things are put together, the result is called a ___mixture___.
2. Mixtures can have different ___combinations___ of solids, liquids, and gases.
3. Some mixtures can be picked ___apart___.

**What mixtures stay mixed?**

4. A mixture that is difficult to take apart is called a ___solution___.
5. When salt is added to water, the salt ___dissolves___ and mixes with the water.
6. Sand and water ___do not___ make a solution.

**Outline, pp. 192–193**
Also available as a student workbook

CHAPTER Vocabulary Name ________ Date ________

**Mixtures**

**Write whether you would need to use a magnet, a filter, evaporation, or your hands in order to take apart each mixture listed below. Some mixtures can be taken apart in more than one way.**

1. salt water ___evaporation___
2. water and sand ___filter, evaporation___
3. iron nails and sand ___hands, magnet___
4. raisins and cornflakes ___hands___
5. iron screws and plastic beads ___hands, magnet___
6. pennies and nickels ___hands___
7. blue paper and white paper ___hands___
8. water and seashells ___filter, hands, evaporation___

**Vocabulary, p. 194**
Also available as a student workbook

## Visual Literacy

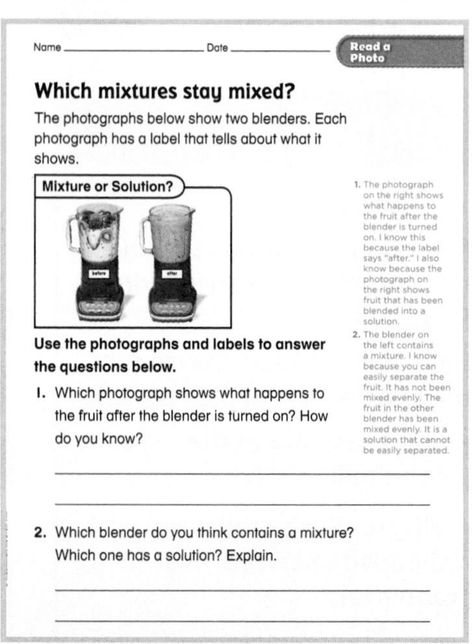

Name ________ Date ________ Read a Photo

**Which mixtures stay mixed?**

The photographs below show two blenders. Each photograph has a label that tells about what it shows.

Mixture or Solution?

**Use the photographs and labels to answer the questions below.**

1. Which photograph shows what happens to the fruit after the blender is turned on? How do you know?
2. Which blender do you think contains a mixture? Which one has a solution? Explain.

1. The photograph on the right shows what happens to the fruit after the blender is turned on. I know this because the label says "after." I also know because the photograph on the right shows fruit that has been blended into a solution.
2. The blender on the left contains a mixture. I know because you can easily separate the fruit. It has not been mixed evenly. The fruit in the other blender has been mixed evenly. It is a solution that cannot be easily separated.

**Read a Photo, p. 31**
Also available as a transparency

# Activity Lab Book

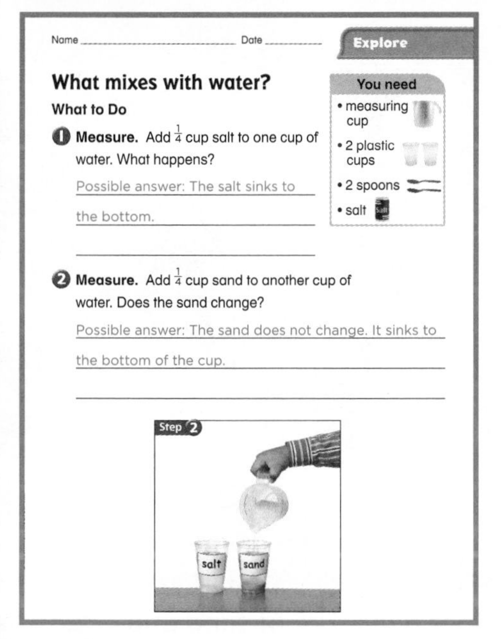

Name ____________ Date ________ **Explore**

### What mixes with water?

**What to Do**

**You need**
- measuring cup
- 2 plastic cups
- 2 spoons
- salt

1. **Measure.** Add $\frac{1}{4}$ cup salt to one cup of water. What happens?
   Possible answer: The salt sinks to the bottom.
2. **Measure.** Add $\frac{1}{4}$ cup sand to another cup of water. Does the sand change?
   Possible answer: The sand does not change. It sinks to the bottom of the cup.



**Explore, pp. 157–158**
Also available as a student workbook

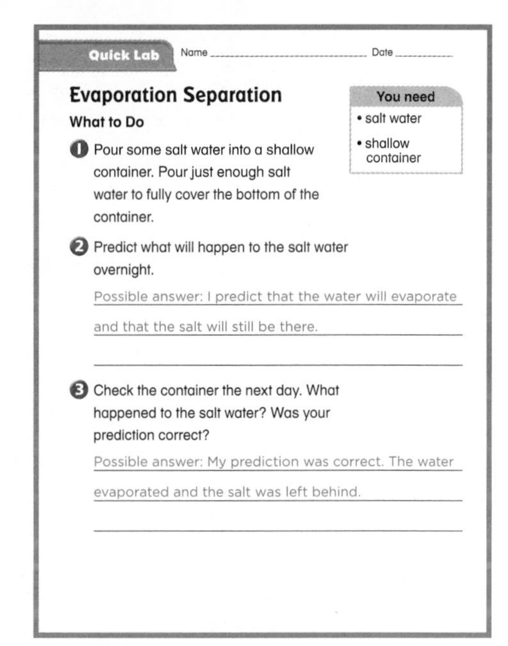

**Quick Lab** Name ____________ Date ________

### Evaporation Separation

**What to Do**

**You need**
- salt water
- shallow container

1. Pour some salt water into a shallow container. Pour just enough salt water to fully cover the bottom of the container.
2. Predict what will happen to the salt water overnight.
   Possible answer: I predict that the water will evaporate and that the salt will still be there.
3. Check the container the next day. What happened to the salt water? Was your prediction correct?
   Possible answer: My prediction was correct. The water evaporated and the salt was left behind.

**Quick Lab, p. 160**
Also available as a student workbook

# Assessment

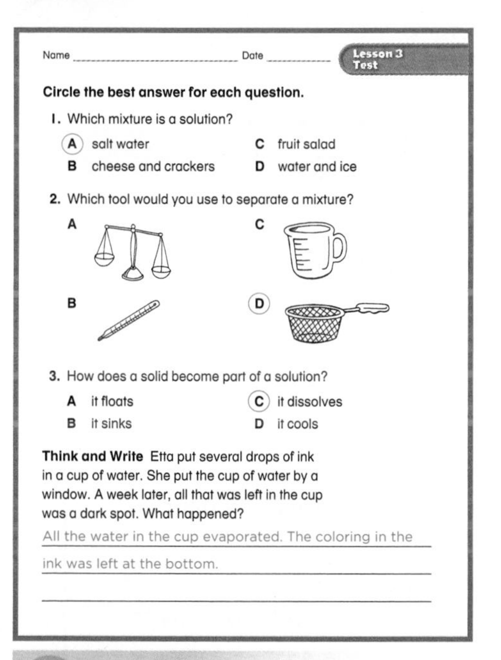

Name ____________ Date ________ **Lesson 3 Test**

**Circle the best answer for each question.**

1. Which mixture is a solution?
   (A) salt water  C fruit salad
   B cheese and crackers  D water and ice
2. Which tool would you use to separate a mixture?
   A  C
   B  (D)
3. How does a solid become part of a solution?
   A it floats  (C) it dissolves
   B it sinks  D it cools

**Think and Write** Etta put several drops of ink in a cup of water. She put the cup of water by a window. A week later, all that was left in the cup was a dark spot. What happened?
All the water in the cup evaporated. The coloring in the ink was left at the bottom.

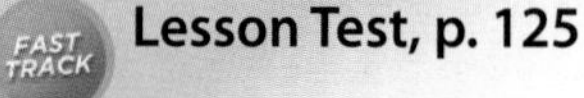


**Lesson Test, p. 125**

## ADDITIONAL RESOURCES

pp. 150–155

*Make a Pizza*

p. 48

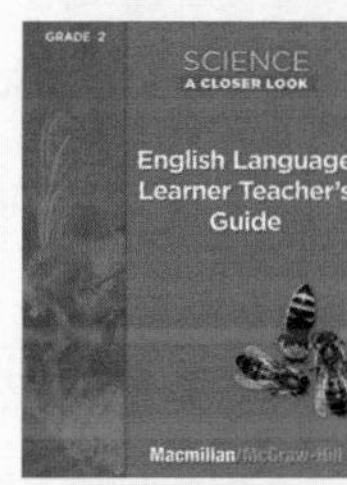


p. 102

p. 31

31

91–100

pp. 99–100

78–80

### Technology

- Science Activity DVD

- TeacherWorks™ Plus CD-ROM

- Classroom Presentation Toolkit CD-ROM
- LOG ON e-Review
- LOG ON NSDL

# Lesson 3 Mixtures

**Objective**

- Observe how solids, liquids, and gases mix.

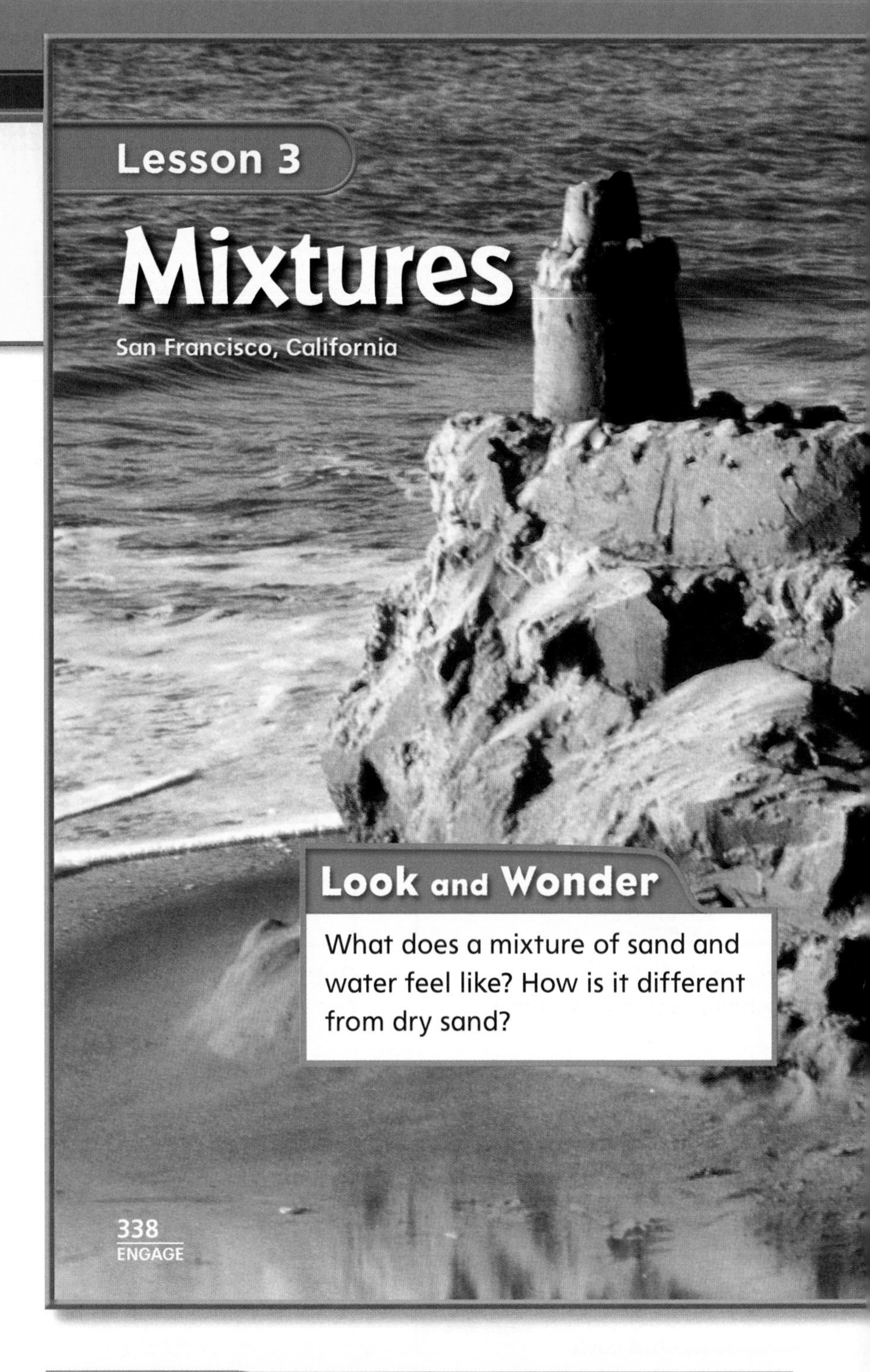

Lesson 3
Mixtures
San Francisco, California
Look and Wonder
What does a mixture of sand and water feel like? How is it different from dry sand?
338
ENGAGE


## 1 Introduce

### ▶ Assess Prior Knowledge

Have children share what they know about how different materials can become mixtures. Ask:

- **What happens when you mix things together?**
- **What happens when you mix liquids and solids?**
- **What happens when you try to take a mixture apart?**

Record children's answers in the What We Know column of the class **KWL** chart.

### Look and Wonder

Read the Look and Wonder questions and discuss children's responses. Then ask:

- **How can a mixture of sand and water be separated?** Possible answers: Leave the sand in a sunny place, so the water will evaporate. Pour the mixture through a strainer to separate most of the sand from the water.
- **Why do you need wet sand to make a sand castle?** Possible answers: Dry sand is free-flowing and does not hold a shape. Wet sand is sticky and can hold together.

### Warm Up

**Start with a Demonstration**

Show children a large glass container full of water. Tell them that the class will make a mixture.

Have a volunteer add sugar, lemon slices, and ice to the water and stir it. Ask:

- **How would you separate the parts of the mixture?** Possible answers: with a strainer; with a colander

Have volunteers pick out the ice and lemon slices by hand. Now ask:

- **How can we separate the water and sugar?** Leave the water in a sunny place until all the water has evaporated. Only the sugar will be left.

### RESOURCES and TECHNOLOGY

- **Activity Lab Book,** pp. 157–159
- **Activity Flipchart,** p. 48
- Science Activity DVD

## Explore — Inquiry Activity

### What mixes with water?

**What to Do**

1. **Measure.** Add $\frac{1}{4}$ cup salt to one cup of water. What happens?
2. **Measure.** Add $\frac{1}{4}$ cup sand to another cup of water. Does the sand change?
3. **Compare.** Stir both mixtures with a spoon. Let them sit. What happens? How are the mixtures different from each other?

**Explore More**

4. **Investigate.** Tell how you could take the sand and the water apart. Can the salt be taken out of the water?

You need
measuring cup
2 plastic cups
2 spoons
Salt
salt


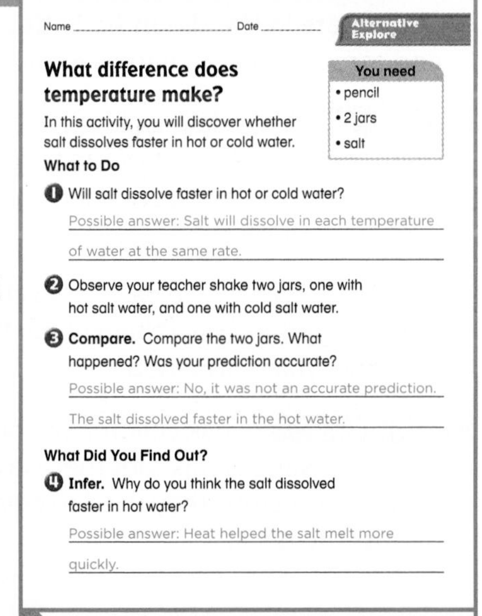




## Explore

 pairs   20 minutes

**Plan Ahead** Provide enough clear cups, measuring cups, sand, salt, and spoons for children to work in pairs.

**Purpose** Children will observe what happens when a material dissolves in a solution.

**Structured Inquiry** What to Do

1. **Measure** Children should notice that some of the salt disappears in the water right away.
2. **Measure** Ask children to compare what they see in the two cups. Children should notice that the sand does not change.
3. **Compare** After stirring, ask children to compare what they see. They should notice that the salt has dissolved while the sand is at the bottom of the cup.

**Guided Inquiry** Explore More

4. **Investigate** Children could use a strainer to separate the sand mixture. They could let the water evaporate to separate the salt mixture. Ask: **What would happen to the water if it was left out for a long period of time?** The water would evaporate and the sand and salt would be left.

**Open Inquiry**

Discuss with children other types of mixtures they are curious about. Give them a variety of solids and liquids, such as oil, rock candy, cereal, and dishwashing liquid, and have them work with the materials to discover more about how solids react when mixed with water.

## Alternative Explore

### What difference does temperature make?

Ask children to predict whether salt will dissolve faster in hot or cold water.

Fill two jars, one with hot water and the other with cold water. Add salt to each jar. Close the jar and shake. After one minute, have the children observe and compare the two jars. Help children discuss possible reasons why salt dissolved faster in hot water.

Name ________ Date ________ Alternative Explore

**What difference does temperature make?**

In this activity, you will discover whether salt dissolves faster in hot or cold water.

You need
- pencil
- 2 jars
- salt

**What to Do**

1. Will salt dissolve faster in hot or cold water?
   Possible answer: Salt will dissolve in each temperature of water at the same rate.
2. Observe your teacher shake two jars, one with hot salt water, and one with cold salt water.
3. **Compare.** Compare the two jars. What happened? Was your prediction accurate?
   Possible answer: No, it was not an accurate prediction. The salt dissolved faster in the hot water.

**What Did You Find Out?**

4. **Infer.** Why do you think the salt dissolved faster in hot water?
   Possible answer: Heat helped the salt melt more quickly.

Activity Lab Book, p. 159

# 2 Teach

## Read Together and Learn

**Reading Skill** Main Ideas and Details The main idea is the most important idea in the reading selection. Details give more information about the main idea.

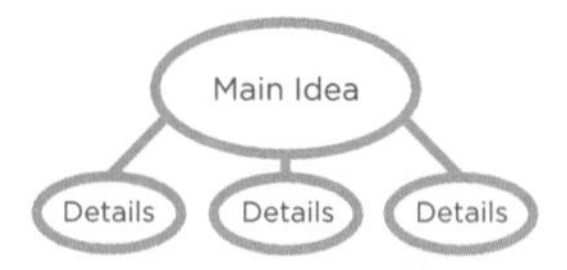


*Graphic Organizer 1,* p. TR3

## What are mixtures?

### FAST TRACK Discuss the Main Idea

**Main Idea** Mixtures are combinations of things, which sometimes can be separated.

Before reading, ask children to list some common mixtures. Discuss whether they are a mixture of two or more solids, or solids and liquids.

Read the text together and ask children who have made papier-mâché to explain what they did. If no one has made papier-mâché, explain the process. Ask:

- **How does the mixture of the solid paper and the liquid glue change?** It hardens as the air dries the glue and it becomes a solid.

### Read Together and Learn

**Essential Question**
How can you make mixtures?

**Vocabulary**
mixture
solution
dissolve

### What are mixtures?

When you put salt into water, yo make a mixture. A **mixture** is two o more things put together. Mixtures can be any combination of solids, liquids, and gases.

When you glue different things to paper, you make a mixture. When you put pieces of clay together, you also make a mixture.

Papier mâché is a mixture of flour, water, and newspaper.

You can cover iten with papier mâché to make things.

340
EXPLAIN

### Science Background

**Mixture** A mixture is a combination of two or more substances. Heterogeneous mixtures, like trail mix, do not look the same throughout. Homogeneous mixtures, such as milk, look the same throughout. If a mixture undergoes a chemical change, such as through the baking process, its parts cannot be separated out again.

See **Science Yellow Pages**, in the Teacher Resources section, for background information.

**Professional Development** For more Science Background and resources from NSDL visit http://nsdl.org/refreshers/science

### ELL Support

**Participate in Hands-on Activities** Give children containers and some solids and liquids to mix together.

**BEGINNING** Help children name each item as they mix them together. Ask the group to repeat the names of the items in each mixture and decide if it is easy or difficult to take the mixture apart.

**INTERMEDIATE** Have each child choose items that they want to mix together. Have them name the items they chose and describe what happened.

**ADVANCED** Ask children to choose three or more items that they want to mix together. Have them describe the sequence of steps they used to create their mixtures and then discuss the results.

ometimes when you mix things
ether, it is easy to pick them apart
ain. You can see the different parts
the mixture. The things in the
ture do not change.

The pencil holder is a mixture made of papier mâché, a can, and buttons. What other mixtures do you see?

What kinds of matter can be used to make a mixture?



## Differentiated Instruction

### Leveled Activities

**EXTRA SUPPORT** Ask children to walk around the classroom and make a list of everything that is a mixture. When they are done, ask them to discuss how they can separate the mixtures they listed.

**ENRICHMENT** Ask children to make a mixture of salt and water. Have children add the salt one spoonful at a time, and keep track of how many tablespoons of salt they add before the salt stops dissolving in the water. Encourage children to repeat the activity using sugar and water, and to compare the results of the two activities in a Venn diagram.

## ▶ Use the Visuals

Have children look at the picture on page 341. Ask:

- **What other objects could have been added to the mixture to decorate the can?** Accept all reasonable responses.
- **What parts of the mixture could be separated?** Possible answers: The buttons could be pulled off. The papier-mâché could be peeled off the can.
- **What parts of the mixture would be difficult to remove?** Possible answers: the paint on the papier-mâché; the glue on the pipe cleaners

## ▶ Develop Vocabulary

**mixture** ***Word Origin*** Write the word *mixture* on the board. Ask children if they notice another word inside the word *mixture*. Circle the word *mix*. Explain that *mix* is the base word of *mixture*. It means, "to make into one substance by stirring." Ask children: **How does the base word *mix* help us understand the word *mixture*?** Mixing or stirring two or more things together into one substance creates a mixture.

## ▶ Explore the Main Idea

ACTIVITY Have children look through magazines to find pictures of things that are mixtures. Have them cut out the pictures and make a "mixture collage."

## ***Quick Check Answer***

Mixtures can be made from solids, liquids, or gases.

### RESOURCES and TECHNOLOGY

- **Reading and Writing,** pp. 192–194
- PuzzleMaker CD-ROM
- Classroom Presentation Toolkit CD-ROM
-  e-Glossary

# Which mixtures stay mixed?

## ▶ Discuss the Main Idea

**Main Idea** Solutions are mixtures that are difficult to separate.

Read the text together and discuss the pictures. Ask:

- **Why is something dissolved in water a solution?** Solutions are mixtures that are difficult to take apart, and when things are dissolved in water they are difficult to take apart.
- **How is a mixture different from a solution?** A mixture can be easily taken apart, but a solution cannot.
- **Do solutions always begin as mixtures? Why?** Yes, because a solution is a kind of mixture. The liquids or solids that make the solution start out as solids, liquids, or gases that are mixed until they are blended or dissolved.

### Read a Photo

Explain to children that the photos show what happens to fruit when it is put in a blender. Ask:

- **What would happen if the blender is turned off before all the fruit is mixed?** There would be a mixture of fruits in the solution.
- **Which of these blenders contains a solution?** the blender on the right

**Answer to Read a Photo** The blender on the right because the mixture cannot be easily taken apart.

## Which mixtures stay mixed?

Sometimes when you mix things, it is not easy to change them back. When you make a shake or a smoothie, you mix different foods together. It is hard to take apart after it has been blended.

**Making a Smoothie**

**Read a Photo**

Which mixture is harder to take apart?



### Classroom Equity

Some children are less likely to participate in classroom discussions and Q & A sessions than others. Foster participation from all children by having each child write their name on a craft stick at the beginning of the year. Place the sticks in a jar on your desk. When you have a question to ask, randomly pull a stick from the jar. Don't replace the sticks until you've made it through all of the children in your class.

**RESOURCES and TECHNOLOGY**

- **Visual Literacy,** p. 31
- Classroom Presentation Toolkit CD-ROM

A **solution** is a mixture that
hard to take apart. Sugar
ıd water make a solution.
ıgar will **dissolve**, or stay
enly mixed in the water.

Sand and water can be
ixed, but they do not make
solution. The sand does not
ay mixed and sinks to the
ottom of the glass.

The drink mix dissolves in the water. ▶

▲ The soapy water is a solution. The dishes are a mixture.

How is a solution a special kind of mixture?



FAST TRACK

## ▶ Develop Vocabulary

**solution** ***Scientific vs. Common Use*** Explain to children that the word *solution* is often used to describe how you solve a problem or get to an answer.

When scientists use the word *solution*, they are talking about "a mixture that is made from a solid or gas that has been dissolved in a liquid." Ask children to use the word *solution* in a sentence using the scientific meaning.

**dissolve** ***Word Origin*** Show children how the parts of a word can help them understand its meaning. Write the word *dissolve* on the board. Explain to children that the Latin prefix *dis-* means "apart" and the Latin verb *solvere* means "to loosen."

Point out that when a solid dissolves in a liquid, it "loosens" and "comes apart" until it mixes evenly throughout the liquid. Have children make a list of solids that dissolve in liquids, such as sugar, salt, and drink mix.

## ▶ Explore the Main Idea

ACTIVITY Have children mix together vegetable oil and water in a jar. Ask them to shake the jar and then decide if the mixture is a solution or not.

Explain that oil and water do not mix. In fact, they will easily separate on their own. Explore further by adding food coloring to the mixture and observe the results.

### Quick Check Answer

Possible answer: A solution is a kind of mixture with parts that cannot be easily separated.

## Differentiated Instruction

### Leveled Questions

**EXTRA SUPPORT** Ask questions such as these to check children's understanding of the material.

- **How are mixtures made?** by putting two or more materials together
- **What is a solution?** a kind of mixture with parts that cannot be easily separated

**ENRICHMENT** Use these types of questions to develop children's higher-order thinking skills.

- **Which types of matter are most likely to make solutions when mixed?** liquids
- **How can a solution be separated?** evaporating the liquid

## How can you take mixtures apart?

### ▶ Discuss the Main Idea

**Main Idea** Some mixtures can be separated with filters, magnets, or evaporation.

After reading together, ask:

- **How does evaporation take water out of a solution?** The water in the solution turns into a gas and floats up into the air.

### ▶ Use the Visuals

Have children look at the picture of the pitcher and filter on page 344. Ask:

- **What would happen if a solution of salt and water was poured into the filter?** It would all pass through the filter.
- **How do you think the filter removes sand from the water and sand mixture?** The holes in the filter are too small for the grains of sand to pass through, so only the water goes through.

### ▶ Develop Vocabulary

To reinforce children's understanding of lesson vocabulary, have children use the words *mixture, dissolve,* and *solution* to describe a simple recipe for hot chocolate.

## How can you take mixtures apart?

Have you ever picked the pretzels out of a snack mix? You were taking apart a mixture. Some mixtures are harder to take apart.

Filters are screens that trap solids but let liquids flow through. Magnets can also help take some mixtures apart. They can be used to separate iron from a mixture.

**Quick La**

**Investigate** hov evaporation helps take a mixture apart.

A filter can help take apart a mixture of sand and water.

A magnet can help take apart a mixtur sand and iron filing



**Quick Lab**

 pairs   15 minutes

**Objective** Identify how evaporation helps separate a solution.

**You need** salt, water, shallow plastic containers

1. Prepare a saltwater solution in advance. Have children pour a small amount of salt water into a shallow container, write their names on it, leave it in a warm, safe place, and draw it.
2. The next day, have them observe and draw the container.
3. Have the children infer what happened to the solution. Ask: **What happened to the water?** it evaporated **What was left?** the salt

**RESOURCES and TECHNOLOGY**

- **School to Home Activities,** pp. 99–100
- **Reading and Writing,** p. 195
- **Activity Lab Book,** p. 160
- **LOG ON e-Review** Narrated Summary and Quiz
- **Exam*View*® *Assessment Suite* CD-ROM**

ome mixtures can be even
der to take apart. Evaporation
be used to take a solution
salt water apart. If you leave
water out to dry, the water
porates. The salt is left behind.

Water has evaporated from the ocean and left salt here.

How do filters help separate mixtures?

## hink, Talk, and Write

1. Main Idea and Details. Describe how different things mix with water.
2. How can you take apart a solution of salt and water?
3. **Essential Question.** How can you make mixtures?

## Health Link

What foods are mixtures? Look for food mixtures in books, magazines, or at the grocery store. Make a list.

**e-Review** Summaries and quizzes online at www.macmillanmh.com





### Quick Check Answer

Filters keep solids from passing through, but allow liquids to pass through easily.

# 3 Close

### Using the KWL Chart

Review the **KWL** chart and ask children to describe what they now know about mixtures and solutions. Record their responses in the What We Learned column of the class **KWL** chart.

### Using the Reading Skill
#### Main Idea and Details

Use the reading skill graphic organizer to identify the main idea and details of the lesson.

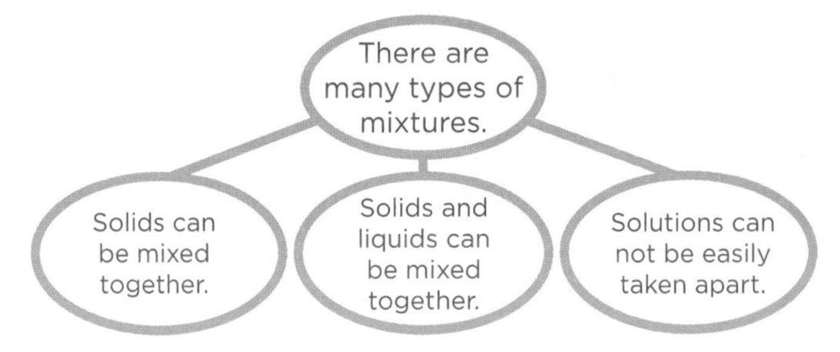

There are many types of mixtures.
Solids can be mixed together.
Solids and liquids can be mixed together.
Solutions can not be easily taken apart.

*Graphic Organizer 1*, p. TR3

### Think, Talk and Write

1. **Main Idea and Details** Possible answers: Some solids and liquids, such as sugar and dishwashing liquid, dissolve in water to become a solution. Some solids and liquids, such as sand and oil, do not dissolve when mixed with water.
2. A saltwater solution can be taken apart by letting the water evaporate, so only the salt remains.
3. **Essential Question** Possible answers: You can make a mixture by mixing together two or more kinds of matter. You can make a mixture with any combination of solids, liquids, or gases.

## Health Link

Discuss the ingredients in different foods to help children realize that many of their favorite foods are mixtures.

## Formative Assessment

### Making a Smoothie

Ask children to describe the steps for making a fruit smoothie like the one pictured on page 342. First, have them explain the parts of the mixture. Next, ask them to describe what would be easy to take apart before it's blended, and what would be difficult to separate after it's blended.

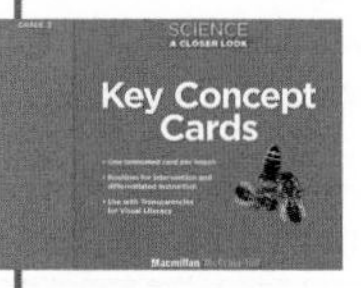


**Key Concept Cards** For student intervention, see the prescribed routine on **Key Concept Card 31.**

# Writing in Science

**Objective**

- Use explanatory writing to record a fruit salad recipe.

## Writing a Recipe

### Talk About It

Read the top paragraph and the recipe on page 346 together with children. Ask:

- **What is a recipe?** A set of step-by-step directions that tells how to make something.
- **Why is trail mix a mixture?** It is made of different things put together.

### Learn About It

Read the Remember box with children. As a class, write the steps in order for making trail mix.

As a demonstration, follow the steps as written by the class. **Be Careful!** Some children may be allergic to peanuts. If this is the case, substitute oat cereal or snack crackers for peanuts. Discuss the result of the demonstration with children. Ask:

- **How would you change the directions?** Possible answer: Add a step to stir the mixture.

Help children modify the directions so the steps are clear.

### Write About It

Read the writing task together. Remind children that recipes have a title, a list of ingredients, and the steps in order. Encourage children to read their recipe aloud to a partner to help them check their work.

**RESOURCES and TECHNOLOGY**

- **Reading and Writing,** pp. 196–197

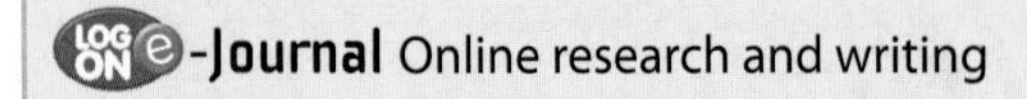
-Journal Online research and writing

---

**Writing in Science**

# Writing a Recipe

A recipe is a set of directions for making something. The steps are explained in order. A recipe can tell you how to make a mixture by adding things together.

Trail Mix Recipe

$\frac{1}{2}$ cup peanuts

$\frac{1}{2}$ cup raisins

$\frac{1}{2}$ cup chocolate chips

$\frac{1}{2}$ cup sunflower seeds

### Write About It

You can write a recipe! Explain how you would use some of the fruit here to make a fruit salad. Tell why it is a mixture.

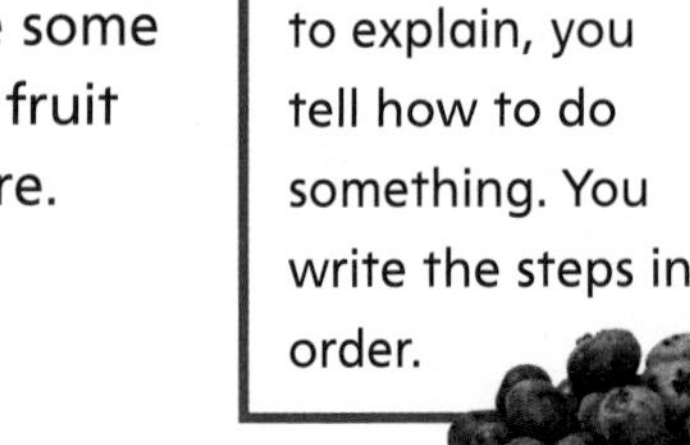

**Remember**

When you write to explain, you tell how to do something. You write the steps in order.

-Journal Write about it online at www.macmillanmh.com

346
EXTEND

## Integrate Writing

**Assembly Instructions**

To tell how to put something together, it is important that the steps are explained in order.

Have children write instructions for making something, such as a toy car made from blocks or a bracelet made from noodles.

Have children trade instructions with a partner and try to follow them. Encourage children to modify instructions as necessary.

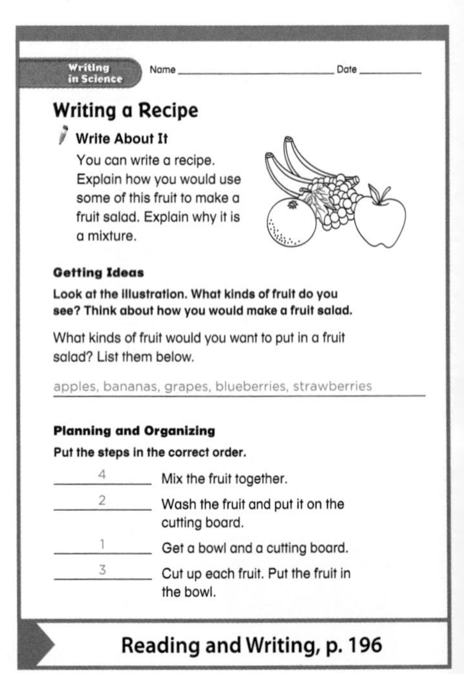

Writing in Science Name ______ Date ______

**Writing a Recipe**

**Write About It**

You can write a recipe. Explain how you would use some of this fruit to make a fruit salad. Explain why it is a mixture.

**Getting Ideas**

**Look at the illustration. What kinds of fruit do you see? Think about how you would make a fruit salad.**

What kinds of fruit would you want to put in a fruit salad? List them below.

apples, bananas, grapes, blueberries, strawberries

**Planning and Organizing**

**Put the steps in the correct order.**

4 Mix the fruit together.

2 Wash the fruit and put it on the cutting board.

1 Get a bowl and a cutting board.

3 Cut up each fruit. Put the fruit in the bowl.

**Reading and Writing, p. 196**

ath in Science

# uffin Math

Aaria and her dad are making muffins. They nt to know how much flour they need to buy.

lve a Problem

e recipe says they will need 2 cups lour to make 12 muffins.

ria and her dad want to make nuffins. How much flour will y need?

**Remember**
Read carefully to know what information you need. You can draw a sketch to help you solve the problem.

347
EXTEND

# Math in Science

## Objective

- Solve a problem about the quantity of ingredients in a recipe.

# Muffin Math

## Talk About It

Read the top text together with children. Ask them to share their baking experiences. Ask:

- **How do you know what you need to bake something?** Possible answer: The recipe tells you.
- **How do you know how much of each ingredient to use?** Possible answer: The recipe tells you how much of each ingredient to use.

## Learn About It

Read the bottom paragraph on page 347 with children. Give children a similar problem using smaller numbers. For example: *A recipe asks for 1 cup of sugar for 5 cookies, but we want to make 10 cookies. How much sugar will we need?*

Read the Remember box together. Discuss with children what information is needed to solve the problem and write it on the board. Ask:

- **How many groups of 5 muffins are there in 10 muffins?** 2
- **How does this information help you to know how many cups of sugar the class will need?** It lets us know that the class will need twice as much sugar to make 10 muffins, or 2 cups of sugar.

## Try It

Have children work in pairs to solve the problem on page 347. Encourage them to make a sketch to help them solve the problem. Have children make up other recipe problems for friends to solve.

## Integrate Math

### Mixed Nuts

Tell children you are making a mixture of nuts. Write the recipe on the board: *1 cup of hazelnuts, 3 cups of walnuts, 5 cups of pecans, 2 cups of almonds*

Ask:

- **How many cups of walnuts are required for this recipe?** 3
- **How many cups of nut mix would we have if we followed this recipe?** 11
- **How many cups of almonds would we buy if we halved the recipe?** 1

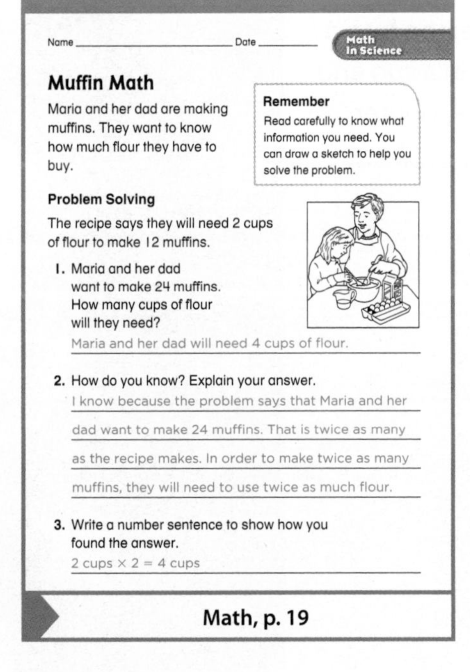

Name ____________ Date ________ Math in Science

**Muffin Math**

Maria and her dad are making muffins. They want to know how much flour they have to buy.

**Remember**
Read carefully to know what information you need. You can draw a sketch to help you solve the problem.

**Problem Solving**

The recipe says they will need 2 cups of flour to make 12 muffins.

1. Maria and her dad want to make 24 muffins. How many cups of flour will they need?
Maria and her dad will need 4 cups of flour.

2. How do you know? Explain your answer.
I know because the problem says that Maria and her dad want to make 24 muffins. That is twice as many as the recipe makes. In order to make twice as many muffins, they will need to use twice as much flour.

3. Write a number sentence to show how you found the answer.
2 cups × 2 = 4 cups

Math, p. 19

**RESOURCES and TECHNOLOGY**

- **Math,** pp. 19–20

# I Read to Review

**Objective**

- Review the concepts of physical and chemical changes with independent reading.

### Buddy Reading

- During small-group instructional time, have pairs of children read the selection to one another.
- Pair weaker readers with stronger partners to ensure success during this work period.
- Ask children to describe things that are similar and different in each picture.

### Independent Reading

- Give children copies of School to Home Activities, pages 101 and 102.
- Have them assemble the pages to make their own books.
- Encourage each child to take his or her book home and read it aloud to a family member.

PAGES **348–349** ▶

Ask: **What caused the chemical change to the muffin mixture?** cooking the muffins

PAGES **350–351** ▶

Ask: **Are the marshmallows part of the solution?** No, they are floating on top.

Ask: **How could you turn the juice pop into a liquid?** Put it in a cup and leave it on the counter.

I Read to Review

## How Things Change

All around me, matter changes. Cutting is a physical change. Even though there are more pieces, this is still an apple. Physical changes do not change what something is.

348

All around me, matter changes. I can make a mixture of chocolate powder and milk. The chocolate milk stays mixed. It is a solution.

350

**RESOURCES and TECHNOLOGY**

- **School to Home Activities**, pp. 101–102
- **Assessment**, pp. 130–131
- **TeacherWorks™ Plus CD-ROM**

All around me, matter changes. Batter cooks in the oven and becomes a muffin. This is a chemical change. The batter has new properties now.
349

All around me, matter changes. I can freeze juice. When something freezes, it changes to a solid.
351


# Performance Assessment

## Separate a Mixture

 individual  25 minutes

**Materials** jars, food coloring, sand, metal screws or other iron objects, water, vegetable oil

1. Show children a mixture of sand and screws. Ask them to draw and write what they would do to separate this mixture. Encourage them to think of a variety of tools that they could use to help them.
2. Show children a small jar of water in which food coloring has been added. Have volunteers shake it up. Children should draw what happens to the water and food coloring and explain in writing whether they can separate the solution ingredients.
3. Show children a jar in which water and vegetable oil are mixed together. Have volunteers shake the jar. Children should draw what happens and write what they would do to separate the mixture.

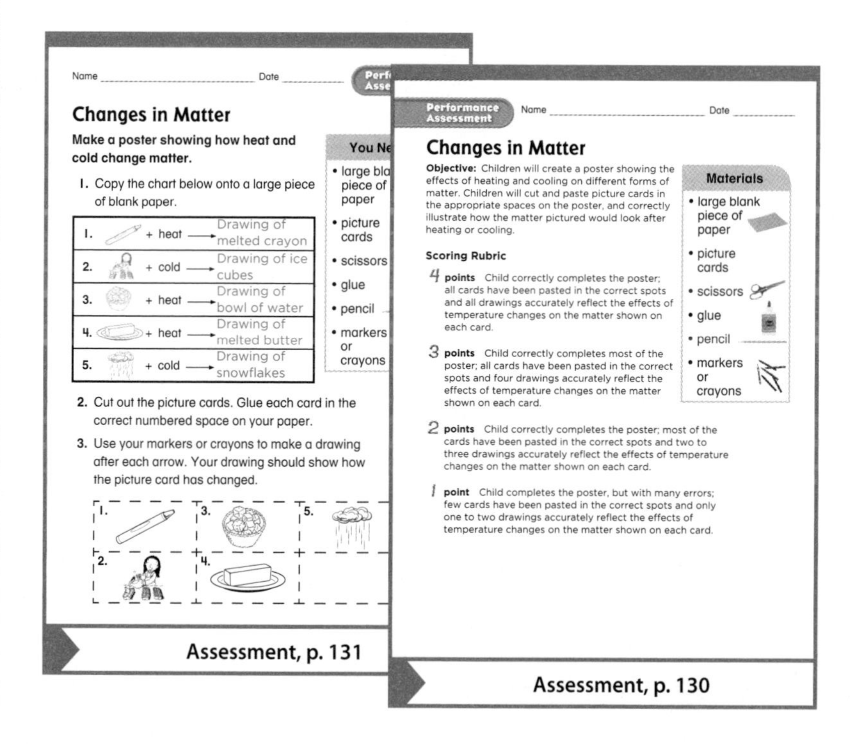

Name ______ Date ______

**Changes in Matter**

**Make a poster showing how heat and cold change matter.**

1. Copy the chart below onto a large piece of blank paper.

| | | |
|---|---|---|
| 1. | + heat → | Drawing of melted crayon |
| 2. | + cold → | Drawing of ice cubes |
| 3. | + heat → | Drawing of bowl of water |
| 4. | + heat → | Drawing of melted butter |
| 5. | + cold → | Drawing of snowflakes |

2. Cut out the picture cards. Glue each card in the correct numbered space on your paper.
3. Use your markers or crayons to make a drawing after each arrow. Your drawing should show how the picture card has changed.

You Ne…
- large bla… piece of paper
- picture cards
- scissors
- glue
- pencil
- markers or crayons

Assessment, p. 131

Performance Assessment Name ______ Date ______

**Changes in Matter**

**Objective:** Children will create a poster showing the effects of heating and cooling on different forms of matter. Children will cut and paste picture cards in the appropriate spaces on the poster, and correctly illustrate how the matter pictured would look after heating or cooling.

**Materials**
- large blank piece of paper
- picture cards
- scissors
- glue
- pencil
- markers or crayons

**Scoring Rubric**

4 **points** Child correctly completes the poster; all cards have been pasted in the correct spots and all drawings accurately reflect the effects of temperature changes on the matter shown on each card.

3 **points** Child correctly completes most of the poster; all cards have been pasted in the correct spots and four drawings accurately reflect the effects of temperature changes on the matter shown on each card.

2 **points** Child correctly completes the poster; most of the cards have been pasted in the correct spots and two to three drawings accurately reflect the effects of temperature changes on the matter shown on each card.

1 **point** Child completes the poster, but with many errors; few cards have been pasted in the correct spots and only one to two drawings accurately reflect the effects of temperature changes on the matter shown on each card.

Assessment, p. 130

# CHAPTER 10 Review

For information on Depth of Knowledge levels, see page 353B.

## ▶ Use the KWL Chart

Review the **KWL** chart that the class made at the beginning of the chapter. Help children compare what they know now about how matter changes with what they knew then. Add any additional information to the What We Learned column of the **KWL** chart.

## ▶ Make a FOLDABLES® Study Guide

Make a large three-tab project Foldable for the class. Divide the class into thirds.

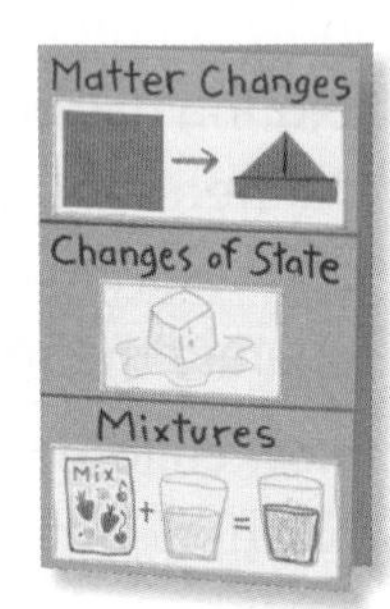


Give large index cards labeled *Physical Changes* and *Chemical Changes* to the Lesson 1 group, and ask them to define each term and give specific, illustrated examples of it.

Ask the Lesson 2 group to create two, two-column charts. One chart should have columns labeled *Matter* and *Heating* and the other should be labeled *Matter* and *Cooling*. Have children fill their charts in by listing and illustrating matter, then describing how it changes when heated or cooled.

Give large index cards labeled *Mixture* and *Solution* to the Lesson 3 group, and ask them to define each term and give specific, illustrated examples of it. Attach the cards under the appropriate tabs.

See page TR38 in the back of the Teacher's Edition for more information on Foldables.

## CHAPTER 10 Review

### Vocabulary
DOK 1

**Use each term once for items 1–6.**

| chemical change |
| --- |
| condense |
| evaporate |
| mixture |
| physical change |
| solution |

1. When wood burns, there is a ________.
2. Water in the air can ________ or change into a liquid.
3. Sugar and water form a mixture that will stay mixed. It is called a ________.
4. Fruit salad is a kind of ________.
5. Tearing paper is a ________.
6. After the snowman melts, the liquid water will turn into a gas, or ________.

352 
LOG ON e-Glossary Words and definitions online at www.macmillanmh...

### Vocabulary

1. chemical change
2. condense
3. solution
4. mixture
5. physical change
6. evaporate

### RESOURCES and TECHNOLOGY

- **Reading and Writing,** pp. 198–199
- **Assessment,** pp. 119–122, 126–129
- Exam*View*® *Assessment Suite* CD-ROM
- PuzzleMaker CD-ROM
-  Vocabulary Games

CHAPTER Vocabulary Name ________ Date ________

**Changes in Matter**

**Fill in the blanks. Use the words in the box.**

| chemical change | evaporation | melts |
| --- | --- | --- |
| condenses | freezes | solution |

1. When matter melts, it changes from a solid to a liquid.
2. A process called evaporation can be used to separate salt from water.
3. A solution is a mixture that is difficult to separate.
4. When matter condenses, it changes from a gas to a liquid.
5. When water freezes, it changes from a liquid to a solid.
6. When a slice of bread is toasted, a chemical change occurs.

**Reading and Writing, p. 198**

Name ________ Date ________ Chapter Test A

**Changes in Matter**

**Write the word or words that best complete each sentence in the spaces below. Words may be used only once.**

| boiling | evaporate | physical change |
| --- | --- | --- |
| chemical change | freeze | solution |
| condense | melting | |
| dissolve | mixture | |

1. A(n) mixture is made when two or more things come together.
2. Heated water that produces bubbles is boiling.
3. Liquids slowly turn to gas as they evaporate.
4. Liquids that freeze change into solids.
5. Making the size or shape of matter different causes a(n) physical change.
6. Solids that mix evenly in a liquid dissolve.
7. Matter that changes into different kind of matter goes through a chemical change.
8. A mixture that is hard to take apart is a(n) solution.
9. Cooling water vapor causes water to condense.
10. Solids change to liquids when they begin melting.

**Assessment, p. 119**

## Science Skills and Ideas

DOK 2

Answer the questions below.

Communicate. Which photo shows a physical change? Which shows a chemical change? What are some other examples of each kind of change?

Predict. What will happen if ice is heated at a high temperature for a long time?

Describe how a solution of sugar and water is different from a mixture of sand and water.

**10.** How can matter change?

e-Review Summaries and quizzes online at www.macmillanmh.com

353

## Science Skills and Ideas

**7. Communicate** Possible answers: The picture of the folded paper on the right shows a physical change. The picture of the rusted bicycle on the left shows a chemical change. Other examples of physical changes are bending clay or wire or melting ice. Other examples of chemical changes are cooking an egg or burning wood.

**8. Predict** Encourage children to complete a graphic organizer like the one shown below.

| What I Predict | What Happens |
|---|---|
| The ice will melt, turn to water, and then evaporate and become a gas. | The ice melts, turns to water, and then evaporates and becomes a gas. |

*Graphic Organizer 3,* p. TR5

**9.** Sugar mixes thoroughly with water and dissolves. This solution cannot be separated easily. Sand does not mix thoroughly with water; it can be easily separated.

**10.** Accept all reasonable answers. Children should address concepts taught in each lesson: physical and chemical changes of matter; how heating and cooling can change matter; and how matter can form mixtures and solutions.

## Summative Assessment and Intervention

**Assessment** provides a summative test for Chapter 10.

**Leveled Readers** may be used to reteach lesson content in an alternative format. Leveled readers deliver chapter content at different readability levels. The back cover of each reader provides comprehension building activities specific to the book content (see page 323).

**Key Concept Cards 29–31** contain prescribed routines for student intervention.

# Performance Assessment

## Matter Changes

**Materials** paper, pencils, crayons

### ▶ Teaching Tips

1. Review how matter can change with the class.
2. Have children fold their sheet of paper in half. Write the words *Physical Changes* and *Chemical Changes* on the board. Instruct children to write one title for each column.
3. Encourage children to draw at least three examples of each type of change.
4. Remind children to write a description for each one.

**Performance Assessment** DOK 3

## Matter Changes

What are ways matter changes?

▶ Fold a sheet of paper in half.

▶ On one side write the words *Physical Changes.* On the other side write the words *Chemical Changes.*

▶ Draw at least three examples of each type of change.

▶ Write a sentence beneath each of your pictures to explain the change you drew.

Physical Chang

Chemical Change

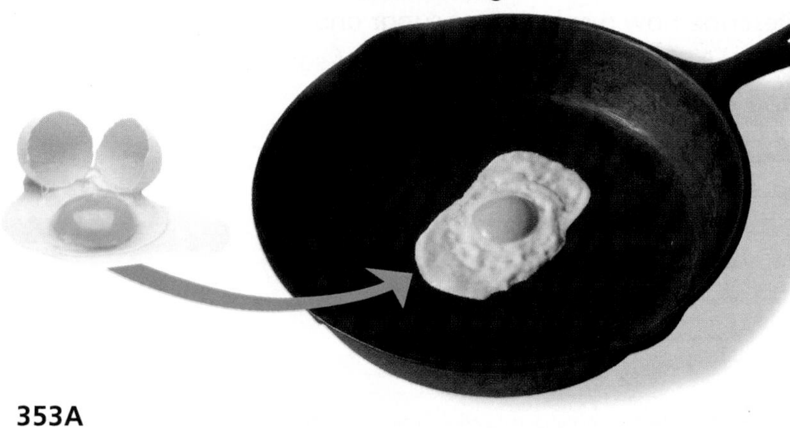

353A

## Scoring Rubric

**4 Points** Child accurately shows three examples of a physical change and three examples of a chemical change with appropriate descriptions.

**3 Points** Child correctly shows four changes (two physical and two chemical) with descriptions.

**2 Points** Child accurately shows one physical and one chemical change with descriptions.

**1 Point** Child draws one type of change with a description of the drawing.

t Preparation

Which of these can change matter into different matter?

**A** folding
**B** tearing
**C** bending
**D** burning
DOK 1

Look at the picture.

What is the first thing that will happen if this is left at room temperature?

**A** The water will evaporate.
**B** The ice will melt.
**C** The water vapor will condense.
**D** The water will freeze.
DOK 2

Which item is a solution?

**A** a fruit salad
**B** a chicken taco
**C** a peanut butter and jelly sandwich
**D** a milk shake
DOK 1

353B

1. D: burning. Folding, tearing, and bending are all physical changes. Only chemical changes, such as burning, can change matter into a different type of matter.
2. B: The ice will melt. Water must melt before it can evaporate or freeze. Water must turn into a gas before it can condense.
3. D: a milk shake. A solution is a type of mixture that is not easily taken apart. Fruit salad, chicken tacos, and a sandwich are all mixtures and can be taken apart.

## Depth of Knowledge

**Level 1 Recall** Level 1 requires memory of a fact, a definition, or a procedure. At this level, there is only one correct answer.

**Level 2 Skill/Concept** Level 2 requires an explanation or the ability to apply a skill. At this level, the answer reflects a deep understanding of the topic.

**Level 3 Strategic Reasoning** Level 3 requires the use of reasoning and analysis, including the use of evidence or supporting information. At this level, there may be more than one correct answer.

**Level 4 Extended Reasoning** Level 4 requires the completion of multiple steps and requires synthesis of information from multiple sources or disciplines. At this level, the answer demonstrates careful planning and complex reasoning.

# Careers in Science

### Objective

- Describe how food chemists use science to make food products.

## Food Chemist

**Genre: Nonfiction** Stories or books about real people and events.

Have children share what they know about food careers. Ask:

- **Do you know any people who work with food?**

Remind children that when they talk about real people and what they do, they are discussing nonfiction.

## Talk About It

Read the text with children. Ask:

- **Why do food chemists need to know about physical science?** Possible answer: They need to know how things change when they are heated, cooled, and mixed.

## Learn About It

Tell children that nutritionists help people choose healthy foods to eat. Explain that chefs study how food changes when it is mixed, heated, and cooled. Ask:

- **How are apples changed when making a pie?** Possible answer: Apples are cut up, mixed with sugar and cinnamon, and baked until soft.

## Write About It

Have children write about a food they would like to see improved and describe what would make it better.

Encourage children to present their improvements to the class.

**RESOURCES and TECHNOLOGY**

-Careers

# Careers in Science

## Food Chemist

Would you like to make your own cereal or flavor of juice? You could become a food chemist. Food chemists explore ways to make new and more delicious foods.

Food chemists learn how to make yogurt smooth. They find out how to keep cereal crunchy. They might find a way to freeze vegetables so they taste fresher. Food chemists have to understand the science of how food products are made.

food chemist

### More Careers to Think About

nutritionist

chef

-Careers at www.macmillanmh.com

## Integrate Writing

### A New Invention

Invite children to be a food chemist for a day. Ask:

- **If you could invent a new food, such as a new kind of ice cream treat, what would it be?**

Have children write about their food creations. Encourage children to describe how their new food would smell, taste, and feel.

Have children describe how the ingredients will change when they are mixed, heated, or cooled.

Encourage children to include unit vocabulary words, such as *solid, liquid, mixture, solution,* and *dissolve* in their writing.

# Motion and Energy

Roller coasters can go more than 160 kilometers per hour.

# UNIT F Materials

**Materials required to complete activities in the Student Edition are listed below. These charts also list materials found in the Deluxe Activity Kit.**

## Non-Consumable Materials

Based on 6 groups

| MATERIALS | QUANTITY NEEDED PER GROUP | KIT QUANTITY | CHAPTER/ LESSON |
|---|---|---|---|
| Book | several | | 11/2 |
| Cardboard | 1 sheet | | 11/2; 12/3 |
| Clock | 1 | | 12/1, 3 |
| Cloth, cotton (12″ x 20″) | small square | 1 | 12/4 |
| Cloth, felt, black | small square | 1 | 12/3 |
| Cloth, flannel (12″ x 20″) | small square | 1 | 12/4 |
| Cloth, white | small square | | 12/3 |
| Cube, wood | 1 | 6 | 11/2, 3 |
| Flashlight | 1 | 6 | 12/3 |
| Light socket, miniature | 1 | 12 | 12/4 |
| Magnet, bar | 1 | 12 | 11/4 |
| Magnet, donut | 1 | 6 | 11/4 |
| Magnet, horseshoe | 2 | 12 | 11/4 |
| Meter stick | 1 | | 11/1 |
| Pail | 1 | | 11/3 |
| Paper clip | 20 | | 11/4; 12/2 |
| Penny | 15 | | 11/3 |
| Prism | 1 | 6 | 12/3 |
| Rolling pin, roller style | 1 | | 11/3 |
| Rope | about 2 ft | | 11/3 |
| Ruler, plastic | 1 | | 11/1, 2, 3; 12/4 |
| Safety goggles | 1 | | 12/2 |
| Scissors | 1 | | 12/2, 4 |
| Stopwatch | 1 | 1 | 11/1, 2 |
| Thermometer | 3 | 18 | 12/1, 3 |
| Toy, car | 1 | 6 | 11/2 |
| Toy, windup | 3 | | 11/1 |
| Tuning fork | 1 | 3 | 12/2 |
| Wire cutter | 1 for teacher | 1 | 12/4 |

## Consumable Materials

Based on 6 groups

| MATERIALS | QUANTITY NEEDED PER GROUP | KIT QUANTITY | CHAPTER/ LESSON |
|---|---|---|---|
| Bag, paper | 1 | 6 | 11/4 |
| Bag, plastic, resealable 6″ x 8″ | 1 | 6 | 11/3 |
| Batteries, size D | 2 | 12 | 12/3, 4 |
| Cup, plastic, 9 oz | 3 | 50 | 12/1, 2 |
| Foil, aluminum | 12-in. sheet | 1 roll | 12/3 |
| Ice cube | 8 | | 12/1 |
| Lamp, miniature | 1 | 12 | 12/4 |
| Markers, assorted colors | several | | 11/3; 12/1 |
| Paper, construction | 1 sheet | | 12/3 |
| Paper, tissue | 1 sheet | | 12/3 |
| Paper, waxed | 12-in. sheet | | 12/3 |
| Pencils, colored | several | | 12/3 |
| Sandpaper | 1 sheet | 6 sheets | 11/2 |
| Soil, potting | 1 c | 2 (8-lb) bags | 12/1 |
| Sticky notes | 4 | | 11/4 |
| String, cotton | several in. | 200 ft | 11/3, 4; 12/2 |
| Tape, masking | small amounts | | 11/1, 2, 3; 12/1 |
| Wire, insulated | 2 (8 in.) | 1 spool | 12/4 |
| Wrap, plastic | 12-in. piece | | 12/3 |

The Big Idea **How do things move?**

Lesson 1 **Position and Motion**
**Essential Question** How can you tell that something has moved?

Lesson 2 **Forces**
**Essential Question** What do forces do?

Lesson 3 **Using Simple Machines**
**Essential Question** How can you use simple machines?

Lesson 4 **Exploring Magnets**
**Essential Question** What are magnets?

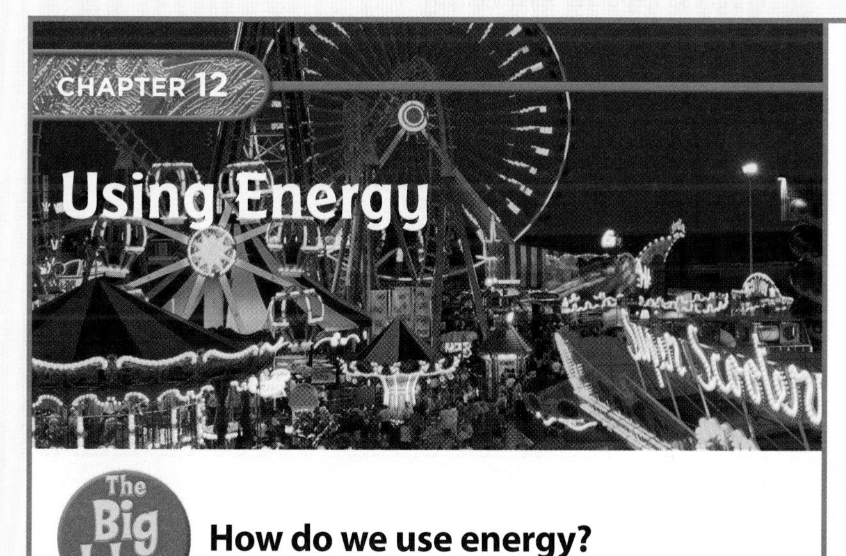


The Big Idea **How do we use energy?**

Lesson 1 **Heat**
**Essential Question** What can heat do to matter?

Lesson 2 **Sound**
**Essential Question** How do we hear sounds?

Lesson 3 **Light**
**Essential Question** How does light help us to see?

Lesson 4 **Exploring Electricity**
**Essential Question** How do we get electricity?

## CHAPTER 11
# How Things Move

### CD-ROM

**TeacherWorks™ Plus CD-ROM**
Interactive Lesson Planner, Teacher's Edition, Worksheets, and Links to Online Resources

**Classroom Presentation Toolkit CD-ROM**

**Exam*View*® *Assessment Suite* CD-ROM**

**PuzzleMaker CD-ROM**

**Science Songs CD** *Tracks 11, 12, 13*

 *Objects in Motion*

### DVD

**Science: Master Teacher DVD Set**

**Science Activity DVD**

### LOG ON www.macmillanmh.com

 **Science in Motion** *How a Ball Changes Directions*

**Online Teacher's Edition**

**Leveled Reader Database**

**Progress Reporter Assessments**

**Professional Development**

**e-Glossary**

**e-Journal**

**e-Review**

**NSDL** National Science Digital Library

## CHAPTER 12
# Using Energy

### CD-ROM

**TeacherWorks™ Plus CD-ROM**
Interactive Lesson Planner, Teacher's Edition, Worksheets, and Links to Online Resources

**Classroom Presentation Toolkit CD-ROM**

**Exam*View*® *Assessment Suite* CD-ROM**

**PuzzleMaker CD-ROM**

**Science Songs CD** *Track 14*

 *Sound*

### DVD

**Science: Master Teacher DVD Set**

**Science Activity DVD**

### LOG ON www.macmillanmh.com

 **Science in Motion** *How We Hear Sound*

**Online Teacher's Edition**

**Leveled Reader Database**

**Progress Reporter Assessments**

**Professional Development**

**e-Glossary**

**e-Journal**

**e-Review**

**NSDL** National Science Digital Library

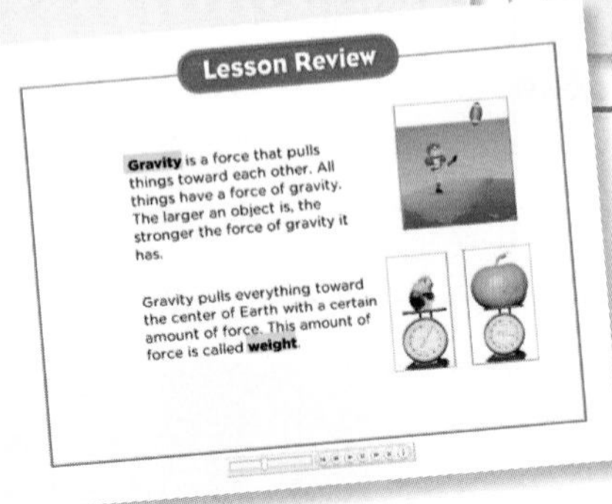


◀ Presentation Toolkit CD-ROM

## CHAPTER 11
## How Things Move

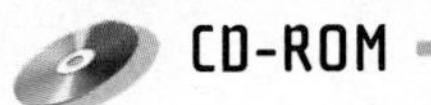 CD-ROM

StudentWorks™ Plus CD-ROM

PuzzleMaker CD-ROM

Science Songs CD *Tracks 11, 12, 13*

 *Objects in Motion*

 DVD

Science Activity DVD

 www.macmillanmh.com

 Science in Motion *How a Ball Changes Directions*

Online Student Edition

Online Vocabulary Games

e-Careers

e-Glossary

e-Journal

e-Review

## CHAPTER 12
## Using Energy

 CD-ROM

StudentWorks™ Plus CD-ROM

PuzzleMaker CD-ROM

Science Songs CD *Track 14*

 *Sound*

 DVD

Science Activity DVD

 www.macmillanmh.com

 Science in Motion *How We Hear Sound*

Online Student Edition

Online Vocabulary Games

e-Careers

e-Glossary

e-Journal

e-Review

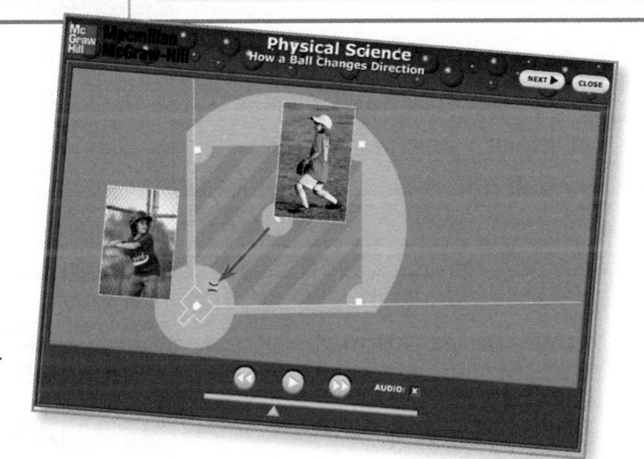


Science in Motion *How a Ball Changes Directions* ▶

# Literature

## Magazine Article

**Objective**

- Understand how bats use sound to locate food.

Literature
National Wildlife Federation
Ranger Rick
Magazine Article
Echolocation
Sounds are all around you, but can sound vibrations help you get dinner? If you are a bat, the answer is yes! Bats use echolocation to move around and catch food.
Bats live in caves. They can not see very well in the dark. They use sound to help them find their way and get food in the dark.
356


## Echolocation

**Genre: Nonfiction** Stories or books about real people and events.

Magazine articles about animals often give interesting information about an animal's behavior. Have children look at the pictures and ask:

- **How do the pictures help the words tell a story?**

Explain that science writers often use charts or illustrations to help the reader understand something new.

### Before Reading

Write the title on the board and underline *echo*. Ask:

- **What does *echo* mean?** Possible answer: a sound that repeats by itself

Explain to children that echolocation is something bats do, and that it has to do with sound. Encourage them to study the compound word for more clues, such as the word *location*. Ask:

- **What might this article tell us about bats?**

Record children's responses on the board.

### During Reading

Read the article paragraph by paragraph. At the end of each paragraph, stop and have one child retell what they learned in that paragraph. Make sure the children use their own language when retelling.

Review the words that relate to sound in the article, such as *high-pitched*, *squeak*, and *vibration*. Ask:

- **What part of the picture shows the vibrations?** the curved lines from the bat to the moth

**RESOURCES and TECHNOLOGY**

- **Reading and Writing,** p. 200
- **ELL Teacher's Guide,** p. 105

### ELL Support

**Visual Representation** Have children identify the bat and moth on page 357. Ask: **What do the blue lines represent? In which direction is the sound vibration moving?** Have them look at the picture and follow the sound vibration with their finger as the last paragraph of the article is read aloud. Help children use context clues for words they may not understand, such as *swoop*.

**BEGINNING** Have children use gestures to show where the sound vibration starts and in which direction it is moving.

**INTERMEDIATE** Ask children to use words and short sentences to answer questions about echolocation.

**ADVANCED** Encourage children to explain in their own words what they learned about echolocation.

from *Ranger Rick*

First, the bat makes a high-pitched squeak. The sound vibrations from the squeak move through the air and hit a moth. The vibrations bounce off the moth and come back to the bat. Then the bat swoops down and catches the moth for a meal.

**Talk About It**
How does sound vibration help animals get food?

357

## After Reading

Point out that the author began the article by talking about something all animals need—food. Ask:

- **Why do bats depend on sound to get food?** They do not see well in the dark.
- **What does the last line of the article have to do with the first line of the article?** Possible answer: The last line answers the question posed in the first line.

## Talk About It

Possible answers: After a bat makes a squeak, the sound vibrations bounce off other animals and return to the bat. The vibrations tell the bat where these animals are so that the bat can find them and eat them.

For children who are having difficulty with the concept of sound vibrations, review the pictures and text. Ask:

- **What color are the sound vibrations of the bat's squeak in the picture?** blue
- **What do the yellow sound vibrations coming from the moth show?** They show the bat's sound bouncing off the moth and returning to the bat.

### More to Read

***Bat Loves the Night,*** by Nicola Davies (Scholastic, 2001).

## Differentiated Instruction

### Leveled Activities

**EXTRA SUPPORT** Have children make a comic strip showing a bat using echolocation. Before they begin writing and drawing their comic strip, ask children to identify the sequence of events described on pages 356 and 357 and write them in a numbered list. Remind children to use either speaking bubbles or labels in each frame of the comic strip.

**ENRICHMENT** Ask children to research other animals that use echolocation, such as whales or dolphins. Invite children to write articles explaining how those animals find food. Encourage them to note how the animal differs from bats by pointing out items such as the animal's habitat, the type of food it eats, and they way it moves. Suggest that children use the article they just read as a model for their own article.

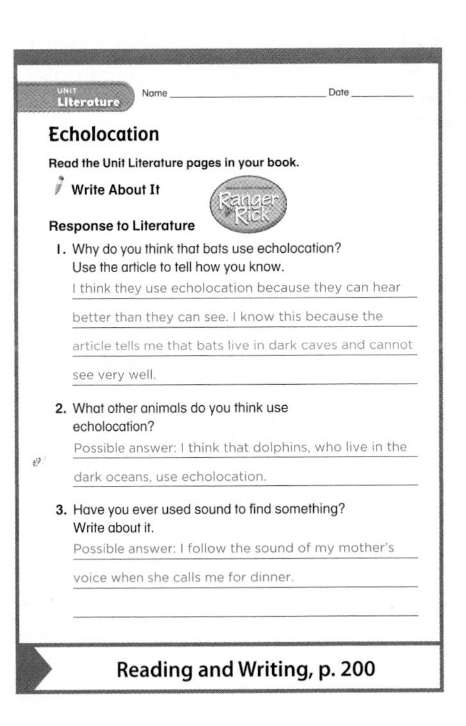

UNIT Literature Name ______ Date ______

**Echolocation**

**Read the Unit Literature pages in your book.**

**Write About It**



**Response to Literature**

1. Why do you think that bats use echolocation? Use the article to tell how you know.
I think they use echolocation because they can hear better than they can see. I know this because the article tells me that bats live in dark caves and cannot see very well.

2. What other animals do you think use echolocation?
Possible answer: I think that dolphins, who live in the dark oceans, use echolocation.

3. Have you ever used sound to find something? Write about it.
Possible answer: I follow the sound of my mother's voice when she calls me for dinner.

**Reading and Writing, p. 200**

# CHAPTER 11 Planner

**Classroom Presentation Toolkit CD-ROM** Lesson Presentations

**TeacherWorks™ Plus CD-ROM** Interactive Lesson Planner, Teacher's Edition, Worksheets, and Online Resources.

| Lesson | OBJECTIVES AND READING SKILLS | VOCABULARY | RESOURCES AND TECHNOLOGY |
|---|---|---|---|
| **1 Position and Motion** PAGES 360–365<br>PACING: 3 days<br>*FAST TRACK:* 1 day | ▪ Describe an object's position in relation to another object.<br>▪ Measure and record changes in an object's position.<br>**Reading Skill** Sequence<br>First → Next → Last<br>*Graphic Organizer 7* | **position**<br>**motion**<br>**speed** | Reading and Writing, pp. 202–205<br>Activity Lab Book, pp. 161–166<br>Visual Literacy, p. 32<br>Transparencies, p. 32<br>**Operation: Science Quest** *Object in Motion*<br>**Science Songs CD** Track 11 |
| **2 Forces** PAGES 366–375<br>PACING: 3 days<br>*FAST TRACK:* 1 day | ▪ Identify a force as a push or a pull.<br>▪ Describe the forces of gravity and friction.<br>**Reading Skill** Cause and Effect<br>Cause → Effect<br>*Graphic Organizer 8* | **force**<br>**gravity**<br>**friction** | Reading and Writing, pp. 206–211<br>Activity Lab Book, pp. 167–170<br>Visual Literacy, p. 33<br>Transparencies, p. 33<br>**Operation: Science Quest** *Object in Motion*<br>**Science Songs CD** Track 12<br>**Science in Motion** *How a Ball Changes Directions* |
| **3 Using Simple Machines** PAGES 376–383<br>PACING: 2 days<br>*FAST TRACK:* 1 day | ▪ Identify simple machines.<br>▪ Discover that simple machines change force to make work easier.<br>**Reading Skill** Summarize<br>Summary<br>*Graphic Organizer 5* | **simple machine**<br>**lever**<br>**fulcrum**<br>**ramp** | Reading and Writing, pp. 212–217<br>Math, pp. 21–22<br>Activity Lab Book, pp. 171–174<br>Visual Literacy, p. 34<br>Transparencies, p. 34<br>**Science Songs CD** Track 13 |
| **4 Exploring Magnets** PAGES 384–389<br>PACING: 2 days<br>*FAST TRACK:* 1 day | ▪ Observe magnets attract and repel objects.<br>▪ Identify magnet poles and explain how they function.<br>**Reading Skill** Problem and Solution<br>Problem → Steps to Solution → Solution<br>*Graphic Organizer 12* | **attract**<br>**poles**<br>**repel** | Reading and Writing, pp. 218–221<br>Activity Lab Book, pp. 175–180<br>Visual Literacy, p. 35<br>Transparencies, p. 35 |

| | | |
|---|---|---|
| **I Read to Review** PAGES 390–393 | **Forces Every Day**<br>▪ Selection for independent reading<br>**Performance Assessment** | **Resources**<br>School to Home Activities, pp. 113–114<br>Assessment, pp. 144–145 |
| **CHAPTER 11 Review** PAGES 394–395, 395A–395B | ▪ Review chapter concepts.<br>**Resources**<br>Assessment, pp. 132–135, 140–143<br>Reading and Writing, pp. 222–223 | **Technology**<br>**ExamView® *Assessment Suite* CD-ROM**<br>**-Review** |

PACING Assumes a day is a 20–25 minute session.

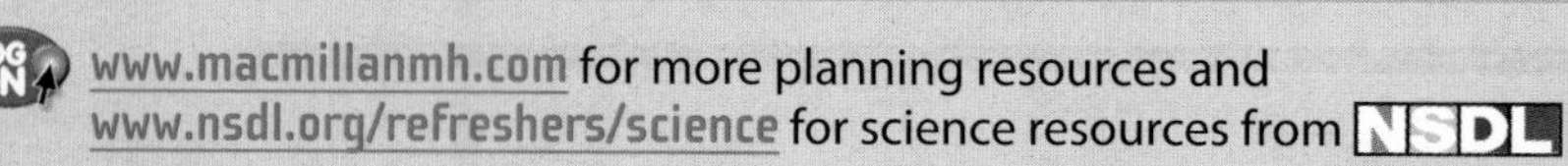

www.macmillanmh.com for more planning resources and
www.nsdl.org/refreshers/science for science resources from NSDL

# Activity Planner

Science Activity DVD Explore Activity demos

Materials included in the Deluxe Activity Kit are listed in *italics*.

## EXPLORE Activities

### Explore *p. 361* PACING: 20 minutes 

**Objective** Describe the position of objects.

**Skills** communicate, draw conclusions

**Materials** classroom objects, paper, pencils

PLAN AHEAD Arrange objects in different locations around the classroom.

### Explore *p. 367* PACING: 20 minutes 

**Objective** Explain how the force applied to an object determines speed.

**Skills** measure, predict

**Materials** *toy cars*, masking tape, rulers

PLAN AHEAD Set up one or more starting lines ahead of time. Group materials for each child ahead of time for distribution.

### Explore *p. 377* PACING: 25 minutes 

**Objective** Observe how a simple machine can make lifting easier.

**Skills** predict

**Materials** markers, masking tape, rulers, fifteen pennies per group, small plastic bags, pencil, paper

PLAN AHEAD Place the supplies needed for each group into plastic bags.

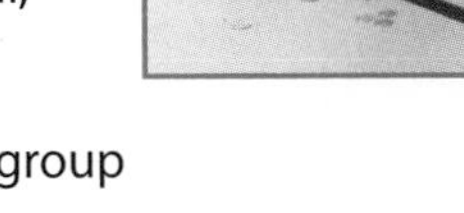

### Explore *p. 385* PACING: 25 minutes 

**Objective** Identify magnetic objects from nonmagnetic objects.

**Skills** predict, classify

**Materials** small objects, paper bags, *string*, pencils, *magnets*

PLAN AHEAD Gather enough small magnetic and nonmagnetic objects for each group.

## QUICK LAB Activities

### Quick Lab *p. 364* PACING: 15 minutes 

**Objective** Compare speeds of walking and hopping.

**Skills** measure, record data, compare

**Materials** *stopwatches*, masking tape, meter sticks, an open space

PLAN AHEAD Clear enough desks to conduct this experiment, or obtain permission to do this activity outdoors.

| | Walk | Hop |
|---|---|---|
| Jane | 7 seconds | 5 seconds |
| Roberto | 7seconds | 4 seconds |

### Quick Lab *p. 371*  PACING: 15 minutes 

**Objective** Discover how friction affects a moving object.

**Skills** record, compare, draw conclusions

**Materials** *stopwatches*, *wood cubes* or other objects with smooth surfaces, cardboard, *sheets of sand paper*, books

PLAN AHEAD Make a ramp ahead of time to show as a model.

### Quick Lab *p. 380* PACING: 15 minutes

**Objective** Investigate how to make and use a pulley.

**Skills** compare

**Materials** roller-style rolling pins, rope, small pails, *wood cubes*

PLAN AHEAD Find pails with strong handles to conduct the experiment.

### Quick Lab *p. 388* 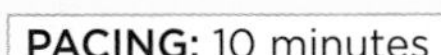 PACING: 10 minutes 

**Objective** Identify the north and south poles of a magnet.

**Skills** draw conclusions

**Materials** *bar magnets*, sticky notes

PLAN AHEAD Review with children how unlike poles attract and like poles repel.

## FOR MORE ACTIVITIES

| Focus on Skills | Be a Scientist | Learning Lab | Everyday Science |
|---|---|---|---|
| Teach the inquiry skill: investigate, p. 50 | Compare the strength of different magnets, p. 54 | Test a magnet's strength, pp. 72–73 | Spin an Egg, p. 65; Gravity, p. 66; Magnets, p. 67 |

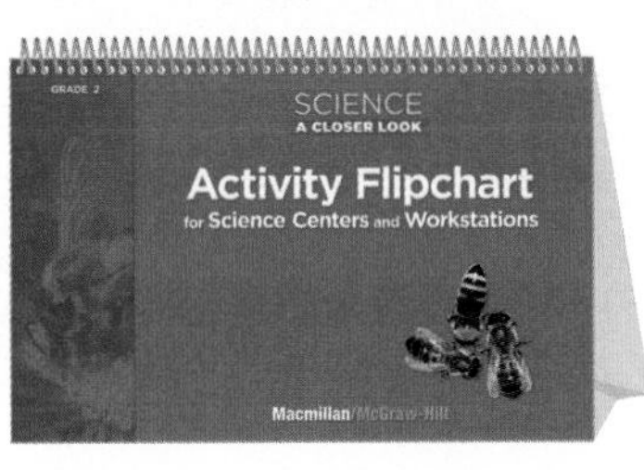


Use the Activities in your work station. See **Activity Lab Book** for more support.

For a comprehensive list of consumable and non-consumable materials, see the back of the unit tab.

# English Language Learner Support

## Technology

For additional language support and vocabulary development, go to www.macmillanmh.com

Science in Motion
*How a Ball Changes Directions*

Vocabulary Games

## ADDITIONAL RESOURCES

pp. 106–115

## Academic Language

English language learners need help in building their understanding of the academic language used in daily instruction and science activities. The following strategies will help to increase children's language proficiency and comprehension of content and instruction words.

### Strategies to Reinforce Academic Language

- **Use Context** Academic language should be explained in the context of the task. Use gestures, expressions, and visuals to support meaning.
- **Use Visuals** Use charts, transparencies, and graphic organizers to explain key labels to help children understand classroom language.
- **Model** Use academic language as you demonstrate the task to help children understand instruction.

### Academic Language Vocabulary Chart

The following chart shows chapter vocabulary and inquiry skills as well as some Spanish cognates. **Vocabulary** words help children comprehend the main ideas. **Inquiry Skills** help children develop questions and perform investigations. **Cognates** are words that are similar in English and Spanish.

| Vocabulary | Inquiry Skills | Cognates | |
|---|---|---|---|
| | | **English** | **Spanish** |
| position, p. 362 | communicate, p. 360 | communicate, p. 360 | *comunicar* |
| motion, p. 363 | draw conclusions, p. 360 | conclusion, p. 360 | *conclusión* |
| speed, p. 364 | measure, p. 367 | position, p. 362 | *posición* |
| force, p. 369 | predict, p. 367 | predict, p. 367 | *predecir* |
| gravity, p. 370 | classify, p. 385 | gravity, p. 370 | *gravedad* |
| friction, p. 371 | | friction, p. 371 | *fricción* |
| simple machine, p. 378 | | simple, p. 378 | *simple* |
| lever, p. 378 | | fulcrum, p . 378 | *fulcro* |
| fulcrum, p. 378 | | ramp, p. 379 | *rampa* |
| ramp, p. 379 | | classify, p. 385 | *clasificar* |
| attract, p. 386 | | attract, p. 386 | *atraer* |
| poles, p. 388 | | poles, p. 388 | *polos* |
| repel, p. 388 | | repel, p. 388 | *repeler* |

## Vocabulary Routine

**Use the routine below to discuss the meaning of each word on the vocabulary list. Use gestures and visuals to model all words.**

**Define** A *lever* is a bar that balances on a point and moves like a seesaw.

**Example** An oar is a *lever*.

**Ask** What is another kind of *lever*?

**Children may respond to questions according to proficiency level with gestures, one-word answers, or phrases.**

### ELL Leveled Reader

**Push or Pull?**
by Debra Lucas

**Summary** People, animals, and machines use push and pull every day.

**Reading Skill**
Compare and Contrast

## Vocabulary Activities

Help children understand how a lever works.

**BEGINNING** Write the words *lever*, *bar*, and *fulcrum*. Use a book, a ruler, and a board eraser to demonstrate how a *lever* works. Ask children to identify the part of the lever that is the *bar*, and which part is the *fulcrum,* or point that does not move. Have children take turns pushing down on one end of the ruler to lift up the book at the other end.

**INTERMEDIATE** Form groups and ask each group to use personal or classroom objects of their choice to build a lever. Have the groups test their levers and use gestures and short sentences to explain how they work.

**ADVANCED** Form groups and give each group a plastic spoon, fork, and knife. Have the groups discuss which utensils are levers and which are not. Then have the groups explain why the spoon and fork are levers but the knife is not.

## Language Transfers

**Grammar Transfer**
Adjectives can stand alone in Spanish, so the pronoun "one" is not used.

*A fork and a shovel are levers.*
*I use the first to eat.*

**Phonics Transfer**
Cantonese and Korean do not have the sound /v/ in *lever*.

CHAPTER 11

# How Things Move

**THE BIG IDEA** How do things move?

**Chapter Preview** Have children take a picture walk and predict what the lessons will be about.

## ▶ Assess Prior Knowledge

Before reading the chapter, create a **KWL** chart with children. Ask the Big Idea question, and then ask:

- How do you know something has moved?
- What causes objects to move?
- How do simple machines help things move?
- How does a magnet move things?

| Motion | | |
|---|---|---|
| What We **K**now | What We **W**ant to Know | What We **L**earned |
| Someone or some thing cause objects to move. | Can objects move by themselves? | |
| Simple machines help us lift things. | What are some simple machines? | |
| Magnets pull some things to them. | | |

Answers shown represent sample student responses.

Follow the **Instructional Plan** at right after assessing children's prior knowledge of chapter content.

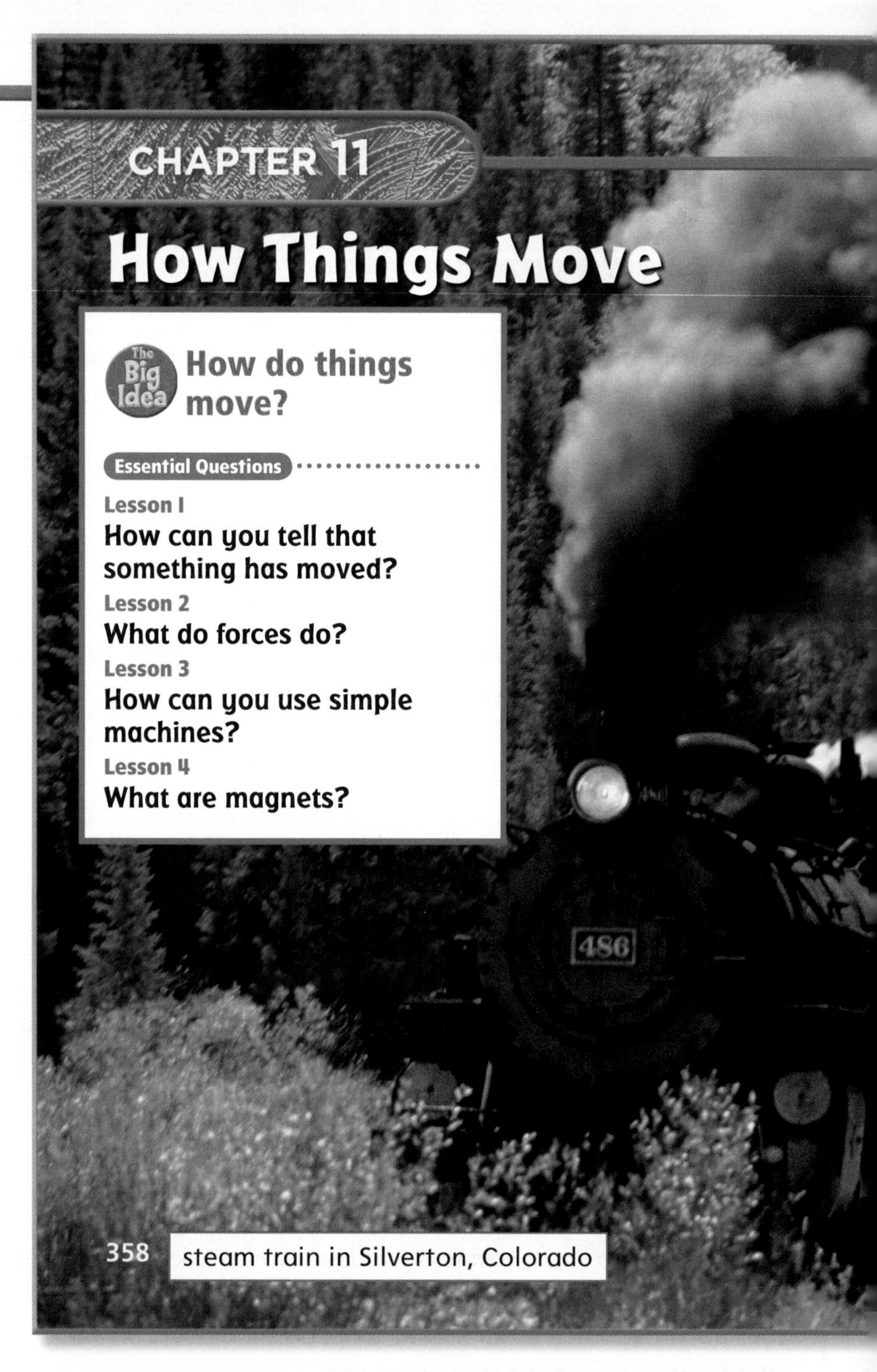

CHAPTER 11
How Things Move
How do things move?
Essential Questions
Lesson 1
How can you tell that something has moved?
Lesson 2
What do forces do?
Lesson 3
How can you use simple machines?
Lesson 4
What are magnets?
486

358 steam train in Silverton, Colorado

### RESOURCES and TECHNOLOGY

- **School to Home Activities,** pp. 103–114
- **Reading and Writing,** pp. 201–223
- **Assessment,** pp. 132–145
- Classroom Presentation Toolkit CD-ROM
- PuzzleMaker CD-ROM

 www.macmillanmh.com

 e-Journal

## Differentiated Instruction

### Instructional Plan

**Chapter Concept** Forces change the way objects move.

**EXTRA SUPPORT** Children who need to describe the motion of objects should cover all of **Lesson 1,** pages 360–365, before continuing with the rest of the chapter.

**ON LEVEL** Children who can describe the motion of objects should explore the topic of speed, **Lesson 1,** pages 364–365, and then can go directly to **Lesson 2,** pages 366–375, to focus on the Chapter 11 concept.

**ENRICHMENT** Children who are ready to enrich their understanding of the Chapter 11 concept may investigate forces with levers and ramps, **Lesson 3,** pages 376–383, and with magnets, **Lesson 4,** pages 384–389.

## Big Idea Vocabulary

**motion** a change in the position of an object (page 363)

**friction** a force that slows down moving things (page 371)

**lever** a simple machine made of a bar that turns around a point (page 378)

**repel** to push away or apart (page 388)

Visit www.macmillanmh.com for online resources.

359

## Big Idea Vocabulary

- Have a volunteer read the **Big Idea Vocabulary** words aloud to the class. Ask children to find one or two of the words in the chapter by using the given page references. Add these words and their definitions to a class Word Wall.
- Encourage children to use the illustrated glossary in the Student Edition's reference section. Guide children to explore the e-**Glossary**, which offers audio pronunciations, definitions, and sentences using the vocabulary words.

## Science Leveled Readers

ALSO ON AUDIO CD
Leveled Reader Library

APPROACHING

**Get Moving!** Look at simple tools and machines that use the forces of push and pull.

**ISBN:** 978-0-02-284645-9

ON LEVEL

**Push or Pull?** People, animals, and machines use push and pull every day.

**ISBN:** 978-0-02-284646-6

BEYOND

**All About Magnets** Learn all about the special features and forces of magnets.

**ISBN:** 978-0-02-284651-0

ELL

**Push or Pull?** Uses sheltered language of On-Level Reader.

**ISBN:** 978-0-02-283463-0

See teaching strategies in the Leveled Reader Teacher's Guide. To order, call 1-800-442-9685.

 **Leveled Reader Database** Online Readers, searchable by topic, reading level, and keywords

Stop Here to

# Plan Your Lesson

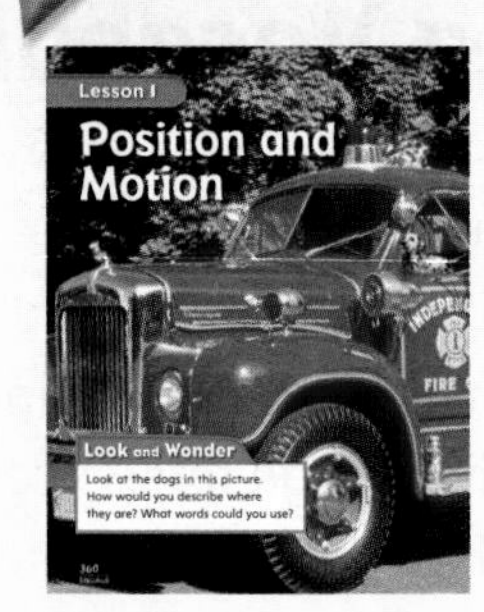


## Lesson 1 Position and Motion

### Essential Question

How can you tell that something has moved?

### Objectives

- Describe an object's position in relation to another object.
- Measure and record changes in an object's position.

**Reading Skill** Sequence

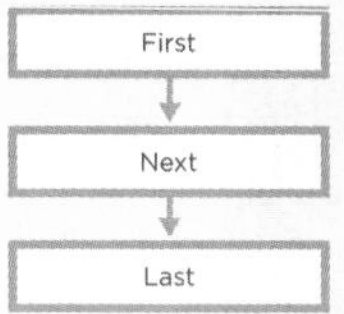


*Graphic Organizer 7*, p. TR9

## FAST TRACK

**Lesson Plan** When time is short, follow the Fast Track and use the essential resources.

**1 Introduce**
**Look and Wonder, p. 360**
Resource **Activity Lab Book, p. 163**

**2 Teach**
**Discuss the Main Idea, p. 362**
Resource **Visual Literacy, p. 32**

**3 Close**
**Think, Talk, and Write, p. 365**
Resource **Assessment, p. 136**

**Professional Development** Look for NSDL to find recommended Science Background resources from the National Science Digital Library.

## Reading and Writing

## Visual Literacy

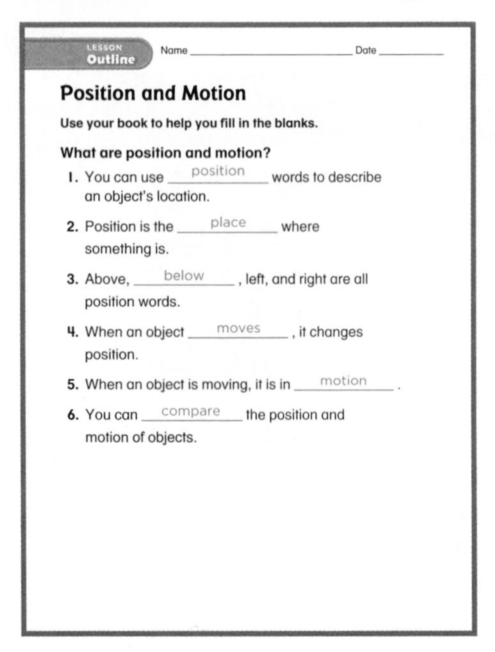

Lesson Outline Name ____________ Date ________

**Position and Motion**

**Use your book to help you fill in the blanks.**

**What are position and motion?**

1. You can use ___position___ words to describe an object's location.
2. Position is the ___place___ where something is.
3. Above, ___below___, left, and right are all position words.
4. When an object ___moves___, it changes position.
5. When an object is moving, it is in ___motion___.
6. You can ___compare___ the position and motion of objects.

**Outline, pp. 202–203**
Also available as a student workbook

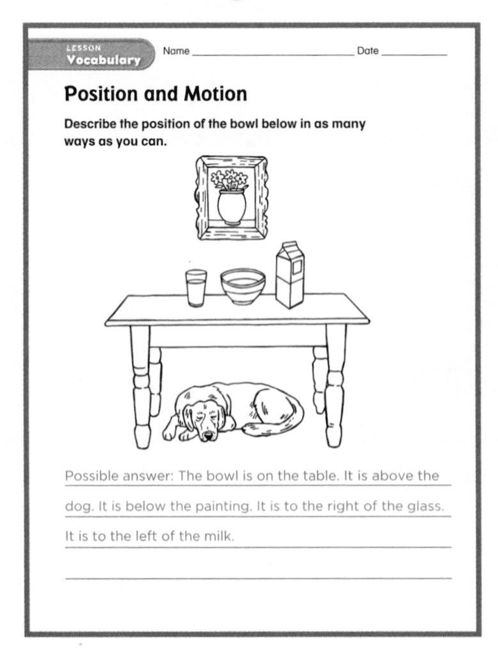

Lesson Vocabulary Name ____________ Date ________

**Position and Motion**

**Describe the position of the bowl below in as many ways as you can.**

Possible answer: The bowl is on the table. It is above the dog. It is below the painting. It is to the right of the glass. It is to the left of the milk.

**Vocabulary, p. 204**
Also available as a student workbook

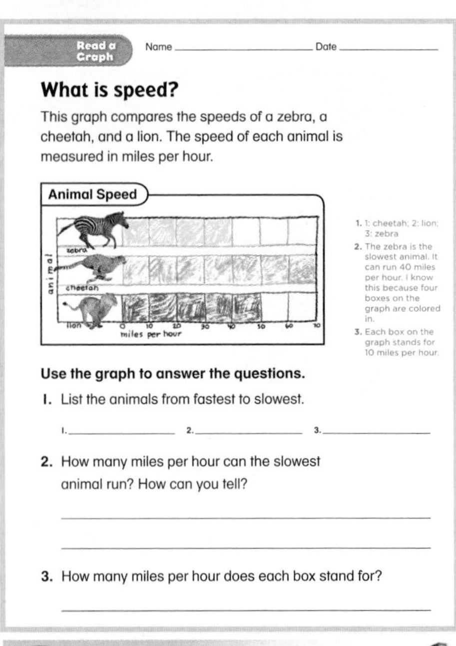

Read a Graph Name ____________ Date ________

**What is speed?**

This graph compares the speeds of a zebra, a cheetah, and a lion. The speed of each animal is measured in miles per hour.



1. 1: cheetah; 2: lion; 3: zebra
2. The zebra is the slowest animal. It can run 40 miles per hour. I know this because four boxes on the graph are colored in.
3. Each box on the graph stands for 10 miles per hour.

**Use the graph to answer the questions.**

1. List the animals from fastest to slowest.
   1. ________ 2. ________ 3. ________
2. How many miles per hour can the slowest animal run? How can you tell?
3. How many miles per hour does each box stand for?

**Read a Graph, p. 32**
Also available as a transparency

## Activity Lab Book

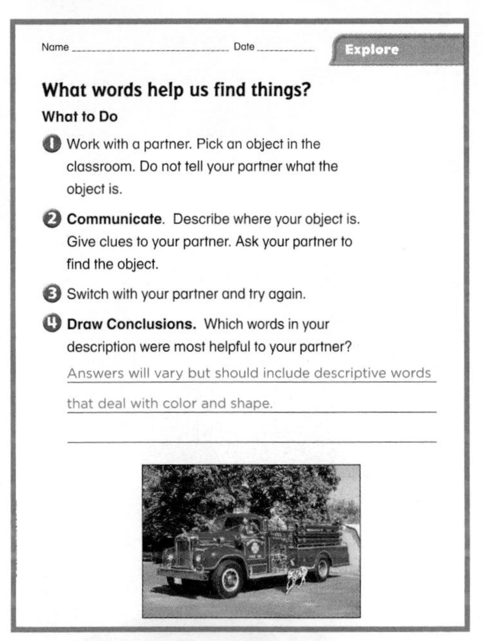

Name ______ Date ______ **Explore**

**What words help us find things?**

**What to Do**

1. Work with a partner. Pick an object in the classroom. Do not tell your partner what the object is.
2. **Communicate.** Describe where your object is. Give clues to your partner. Ask your partner to find the object.
3. Switch with your partner and try again.
4. **Draw Conclusions.** Which words in your description were most helpful to your partner?
   Answers will vary but should include descriptive words that deal with color and shape.

**Explore, pp. 161–162**
Also available as a student workbook

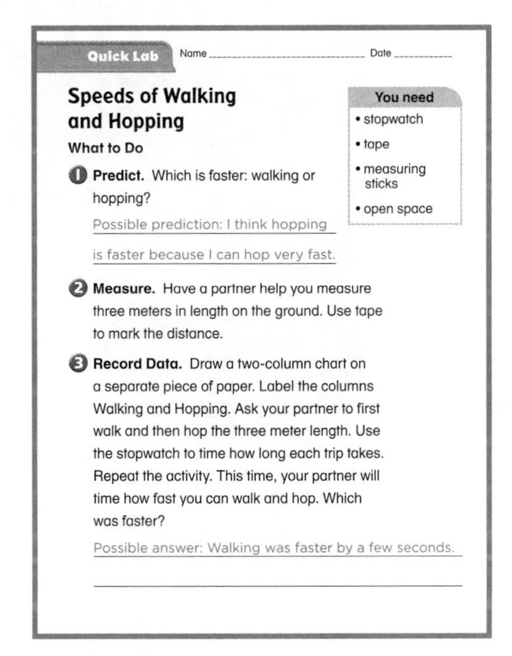

**Quick Lab** Name ______ Date ______

**Speeds of Walking and Hopping**

**You need**
- stopwatch
- tape
- measuring sticks
- open space

**What to Do**

1. **Predict.** Which is faster: walking or hopping?
   Possible prediction: I think hopping is faster because I can hop very fast.
2. **Measure.** Have a partner help you measure three meters in length on the ground. Use tape to mark the distance.
3. **Record Data.** Draw a two-column chart on a separate piece of paper. Label the columns Walking and Hopping. Ask your partner to first walk and then hop the three meter length. Use the stopwatch to time how long each trip takes. Repeat the activity. This time, your partner will time how fast you can walk and hop. Which was faster?
   Possible answer: Walking was faster by a few seconds.

**Quick Lab, p. 164**
Also available as a student workbook

## Assessment

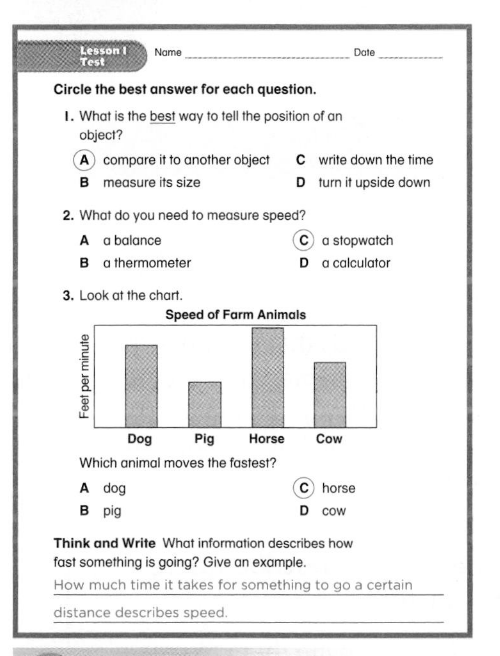

**Lesson 1 Test** Name ______ Date ______

**Circle the best answer for each question.**

1. What is the best way to tell the position of an object?
   - (A) compare it to another object
   - B measure its size
   - C write down the time
   - D turn it upside down
2. What do you need to measure speed?
   - A a balance
   - B a thermometer
   - (C) a stopwatch
   - D a calculator
3. Look at the chart.

   **Speed of Farm Animals** (Feet per minute: Dog, Pig, Horse, Cow)

   Which animal moves the fastest?
   - A dog
   - B pig
   - (C) horse
   - D cow

**Think and Write** What information describes how fast something is going? Give an example.
How much time it takes for something to go a certain distance describes speed.

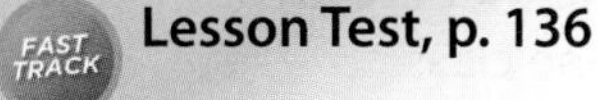

**Lesson Test, p. 136**

# ADDITIONAL RESOURCES

**pp. 158–161**

***Push or Pull?***

**pp. 49–50**

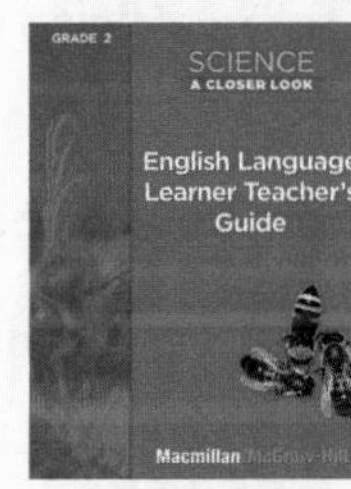


**p. 110**

**p. 32**

**32**

**101–110**

**pp. 105–106**

**81–83**

### Technology

- **Science Activity DVD**
- **TeacherWorks™ Plus CD-ROM**
- **Classroom Presentation Toolkit CD-ROM**
- SCIENCE QUEST *Objects in Motion*
- **Science Songs CD** Track 11

**-Review**

# Lesson 1 Position and Motion

### Objectives

- Describe an object's position in relation to another object.
- Measure and record changes in an object's position.

## 1 Introduce

### ▶ Assess Prior Knowledge

Have children share what they know about position and motion. Ask:

- **Why is it important to be able to describe where something is located?**
- **How can you tell when something has moved?**
- **How do you know when something moves faster than something else?**

Record children's answers in the What We Know column of the class **KWL** chart.

### Look and Wonder

Read the Look and Wonder statement and questions about position and motion.

Invite children to share their responses to the questions. Ask:

- **Which dog is at the highest point in the picture?** Have children point to the photo.
- **Find the dog that is moving. How would you describe the way the dog is moving?** Possible answer: The dog on the ground seems like it is about to run.

Write children's responses on the class **KWL** chart and note any misconceptions that they may have.

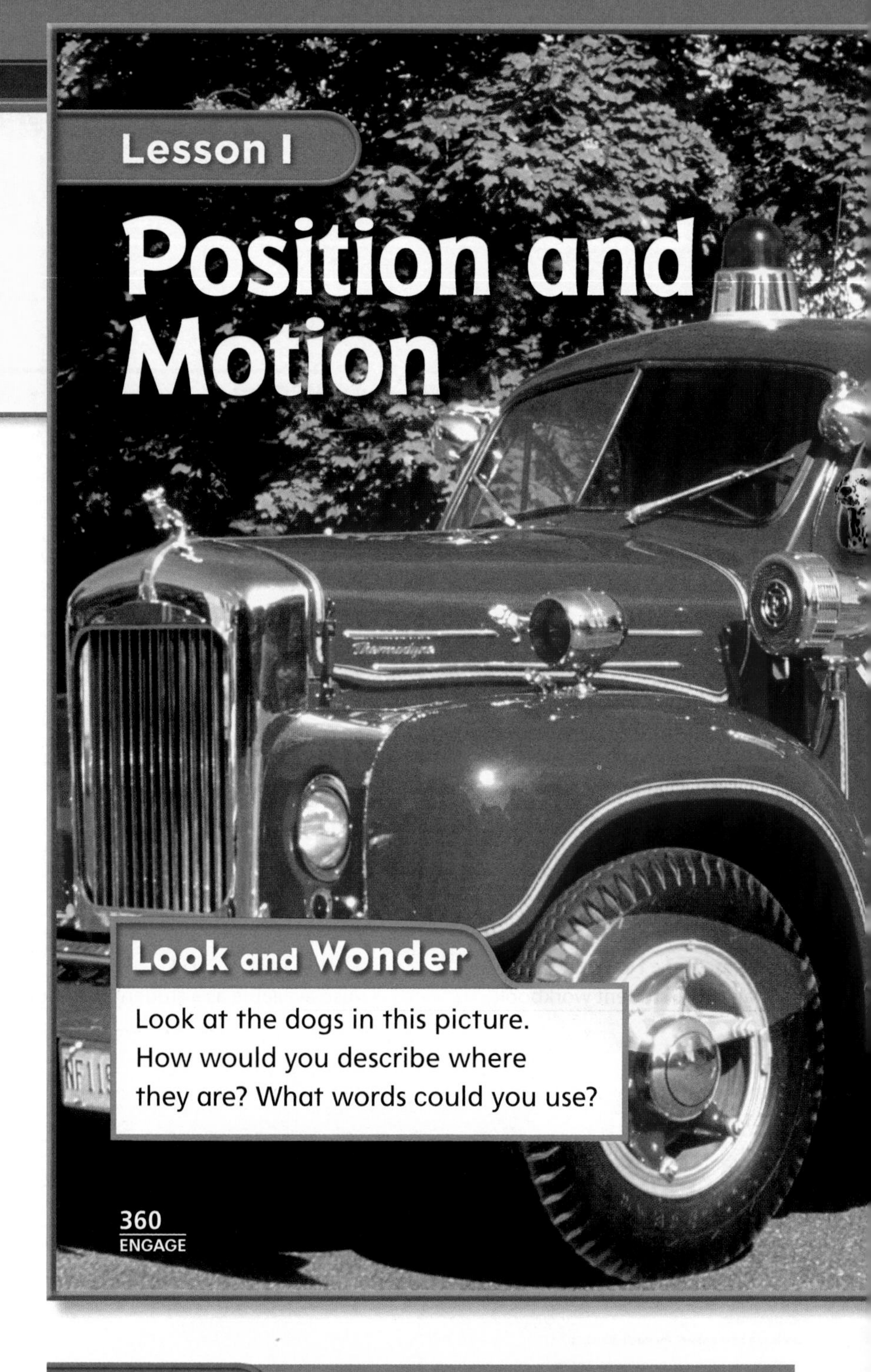

Lesson 1
Position and Motion
Look and Wonder
Look at the dogs in this picture. How would you describe where they are? What words could you use?
360
ENGAGE


### Warm Up

**Start with a Book**

Introduce a book from the ***I Spy*** series, such as ***I Spy: A Balloon,*** by Jean Marzollow and Walter Wick (Scholastic, 2006).

Before reading, generate a class list of words that can be used to describe the position of an object. Tell children that they can use words from the list to tell where things are in the photographs in the book. Read the story and have children use position words to answer the riddles.

After reading, ask:

- **What other position words helped you describe where something was located in the pictures?**

Add children's responses to the class list. Invite them to create their own pictures and riddles to share with the class.

**RESOURCES and TECHNOLOGY**

- **Activity Lab Book,** pp. 161–163
- **Activity Flipchart,** p. 49
- Science Activity DVD

## Explore — Inquiry Activity

### What words help us find things?

**What to Do**

1. Work with a partner. Pick an object in the classroom. Do not tell your partner what the object is.
2. **Communicate.** Describe where your object is. Give clues to your partner. Ask your partner to find the object.
3. Switch with your partner and try again.
4. **Draw Conclusions.** Which words in your description were most helpful to your partner?

**Explore More**

5. **Communicate.** Draw a picture and write directions to find an object in your picture. Then switch with a partner.

361
EXPLORE

## Alternative Explore

### Can you get there?

Have children work in pairs. Ask each child to choose a secret spot and write it on a piece of paper.

Have children write a three-step set of directions to help their partner locate the secret spot.

After they write, ask children to discuss whether the written directions were helpful. If the directions were not helpful, ask children to work together to develop another set of directions.

Name ______________ Date ________ Alternative Explore

**Can you get there?**

In this activity, you will create and follow directions that lead to a mystery place.

You need
- pencil

**What to Do**

1. Choose a mystery place for a partner to find. Write three directions to tell how to move to the place. Use position words.
   Possible answer: First, walk to the front of the teacher's desk. Next, turn left. Finally, walk straight ahead.
2. Read the clues to your partner. Ask your partner to follow the directions.
3. After your partner finds the mystery place, the activity can be repeated. This time your partner can choose the mystery place.
4. Which words helped you to find the place?
   Possible answer: to the right of, in front of, under the table, on top of the chair

Activity Lab Book, p. 163

## Explore

pairs · 20 minutes

**Plan Ahead** Arrange objects in different locations in the classroom.

**Purpose** This activity will help children develop vocabulary necessary to describe position. They will use their observational skills to draw conclusions about where things are positioned.

**Structured Inquiry** What to Do

Create a class list of position words with children. Model the activity by describing the location of an unnamed object in the classroom. Use words from the class list and other position words to demonstrate different ways a location can be described. Have children guess the object.

1. After children select an object, encourage them to look at the things that are surrounding the object in the classroom. Tell them that these items can help to describe the location of the object.
2. **Communicate** Encourage children to use a variety of words to describe the position of their object to their partner. Remind them to refer to the class list of position words for ideas.
3. After children have switched roles once, repeat the activity, and encourage them to use different words each time.
4. **Draw Conclusions** Help children use words, such as: *top, above, bottom, next to, under,* or *below.*

**Guided Inquiry** Explore More

5. **Communicate** Discuss situations in which a person might need to give or understand directions. Ask: **What words are most helpful to direct someone to find an object?**

**Open Inquiry**

Have children talk with a partner about other reasons it is important to be able to explain the position of something. Ask them what people do to find out how to get to a place. Encourage them to think of questions they have about materials people use to get where they need to go. Provide print and non-print resources to help them find answers to some of their questions.

# 2 Teach

## Read Together and Learn

**Reading Skill** Sequence The order in which things happen.

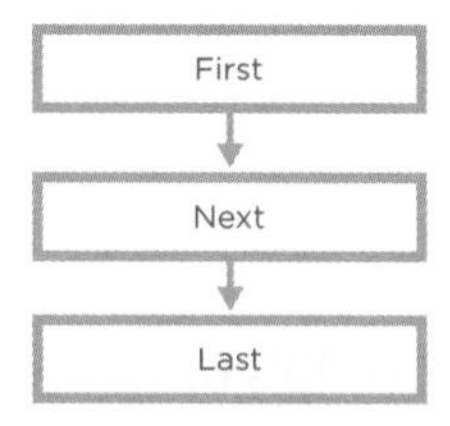


*Graphic Organizer 7,* p.TR9

## What are position and motion?

### Discuss the Main Idea

FAST TRACK

**Main Idea** The position of an object can be described by using words that relate it to another nearby object.

After reading together, ask:

- **How would you describe the position of the train in the photo?** Possible answers: to the right of the dog on the stool; behind the cat in the wagon
- **How can you describe the position of something that has moved?** compare it to other objects
- **What words would you use to describe the position of the race car?** Possible answers: under, on, between, next to, beside

### Read Together and Learn

**Essential Question**

How can you tell that something has moved?

**Vocabulary**

position
motion
speed

### What are position and motion?

**Position** is the place where something is. You can tell the position of an object by comparing it to the positions of other objects.

You can use words such as above or below to describe where things are. You can also use the words in, on, under, next to, near, left, or right.

**Where are the dog and the cat? What sentences can you make about their positions?**

362
EXPLAIN

### Science Background

**Position** An object's position is described by comparing its location to the position of one or more other objects. Many descriptions of position rely on words, such as *in, over, between,* and *above* that describe relationships between objects. The relationship between two objects can also be described in terms of direction and distance.

See **Science Yellow Pages**, in the Teacher Resources section, for background information.

**Professional Development** For more Science Background and resources from NSDL visit http://nsdl.org/refreshers/science

### ELL Support

**Act Out** Demonstrate a simple motion, such as jumping, clapping, or hopping, so that children can identify, draw, and compare the different movements that are acted out.

**BEGINNING** Invite children to guess the movement made. When someone guesses correctly, have everyone act out the movement and identify the movement out loud.

**INTERMEDIATE** Have children draw a picture of a movement and label it. Invite children to act out the movements illustrated by other children.

**ADVANCED** Ask a volunteer to act out a movement. Ask children to identify and imitate the movement. Repeat the activity with another volunteer. Ask children to describe how the two movements are alike and different.

When something moves, its
›sition changes. **Motion** is a change
the position of an object. Some
ays objects move are up, down,
ound, sideways, or zigzag. You can
scribe an object's motion by telling
›w its position changed.

How can you describe where
an object is and how it moves?

jump up

leap over

zag

around and through

363
EXPLAIN

## Use the Visuals

Look at the pictures on pages 362 and 363 and read the captions. Ask:

- **What words could you use to direct a friend to find the toy train on page 362?** Possible answers: behind the cat in the wagon; to the right of the rug
- **What do you see in the pictures of the dogs that help describe their motions?** Possible answer: The dogs' motions are related to the objects.

## Develop Vocabulary

**position** ***Word Origin*** Explain that the word *position* comes from the Latin word *positio*, which means "to place." Write *position* on the board and underline *positio*. Explain that *position* is "a place that is occupied by a person or thing," or "a location." Have children use *position* in a sentence to describe where an object is in the classroom.

**motion** ***Scientific vs. Common Use*** In common use *motion* can be used as an idiom, for example: *She went through the motions, but her heart was not in it.* Explain to children that *motion* used in the sentence means "a lack of interest." In science, *motion* means "a change in position or place." Ask children to give examples of *motions* they do every day, such as climbing stairs, turning the pages in a book, and drinking from a glass. Play a game in which a volunteer says *motion* and children move and *no motion* and children are still.

## Differentiated Instruction

### Leveled Activities

**EXTRA SUPPORT** Have children play a variation of the game Mother, May I? Substitute a classmate's name for *Mother*. Vary the commands to include motions such as jumping, tiptoeing, and hopping.

**ENRICHMENT** Ask children to draw a basic map of the classroom. Next, on a separate piece of paper, have children write descriptions of where objects, such as a ball and a book, could be placed. For example: *Put a red book in the center of the teacher's desk.* Invite children to exchange maps and draw the objects in the positions described by their classmates.

## Quick Check Answer

Possible answers: You can describe where an object is by noticing what other objects are nearby and using direction words, such as *left, right, center, top, bottom, in front of,* or *in back of*. To show how an object moves use words like *jump, leap, zigzag,* and *around*.

### RESOURCES and TECHNOLOGY

- **Reading and Writing,** pp. 202–204
- **PuzzleMaker CD-ROM**
- **Classroom Presentation Toolkit CD-ROM**
- SCIENCE QUEST *Objects in Motion*
- **Science Songs CD** Track 11
- LOG ON **e-Glossary**

## What is speed?

### ▶ Discuss the Main Idea

**Main Idea** Speed can be determined by measuring how far a thing moves in a certain amount of time.

Read the blue question at the top of page 364 and discuss with children the speeds of different things. Ask:

- **How would you describe the speed of walking compared to running?** Possible answer: Walking is slower than running.

**Read a Graph**

Have children look at the bar graph on page 365 and compare the speeds of the animals. Ask:

- **How fast can a zebra run?** 40 mph
- **How can you compare the speed of each animal?** Possible answer: Measure how fast each goes and subtract the slower speed from the faster speed.

**Answer to Read a Graph** cheetah

### ▶ Develop Vocabulary

**speed** ***Scientific vs. Common Use*** *Speed* in common usage describes a person's skills or character. For example: *I am not up to speed with the rest of the team.* In science, *speed* is a specific characteristic that can be measured in units, such as miles per hour or meters per minute. Have children make a two-column chart with the labels: *Animals with Fast Speeds* and *Animals with Slow Speeds.* Ask children to write sentences to compare the speeds of a fast animal and a slow animal.

**RESOURCES and TECHNOLOGY**

- **School to Home Activities,** pp. 105–106
- **Reading and Writing,** p. 205
- **Activity Lab Book,** p. 164
- **Visual Literacy,** p. 32
- SCIENCE QUEST *Objects in Motion*
- LOG ON e**-Review** Narrated Summary and Quiz
- **Exam*View*® *Assessment Suite* CD-ROM**

### What is speed?

Some things, such as snails, move slowly. Others, such as cheetahs, move quickly. **Speed** is how far something moves in a certain amount of time.

**Quick Lab**

Measure three meter in the classroom. Wa and then hop the distance. Record the time it took for each.

A cheetah's speed can be measured with a stopwatch and a tape measure. The fastest objects move farthest in a certain amount of time.

364
EXPLAIN

**Quick Lab**

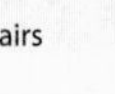
pairs

15 minutes

**Objective** Compare speeds of walking and hopping.

**You need** stopwatches, masking tape, meter sticks, open space

1. Have children measure three meters in length and mark it with tape on the ground.
2. Ask one child to walk the length as their partner uses the stopwatch to clock the time. Have children record data on a two-column chart with the labels: *Walk* and *Hop.* Ask the same child to hop the distance and record the speed. Repeat the activity, allowing the pairs to switch roles.
3. Invite children to compare their walking and hopping speeds.

| | Walk | Hop |
|---|---|---|
| Jane | 7 seconds | 5 seconds |
| Roberto | 7seconds | 4 seconds |

**Animal Speed**

zebra
cheetah
lion
0 10 20 30 40 50 60 70
miles per hour

**Read a Graph**

Which animal is the fastest?

What are some objects that move at high speeds?

## Think, Talk, and Write

1. **Sequence.** Describe the order of the objects you would pass as you move from your desk to your teacher's desk.
2. What words can you use to describe motion?
3. **Essential Question.** How can you tell that something has moved?

**Social Studies Link**

Make a map of your classroom. Draw yourself on the map and describe your position.

**e-Review** Summaries and quizzes online at www.macmillanmh.com

365
EVALUATE

## Formative Assessment

**Travelogue**

Have children draw three pictures that show the sequence of their trip to school. Under each drawing, have children write a sentence to describe the pictured position and motion.

I start my trip when I step into the bus.
The bus is moving fast.
SCHOOL
I walk quickly to school.

**Key Concept Cards** For student intervention, see the prescribed routine on **Key Concept Card 32.**

### *Quick Check Answer*

Possible answers: jet plane; rocket; cheetah

# 3 Close

### Using the KWL Chart

Review with children what they have learned about position and motion. Record their responses in the What We Learned column of the class **KWL** chart.

### Using the Reading Skill Sequence

Have children use the reading skill graphic organizer to describe the sequence of motions made when singing "The Itsy, Bitsy Spider."

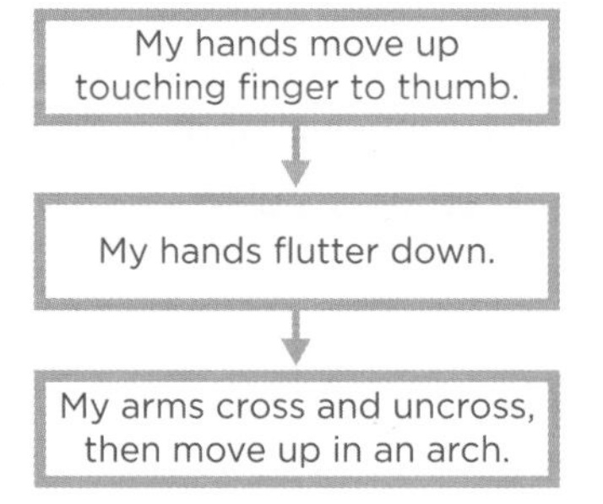

My hands move up touching finger to thumb.
My hands flutter down.
My arms cross and uncross, then move up in an arch.


*Graphic Organizer 7,* p. TR9

### FAST TRACK Think, Talk, and Write

1. **Sequence** Answers will vary, depending on the child's location.
2. Possible answers: straight, zigzag, in a circle
3. **Essential Question** Possible answer: You can look at its position now and its position in the past. If the object has moved, its position will be different.

## Social Studies Link

Encourage children to include details on their maps. They should indicate their position on the map and include pictures of objects used to describe their position. After children have related their position to different things in the room, invite them to move to a different location and describe their position from a new viewpoint.

## Focus on Skills

Activity Flipchart, p. 50

### Objective

- Investigate the speeds of different objects.

**You need** masking tape, rulers, windup toys, stopwatches

**EXTEND** As children investigate the time it takes for toys to travel a measured distance, they will understand that the process of investigating requires a plan.

### Inquiry Skill: Investigate

### ▶ Learn It

Read the Learn It section on page 50 together. Explain to children that investigating is making a plan in order to test an idea.

Ask:

- **What do Joe and Pat want to investigate?** which boy is the faster runner
- **What is their plan for the investigation?** Possible answer: They measure where they will run, mark it, and measure the time it takes for each boy to run.

Point out that a chart is a good way to organize information in an investigation because it records the information in a way that makes it easy to read and compare.

## Focus on Skills

### Inquiry Skill: Investigate

When you **investigate**, you make a plan and test it out.

### ▶ Learn It

Joe and Pat will run in a race. They want to find their speeds. They make a plan.

First, they measure 20 meters. They make a start and a finish line. Next, they measure the time it took them to run the distance.

Look at the chart. Who is faster?

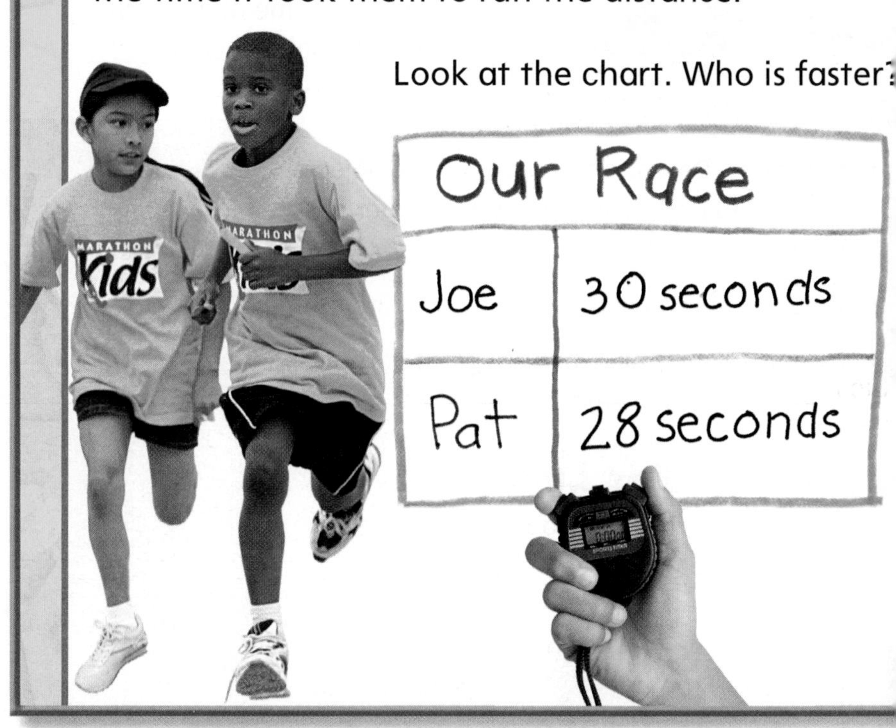


## Integrate Math

**Investigating Speed and Distance**

Tell children that a farmer held a race to investigate the speeds of his favorite chicken and beloved pig. Charles the Chicken strutted 2 miles in 30 minutes. Paddy the Pig waddled 1 mile in the same amount of time. Ask:

- **If the animals continued to move at the same speeds, how far would Charles strut in 60 minutes? How far would Paddy waddle in 60 minutes?** Charles would strut 4 miles, and Paddy would waddle 2 miles.
- **What is your plan to find the answers?** Possible answer: Look at total miles per 30 minutes for Charles and Paddy and double it.

## Skill Builder

### Try It

Which toy moves fastest? Make a
n to find out. Then test your plan.

Put tape on the floor to make a start line. Measure how far away your finish line will be. Mark it with tape.

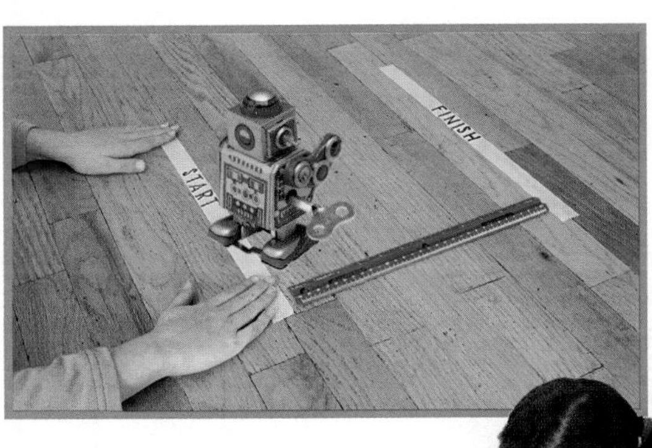


Use a stopwatch to find out how long it took each toy to go the distance. Record the times.

Which toy was fastest?

**You need**

masking tape

ruler

windup toys

stopwatch

Use with Chapter 11, Lesson 1 50

## ▶ Try It

Read the Try It section on page 50 with the class.

1. Demonstrate with children how to use a ruler to measure how far the finishing line will be. Explain why it is important for the starting line and finishing line to be parallel to one another.
2. Model how to use the stopwatch for children. Have them practice using stopwatches before doing the experiment. Ask children to predict the time it will take for each toy to get to the finish line.
3. Have children make a chart like the one shown in the Learn It section to compare the times it took the toys to go the distance. Ask: **Why was the time it took for each toy to move the distance the same or different?**

## ▶ Apply It

Have children work in small groups to investigate the speed of each child in the group in a skipping race. Ask children to make a plan for the race.

Go to an open area of the school yard or gym and ask children to mark off a 20 meter distance with masking tape. Have children take turns to participate in the race, use the stopwatch to time the race, and record the speeds.

Encourage children to test other movements, such as hopping and jumping. If there is a sloped area in the schoolyard, children can test skipping up that incline.

Have children use a chart similar to the one shown on page 50 to record their results.

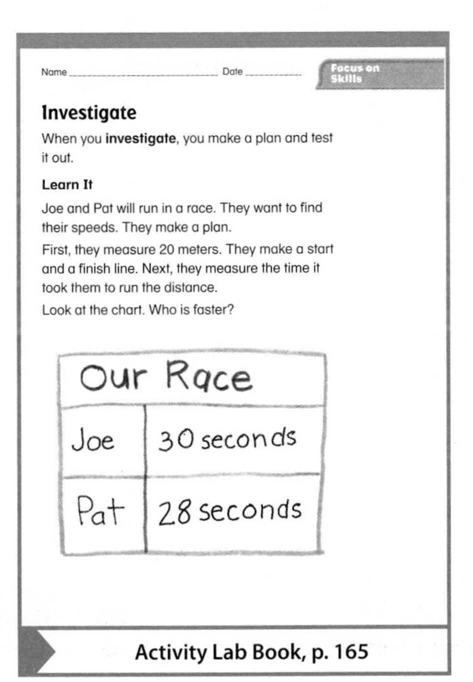

Name ________________ Date ________ Focus on Skills

**Investigate**

When you **investigate**, you make a plan and test it out.

**Learn It**

Joe and Pat will run in a race. They want to find their speeds. They make a plan.

First, they measure 20 meters. They make a start and a finish line. Next, they measure the time it took them to run the distance.

Look at the chart. Who is faster?

| Our Race | |
|---|---|
| Joe | 30 seconds |
| Pat | 28 seconds |

Activity Lab Book, p. 165

### RESOURCES and TECHNOLOGY

- **Activity Lab Book**, pp. 165–166
- **TeacherWorks™ Plus CD-ROM**

# Plan Your Lesson

## Lesson 2 Forces

### Essential Question

What do forces do?

### Objectives

- Identify a force as a push or a pull.
- Describe the forces of gravity and friction.

**Reading Skill** Cause and Effect

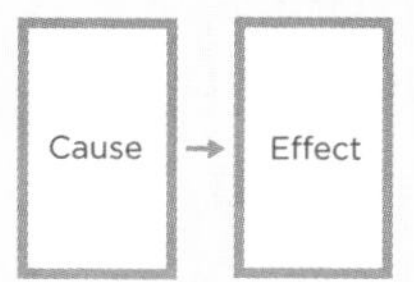


*Graphic Organizer 8*, p. TR10

## FAST TRACK

**Lesson Plan** When time is short, follow the Fast Track and use the essential resources.

**1 Introduce**

**Look and Wonder, p. 366**

Resource **Activity Lab Book, p. 169**

**2 Teach**

**Discuss the Main Idea, p. 368**

**Use the Visuals, p. 370**

Resource **Visual Literacy, p. 33**

**3 Close**

**Think, Talk, and Write, p. 373**

Resource **Assessment, p. 137**

**Professional Development** Look for NSDL to find recommended Science Background resources from the National Science Digital Library.

## Reading and Writing

LESSON Outline Name ______ Date ______

**Forces**

**Use your book to help you fill in the blanks.**

**What makes things move?**

1. It takes a push or a pull to make something move.
2. A push or pull is a force.
3. To push something, you move it away from you.
4. To pull something, you move it toward you.

**What are some forces?**

5. When you throw a ball in the air, gravity pulls it back to Earth.
6. Gravity is a force that pulls things to Earth.
7. One measure of gravity is weight.
8. Weight is how much force it takes to pull something to Earth.

**Outline, pp. 206–207**

Also available as a student workbook

LESSON Vocabulary Name ______ Date ______

**Forces**

**Answer each riddle. Then find each word in the word search.**

1. I am a force that slows down moving things. What am I? friction
2. I am a force that pulls things to Earth. What am I? gravity
3. To put an object in motion, you must use me. What am I? force
4. I am the amount of force Earth pulls on an object. What am I? weight
5. To move an object closer to you, you must use me. What am I? pull

| | | | | | | | | |
|---|---|---|---|---|---|---|---|---|
| f | r | i | c | t | i | o | n | d | w |
| o | l | m | s | h | i | e | h | g | e |
| r | g | r | a | v | i | t | y | c | i |
| c | a | t | v | m | p | s | t | u | g |
| e | m | n | x | y | r | l | m | e | h |
| p | u | l | l | n | z | c | b | o | t |

**Vocabulary, p. 208**

Also available as a student workbook

## Visual Literacy

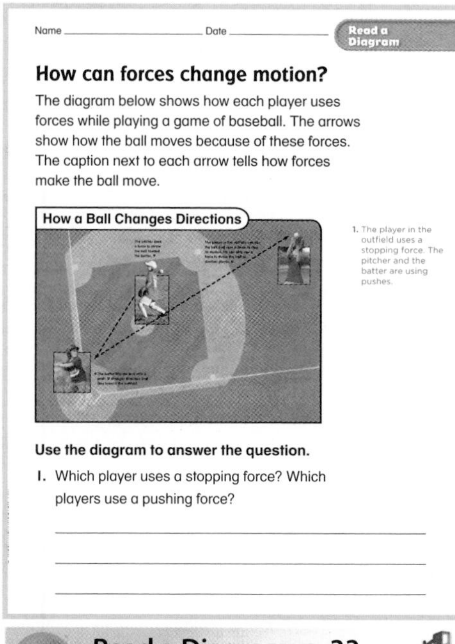
Name ______ Date ______ Read a Diagram

**How can forces change motion?**

The diagram below shows how each player uses forces while playing a game of baseball. The arrows show how the ball moves because of these forces. The caption next to each arrow tells how forces make the ball move.

**How a Ball Changes Directions**

1. The player in the outfield uses a stopping force. The pitcher and the batter are using pushes.

**Use the diagram to answer the question.**

1. Which player uses a stopping force? Which players use a pushing force?

**Read a Diagram, p. 33**

Also available as a transparency

## Activity Lab Book

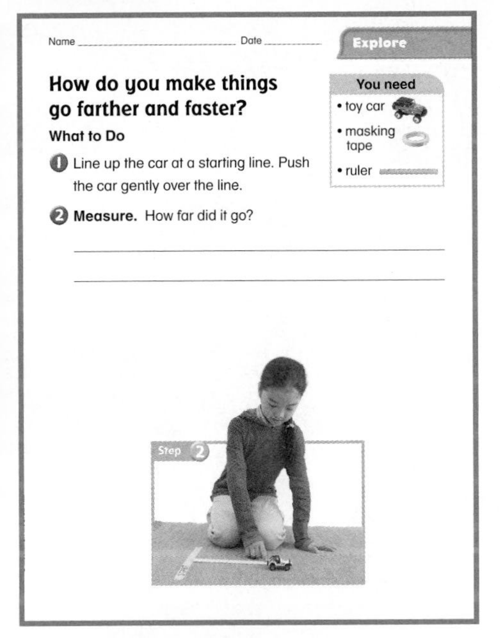

Name ______ Date ______ **Explore**

**How do you make things go farther and faster?**

**What to Do**

1. Line up the car at a starting line. Push the car gently over the line.
2. **Measure.** How far did it go?

**You need**
- toy car
- masking tape
- ruler

Step 2

**Explore, pp. 167–168**
Also available as a student workbook

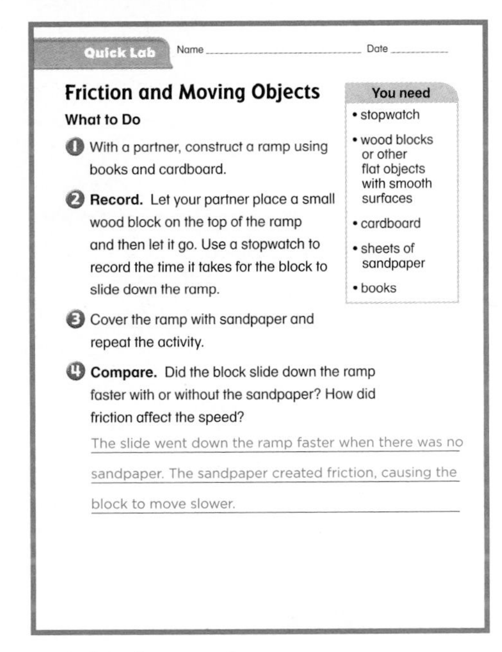

**Quick Lab** Name ______ Date ______

**Friction and Moving Objects**

**What to Do**

1. With a partner, construct a ramp using books and cardboard.
2. **Record.** Let your partner place a small wood block on the top of the ramp and then let it go. Use a stopwatch to record the time it takes for the block to slide down the ramp.
3. Cover the ramp with sandpaper and repeat the activity.
4. **Compare.** Did the block slide down the ramp faster with or without the sandpaper? How did friction affect the speed?
The slide went down the ramp faster when there was no sandpaper. The sandpaper created friction, causing the block to move slower.

**You need**
- stopwatch
- wood blocks or other flat objects with smooth surfaces
- cardboard
- sheets of sandpaper
- books

**Quick Lab, p. 170**
Also available as a student workbook

## Assessment

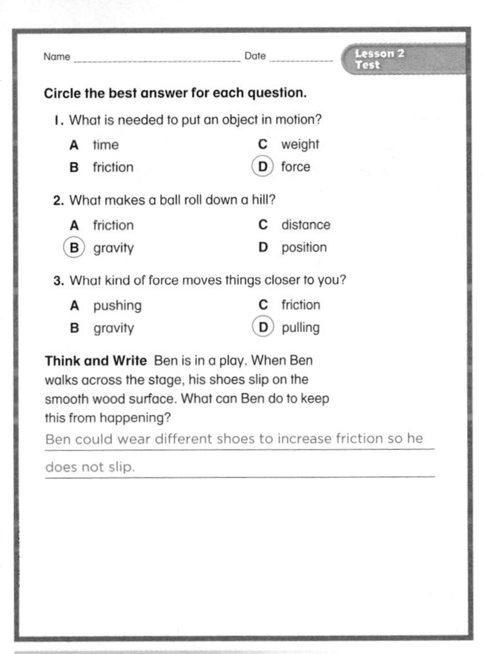

Name ______ Date ______ **Lesson 2 Test**

**Circle the best answer for each question.**

1. What is needed to put an object in motion?
   A time C weight
   B friction (D) force
2. What makes a ball roll down a hill?
   A friction C distance
   (B) gravity D position
3. What kind of force moves things closer to you?
   A pushing C friction
   B gravity (D) pulling

**Think and Write** Ben is in a play. When Ben walks across the stage, his shoes slip on the smooth wood surface. What can Ben do to keep this from happening?
Ben could wear different shoes to increase friction so he does not slip.

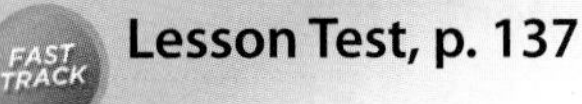
FAST TRACK **Lesson Test, p. 137**

## ADDITIONAL RESOURCES

**pp. 162–167**

***Push or Pull?***

**p. 51**

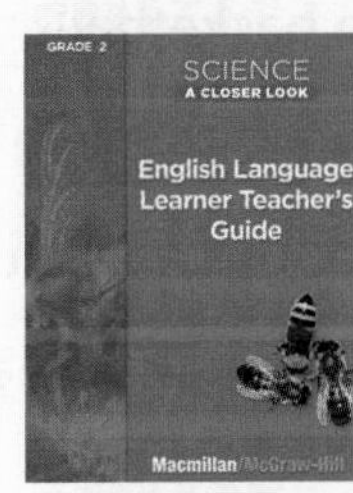


**p. 111**

**p. 33**

**33**

**101–110**

**pp. 107–108**

**84–86**

### Technology

- **Science Activity DVD**
- **TeacherWorks™ Plus CD-ROM**
- **Classroom Presentation Toolkit CD-ROM**
- SCIENCE QUEST *Objects in Motion*
- LOG ON 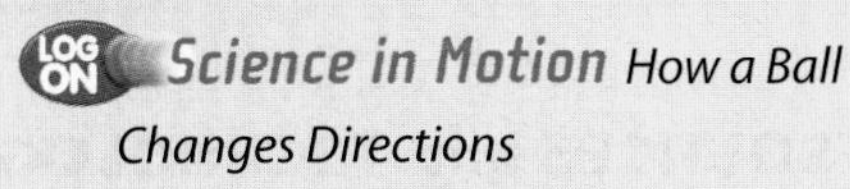 ***Science in Motion*** *How a Ball Changes Directions*
- **Science Songs CD** Track 12
-  LOG ON **e-Review**
-  LOG ON **NSDL**

# Lesson 2 Forces

## Objectives

- Identify a force as a push or a pull.
- Describe the forces of gravity and friction.

# 1 Introduce

### ▶ Assess Prior Knowledge

Have children share what they know about forces. They may understand the concept even if they do not know the word. Ask:

- **How do you move a basketball?**
- **What are different ways to make something move?**
- **How can you change the way something moves?**

Record children's answers in the What We Know column of the class **KWL** chart.

## Look and Wonder

Read the Look and Wonder questions. Invite children to share their responses to the questions.

Encourage children to look at the picture and discuss what they see. If necessary, explain that the pictured yellow vehicle is a bobsled, which is a very fast sled that fits several people. Note that the bobsled travels across ice on runners. Ask:

- **How are the athletes making the bobsled move?** They are running and pushing it.
- **How could they make it go faster?** Possible answers: They could push harder. More people could push.
- **If the bobsled did not have runners would it be harder to move?** Possible answer: Yes, because the runners help it to slide.

Write children's responses on the class **KWL** chart and note any misconceptions that they may have.

**RESOURCES and TECHNOLOGY**

- **Activity Lab Book,** pp. 167–169
- **Activity Flipchart,** p. 51
- Science Activity DVD

Lesson 2
Forces
Look and Wonder
How can you make something move?
How can you make it move farther?
366
ENGAGE


## Warm Up

**Start with a Book**

Read the book ***Jump, Frog, Jump,*** by Robert Kalan (HarperCollins, 2003) with the class. Review the activities of each animal in the story. Ask:

- **What kind of force is a jump?** a push
- **What would you do to jump higher?** push harder

Have a volunteer demonstrate a jump. Ask the class to observe how the jumper pushed off of the ground, leaped into the air, and was pulled back to the ground.

Encourage children to think of other movements that use a pushing and a pulling force, such as skipping or hopping. Have volunteers demonstrate and describe each action.

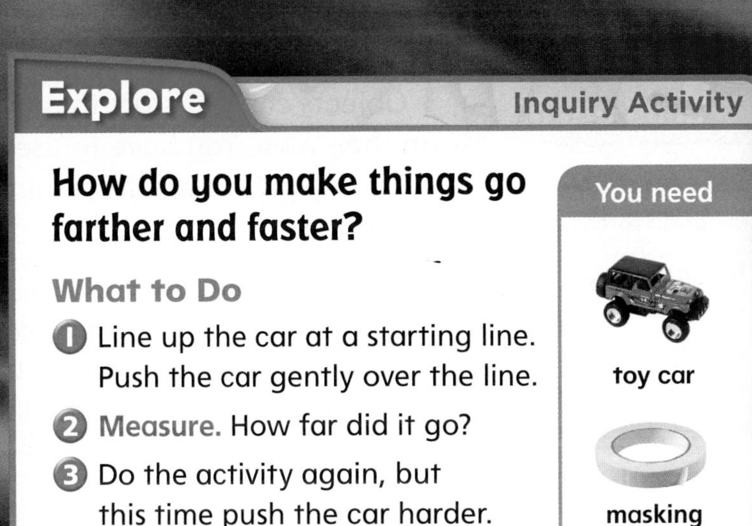

## Explore — Inquiry Activity

### How do you make things go farther and faster?

**What to Do**

1. Line up the car at a starting line. Push the car gently over the line.
2. **Measure.** How far did it go?
3. Do the activity again, but this time push the car harder. Observe what happens.

**Explore More**

4. **Predict.** What might happen if you pulled the car toward you with your hands? Would it go as far?

**You need**

- toy car
- masking tape
- ruler

Step 2

367
EXPLORE

## Explore

 individual  20 minutes

**Plan Ahead** Set up the starting line ahead of time. Consider creating multiple starting lines, so each child has enough space to work. Group materials (toy car, ruler, pencil, and paper) for each child ahead of time for distribution.

**Purpose** Children use their observational skills to draw conclusions about how different amounts of force can affect an object.

**Structured Inquiry** What to Do

Have children describe the different ways they can make toy cars move. Explain that for this experiment, they want to control the speed of the toy car by moving it with different amounts of force.

1. Model what a gentle force is by lightly poking the car across the line.
2. **Measure** Demonstrate how to mark the end of the ruler with a finger and move it to the other side of the finger to measure longer distances. Have children record the measurement.
3. Encourage children to describe what they see as well as measure the distances. Ask: **Which force made the car move farther? Why?**

**Guided Inquiry** Explore More

4. **Predict** Have children predict how pulling the car might be different from pushing the car. Ask: **Do you think it will be easier to pull the toy car than to push it? Why?**

**Open Inquiry**

Encourage children to think of other things that can affect an object's speed or the distance that it moves.

For example, children may want to know how other objects, such as balls, marbles, or wooden blocks are affected by different amounts of force.

Supply a variety of different objects for children to test. Ask them to compare the results of the different objects. Encourage children to use charts to record their information.

## Alternative Explore

### How much force is needed?

Help children compare the force needed to pull two different loads: one book and two books.

Model how to secure the book or books to a rubber band and pull the band until the books move.

Have children measure the lengths of the stretched rubber bands. Ask:

- **Why was the rubber band longer when it pulled two books?**

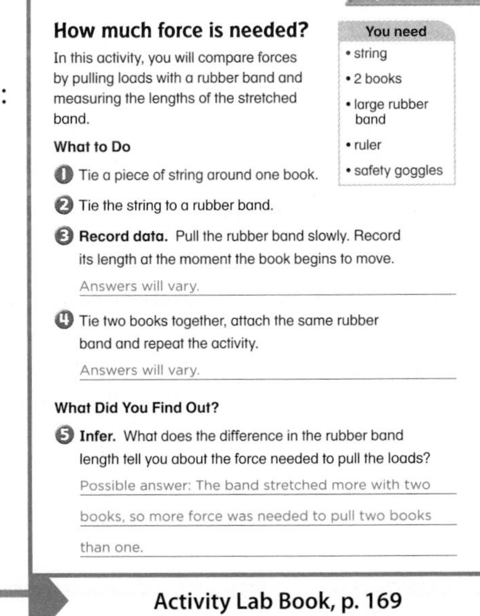

Name ____________ Date ________ Alternative Explore

**How much force is needed?**

In this activity, you will compare forces by pulling loads with a rubber band and measuring the lengths of the stretched band.

You need
- string
- 2 books
- large rubber band
- ruler
- safety goggles

**What to Do**

1. Tie a piece of string around one book.
2. Tie the string to a rubber band.
3. **Record data.** Pull the rubber band slowly. Record its length at the moment the book begins to move.
Answers will vary.
4. Tie two books together, attach the same rubber band and repeat the activity.
Answers will vary.

**What Did You Find Out?**

5. **Infer.** What does the difference in the rubber band length tell you about the force needed to pull the loads?
Possible answer: The band stretched more with two books, so more force was needed to pull two books than one.

Activity Lab Book, p. 169

# 2 Teach

## Read Together and Learn

**Reading Skill** Cause and Effect A cause is why an event happens. An effect is the event that happens.

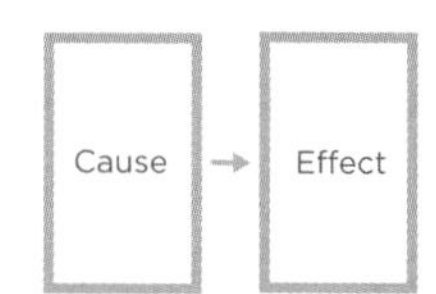


*Graphic Organizer 8,* p. TR10

## What makes things move?

### FAST TRACK Discuss the Main Idea

**Main Idea** Pushes and pulls are forces that put things in motion.

Before reading, discuss with children things they push and pull every day.

After reading the text together, ask:

- **How are pushes and pulls alike?** Possible answers: They are both forces. They both make things move.
- **How are pushes and pulls different?** Possible answer: Pushes move things away and pulls move things closer.

**Read Together and Learn**

**Essential Question**
What do forces do?

**Vocabulary**
force
gravity
friction

## What makes things move?

Objects can not start to move on their own. You have to use a push or a pull to put something in motion.

When you play soccer, you kick the ball to move it across the field. Your kick is a push. If you do not kick the ball, it will stay in the same place.

A stronger kick will make an object move farther.

368
EXPLAIN

## Science Background

**Forces and Motion** Inertia is the resistance an object has to a change in its state of motion. All materials resist changes to their state of motion. If two forces act on an object from different directions, it can cause a state of equilibrium. Isaac Newton's concept of inertia is stated in his *Laws of Motion*: "Unless acted upon by an unbalanced force, an object will maintain a constant velocity."

See **Science Yellow Pages**, in the Teacher Resources section, for background information.

**Professional Development** For more Science Background and resources from NSDL visit http://nsdl.org/refreshers/science

## ELL Support

**Use Pictures** Show children pictures of people doing different physical activities.

**BEGINNING** For each picture, say the name of the activity and have children repeat it. If appropriate, invite children to act out the motion. Discuss whether the motion is a push or pull.

**INTERMEDIATE** Have children identify each activity and describe in a simple statement what the person is doing. Ask children to identify push or pull in the activity.

**ADVANCED** For each picture, have children name the activity and explain the kind and amount of force shown.

A push or pull is called a **force**.
you push something, it will
ove away from you.
you pull it, it will
ove closer to you.

A kick is a kind of push.
pening a drawer is a kind of pull.
u can move different objects
ith different amounts of force.

What is making the cart move?

Why do we need forces?

Both groups are pulling the rope. Why does it not move?



## Differentiated Instruction

### Leveled Questions

EXTRA SUPPORT Ask questions such as these to check children's understanding of the material.

- **What happens when you pull on a drawer?** It opens.
- **What happens when you push on a drawer?** It closes.

ENRICHMENT Use these types of questions to develop children's higher-order thinking skills.

- **What happens if you pull a door on one side, and someone else is pulling the door on the other side with the same amount of force?** The door will not move.
- **What does a ball thrown a long distance tell you about the force used to pitch it?** Possible answer: A large push was used to throw the ball.

## ▶ Address Misconceptions

Children might think that a larger object exerts a greater force than a smaller object. Explain that a force is the amount of effort given to a push or a pull. For example, a large object, like a foot, can push gently and exert a light force. A small object, like a finger, can push hard and exert a strong force.

## ▶ Use the Visuals

Have children look at the photographs and read the captions on pages 368 and 369. Ask children to use the words *push* and *pull* to describe what the people in the photographs are doing. Ask:

- **What kind of force will make the soccer ball go higher?** a larger push
- **What other force could be used to move the cart?** a pull
- **Which photograph shows a pulling force?** The photograph of children tugging a rope.
- **What would happen if one team let go of the rope?** The other team would continue pulling and fall.

## ▶ Develop Vocabulary

**force** *Word Origin* *Force* comes from the Latin word *fortis*, meaning "strong." Have children describe situations in which a strong force is used, such as opening a stuck window. Make sure children use the word *force* correctly in the description.

## *Quick Check Answer*

Possible answer: to make things move

### RESOURCES and TECHNOLOGY

- **Reading and Writing,** pp. 206–208
- **PuzzleMaker CD-ROM**
- **Classroom Presentation Toolkit CD-ROM**
- 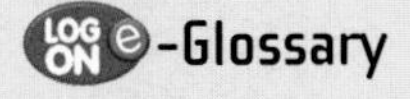 *Objects in Motion*
- **Science Songs CD** Track 12
- **e-Glossary**

# What are some forces?

## ▶ Discuss the Main Idea

**Main Idea** Friction is a force that slows moving objects, and gravity is a force that pulls objects to Earth.

Read the question and ask children to discuss the forces of push and pull.

After reading together, ask:

- **How does gravity affect your everyday life?** Possible answers: It keeps me from floating off the ground. It lets me play soccer because the ball doesn't fly away.
- **How does friction help us?** Possible answer: It helps us to come to a stop when we are skating.

## FAST TRACK Use the Visuals

Have children look at the photographs and read the captions on pages 370 and 371. Ask:

- **What forces are acting on the ball?** gravity and friction
- **In the photograph of the soccer ball, what causes friction that makes the ball slow down?** the grass
- **How does the brake on a skate help a child slow down?** Possible answer: The brake drags on the ground, causing friction that makes the person stop.

## ▶ Address Misconceptions

Children may think that Earth is the only planet with gravity because they may associate photographs of astronauts floating in space with conditions on other planets.

**FACT All planets have gravity.** Display photographs of Mars Rover on Mars and discuss how gravity is holding the rover and the rocks around it in place.

### What are some forces?

When you let go of a ball, it falls. **Gravity** is a force that pulls down on everything on Earth. When you jump up in the air, gravity pulls you back down to the ground. Gravity pulls on objects through solids, liquids, and gases. The amount of force that pulls something down toward Earth is called its weight.

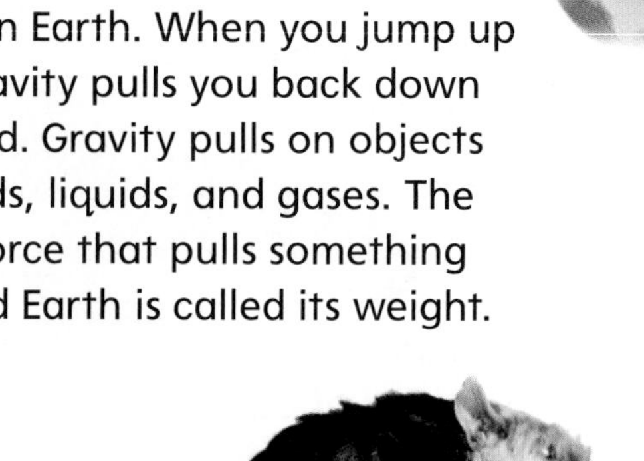

▲ **Why is the ball falling? What do you think will happen to the dog?**

**FACT** All planets have gravity.

370 EXPLAIN

## Quick Lab

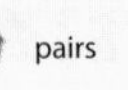
pairs

15 minutes

**Objective** Discover how friction affects a moving object.

**You need** stopwatches, wood cubes or other objects with smooth surfaces, cardboard, sheets of sandpaper, books

1. Have pairs make a ramp using books and cardboard.
2. Have one child release a small wood block from the top of the ramp. The other child will use a stopwatch to record the time it takes to slide down the ramp.
3. Have children cover the ramp with sandpaper and repeat the activity. Ask children to compare the speeds and draw conclusions about how friction affects speed.

When you skate, you drag a
ıbber stopper on the ground to
ɔp. The dragging causes friction.
**ction** is a force that slows down
ɔving things. Friction happens
ıen two things rub together.

There is usually more friction on
ugh surfaces than on smooth
es. It is usually harder to push or
ll something on a rough surface
an on a smooth surface.

How are gravity and friction alike?

## Quick Lab

Slide a wooden block on different slanted surfaces. **Compare** how friction affects the speed of the block.

Dragging the rubber stopper on the ground causes friction. This slows the skater down.

The ball falls to the grass and rolls. Friction makes the ball slow down and stop. ▼

371
EXPLAIN

## Classroom Equity

Materials, such as posters and handouts, can introduce gender bias by portraying men and women in stereotypical roles. As you teach about forces, be careful about using only "male activities" as examples. If you use such an example, switch the gender of the person involved—for example, use a female auto mechanic. If possible, invite one from your community to the class, and have children ask their own "interview" questions about her work.

## ▶ Develop Vocabulary

**gravity** Demonstrate *gravity* to children by dropping small objects, such as erasers and paper clips, to the ground. Ask: **What force pulls an object to the ground?** gravity
Give children an example to show why they may be grateful for gravity. For example: *Gravity makes rain fall on my roof and I like the sound it makes.* Have children write a sentence about why they are grateful for gravity, and collect them in a class book.

**friction** ***Scientific vs. Common Use*** Explain that in common use, *friction* can be used to describe "a conflict or disagreement." For example: *There was* friction *between the two friends because they did not agree on which game to play.*
Scientists use the word *friction* to describe "a force that slows a moving object." Ask children to list things that can cause *friction,* such as sandpaper, rubber stopper, and grass. Have children use one of the words on the list in a sentence to describe *friction*.

### ✓ Quick Check Answer

They are both forces.

### RESOURCES and TECHNOLOGY

- **Activity Lab Book,** p. 170
- **Classroom Presentation Toolkit CD-ROM**
- 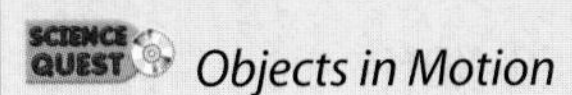 *Objects in Motion*

# How can forces change motion?

## ▶ Discuss the Main Idea

**Main Idea** Forces can change the way things move.

Read the blue question at the top of page 372 and discuss various ways objects in motion can change direction.

### Read a Diagram

Explain that the diagram shows how forces can change the speed and direction of a ball. After children read the captions, have them follow the direction of the arrows to see how the ball is traveling. Ask:

- **What force is making the ball move from the pitcher to the batter?** Possible answer: The pitcher is pushing or throwing the ball.
- **In the diagram, how many times does the ball change direction?** once

**Answer to Read a Diagram** pushes, pulls, and friction

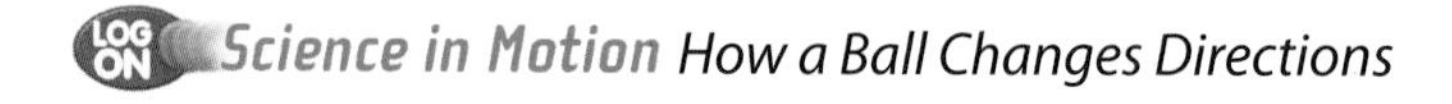
LOG ON *Science in Motion* *How a Ball Changes Directions*

## ▶ Develop Vocabulary

To review the words *force, gravity,* and *friction* with children, have them create a three-column chart using these words as labels. Under each word have children write a sentence using the word *ball* and the vocabulary word. For example: *I used a pushing force to hit the ball. Gravity pulled the ball to the ground. Friction slowed the motion of the ball.*

### How can forces change motion?

You know that forces can change how things move. Forces can make things start moving, speed up, slow down, and stop. Forces can make things change direction too. In softball, the players use forces to change the direction of the ball.

✓ **Think about a sport that uses a ball. How does the ball change direction?**

**How a Ball Changes Direction**

The pitcher u[ses] a force to thr[ow] the ball towa[rd] the batter. ▼

◀ The batter hits the ball with a push. It changes direction and flies toward the outfield.

372
EXPLAIN

## RESOURCES and TECHNOLOGY

- **School to Home Activities,** pp. 107–108
- **Reading and Writing,** p. 209
- **Visual Literacy,** p. 33
- SCIENCE QUEST *Objects in Motion*
- LOG ON *Science in Motion* *How a Ball Changes Directions*
- LOG ON **e-Review** Narrated Summary and Quiz
- ***ExamView® Assessment Suite* CD-ROM**

## Differentiated Instruction

### Leveled Activities

**EXTRA SUPPORT** Use a ball to demonstrate different forces. Ask a volunteer to gently toss a ball to another child. Have the volunteer identify the force used to throw the ball. push Ask the volunteer catching the ball to identify the force that stopped the ball. friction

**ENRICHMENT** Have children draw a diagram, showing how the ball in the diagram on pages 372 and 373 might be completed during play. For example, children might show the ball being thrown to one of the bases or to home plate. Ask children to write captions to describe the forces and the direction in which the ball travels.

The player in the outfield catches the ball and uses a force to stop its motion. He can also use a force to throw the ball to another player. ▶

**Read a Diagram**

What kind of forces do the players use?

*Science in Motion* Watch forces work at www.macmillanmh.com

## Think, Talk, and Write

1. **Cause and Effect.** What happens when you put more force on an object?
2. Why is it hard to push objects on some surfaces?
3. **Essential Question.** What do forces do?

### Social Studies Link

Learn about a sport played in another country. Describe the pushes and pulls in this sport.

**e-Review** Summaries and quizzes online at www.macmillanmh.com



## Formative Assessment

### Effects of Forces

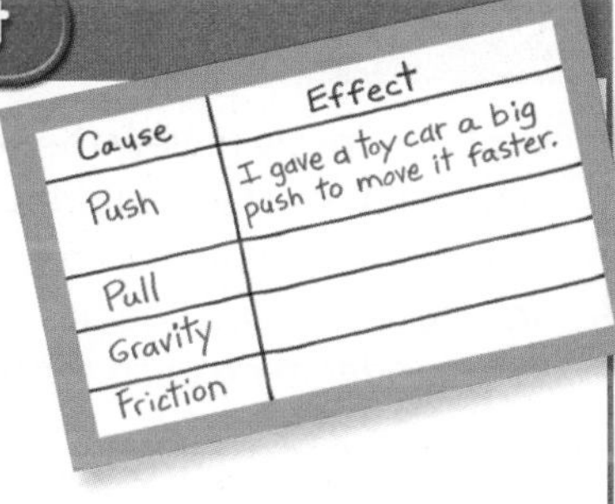


Have children label the long side of a piece of paper to make a two-column chart. The first column should be labeled *Cause* and the second column should be wider and labeled *Effect*.

Under the label *Cause* have children write the four forces described in this lesson: *push, pull, gravity,* and *friction*. On the right side, ask them to write a sentence that describes the effect of that force on an object.

**Key Concept Cards** For student intervention, see the prescribed routine on **Key Concept Card 33.**

### ✓ Quick Check Answer

Possible answers: In football, the ball is thrown or kicked using a push in one direction. The other team intercepts the ball. They stop it by using friction, and run with it in the opposite direction.

# 3 Close

### ▶ Using the KWL Chart

Review with children what they have learned about forces. Record their responses in the What We Learned column of the class **KWL** chart.

### ▶ Using the Reading Skill
### Cause and Effect

Use the reading skill graphic organizer to identify causes and effects in the lesson. Ask: **How does dragging the rubber stopper of a skate affect the skater?**

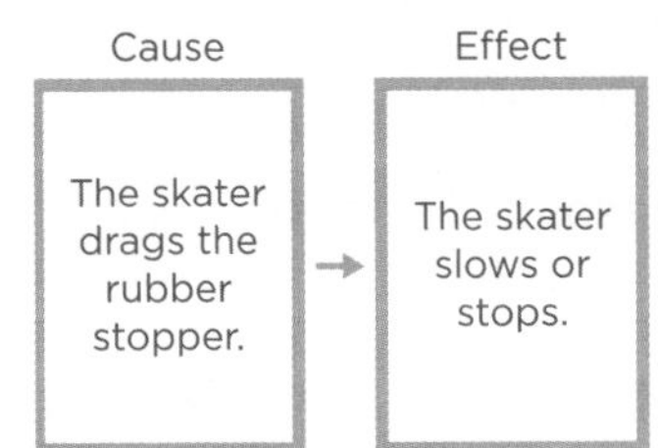


*Graphic Organizer 8,* p. TR10

## FAST TRACK Think, Talk, and Write

1. **Cause and Effect** Possible answers: It can move farther. It can move faster. It can move both farther and faster.
2. Possible answers: Friction causes objects to rub together and slow down. It is harder to move objects on rough surfaces, because there is more friction.
3. **Essential Question** Friction can stop or change the direction of objects that are moving. Pushes or pulls can also cause an object that is still to move. Gravity pulls us down so we stay close to Earth.

## Social Studies Link

Have resources available for children to research other sports, such as cricket or rugby. Encourage children to draw diagrams similar to one on pages 372 and 373 to show how players or game pieces move.

# Reading in Science

**Objective**

- Identify the effects of gravity on stars and Earth.

## Meet Héctor Arce

**Genre: Nonfiction** Stories or books about real people and events.

Have children read the title and look at the photos. Read the captions together. Ask:

- **Who is the person you will read about?** Héctor Arce
- **What do the photographs tell you about Héctor Arce?** Possible answer: He studies planets and stars.

## Before Reading

Review the concept of gravity with children by holding an eraser or other object in your hand. Ask:

- **What will happen to the eraser if I let it go?** It will fall to the ground.
- **What do we call the force that makes an object fall?** gravity

## During Reading

Read the text together. Have children look for the words *astrophysicist* and *gravity* as they read.

- **What does this story tell you about gravity on planets, moons, and stars?** Gravity is everywhere.
- **How does gravity create a star?** Gravity pulls clouds of dust and gas together, and makes the center so hot it lights up.

**RESOURCES and TECHNOLOGY**

- **Reading and Writing,** pp. 210–211

Real World **Reading in Science**

# Meet Héctor Arce

Héctor Arce is a scientist at the American Museum of Natural History. Héctor studies how stars form. When gravity pulls together huge clouds of gas and dust, stars form. Gravity makes their centers so hot that they light up. This is why stars shine in our night sky.

Gravity is the force that keeps you on Earth. You may not be able to see gravity, but it is all around you. In fact, it is everywhere! There is gravity on planets, moons, and stars. How powerful is gravity? It is powerful enough to create a star!

**Héctor Arce is an astrophysicist, a scientist who studies the planets, moons, and stars.**

374
EXTEND

## ELL Support

**Explain** Closely review the second paragraph of the article with children. Help children break into steps the formation of a star. Help them restate the sequence in their own words, using *first, next,* and *last*. Use gestures and simple drawings to help them understand each sentence.

**BEGINNING** Encourage children to use gestures or one- and two-word responses to modify questions about how stars form.

**INTERMEDIATE** Ask children to use short sentences to explain how stars form.

**ADVANCED** Have children use their own words to explain how stars form and why they shine brightly.

## ▶ Address Misconceptions

Children may believe that people can fall off the bottom of Earth. However, there is no bottom of Earth. Gravity pulls toward the center of the planet, so everything on the surface remains there because of gravity. The direction of *down* does not change if a person is in California, Australia, or the South Pole. *Down* is always toward the center of Earth. In fact, the closer a person gets to the center of Earth, the less gravity he or she would feel. At the very center, there's no gravity at all!

## After Reading

Discuss the article with children. Ask:

- **Why is gravity important to Héctor Arce's research?** Possible answer: Héctor studies how gravity causes stars to form.

Draw Graphic Organizer 8 on chart paper. In the left box write: *Gravity*. Ask:

- **What things happen because of gravity?**

Remind children that a cause is why an event happens, and an effect is the event that happens. Ask children to fill in an effect of gravity for the right-hand box.

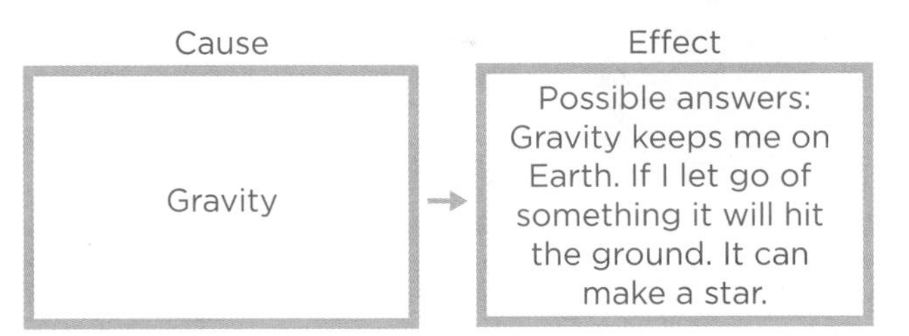

Cause
Effect
Gravity
Possible answers: Gravity keeps me on Earth. If I let go of something it will hit the ground. It can make a star.

*Graphic Organizer 8,* p. TR10

## Extended Reading

### Understanding Prefixes

Write the word *astrophysicist* on the board and circle *astro*. Explain that *astro* comes from the Greek word that means "star." Tell children that words with this prefix always have something to do with stars.

Write *astronaut* and *astronomy* on the board. Have children suggest meanings of these words and check them in a dictionary.

Reading in Science Name ______ Date ______

**Meet Hector Arce**

Read the Reading in Science pages in your book. As you read, keep track of what happens and why. Record the causes and effects you read about in the chart below. Remember, a cause is why something happens. An effect is the thing that happens. Sometimes, one cause can have many effects.

| Cause | Effect |
| --- | --- |
| Gravity | It keeps living things and objects on Earth. |
| Gravity | It pulls together huge clouds of gas and dust to form stars. |
| Gravity | It makes the centers of stars hot enough to glow. |

**Reading and Writing, p. 210**

Héctor uses a telescope, like the one in this building, to get a closer look at stars.
## Talk About It
How do stars form?

Gravity pulls together huge clouds of gas and dust.

If children are having difficulty understanding how stars are formed, ask them to read the first paragraph again. Have them identify what happens first. Direct their attention to the picture of stars on page 375. Explain to children that the dust and gases pulling together may look like the picture.

Have children describe what happens next in the formation of a star. Ask children to explain how gravity helps form a star.

# Stop Here to Plan Your Lesson

Lesson 3
Using Simple Machines
Look and Wonder
Have you ever used a shovel? How does it make digging easier?


## Lesson 3 Using Simple Machines

### Essential Question

How can you use simple machines?

### Objectives

- Identify simple machines.
- Discover that simple machines change force to make work easier.

**Reading Skill** Summarize

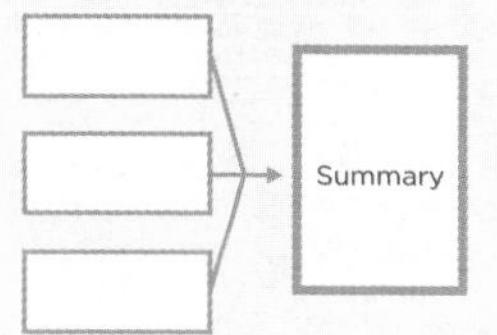


*Graphic Organizer 5*, p. TR7

## FAST TRACK

**Lesson Plan** When time is short, follow the Fast Track and use the essential resources.

**1 Introduce**
**Look and Wonder, p. 376**
Resource **Activity Lab Book, p. 173**

**2 Teach**
**Develop Vocabulary, p. 379**
Resource **Visual Literacy, p. 34**

**3 Close**
**Think, Talk, and Write, p. 381**
Resource **Assessment, p. 138**

**Professional Development** Look for **NSDL** to find recommended Science Background resources from the National Science Digital Library.

## Reading and Writing

## Visual Literacy

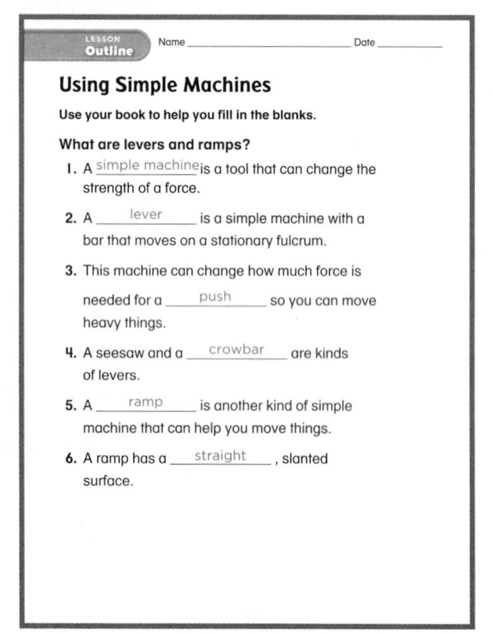

Lesson Outline Name ______ Date ______

**Using Simple Machines**

Use your book to help you fill in the blanks.

**What are levers and ramps?**

1. A simple machine is a tool that can change the strength of a force.
2. A lever is a simple machine with a bar that moves on a stationary fulcrum.
3. This machine can change how much force is needed for a push so you can move heavy things.
4. A seesaw and a crowbar are kinds of levers.
5. A ramp is another kind of simple machine that can help you move things.
6. A ramp has a straight, slanted surface.

**Outline, pp. 212–213**
Also available as a student workbook

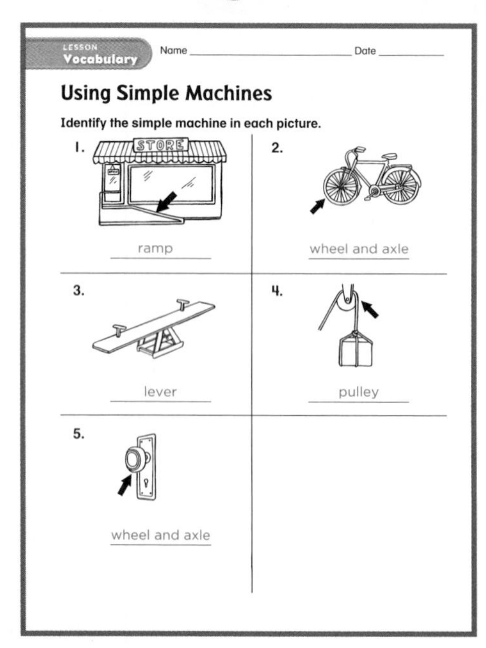

Lesson Vocabulary Name ______ Date ______

**Using Simple Machines**

Identify the simple machine in each picture.

1. ramp
2. wheel and axle
3. lever
4. pulley
5. wheel and axle

**Vocabulary, p. 214**
Also available as a student workbook

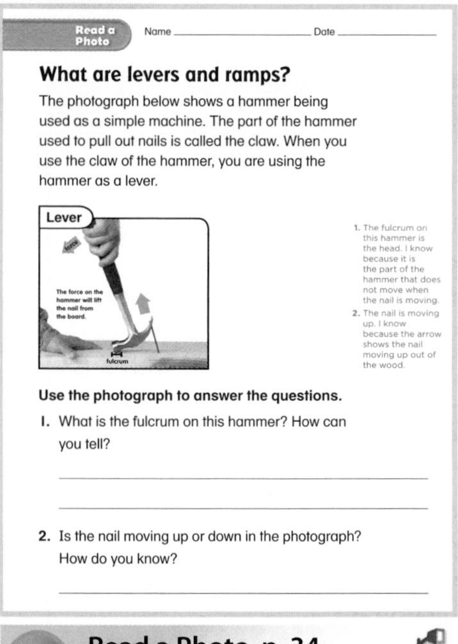

Read a Photo Name ______ Date ______

**What are levers and ramps?**

The photograph below shows a hammer being used as a simple machine. The part of the hammer used to pull out nails is called the claw. When you use the claw of the hammer, you are using the hammer as a lever.



1. The fulcrum on this hammer is the head. I know because it is the part of the hammer that does not move when the nail is moving.
2. The nail is moving up. I know because the arrow shows the nail moving up out of the wood.

Use the photograph to answer the questions.

1. What is the fulcrum on this hammer? How can you tell?
2. Is the nail moving up or down in the photograph? How do you know?

**Read a Photo, p. 34**
Also available as a transparency

The force on the hammer will lift the nail from the board.

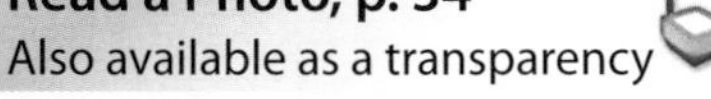

# Activity Lab Book

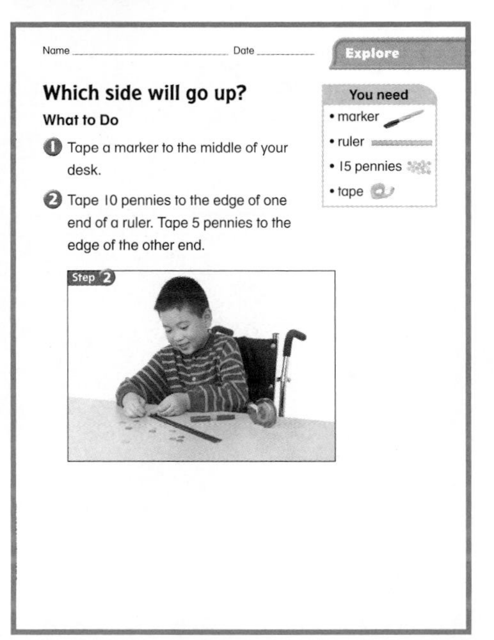

Name ______ Date ______ Explore

**Which side will go up?**

**What to Do**

1. Tape a marker to the middle of your desk.
2. Tape 10 pennies to the edge of one end of a ruler. Tape 5 pennies to the edge of the other end.

You need
- marker
- ruler
- 15 pennies
- tape

Step 2

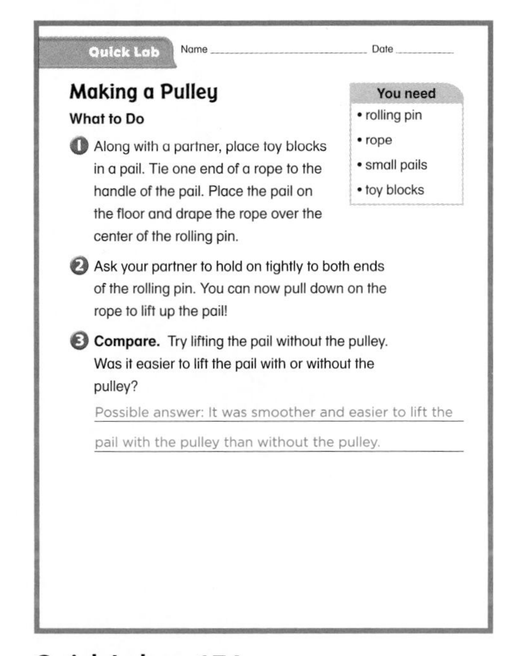

Quick Lab Name ______ Date ______

**Making a Pulley**

**What to Do**

1. Along with a partner, place toy blocks in a pail. Tie one end of a rope to the handle of the pail. Place the pail on the floor and drape the rope over the center of the rolling pin.
2. Ask your partner to hold on tightly to both ends of the rolling pin. You can now pull down on the rope to lift up the pail!
3. **Compare.** Try lifting the pail without the pulley. Was it easier to lift the pail with or without the pulley?

Possible answer: It was smoother and easier to lift the pail with the pulley than without the pulley.

You need
- rolling pin
- rope
- small pails
- toy blocks

**Explore, pp. 171–172**
Also available as a student workbook

**Quick Lab, p. 174**
Also available as a student workbook

# Assessment

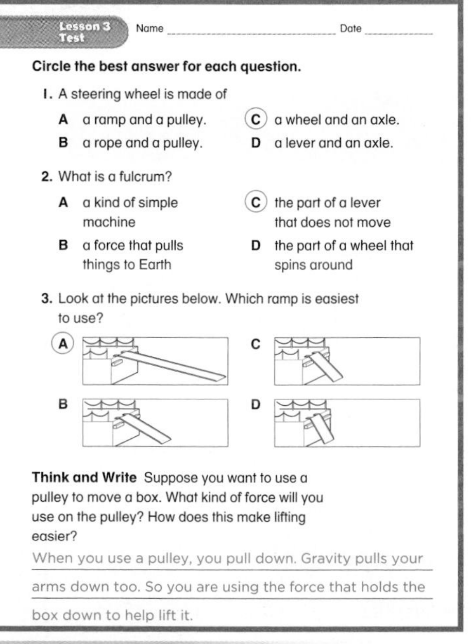

Lesson 3 Test Name ______ Date ______

**Circle the best answer for each question.**

1. A steering wheel is made of
   - A a ramp and a pulley.
   - B a rope and a pulley.
   - (C) a wheel and an axle.
   - D a lever and an axle.
2. What is a fulcrum?
   - A a kind of simple machine
   - B a force that pulls things to Earth
   - (C) the part of a lever that does not move
   - D the part of a wheel that spins around
3. Look at the pictures below. Which ramp is easiest to use?
   - (A)
   - B
   - C
   - D

**Think and Write** Suppose you want to use a pulley to move a box. What kind of force will you use on the pulley? How does this make lifting easier?

When you use a pulley, you pull down. Gravity pulls your arms down too. So you are using the force that holds the box down to help lift it.

**Lesson Test, p. 138**

# ADDITIONAL RESOURCES

**pp. 168–171**

***Push or Pull?***

**p. 52**

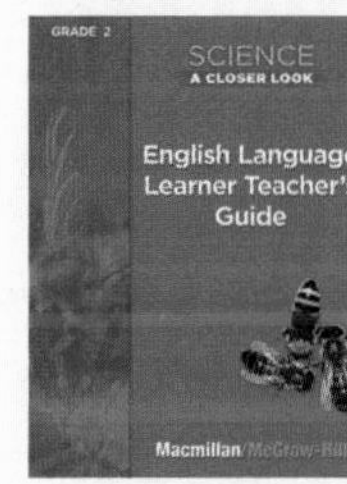


**p. 112**

**p. 34**

**34**

**101–110**

**pp. 109–110**

**87–90**

## Technology

- Science Activity DVD
- TeacherWorks™ Plus CD-ROM
- Classroom Presentation Toolkit CD-ROM
- Science Songs CD Track 13
-  e-Review
-  NSDL

# Lesson 3 Using Simple Machines

## Objectives

- Identify simple machines.
- Discover that simple machines change force to make work easier.

# 1 Introduce

## ▶ Assess Prior Knowledge

Evaluate children's knowledge of simple machines. Ask:

- **What tools have you used?**
- **How did the tools help you?**

Record children's answers in the What We Know column of the class **KWL** chart.

FAST TRACK

## Look and Wonder

Have a volunteer read the Look and Wonder questions about using a shovel. Invite children to share their responses to the questions. Ask:

- **Why would it be harder to dig the soil without the shovel?** Possible answer: The shovel holds more than my hands, so it would take more time and strength to dig without it.
- **What force is the person using to dig up the soil?** Possible answer: The person is pushing down.

Write children's responses on the class **KWL** chart and note any misconceptions that they may have.

Lesson 3
Using Simple Machines
Look and Wonder
Have you ever used a shovel?
How does it make digging easier?
376
ENGAGE

**RESOURCES and TECHNOLOGY**

- **Activity Lab Book,** pp. 171–173
- **Activity Flipchart,** p. 52
- Science Activity DVD

## Warm Up

**Start with a Book**

Preview the book ***Simple Machines,*** by Allan Fowler (Children's Press, 2001) by looking at the pictures. Have children raise their hands if they are familiar with any of the simple machines in the pictures.

Read the book aloud.

Distribute index cards to children and encourage them to write a list of facts about one of the simple machines on each card. Ask:

- **How does the simple machine help people?**
- **How does the simple machine work?**

Have children exchange and read each other's cards.

## Explore — Inquiry Activity

### Which side will go up?

**What to Do**

1. Tape a marker to the middle of your desk.
2. Tape 10 pennies to the edge of one end of a ruler. Tape 5 pennies to the edge of the other end.
3. **Predict.** What will happen if you put the middle of the ruler on the marker? Which side will lift up? Try it. Was your prediction correct?

**Explore More**

4. Try to move the ruler so that 5 pennies can lift 10 pennies. Where did you need to move the ruler?

**You need**

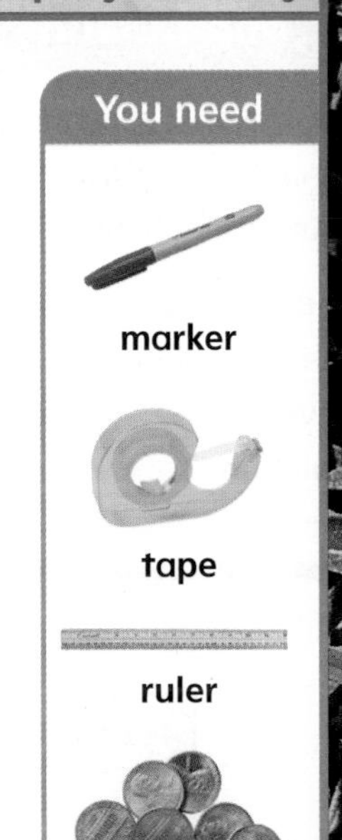

377
EXPLORE

## Alternative Explore

### How can you make a lever?

Tell children they will make and use a lever to lift a heavy book. Explain that people have lifted heavy objects in this same way for thousands of years.

Have children tape a marker to their desk. Ask them to place a ruler on the marker with the book at one end of the ruler.

Ask children to position the ruler at different points to observe where the lever lifts the book the highest.

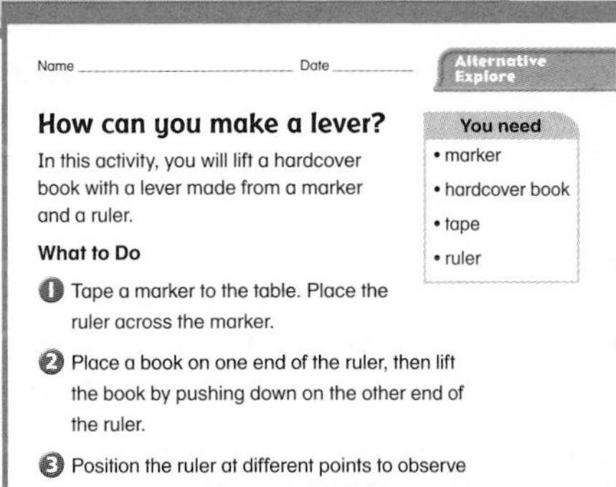

Name ______________ Date ________ Alternative Explore

**How can you make a lever?**

In this activity, you will lift a hardcover book with a lever made from a marker and a ruler.

You need
- marker
- hardcover book
- tape
- ruler

**What to Do**

1. Tape a marker to the table. Place the ruler across the marker.
2. Place a book on one end of the ruler, then lift the book by pushing down on the other end of the ruler.
3. Position the ruler at different points to observe where the lever lifts the book highest.

**What Did You Find Out?**

4. **Observe.** Where was the best place to position the ruler?
Possible answer: When the distance between the book and the marker was greatest, it was easiest to lift the book.

Activity Lab Book, p. 173

## Explore

 pairs  25 minutes 

**Plan Ahead** Have 15 pennies, a ruler, marker, and tape ready in a small plastic bag for each pair. Masking tape will be easier to remove from pennies than clear tape. If possible, distribute the pieces of tape to children.

**Purpose** This activity supports children's observational skills and will help them understand how predictions can be tested through experimentation.

### Structured Inquiry What to Do

Explain to children that their models will be similar to a seesaw. Ask: **What happens when you are on a seesaw?** One person goes up and the other goes down.

1. Have children securely tape the ends of the marker to the desk.
2. Model how to tape the pennies to the ruler so pennies do not tumble during the activity.
3. **Predict** Before children put the middle of the ruler on the marker, ask them to predict which side will lift up and have them record their predictions. Show the middle number on the ruler to make sure children know where the middle of the ruler is.

### Guided Inquiry Explore More

4. Ask: **How does moving the ruler relative to the marker affect lifting the pennies?** Possible answers: When the marker is far away from five pennies, five pennies can lift ten pennies. Less force can lift the heavier side if the marker is closer to it.

### Open Inquiry

Encourage children to explore levers further. Have them share additional research questions. If they need a prompt, ask: **What kind of lever would lift a heavy object such as a dog or a person? What other kinds of tools are levers?**

Help children make a plan to find answers to their questions. Provide resources to help children conduct their research.

# 2 Teach

## Read Together and Learn

**Reading Skill** Summarize To retell the most important ideas from the reading selection.

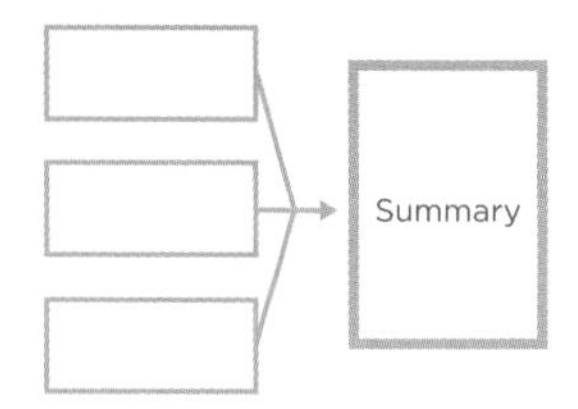
Summary

*Graphic Organizer 5,* p. TR7

## What are levers and ramps?

### ▶ Discuss the Main Idea

**Main Idea** Levers and ramps are simple machines that make work easier.

Before reading, have children discuss what they know about ramps and levers and how they are used. After reading, discuss with children how levers and ramps help people do work. Ask:

- **Why are ramps and levers important tools?** Possible answer: They help make work easier by letting people use less strength, or force, to do work.

**Read Together and Learn**

**Essential Question**
How can you use simple machines?

**Vocabulary**
simple machine
lever
fulcrum
ramp

### What are levers and ramps?

A **simple machine** is a tool that changes the size or the direction of a force. A simple machine can make work easier.

A **lever** is a bar that moves against an unmoving point. The unmoving point that a lever moves against is called a **fulcrum**. Shovels and seesaws are levers. When you push down on one side of the lever, the other side moves up.

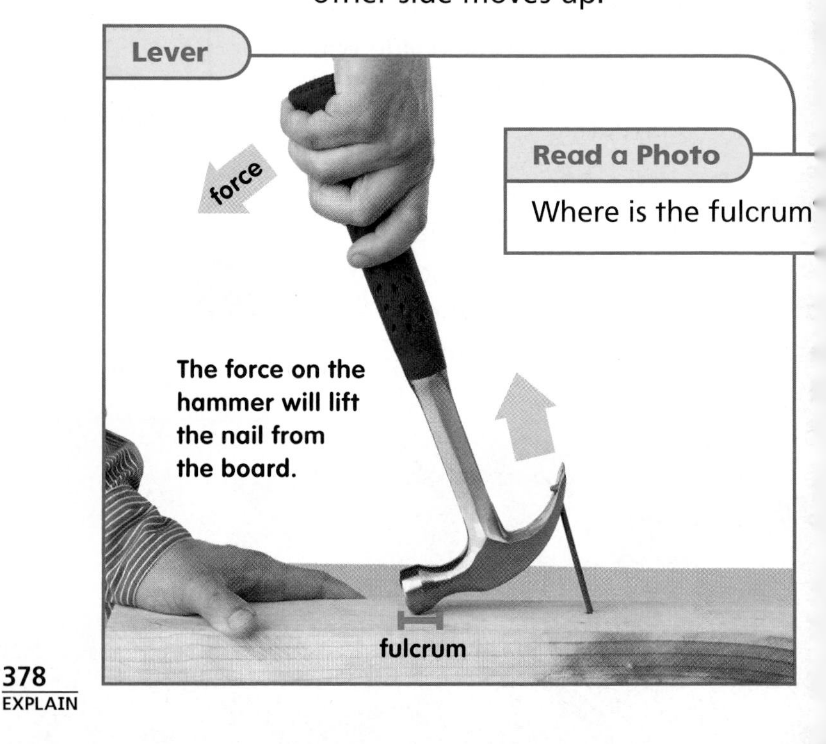
Lever
force
The force on the hammer will lift the nail from the board.
fulcrum
Read a Photo
Where is the fulcrum

378
EXPLAIN

## Science Background

**Simple Machines** Simple machines help people do work against an opposing force, such as gravity or friction. The lever, ramp, and wheel/axle decrease the amount of the force needed to do work. The force placed on the wheel is multiplied on the axle. A simple pulley changes the direction of the force applied to it. Other simple machines are the wedge and screw.

See **Science Yellow Pages**, in the Teacher Resources section, for background information.

**Professional Development** For more Science Background and resources from NSDL visit http://nsdl.org/refreshers/science

## ELL Support

**Participate in Hands-On Activities** Take children on a walk around the school or play yard to find ramps and levers.

**BEGINNING** Have children point to or name the simple machine and say whether it is a ramp or a lever.

**INTERMEDIATE** Ask children to use a phrase or short sentence to name the ramp or lever and to tell the work it helps to do.

**ADVANCED** Have children name the simple machine and describe in full sentences how it makes work easier.

ou use less force to push a heavy
ox up a ramp than to lift the box.

Another kind of simple machine is a
amp. A **ramp** is a surface that is straight
nd slanted. Ramps can be used to move an
bject from one place to another. Pushing
omething up a ramp is easier than lifting it.
ess force is needed to move something on a
ng, low ramp than on a short, steep ramp.

How do a lever and a ramp make work easier?



## Differentiated Instruction

### Leveled Activities

**EXTRA SUPPORT** Give children spatulas and have them lift small objects with them. Ask:

- **What kind of simple machine is the spatula?** a lever
- **Why is a spatula helpful in the kitchen?** Possible answer: It can lift very hot food off a tray without having to touch the food.

**ENRICHMENT** Have children use a heavy piece of cardboard (about 12" x 4") and a stack of several blocks to make a ramp. Encourage children to experiment with different degrees of the angle to the slant to observe how the angle affects the amount of force required to push objects of different weights down the ramp.

### Read a Photo

Explain to children how the arrows in the diagram show the direction of the push and pull forces. Have children look at the diagram and read the labels. Ask:

- **What kind of force does it take to remove the nail?** a pull or lift

**Answer to Read a Photo** It is the part of the wood board that the head of the hammer is pushing against.

## FAST TRACK Develop Vocabulary

**simple machine** Explain that a *machine* is "something that makes a job easier to do." Remind children that *simple* means "plain and uncomplicated." Have them put these definitions together to define a *simple machine*.

**lever** *Word Origin* Explain that the word *lever* comes from the Latin words *lavare,* which means "to raise," and *levis* which means "light." Ask children to think about the work of levers and explain how the the words *raise* and *light* are related to *levers*.

**fulcrum** *Word Origin* *Fulcrum* comes from the Latin word *fulcire,* which means "support." Have children put their elbow on the desk and lift a book. Explain that their arm is a lever and the elbow is the *fulcrum*. Ask children to explain how the fulcrum supports the arm lifting the book.

**ramp** *Word Origin* *Ramp* comes from the Old French verb *ramper,* meaning "to climb." Ask children to identify where they have seen ramps and to describe how the ramp was used. Have them explain how ramps are related to climbing.

### Quick Check Answer

They let people use less force to do work.

### RESOURCES and TECHNOLOGY

- **Reading and Writing,** pp. 212–214
- **Visual Literacy,** p. 34
- PuzzleMaker CD-ROM
- Classroom Presentation Toolkit CD-ROM
- Science Songs CD Track 13
-  e-Glossary

# What are other simple machines?

## ▶ Discuss the Main Idea

**Main Idea** The pulley and the wheel/axle are simple machines that help move things.

Read the text together and discuss how children use simple machines every day. Ask:

- **What things that have a wheel and axle did you use today?** Possible answers: car, bus, bicycle, door
- **Why is a pulley helpful?** Possible answer: It can help lift an object up instead of carrying it up stairs.

## ▶ Use the Visuals

Have children look at the photographs and read the captions on pages 380 and 381.

Ask:

- **What does the wheel and axle do on the monster truck?** Possible answer: It helps move the truck forward.
- **What is the force applied to the pulley?** a pull
- **What is the pulley doing?** Possible answer: changing the direction of the force so the pig can be pulled up

## ▶ Develop Vocabulary

Review the lesson vocabulary with children. Draw a word web for *simple machines* on the board. Have children copy the web and identify different kinds of simple machines. Ask them to list whether a push or pull force is used for the simple machine and how the tool is used.

### RESOURCES and TECHNOLOGY

- **School to Home Activities,** pp. 109–110
- **Reading and Writing,** p. 215
- **Activity Lab Book,** p. 174
- **e-Review** Narrated Summary and Quiz
- **ExamView® *Assessment Suite* CD-ROM**

## What are other simple machines?

A bicycle uses a simple machine called a wheel and axle. A wheel and axle is made of a wheel and a bar, or axle. The bar is connected to the center of the wheel. When the wheel turns, the bar turns too.

A doorknob and a steering wheel each use a wheel and axle. Each axle on a car or bus has two wheels attached.

**Where is the axle on this monster truck?**

380
EXPLAIN

### Quick Lab

**Investigate** how t… make a pulley. Us… the pulley to lift a pail filled with blocks.

### Quick Lab

 pairs  15 minutes

**Objective** Investigate how to make and use a pulley.

**You need** roller-style rolling pins, rope, small pails, wood cubes

1. Have children place wood cubes in the pail and tie one end of the rope to the handle. Place the pail on the floor and drape the rope over the center of the rolling pin.
2. Have one child securely hold both handles of the rolling pin. Ask the other child to pull down on the rope to lift the pail.
3. Have children lift the pail without using the pulley. Ask them to compare using a pulley with not using a pulley.

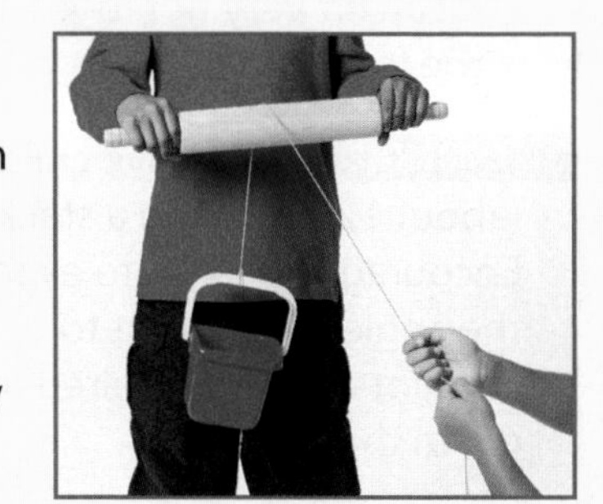

A pulley is also a simple
ıchine. A pulley is made with
·ope that moves around a
ıeel. When you attach a pulley
a object, you can change the
·ection of the force on the
ject. A pulley can help lift
object up high.

When might it be helpful to use a pulley?

When you pull the rope down, the pig in the pail goes up to the pulley.

## Think, Talk, and Write

1. **Summarize.** How do simple machines make our lives easier?
2. What are some kinds of simple machines?
3. **Essential Question.** How can you use simple machines?

### Math Link

Make a tally chart of simple machines used at home and school.

LOG ON e-Review Summaries and quizzes online at www.macmillanmh.com

381
EVALUATE

### Quick Check Answer

Possible answer: It would be helpful to use a pulley to move an object higher without needing to use a ladder.

# 3 Close

## ▶ Using the KWL Chart

Review with children what they have learned about simple machines. Record their responses in the What We Learned column of the class **KWL** chart.

## ▶ Using the Reading Skill
### Summarize

Use the reading skill graphic organizer to summarize the lesson.

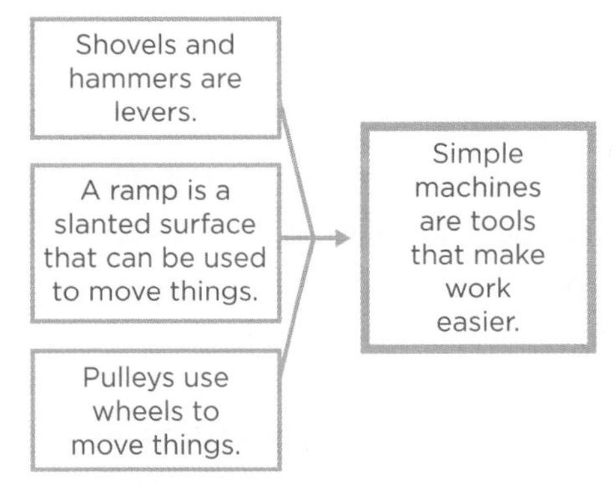

Shovels and hammers are levers.
A ramp is a slanted surface that can be used to move things.
Pulleys use wheels to move things.
Simple machines are tools that make work easier.

*Graphic Organizer 5*, p. TR7

## FAST TRACK Think, Talk, and Write

1. **Summarize** Possible answers: Simple machines can help move things and lift things to make work easier.
2. Possible answers: ramps, levers, wheels/axles, pulleys
3. **Essential Question** Possible answers: I use pulleys to close my curtains. I use a ramp to help lift heavy items up. I use a lever when I play on a seesaw.

## Formative Assessment

### It's Simple

Have children fold a piece of paper into quarters. Ask them to draw a different kind of simple machine in each box. Encourage them to choose simple machines they have used. Under each picture, have children write a sentence describing how the simple machine is used.

The wheels and axles turn the wheels when I push on the pedals.

**Key Concept Cards** For student intervention, see the prescribed routine on **Key Concept Card 34.**

## Math Link

Have children make a list of the simple machines they found at home and at school. From the list, children can create a chart with the simple machines listed in rows and tally marks next to each simple machine.

# Writing in Science

**Objective**

- Use expository writing to describe how penguins slide on ice.

## Slip and Slide

### Talk About It

Have children study the photograph and describe how the penguins are moving. Ask:

- **How does the ice help the penguin move?** Ice is slippery so it is easy for a penguin to move.

### Learn About It

Read together the top paragraph and Remember box.

Remind children that authors use many descriptions to explain something. Encourage children to write about the penguins, their movements, and the environment using descriptive words. Ask:

- **What kind of information does the photograph show you?** Possible answers: the time of day; how the penguins move; the weather

### Write About It

Before writing, review with children the forces of push, pull, and friction. Have children brainstorm a list of reasons why penguins can slide on the ice.

Guide children by taking a reason from the list and modeling a complete sentence. Have children write their fact sentences on index cards. Ask children to trade their fact cards with classmates.

**RESOURCES and TECHNOLOGY**

**Reading and Writing,** pp. 216–217

e-Journal Online research and writing

# Slip and Slide

Have you ever walked on ice? It is smooth and slippery! Sometimes penguins slide on their bellies to move.

### Write About It

Explain why penguins can slide on the ice. Think about what you learned about forces. Make sure to explain why ice is slippery.

**Remember**

When you write to give information, you give facts.

e-Journal Write about it online at www.macmillanmh.com



## Integrate Writing

**Compare and Contrast**

Display Photo Sorting Cards 103–106. Have children describe the movements in each photo.

Ask children to compare the movements and explain how they are alike or different. Ask:

- **How is the motion of children playing a game of tug different than the motion of the people with the cart?** The children playing a game are pulling and the people with the cart are pushing.

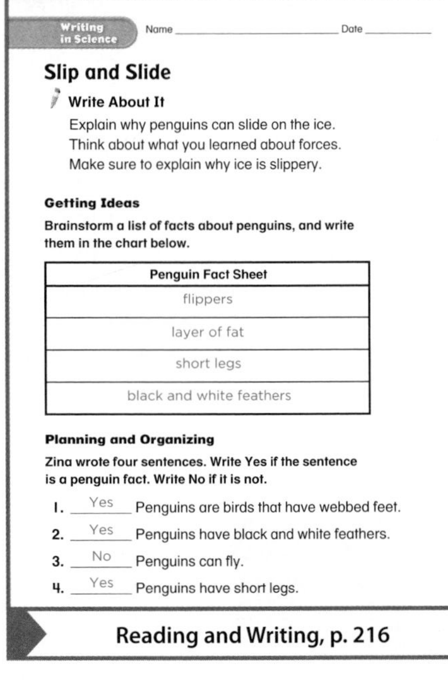
Writing in Science Name ____ Date ____

**Slip and Slide**

**Write About It**

Explain why penguins can slide on the ice. Think about what you learned about forces. Make sure to explain why ice is slippery.

**Getting Ideas**

Brainstorm a list of facts about penguins, and write them in the chart below.

| Penguin Fact Sheet |
| --- |
| flippers |
| layer of fat |
| short legs |
| black and white feathers |

**Planning and Organizing**

Zina wrote four sentences. Write Yes if the sentence is a penguin fact. Write No if it is not.

1. Yes Penguins are birds that have webbed feet.
2. Yes Penguins have black and white feathers.
3. No Penguins can fly.
4. Yes Penguins have short legs.

Reading and Writing, p. 216

ath in Science

# How Far Did It Move?

These students are playing softball.
They want to know how far the ball moved.

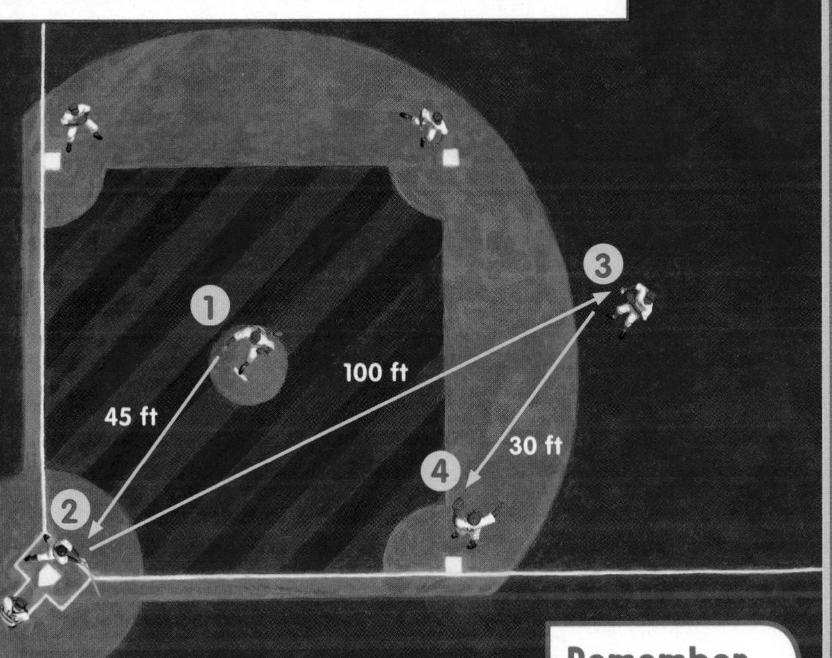


**Remember**
Add the 1s first.
Then add the 10s. Then add the 100s.

**dd Measurements**
ld the distances the ball moved. How r did it go? How many times did the ll change direction? Make up your own ath problem about the softball game.



# Math in Science

**Objective**

- Add three-digit numbers.

## How Far Did It Move?

### Talk About It

Have a volunteer explain the basic idea of a softball game, and then ask children to examine the diagram.

Remind children that the circled numbers in the diagram show the order in which the ball traveled. The arrows show the direction the ball was thrown. Point to the labels and explain that they indicate the distance the ball traveled between two players.

Read the introduction together. Ask:

- **Where did the ball start?** at the pitcher's mound; number 1 on the diagram
- **How many players touched the ball?** 4

### Learn About It

Read the problem paragraph together. Help children come up with a strategy to answer the questions. Ask:

- **How can you find the distances?** They are the numbers beside the arrows.
- **What is the best way to find the different directions?** look at the arrows

Remind children that they can use a different operation to create another math problem from the diagram.

### Try It

Help children list ideas for their math problems. For example, find the number of times the ball was thrown (not batted) and make a number sentence to show the total distances the ball was pitched.

## Integrate Math

**Map Math**

Have children make a map of the classroom. Ask them to use a ruler to measure the distances between objects. Encourage them to use the softball diagram as a model.

Encourage children to make math problems for their classmates to solve, such as the total distance between the computer center, the math center, and the library. Have children think about using addition or subtraction in their problems.

Name ____________ Date ________ Math in Science

**How Far Did It Move?**

These students are playing softball. They want to know how far the ball moved.

**Remember**
Add the 1s first.
Then add the 10s.
Then add the 100s.

**Add Measurements**

1. Add the distances that the ball moved. How far did it go?
   45 ft + 100 ft + 30 ft = 175 ft; the ball moved 175 feet.
2. How many times did the ball change direction?
   The ball changed direction twice.
3. Now write your own math problem about the softball game. Then solve it.
   Answers will vary.

**Math, p. 21**

**RESOURCES and TECHNOLOGY**

- **Math,** pp. 21–22

Stop Here to

# Plan Your Lesson

Lesson 4
Exploring Magnets
Look and Wonder
Why does the magnet pull some of these objects and not others?


## Lesson 4 Exploring Magnets

### Essential Question

What are magnets?

### Objectives

- Observe magnets attract and repel objects.
- Identify magnet poles and explain how they function.

**Reading Skill** Problem and Solution

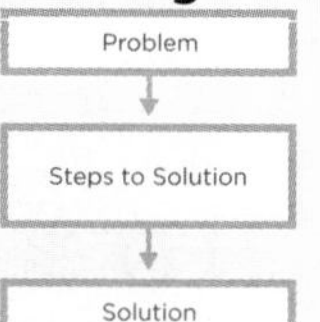


*Graphic Organizer 12*, p. TR14

## FAST TRACK

**Lesson Plan** When time is short, follow the Fast Track and use the essential resources.

**1 Introduce**
**Look and Wonder, p. 384**
Resource **Activity Lab Book, p. 177**

**2 Teach**
**Read a Chart, p. 387**
Resource **Visual Literacy, p. 35**

**3 Close**
**Think, Talk, and Write, p. 389**
Resource **Assessment, p. 139**

**Professional Development** Look for NSDL to find recommended Science Background resources from the National Science Digital Library.

## Reading and Writing

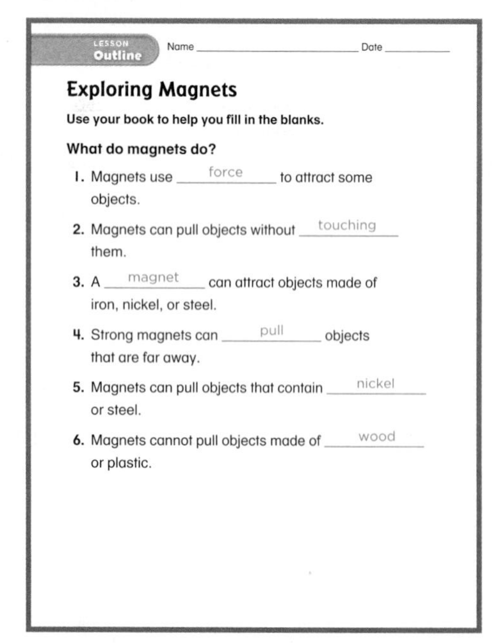

LESSON Outline Name ______ Date ______

**Exploring Magnets**

**Use your book to help you fill in the blanks.**

**What do magnets do?**

1. Magnets use ___force___ to attract some objects.
2. Magnets can pull objects without ___touching___ them.
3. A ___magnet___ can attract objects made of iron, nickel, or steel.
4. Strong magnets can ___pull___ objects that are far away.
5. Magnets can pull objects that contain ___nickel___ or steel.
6. Magnets cannot pull objects made of ___wood___ or plastic.

**Outline, pp. 218–219**
Also available as a student workbook

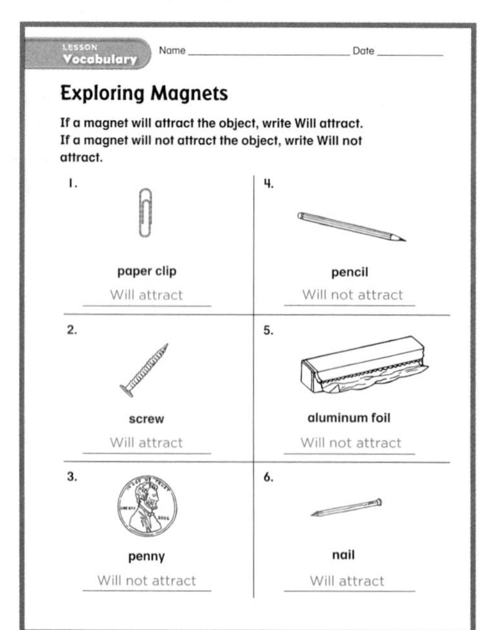

LESSON Vocabulary Name ______ Date ______

**Exploring Magnets**

**If a magnet will attract the object, write Will attract. If a magnet will not attract the object, write Will not attract.**

1. paper clip — Will attract
2. screw — Will attract
3. penny — Will not attract
4. pencil — Will not attract
5. aluminum foil — Will not attract
6. nail — Will attract

**Vocabulary, p. 220**
Also available as a student workbook

## Visual Literacy

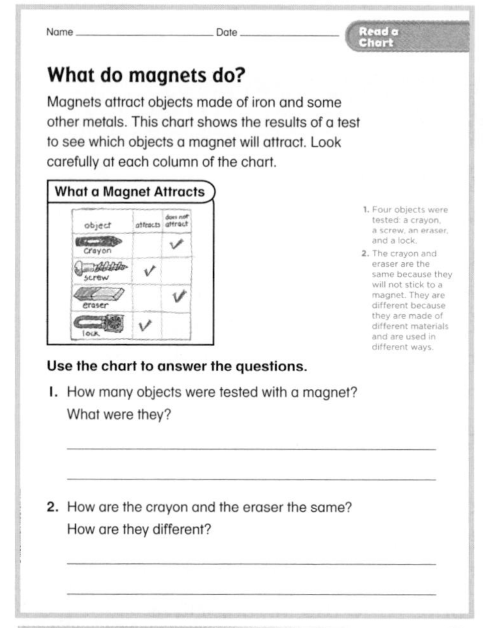

Name ______ Date ______ Read a Chart

**What do magnets do?**

Magnets attract objects made of iron and some other metals. This chart shows the results of a test to see which objects a magnet will attract. Look carefully at each column of the chart.

**What a Magnet Attracts**

| object | attracts | does not attract |
|---|---|---|
| crayon | | ✓ |
| screw | ✓ | |
| eraser | | ✓ |
| lock | ✓ | |

1. Four objects were tested: a crayon, a screw, an eraser, and a lock.
2. The crayon and eraser are the same because they will not stick to a magnet. They are different because they are made of different materials and are used in different ways.

**Use the chart to answer the questions.**

1. How many objects were tested with a magnet? What were they?
2. How are the crayon and the eraser the same? How are they different?

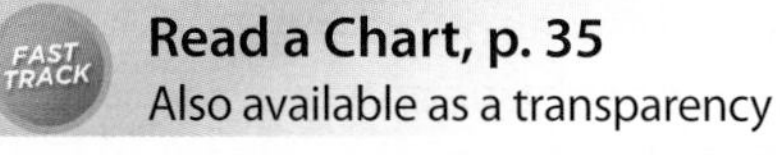

**Read a Chart, p. 35**
Also available as a transparency

# Activity Lab Book

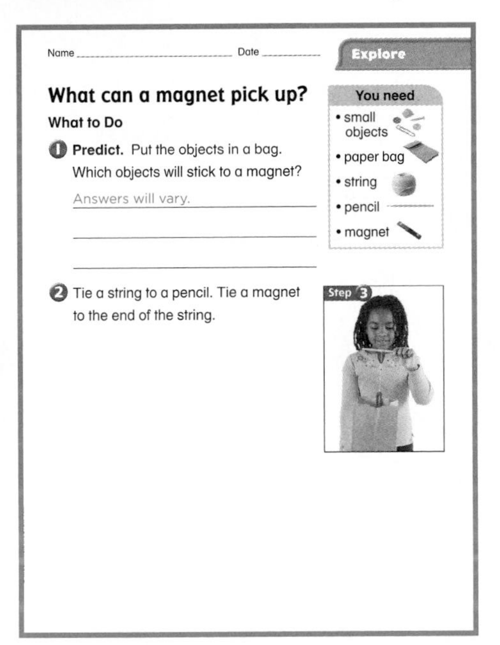

Name ________ Date ________ **Explore**

**What can a magnet pick up?**

**What to Do**

**You need**
- small objects
- paper bag
- string
- pencil
- magnet

1. **Predict.** Put the objects in a bag. Which objects will stick to a magnet?
   Answers will vary.
2. Tie a string to a pencil. Tie a magnet to the end of the string.

Step 3

**Explore, pp. 175–176**
Also available as a student workbook

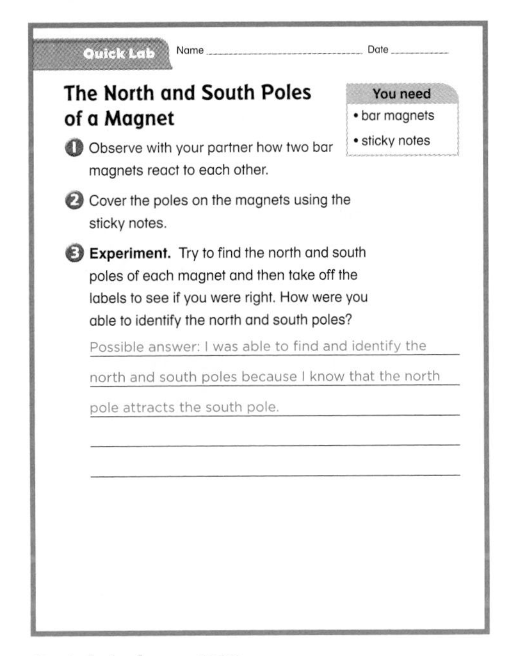

**Quick Lab** Name ________ Date ________

**The North and South Poles of a Magnet**

**You need**
- bar magnets
- sticky notes

1. Observe with your partner how two bar magnets react to each other.
2. Cover the poles on the magnets using the sticky notes.
3. **Experiment.** Try to find the north and south poles of each magnet and then take off the labels to see if you were right. How were you able to identify the north and south poles?
   Possible answer: I was able to find and identify the north and south poles because I know that the north pole attracts the south pole.

**Quick Lab, p. 178**
Also available as a student workbook

# Assessment

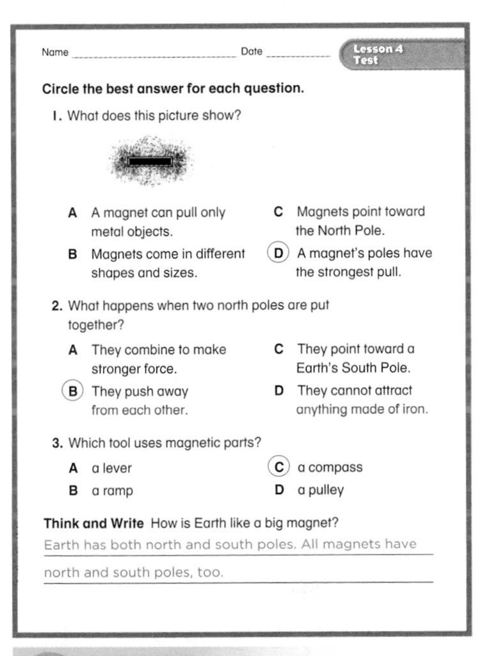

Name ________ Date ________ **Lesson 4 Test**

**Circle the best answer for each question.**

1. What does this picture show?
   - A A magnet can pull only metal objects.
   - B Magnets come in different shapes and sizes.
   - C Magnets point toward the North Pole.
   - (D) A magnet's poles have the strongest pull.
2. What happens when two north poles are put together?
   - A They combine to make stronger force.
   - (B) They push away from each other.
   - C They point toward a Earth's South Pole.
   - D They cannot attract anything made of iron.
3. Which tool uses magnetic parts?
   - A a lever
   - B a ramp
   - (C) a compass
   - D a pulley

**Think and Write** How is Earth like a big magnet?
Earth has both north and south poles. All magnets have north and south poles, too.

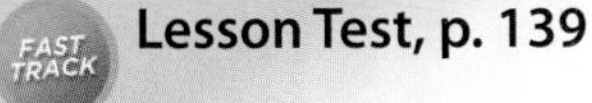


**Lesson Test, p. 139**

# ADDITIONAL RESOURCES

**pp. 172–175**

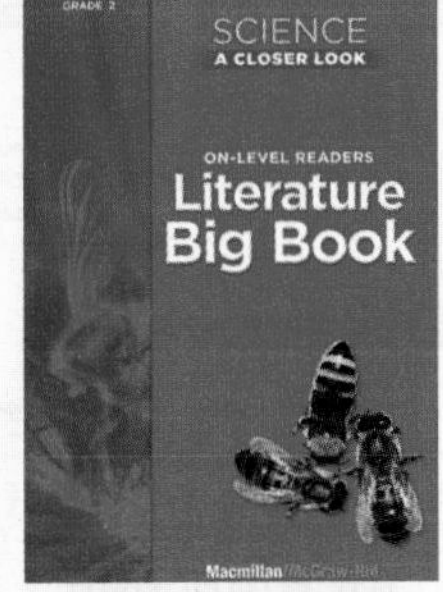


***Push or Pull?***

**pp. 53–54**

**p. 113**

Transparencies for Visual Literacy

**p. 35**

**35**

**101–110**

**pp. 111–112**

**91–93**

## Technology

- Science Activity DVD
- TeacherWorks™ Plus CD-ROM
- Classroom Presentation Toolkit CD-ROM

-Review

NSDL

# Lesson 4 Exploring Magnets

## Objectives

- Observe magnets attract and repel objects.
- Identify magnet poles and explain how they function.

# 1 Introduce

## ▶ Assess Prior Knowledge

Have children share what they know about magnets.

Ask:

- **When have you used magnets at home or school?**
- **Why are magnets helpful?**
- **What things are attracted to a magnet?**

Record children's answers in the the What We Know column of the class **KWL** chart.

## Look and Wonder

Read and discuss the Look and Wonder question. Encourage children to share their responses to the question. Ask:

- **What are the materials used to make some of the objects in the picture?** Possible answers: metal, rubber, plastic
- **What do the objects that are sticking to the magnet have in common?** Possible answer: They are all made of metal.

### RESOURCES and TECHNOLOGY

- **Activity Lab Book,** pp. 175–177
- **Activity Flipchart,** p. 53
- **Science Activity DVD**

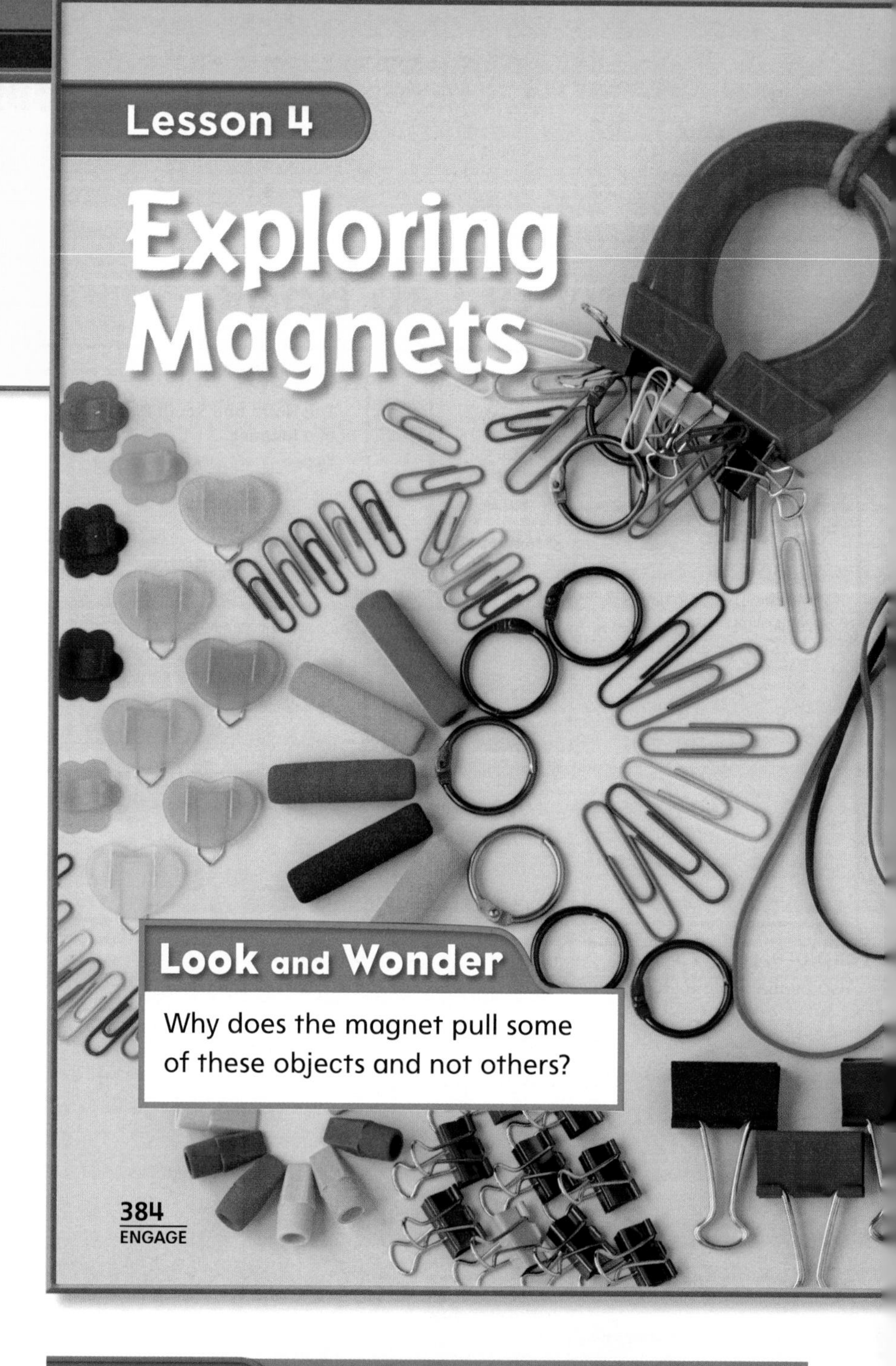

Lesson 4
Exploring Magnets
Look and Wonder
Why does the magnet pull some of these objects and not others?
384
ENGAGE


## Warm Up

### Start with a Book

Before reading ***What Makes a Magnet,*** by Franklyn M. Branley (HarperCollins, 1996), show children the cover of the book and invite them to share their predictions about the story.

Read the book and discuss what a magnet does. After reading the book, help children use the information they learned from the book to identify two objects in the classroom that they predict are magnetic.

Distribute magnets to children. Give them an opportunity to experiment with magnets and see whether their predictions were correct. Ask:

- **What made you think the object would be attracted to a magnet?**

## Explore

Inquiry Activity

### What can a magnet pick up?

**What to Do**

1. **Predict.** Put the objects in a bag. Which objects will stick to a magnet?
2. Tie a string to a pencil. Tie a magnet to the end of the string.
3. Use the magnet to pull objects out of the bag.

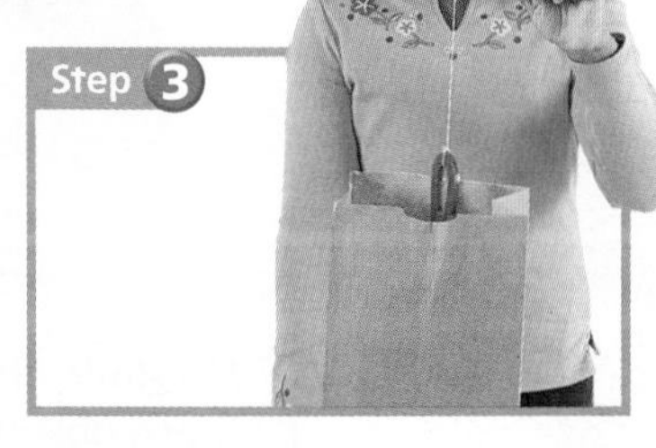


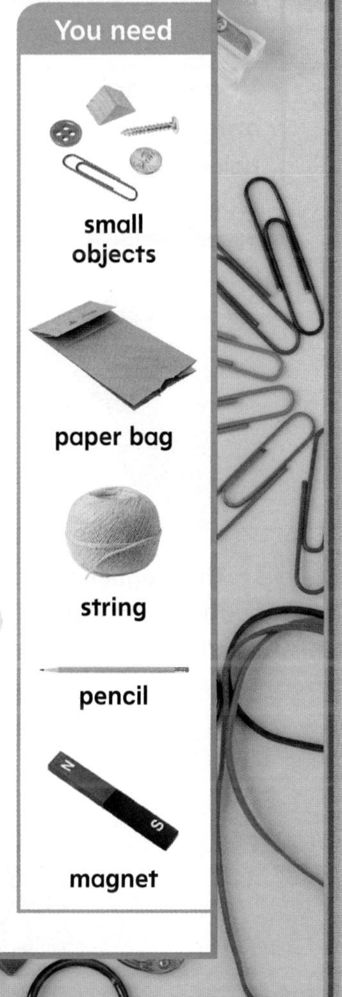

You need
small objects
paper bag
string
pencil
magnet


**Explore More**

4. **Classify.** How are the things that stick to the magnet alike?



## Alternative Explore

### What metals are attracted to magnets?

Collect aluminum cans and magnetic objects. Explain that not all metals are magnetic. For example, aluminum is not magnetic. Recycling centers sort aluminum from other metals.

Have children use a magnet to test each object. Ask them to separate the aluminum objects from the magnetic metals.

Name ______ Date ______ Alternative Explore

**What metals are attracted to magnets?**

In this activity, you will separate non-magnetic aluminum from magnetic metals by sorting metal objects for recycling.

You need
- metal objects such as soda cans and aluminum foil
- magnet

**What to Do**

1. Work with a partner. Gather several pieces of metal and aluminum provided by your teacher.
2. **Investigate.** Test each object to see if it can be pulled by a magnet.
3. **Classify.** Sort the objects that can be pulled by the magnet in one group. Put the ones that cannot be pulled by a magnet in another group.

**What Did You Find Out?**

4. Which objects cannot be pulled by a magnet?

Possible answer: The soda cans and aluminum foil cannot be pulled by a magnet.

Activity Lab Book, p. 177

## Explore

small groups

25 minutes

**Plan Ahead** Find enough small magnetic objects, such as paper clips and iron screws, so each child can "fish out" something magnetic. Gather nonmagnetic objects, such as erasers, paper, crayons, pencils. Use strong paper bags that will not rip easily. Precut the string (about 6–8 inches in length) for each group.

**Purpose** This activity will allow children to test their predictions by experimenting with an assortment of objects. Through experimentation, children will be able to classify which objects are attracted to magnets.

**Structured Inquiry** What to Do

Make a model magnetic "fishing" rod to show to the children. Ask: **What kinds of "fish" might be caught with the rod?**

1. **Predict** Before children make their own bags, show them a model bag and the objects in it. Have them predict which objects will be magnetic. Record their predictions on chart paper.
2. Make sure children tie the string in the center of the magnet to balance it properly.
3. Have children take turns using the magnetic fishing rod to fish something out of the bag. Remind them to use only the rod and not their hands. As they fish out magnetic objects, have children put them in a pile.

**Guided Inquiry** Explore More

4. **Classify** Ask children to think about how the objects that are attracted to the magnet are alike. Have children remove the objects from the bag that did not stick to the magnet and put them on their desks in a pile. Ask: **How are the objects that did not stick to the magnet alike?**

**Open Inquiry**

Encourage children to explore further by researching how magnets are used. For example, children can investigate machines or devices that use magnets to sort objects. Have resource materials available for children to conduct their research.

# 2 Teach

## Read Together and Learn

**Reading Skill** Problem and Solution A problem is what needs to be done, found out, or changed. A solution fixes the problem.

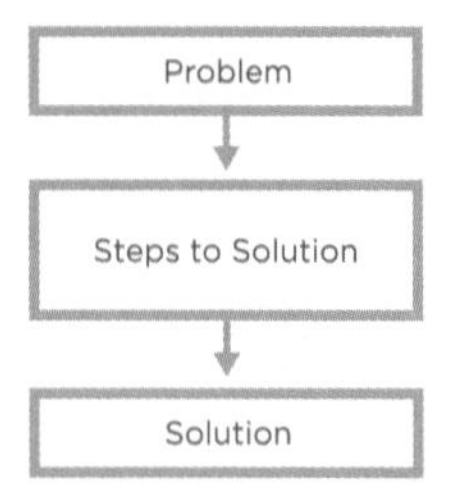


*Graphic Organizer 12,* p. TR14

## What do magnets do?

### ▶ Discuss the Main Idea

**Main Idea** Magnets attract objects made with nickel or iron, such as steel.

Before reading, ask children to describe magnets.

After reading together, ask:

- **What are the magnets in the photographs doing?** Possible answer: They are pulling metal things.
- **Why can a magnet attract a steel paper clip through a piece of paper?** because magnets attract objects through solids and paper is a solid

**Read Together and Learn**

**Essential Question**
What are magnets?

**Vocabulary**
attract
poles
repel

## What do magnets do?

A magnet can **attract**, or pull, some objects. Magnets attract objects through solids, liquids, and gases. A very strong magnet can pu objects from far away. The farther a magnet is from the object, the weaker the magnet's pull will be.

Many magnets contain iron. Magnets attract objects made with iron, including steel. They can also attract objects containing nickel.

The magnet pulls the paper clip without touching it.

Magnets are holding these objects in place.



386
EXPLAIN

### Science Background

**Magnetism** Magnetism is a force that pulls objects made with iron or nickel toward magnets. Magnetic items are attracted to a magnet's poles, or ends. This is where a magnet has the strongest force. Horseshoe and bar magnets have poles at each end, but a doughnut magnet has one pole on its outer surface and the other pole on its inner one.

See **Science Yellow Pages**, in the Teacher Resources section, for background information.

**Professional Development** For more Science Background and resources from NSDL visit http://nsdl.org/refreshers/science

### ELL Support

**Classify** Hold up different objects. Invite children to predict whether a magnet will attract or repel them. Have volunteers take turns testing each object with a magnet.

**BEGINNING** Help children name each object and then predict whether a magnet will pull the object.

**INTERMEDIATE** Have children name each object and use a short phrase to indicate whether it will be attracted to a magnet.

**ADVANCED** Have children explain, in their own words, the reasons for their predictions, using full sentences and vocabulary related to forces and magnets.

There are many objects that agnets can not attract. These clude plastic, wood, and some etals. Walk around your classroom ith a magnet. See what the magnet tracts and what it does not.

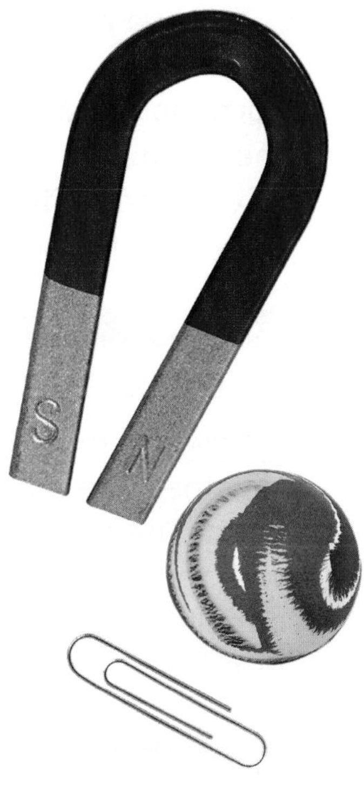

### 'hat a Magnet Attracts

| object | attracts | does not attract |
|---|---|---|
| crayon | | ✓ |
| screw | ✓ | |
| eraser | | ✓ |
| lock | ✓ | |

**Read a Chart**

Which objects will stick to a magnet?

✓ Will a magnet attract a button? Why or why not?

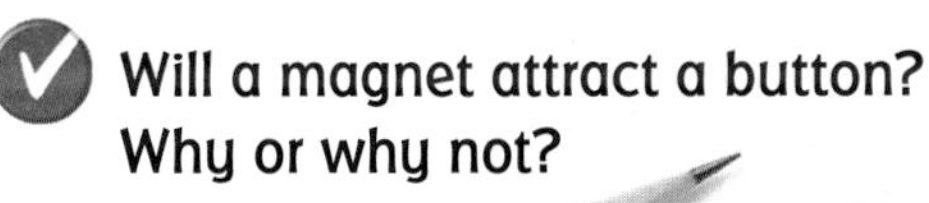

387
EXPLAIN

## Differentiated Instruction

**Leveled Activities**

**EXTRA SUPPORT** Provide magnets for children to use to explore the classroom for magnetic objects. Hang a piece of chart paper up where children can reach it. Have them record the name of each object they test and whether or not it is magnetic. Be sure children do not test the classroom computer.

**ENRICHMENT** Have children compare the strength of different areas on a magnet by making a paper clip chain, one clip at a time, on different parts of the magnet. Ask children to write about what they discovered.



FAST TRACK **Read a Chart**

Review the titles and format of the chart with children. Point out that the objects are listed in the rows of the chart. Ask:

- **What do you see that gives you information about what a magnet attracts?** Possible answer: There are checks in the *Attracts* column next to the screw and lock.
- **What do you see that helps you know what a magnet does not attract?** Possible answer: The crayon and eraser rows have a check under the *Does Not Attract* column.

**Answer to Read a Chart** screw, lock

## ▶ Develop Vocabulary

**attract** ***Scientific vs. Common Use*** In common usage, the word *attract* means "to arouse the interest or attention of." For example: *The new playground will attract lots of children.* Help children describe how *attract* was used in the sentence.

Explain to children that scientists use the word *attract* to describe "the connecting pull of a magnet and a magnetic object." Have children use *attract* to describe how magnets work.

## ▶ Explore the Main Idea

**ACTIVITY** Explain that iron is a magnetic mineral that is found in the rock *magnetite*. If possible, get a piece of *magnetite,* also called *lodestone*, and other small rocks. Have children use a magnet to classify rocks that do and do not contain the mineral *magnetite*.

### ✓ *Quick Check Answer*

Maybe; only if the button is made with iron or nickel parts

**RESOURCES and TECHNOLOGY**

- **Reading and Writing,** pp. 218–220
- **Visual Literacy,** p. 35
- **PuzzleMaker CD-ROM**
- **Classroom Presentation Toolkit CD-ROM**
- LOG ON e-Glossary

## What are poles?

### ▶ Discuss the Main Idea

**Main Idea** Magnets have north and south poles.

Read the text together and discuss how magnets attract or repel one another.

### ▶ Use the Visuals

Encourage children to describe what they see in the illustrations on page 388. Ask:

- **What do the arrows in the illustrations show?** Possible answer: The arrows pointing to one another show how magnets attract. The arrows pointing in opposite directions show how magnets repel.

Direct children to the photograph on page 388. Ask:

- **Why are there more pieces of iron at the ends of the magnet than in the middle?** Possible answers: The magnet is stronger at the poles. The north and south poles of the magnet are stronger than the middle of the magnet.

### ▶ Develop Vocabulary

**poles** *Word Origin* *Pole* comes from the Latin *polus,* meaning "pole of the heavens" and from the Greek *polos,* meaning "axis of the sphere." Write each word on the board and discuss how the meanings and word structures are similar. Ask children to use the word *poles* in a sentence to describe where they are located on a magnet

**repel** *Word Origin* Explain that *repel* comes from the Latin word *pellere,* meaning "to drive." Point out that the prefix *re-* means "back" or "again," so the word *repel* means "drive back." Have children try to put the same poles of two magnets together. Ask them to explain what happens, using the word *repel.*

**RESOURCES and TECHNOLOGY**

- **School to Home Activities,** pp. 111–112
- **Reading and Writing,** p. 221
- **Activity Lab Book,** p. 178
- LOG ON **e-Review** Narrated Summary and Quiz
- **ExamView® *Assessment Suite* CD-ROM**

## What are poles?

The two ends of a magnet are its **poles**. Every magnet has a north pole and a south pole. Put the north pole of one magnet next to the south pole of another. They will attract each other.

Now put the two south poles together. They will **repel**, or push apart. The same thing will happen with the two north poles. The push or pull of a magnet is strongest at its poles.

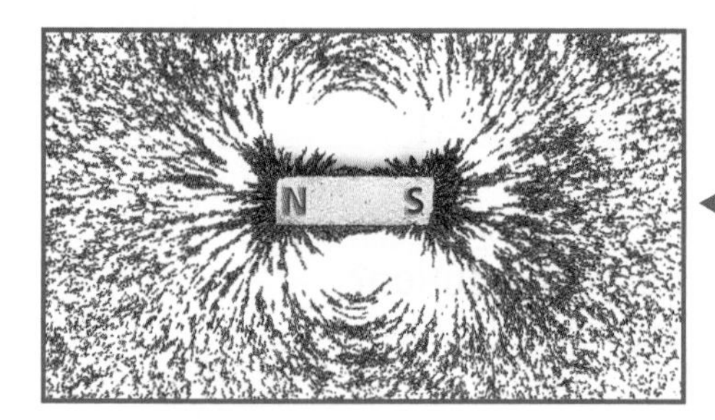


◀ This magnet attracts tiny pieces of iron.

**Quick Lab**

Cover the labels o[n] two bar magnets. **Investigate** to fi[nd] which poles are alike and which are different.



**FACT** Some magnets are much stronger than others.

## Quick Lab

 pairs  10 minutes

**Objective** Identify the north and south poles of a magnet.

**You need** bar magnets, sticky notes

1. Distribute two bar magnets and four sticky notes to each pair.
2. Have children cover the poles on their magnets using the sticky notes or labels.
3. Have children experiment to find the north poles and the south poles of each bar magnet. Ask children to uncover the labels to see if they were correct.

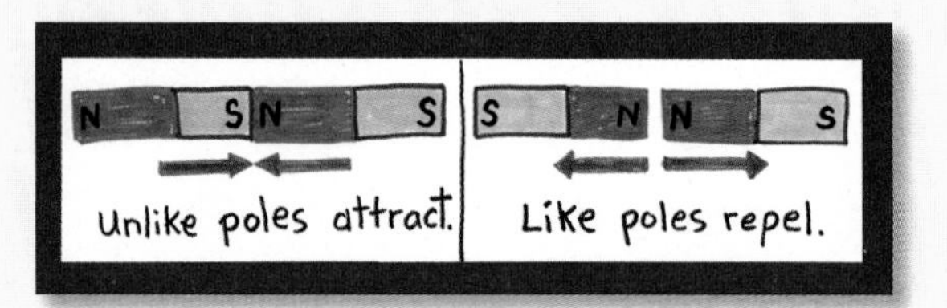

Earth acts like a big magnet.
ke every magnet, it has north
nd south poles.

A compass is a magnet that is
ee to spin. The north pole in
e magnet points toward Earth's
orth Pole.

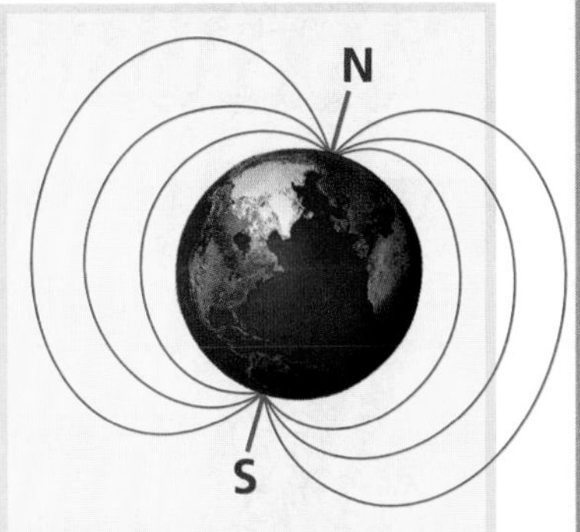


Earth has a magnetic force around the North and South Poles.

The needle inside a compass is a magnet that points to Earth's North Pole.

Where is the pull of a magnet strongest?

## Think, Talk, and Write

1. **Problem and Solution.** Two magnets repel each other. How can you make them stick together?
2. What will a magnet attract?
3. **Essential Question.** What are magnets?

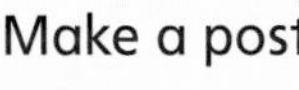

### Art Link

Make a poster that shows how people use magnets.

LOG ON e-Review Summaries and quizzes online at www.macmillanmh.com

389
EVALUATE

## Quick Check Answer

at the poles

# 3 Close

### ▶ Using the KWL Chart

Review with children what they have learned about magnets. Record their responses in the What We Learned column of the class **KWL** chart.

### ▶ Using the Reading Skill
#### Problem and Solution

Use the reading skill graphic organizer to solve a problem about a magnet. Ask: **What could you do to find the south pole of an unmarked bar magnet?**

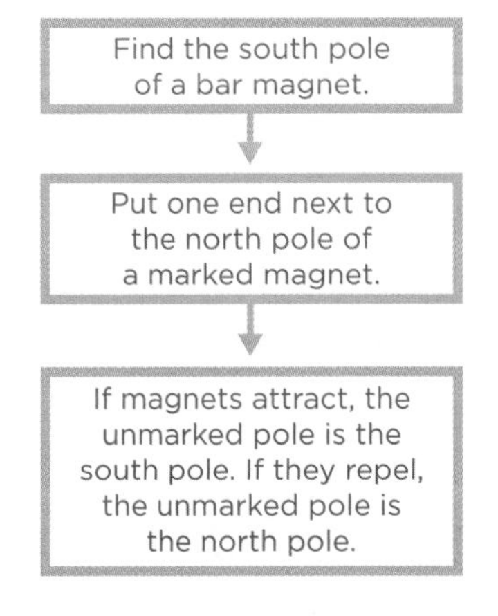

Find the south pole of a bar magnet.
Put one end next to the north pole of a marked magnet.
If magnets attract, the unmarked pole is the south pole. If they repel, the unmarked pole is the north pole.


*Graphic Organizer 12,* p. TR14

## Formative Assessment

### Attract and Repel

Have children cut a piece of paper in two lengthwise.

On one piece of paper, ask them to draw a picture of two attracting magnets. On the other piece of paper, have them draw repelling magnets.

Have them label the poles of the magnets. Encourage them to write a sentence to describe why the magnets attract and why they repel.

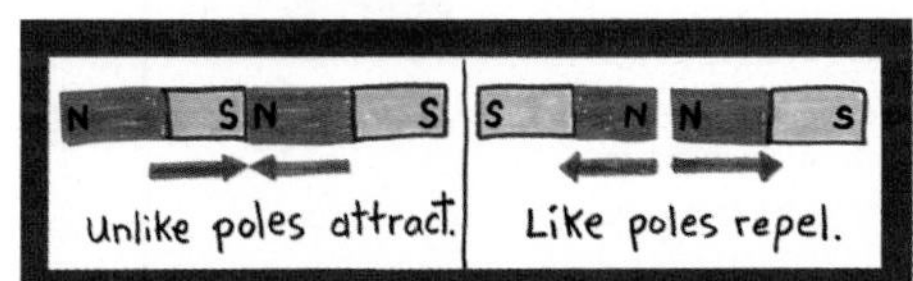

Key Concept Cards For student intervention, see the prescribed routine on **Key Concept Card 35.**

## FAST TRACK Think, Talk and Write

1. **Problem and Solution** Possible answer: Turn one magnet so that opposite poles are near each other.
2. Possible answer: Magnets attract objects made with iron or nickel, such as paper clips.
3. **Essential Question** Possible answer: A magnet is a piece of metal that attracts or pulls objects made with iron or nickel.

## Art Link

Collect reference books, such as ***What Magnets Can Do,*** by Allan Fowler (Children's Press, 1995), and ***Science All Around Me: Magnets,*** by Karen Bryant-Mole (Heinemann, 1998), to help children create their posters.

## Be a Scientist

pairs | 30 minutes

Use Your

GRADE 2
SCIENCE A CLOSER LOOK
Activity Flipchart for Science Centers and Workstations
Macmillan/McGraw-Hill

Activity Flipchart, p. 54

**Skills** record data, communicate, investigate

### Objective

- Record the results of an experiment in a bar graph.

**You need** paper clips, magnets

**Plan Ahead** Test paper clips ahead of time to check that they are magnetic. Make copies of 1-inch graph paper for children to use to make their bar graphs.

**EXTEND** Children will experiment to find out how many paper clips will hang from magnets.

## Be a Scientist

**You need**

paper clips

magnets

### How can you compare the strength of different magnets?

Find out how many paper clips each of the magnets can attract.

**What to Do**

1. Hang a paper clip from a magnet. Keep adding more clips in a line until no more will stick.

**Structured Inquiry** What to Do

### How can you compare the strength of different magnets?

1. Explain to children that they are going to test the strength of different magnets by seeing how many paper clips each magnet can hold. Model for children how to add one paper clip at a time to form a line.
2. **Record Data** Have children record how many paper clips hang from each magnet. They may use tally marks or numbers.
3. Have children repeat the activity using bar, ring, and horseshoe magnets.

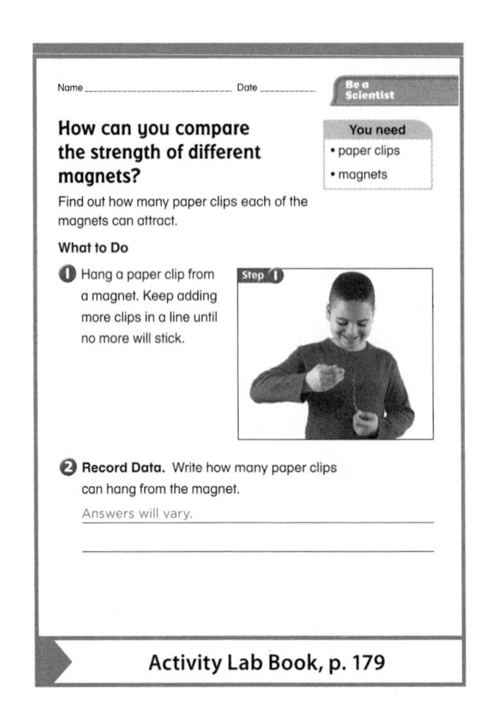

Name ______ Date ______ Be a Scientist

**How can you compare the strength of different magnets?**

You need
- paper clips
- magnets

Find out how many paper clips each of the magnets can attract.

What to Do

1. Hang a paper clip from a magnet. Keep adding more clips in a line until no more will stick.

Step 1

2. **Record Data.** Write how many paper clips can hang from the magnet.

Answers will vary.

Activity Lab Book, p. 179

**RESOURCES and TECHNOLOGY**

- **Activity Lab Book,** pp. 179–180
- TeacherWorks™ Plus CD-ROM

Inquiry Investigation

2 **Record Data.** Write how many paper clips can hang from the magnet.

3 Repeat the steps using different magnets.

4 **Communicate.** Make a bar graph to show the strengths of your magnets.

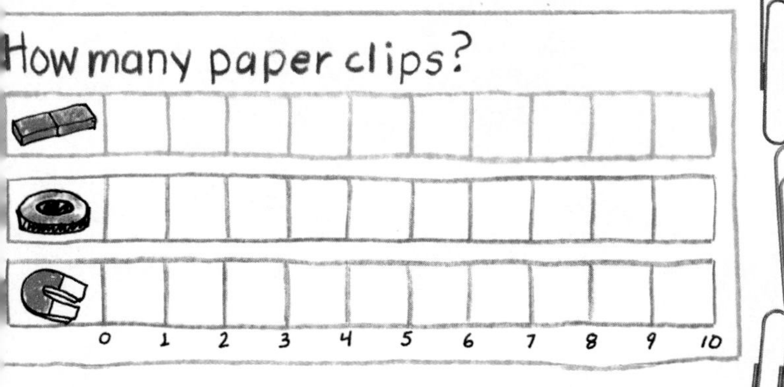


**Investigate More**

**Investigate.** How many paper clips can you pick up with two magnets? Find a way to attach two magnets and try it out.

Use with Chapter 11, Lesson 4 54

④ **Communicate** When children are finished, distribute graph paper. Discuss and model how children can use the graph paper to make a bar graph showing their results. Ask: **What does your bar graph tell you about the different magnets? Why do you think you got the results you did from the experiment?**

**Guided Inquiry** **Investigate More**

**Investigate** Have children make predictions about the number of paper clips that can be picked up with two magnets before they conduct their experiments. Ask: **Will two magnets hold more paper clips than a single magnet? Which parts of the magnets will you attach to one another?**

Have children explain their results using the two magnets together.

Encourage children to add the results of this experiment to their bar graphs.

**Open Inquiry**

Have children find out how magnets can be used in different ways. Ask: **What machines or devices use magnets?**

Ask children to work in pairs to research a machine or device that uses a magnet, such as a can opener, compass, cabinet door, or magnet crane. Have children find out how the magnets are used to make the machine or device work. Encourage children to write a report about their machine or device and share it with the class.

## Integrate Writing

### Using a Compass

Explain to children that a compass is an instrument used for finding directions on Earth. Display a magnet and point to the needle. Tell children that the needle is magnetized. Ask:

- **Why is it helpful that the compass needle is a magnet?** Possible answers: Earth has a magnetic force around its poles. The magnet will line up with the Earth's poles to show direction.
- **Where does the compass needle point to?** north
- **Why is a compass a useful tool?** Possible answer: It helps people find their way when they are lost.

Ask children to investigate the questions and write a story about how a compass works. Encourage children to draw illustrations to accompany their stories.

# I Read to Review

**Objective**

- Review simple machines, forces, and movement with independent reading.

### Buddy Reading

- During small-group instructional time, have children work in pairs to read the selection to one another.
- Pair weaker readers with stronger partners to ensure success during this work period.

### Independent Reading

- Give children copies of School to Home Activities, pages 113 and 114. Have them assemble the pages to make their own book.
- Encourage children to take their book home and read it aloud to a family member.

PAGES **390–391** ▶

Ask: **Which forces move things in the story?**
Possible answers: Pushes and pulls move the shopping cart; gravity moves the egg.

I Read to Review
Forces Every Day
Push! I use forces all day long. I push the heavy cart to make it move. A push is a force. I pull on the cart to make it stop. A pull is a force too.
390


PAGES **392–393** ▶

Ask: **Which tool in the picture is a simple machine?**
Possible answer: The spatula is a lever.
Ask: **What type of metal is in the refrigerator door?**
iron

Flip! I flip my pancakes. Simple machines can make work easier. A lever can change the direction of a force. The force makes the pancake fly through the air.

**RESOURCES and TECHNOLOGY**

- **School to Home Activities,** pp. 113–114
- **Assessment,** pp. 144–145
- **TeacherWorks™ Plus CD-ROM**

Plop! I crack an egg. A force called gravity makes the egg fall into the bowl. Objects move with different motions. I stir the pancake batter around and around.

391

Yum! The pancake lands on my plate. Other things use pushes and pulls too. A magnet keeps my shopping list on the refrigerator. It is time to buy more eggs and milk!

393

# Performance Assessment

## Forces at Work

 individual  40 minutes

**Materials** a picture of people using simple tools in a garden, drawing paper, colored pencils

1. Show children pictures of people working in a park or a garden, with shovels, rakes, carts, and other tools. Point out to children that simple tools are used in everyday life to help complete many chores.
2. Have children think of other activities where simple machines are used. Ask: **What kinds of simple machines are used for cooking?**
3. Ask children to write a story about how simple machines are used for a specific activity. Remind children that simple machines are used in games and recreational activities.
4. Distribute drawing paper and colored pencils. Encourage children to make a drawing for the cover of their stories. Encourage children to share their stories with their classmates.

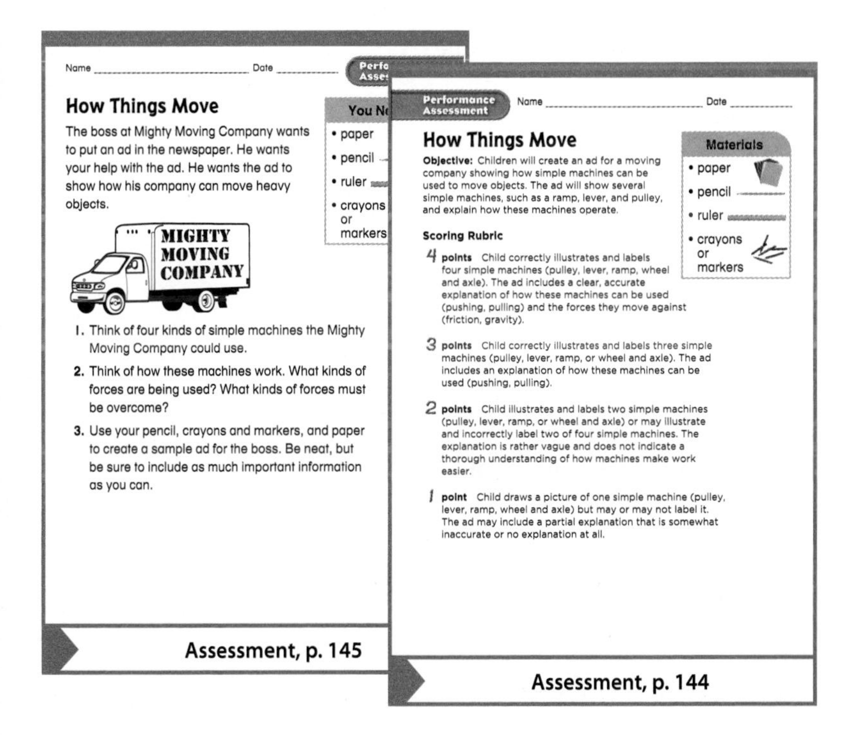

Name ______________ Date ________

### How Things Move

The boss at Mighty Moving Company wants to put an ad in the newspaper. He wants your help with the ad. He wants the ad to show how his company can move heavy objects.

**You N…**
- paper
- pencil
- ruler
- crayons or markers

MIGHTY MOVING COMPANY

1. Think of four kinds of simple machines the Mighty Moving Company could use.
2. Think of how these machines work. What kinds of forces are being used? What kinds of forces must be overcome?
3. Use your pencil, crayons and markers, and paper to create a sample ad for the boss. Be neat, but be sure to include as much important information as you can.

Assessment, p. 145

Performance Assessment

Name ______________ Date ________

### How Things Move

**Objective:** Children will create an ad for a moving company showing how simple machines can be used to move objects. The ad will show several simple machines, such as a ramp, lever, and pulley, and explain how these machines operate.

**Materials**
- paper
- pencil
- ruler
- crayons or markers

**Scoring Rubric**

4 **points** Child correctly illustrates and labels four simple machines (pulley, lever, ramp, wheel and axle). The ad includes a clear, accurate explanation of how these machines can be used (pushing, pulling) and the forces they move against (friction, gravity).

3 **points** Child correctly illustrates and labels three simple machines (pulley, lever, ramp, or wheel and axle). The ad includes an explanation of how these machines can be used (pushing, pulling).

2 **points** Child illustrates and labels two simple machines (pulley, lever, ramp, or wheel and axle) or may illustrate and incorrectly label two of four simple machines. The explanation is rather vague and does not indicate a thorough understanding of how machines make work easier.

1 **point** Child draws a picture of one simple machine (pulley, lever, ramp, wheel and axle) but may or may not label it. The ad may include a partial explanation that is somewhat inaccurate or no explanation at all.

Assessment, p. 144

# CHAPTER 11 Review

For information on Depth of Knowledge levels, see page 395B.

## ▶ Use the KWL Chart

Review the **KWL** chart that the class made at the beginning of the chapter. Help children compare what they know about how things move now with what they knew then. Add any additional information to the What We Learned column of the **KWL** chart.

## ▶ Make a FOLDABLES Study Guide

Make a large four-door Foldables for the class. Divide the class into four groups.

Ask the Lesson 1 group to make a three-column chart in which they list words that describe position, motion, and speed and give an example of each.

Have the Lesson 2 group make a three-column chart listing types of forces, an example of the force, and how the force changes motion.

Ask the Lesson 3 group to make a three-column chart in which they list simple machines, an illustrated example of each, and how it makes work easier.

Give the Lesson 4 group three large index cards. Have them do the following: on first card, draw a diagram of a magnet; on second card, list things magnets attract and repel; on last card, list magnet facts.

Paste each group's work under the appropriate door. See page TR38 in the back of the Teacher's Edition for more information on Foldables.

## Vocabulary

1. friction
2. ramp
3. position
4. gravity
5. speed
6. lever

## RESOURCES and TECHNOLOGY

- **Reading and Writing,** pp. 222–223
- **Assessment,** pp. 132–135, 140–143
- ExamView® *Assessment Suite* CD-ROM
- PuzzleMaker CD-ROM
- Vocabulary Games

## CHAPTER 11 Review

### Vocabulary
DOK 1

**Use each word once for items 1–6.**

friction
gravity
lever
position
ramp
speed

1. When two objects rub together, they can be slowed down by ________.
2. A simple machine that makes it easier to push an object to a higher level is a ________.

3. We can tell where an object is by its ________.
4. Objects fall to the floor because of a force called ________.
5. How far an object moves in a period of time is called ________.
6. A simple machine that moves against a fulcrum is called a ________.

394 e-Glossary Words and definitions online at www.macmillanmh.com

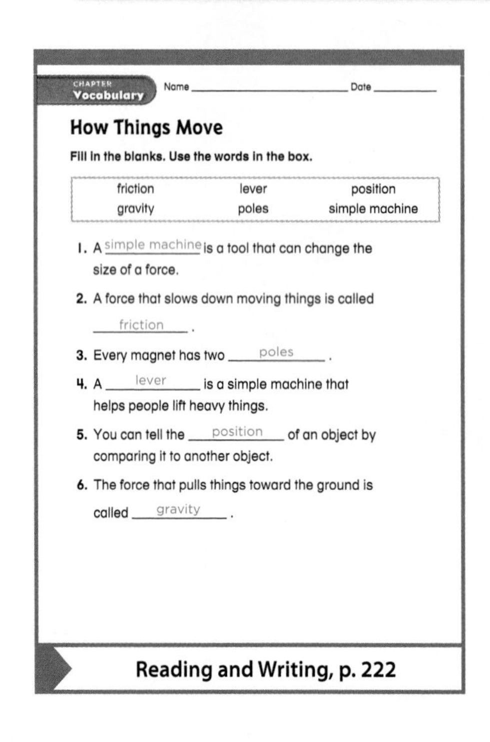
Chapter Vocabulary Name ________ Date ________

**How Things Move**

**Fill in the blanks. Use the words in the box.**

| friction | lever | position |
|---|---|---|
| gravity | poles | simple machine |

1. A simple machine is a tool that can change the size of a force.
2. A force that slows down moving things is called friction.
3. Every magnet has two poles.
4. A lever is a simple machine that helps people lift heavy things.
5. You can tell the position of an object by comparing it to another object.
6. The force that pulls things toward the ground is called gravity.

Reading and Writing, p. 222

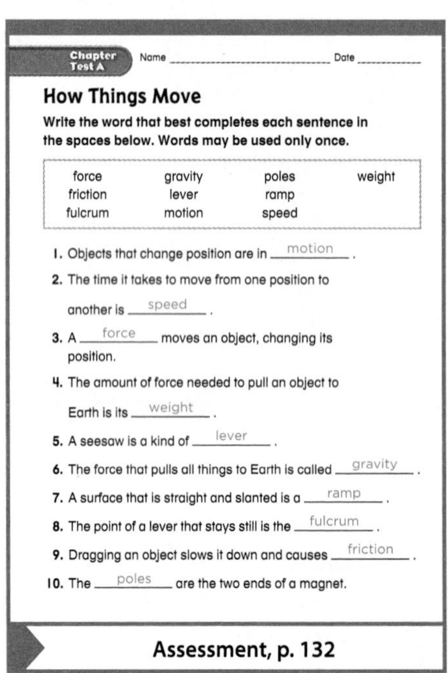
Chapter Test A Name ________ Date ________

**How Things Move**

**Write the word that best completes each sentence in the spaces below. Words may be used only once.**

| force | gravity | poles | weight |
|---|---|---|---|
| friction | lever | ramp | |
| fulcrum | motion | speed | |

1. Objects that change position are in motion.
2. The time it takes to move from one position to another is speed.
3. A force moves an object, changing its position.
4. The amount of force needed to pull an object to Earth is its weight.
5. A seesaw is a kind of lever.
6. The force that pulls all things to Earth is called gravity.
7. A surface that is straight and slanted is a ramp.
8. The point of a lever that stays still is the fulcrum.
9. Dragging an object slows it down and causes friction.
10. The poles are the two ends of a magnet.

Assessment, p. 132

## Science Skills and Ideas

DOK 2

Answer the questions below.

7. Summarize. Describe the position of the blue paper.

8. Investigate. What can help you move a heavy object?

9. Describe some of the simple machines in this picture and how they work.

10. How do things move?

LOG ON e-Review Summaries and quizzes online at www.macmillanmh.com

395

## Science Skills and Ideas

**7. Summarize** Encourage children to complete the summarize graphic organizer.

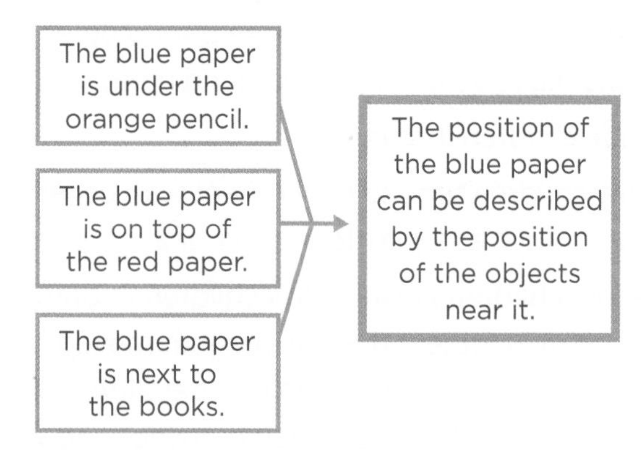

The blue paper is under the orange pencil.
The blue paper is on top of the red paper.
The blue paper is next to the books.
The position of the blue paper can be described by the position of the objects near it.


*Graphic Organizer 5,* p. TR7

**8. Investigate** Possible answers: levers, pulleys, ramps, wheel and axle

**9.** Possible answers: A shovel is a lever that is used to dig; you push down on the handle and the other end moves up. A wagon has wheels and axles; it is pulled or pushed to move things easily.

**10.** Accept all reasonable responses. Children should address concepts taught in each lesson: You can describe how an object moves by talking about the things around it; things are moved or stopped by forces, such as a push, pull, or gravity; simple machines can change the direction or strength of a force; magnets can attract or repel metals that contain iron or nickel.

## Summative Assessment and Intervention

**Assessment** provides a summative test for Chapter 11.

**Leveled Readers** may be used to reteach lesson content in an alternative format. Leveled readers deliver chapter content at different readability levels. The back cover of each reader provides comprehension building activities specific to the book content (see page 359).

**Key Concept Cards 32–35** contain prescribed routines for student intervention.

## Magnet Maze

**Materials** paper, crayons, paper clips, magnets

### ▶ Teaching Tips

1. As an option to save time, make photocopies of a simple maze before class and hand it out to children.
2. Divide children into pairs.
3. Allow children time to experiment with other objects, such as a plastic checker or another magnet.
4. Discuss children's results. Ask children to draw a picture of an object that the magnet moved, and a picture of an object that the magnet didn't affect. Encourage them to label their drawings and write a few sentences describing what happened.

Performance Assessment DOK 2

## Magnet Maze

- Draw a maze on a piece of paper. Put a magnet under the paper to move a paper clip through the maze. Have a partner time how long it takes to finish the maze.
- Move the magnet away from the paper and try the maze again. Why do you think it took longer to finish the maze?
- What would happen if you used a plastic checker instead of a paper clip? Why?
- What other objects could you use in your magnet maze?

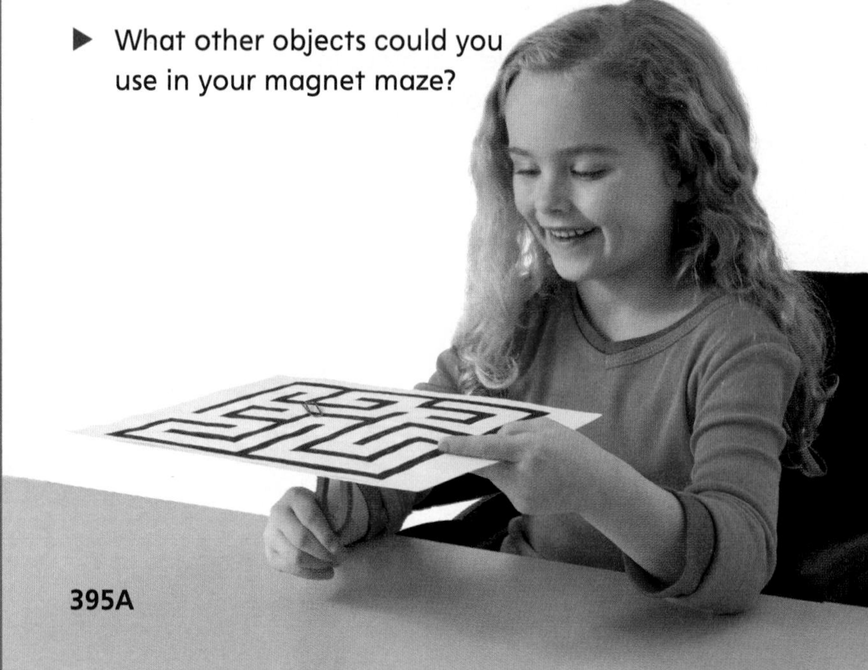

395A

### Scoring Rubric

**4 Points** Child draws and labels two pictures correctly, and writes at least three complete sentences describing what a magnet attracts and why.

**3 Points** Child draws one picture but does not label it, and writes at least two complete sentences describing what a magnet attracts and why.

**2 Points** Child draws one picture and writes at least one complete sentence describing what a magnet attracts and why.

**1 Point** Child's pictures and writing do not reflect an understanding of magnets.

## est Preparation

This graph shows how fast a train, a car, a plane, and a boat can move.

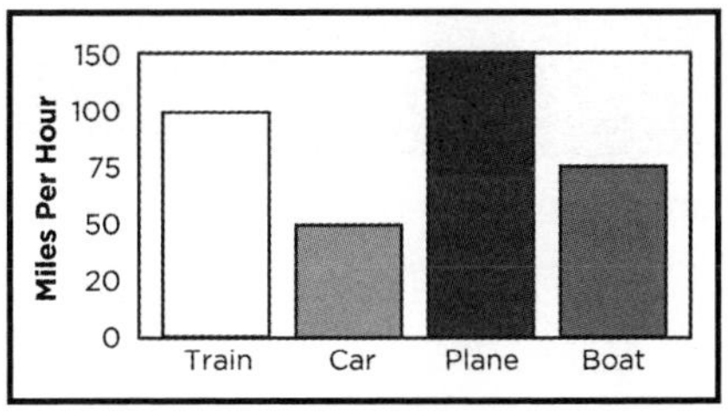

Miles Per Hour

150
100
75
50
20
0

Train Car Plane Boat

Which moves the fastest?

**A** a train
**B** a car
**C** a plane
**D** a boat
DOK 2

Look at the picture.

What force is the boy using to close the door?

**A** gravity
**B** friction
**C** pulling
**D** pushing
DOK 1

Which item will not be attracted by a magnet?

**A** a paper clip
**B** scissors
**C** a rubber band
**D** a safety pin
DOK 1

395B

## Depth of Knowledge

**Level 1 Recall** Level 1 requires memory of a fact, a definition, or a procedure. At this level, there is only one correct answer.

**Level 2 Skill/Concept** Level 2 requires an explanation or the ability to apply a skill. At this level, the answer reflects a deep understanding of the topic.

**Level 3 Strategic Reasoning** Level 3 requires the use of reasoning and analysis, including the use of evidence or supporting information. At this level, there may be more than one correct answer.

**Level 4 Extended Reasoning** Level 4 requires the completion of multiple steps and requires synthesis of information from multiple sources or disciplines. At this level, the answer demonstrates careful planning and complex reasoning.

1. C: a plane. The bars in the graph for train, car, and boat are all shorter than the bar for plane.
2. D: pushing. Gravity and friction are forces that occur without a person causing them. If the boy was pulling, the door would look like it was moving toward him.
3. C: a rubber band. Paper clips, scissors, and safety pins are usually made of steel (which contains iron) and would probably be attracted by a magnet.

Classroom Presentation Toolkit CD-ROM Lesson Presentations

# CHAPTER 12 Planner

TeacherWorks™ Plus CD-ROM Interactive Lesson Planner, Teacher's Edition, Worksheets, and Online Resources.

| Lesson | OBJECTIVES AND READING SKILLS | VOCABULARY | RESOURCES AND TECHNOLOGY |
|---|---|---|---|
| **1 Heat** PAGES **398–403**<br>PACING: 2 days<br>*FAST TRACK:* 1 day | ▪ Recognize that the Sun supplies heat and energy to Earth.<br>**Reading Skill** Main Idea and Details<br>*Graphic Organizer 1* (Main Idea; Details, Details, Details) | **heat**<br>**fuel** | Reading and Writing, pp. 225–228<br>Activity Lab Book, pp. 181–186<br>Visual Literacy, p. 36<br>Transparencies, p. 36 |
| **2 Sound** PAGES **404–413**<br>PACING: 3 days<br>*FAST TRACK:* 1 day | ▪ Discover how different sounds are produced.<br>▪ Describe the volume and pitch of sounds.<br>**Reading Skill** Problem and Solution<br>*Graphic Organizer 12* (Problem; Steps to Solution; Solution) | **sound**<br>**vibrate**<br>**pitch** | Reading and Writing, pp. 229–234<br>Math, pp. 23–24<br>Activity Lab Book, pp. 187–190<br>Visual Literacy, p. 37<br>Transparencies, p. 37<br>**Operation: Science Quest** *Sound*<br>**Science Songs CD** Track 14<br>**Science in Motion** *How We Hear Sound* |
| **3 Light** PAGES **414–419**<br>PACING: 2 days<br>*FAST TRACK:* 1 day | ▪ Identify the composition and properties of light.<br>**Reading Skill** Sequence<br>*Graphic Organizer 7* (First; Next; Last) | **light**<br>**reflect** | Reading and Writing, pp. 235–238<br>Activity Lab Book, pp. 191–196<br>Visual Literacy, p. 38<br>Transparencies, p. 38 |
| **4 Exploring Electricity** PAGES **420–427**<br>PACING: 3 days<br>*FAST TRACK:* 1 day | ▪ Identify forms of electricity and their uses.<br>**Reading Skill** Cause and Effect<br>*Graphic Organizer 8* (Cause → Effect) | **current electricity**<br>**circuit**<br>**static electricity** | Reading and Writing, pp. 239–244<br>Activity Lab Book, pp. 197–200<br>Visual Literacy, p. 39<br>Transparencies, p. 39 |

**I Read to Review** PAGES **428–431**

**Energy Poem**
- Selection for independent reading

**Performance Assessment**

**Resources**
- School to Home Activities, pp. 125–126
- Assessment, pp. 158–159

**CHAPTER 12 Review** PAGES **432–433, 433A–433B**

- Review chapter concepts.

**Resources**
- Assessment, pp. 146–149, 154–157
- Reading and Writing, pp. 245–246

**Technology**
- Exam*View*® *Assessment Suite* CD-ROM
- e-Review

PACING Assumes a day is a 20–25 minute session.

www.macmillanmh.com for more planning resources and www.nsdl.org/refreshers/science for science resources from NSDL

Science Activity DVD Explore Activity demos

Materials included in the Deluxe Activity Kit are listed in *italics*.

## EXPLORE Activities

### Explore p. 399 PACING: 15 minutes 

**Objective** Observe how ice melts.

**Skills** predict

**Materials** ice cubes, two *plastic cups* per grouping, *stopwatches* or clocks, pencils, paper

PLAN AHEAD If there is no source of sunlight in the classroom, use a bright lamp.

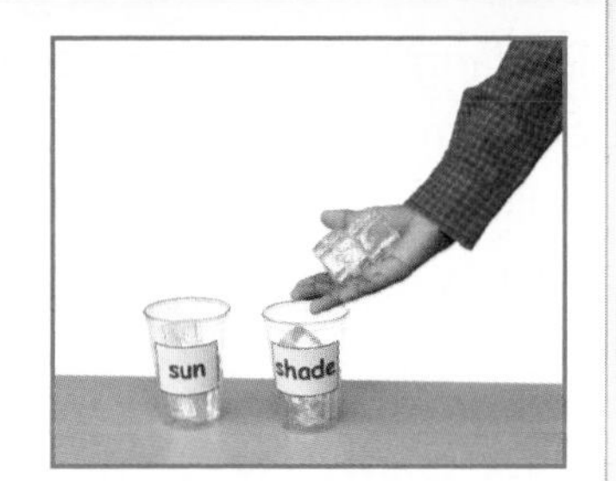


### Explore p. 405 PACING: 30 minutes 

**Objective** Observe that vibrations create sound.

**Skills** observe, predict

**Materials** *string*, *paper cups*, goggles, paper clips, scissors or other sharp object

PLAN AHEAD Collect enough paper cups, pre-cut string, and paper clips for every group.

### Explore p. 415 PACING: 15 minutes 

**Objective** Compare how different materials allow light to pass through them.

**Skills** predict, observe, compare

**Materials** *flashlights*, cardboard, plastic wrap, various items

PLAN AHEAD Provide enough flashlights and materials for pairs. Check flashlights to make sure each emits a strong beam.

### Explore p. 421 PACING: 30 minutes 

**Objective** Learn how to make a closed circuit.

**Skills** predict, record data

**Materials** *insulated wire*, *D cell batteries*, *light bulbs*, *miniature light sockets*, red and blue dot stickers

PLAN AHEAD If you use a spool of wire, you will have to cut the wires and strip about a half inch of plastic coating for proper contact with the battery.

## QUICK LAB Activities

### Quick Lab p. 403 PACING: 15 minutes 

**Objective** Compare the temperature of soil, water, and air.

**Skills** predict, measure, record, compare

**Materials** *thermometers*, three *plastic cups* per grouping, water, *soil*, crayons, paper

PLAN AHEAD Pre-measure soil into cups before class.

Temperature Chart

| air | soil | water |
| --- | --- | --- |
| 70°F | 62°F | 68°F |

### Quick Lab p. 407 PACING: 15 minutes 

**Objective** Observe sound energy.

**Skills** observe

**Materials** *tuning forks*, *plastic cups*, water

PLAN AHEAD If there is no sink in the classroom, have a pitcher filled with water readily available to fill cups. Clear plastic cups will allow children to observe the ripples more clearly.

### Quick Lab p. 418 PACING: 15 minutes 

**Objective** Observe what happens to light as it passes through a prism.

**Skills** observe, record

**Materials** *prisms*, a light source, colored pencils, paper

PLAN AHEAD The same flashlights that were used for the Explore Activity may be used for Quick Lab. Dim lights in classroom for maximum effectiveness.

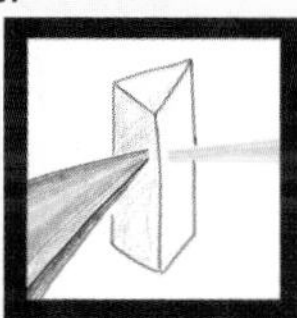

### Quick Lab p. 424 PACING: 10 minutes 

**Objective** Use static electricity to move tissue paper.

**Skills** observe

**Materials** plastic rulers, *tissue paper*, scissors, pieces of *cotton* or *flannel cloth*

PLAN AHEAD This experiment will work best in cool, dry air. Check the weather ahead of time and plan accordingly.

## FOR MORE ACTIVITIES

| Focus on Skills | Be a Scientist |
| --- | --- |
| Teach the inquiry skill: draw conclusions, p. 56 | How does sunlight affect the temperature of light and dark objects?, p. 59 |

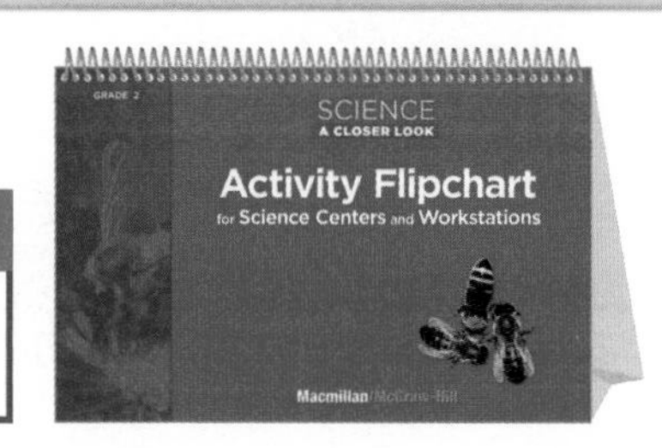


Use the Activities in your work station. See **Activity Lab Book** for more support.

For a comprehensive list of consumable and non-consumable materials, see the back of the unit tab.

# English Language Learner Support

## Technology

For additional language support and vocabulary development, go to www.macmillanmh.com

*Science in Motion*

*How We Hear Sound*

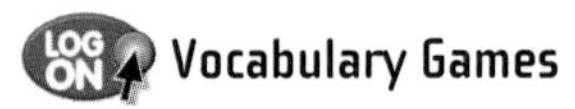

Vocabulary Games

## ADDITIONAL RESOURCES

pp. 116–125

## Academic Language

English language learners need help in building their understanding of the academic language used in daily instruction and science activities. The following strategies will help to increase children's language proficiency and comprehension of content and instruction words.

### Strategies to Reinforce Academic Language

- **Use Context** Academic language should be explained in the context of the task. Use gestures, expressions, and visuals to support meaning.
- **Use Visuals** Use charts, transparencies, and graphic organizers to explain key labels to help children understand classroom language.
- **Model** Use academic language as you demonstrate the task to help children understand instruction.

### Academic Language Vocabulary Chart

The following chart shows chapter vocabulary and inquiry skills as well as some Spanish cognates. **Vocabulary** words help children comprehend the main ideas. **Inquiry Skills** help children develop questions and perform investigations. **Cognates** are words that are similar in English and Spanish.

| Vocabulary | Inquiry Skills | Cognates: English | Cognates: Spanish |
|---|---|---|---|
| heat, p. 400 | predict, p. 399 | predict, p. 399 | *predecir* |
| fuel, p. 401 | observe, p. 405 | observe, p. 405 | *observar* |
| sound, p. 406 | compare, p. 415 | vibrate, p. 407 | *vibrar* |
| vibrate, p. 407 | record data, p. 421 | compare, p. 415 | *comparar* |
| pitch, p. 409 | | reflect, p. 416 | *reflejar* |
| light, p. 416 | | current electricity, p. 422 | *corriente eléctrica* |
| reflect, p. 416 | | circuit, p. 422 | *circuito* |
| current electricity, p. 422 | | | |
| circuit, p. 422 | | | |
| static electricity, p. 424 | | | |

## Vocabulary Routine

**Use the routine below to discuss the meaning of each word on the vocabulary list. Use gestures and visuals to model all words.**

**Define** To *vibrate* means to move back and forth quickly.

**Example** Sound is made when something *vibrates*.

**Ask** What *vibrates* in your throat when you make a sound?

**Children may respond to questions according to proficiency level with gestures, one-word answers, or phrases.**

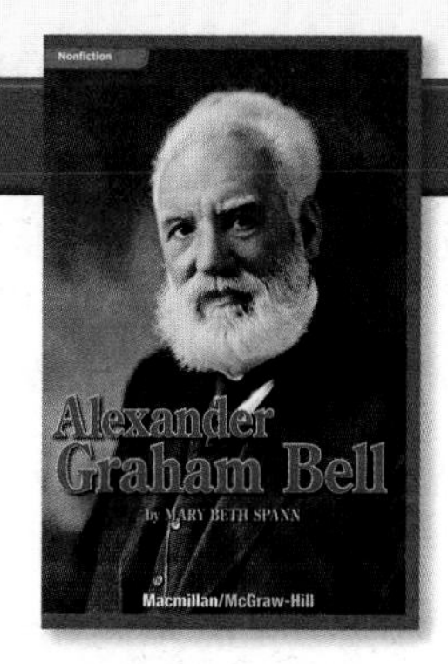


**ELL Leveled Reader**

**Alexander Graham Bell**
by Mary Beth Spann

**Summary** Read all about this brilliant man and his inventions.

**Reading Skill**
Summarize

## Vocabulary Activities

Help children understand how sound is made.

**BEGINNING** Write the word *vibrate*. Have children read it with you. Then have children listen as you shake a large sheet of paper vigorously. Explain how shaking the paper causes it and the air around it to vibrate and create sound. Encourage children to describe the sound.

**INTERMEDIATE** Form groups. Encourage the groups to make a list of sounds they hear on the way to school. Then have the groups classify the sounds listed as *loud* or *soft* and *high* or *low*. Have groups exchange lists and compare whether the sounds listed are the same or different.

**ADVANCED** Turn to the diagram on pages 406–407. Encourage children to explain why the guitar strings make a sound and how the air carries the sound to your eardrum. Encourage children to describe how you make sounds with other string instruments (piano, violin).

## Language Transfers

**Grammar Transfer**
In Cantonese and Korean, adverbs usually come before verbs.
*When the strings quickly vibrate, they make sound.*

**Phonics Transfer**
In Spanish, most speakers do not differentiate between the sounds /v/ and /b/.

# CHAPTER 12

## Using Energy

 **THE BIG IDEA** How do we use energy?

**Chapter Preview** Have children take a chapter picture walk and predict what the lessons will be about.

### ▶ Assess Prior Knowledge

Before reading the chapter, create a **KWL** chart with children. Ask the Big Idea question and then ask:

- What kind of energy comes from the Sun?
- What types of sounds are there?
- What can light pass through?
- How do people use electricity?

| Energy | | |
|---|---|---|
| What We Know | What We Want to Know | What We Learned |
| The Sun heats Earth. | Does the Sun make other kinds of energy? | |
| Car horns are loud sounds. | Are there sounds people can't hear? | |
| Electricity turns on lights. | Where does electricity come from? | |

Answers shown represent sample student responses.

Follow the **Instructional Plan** at right after assessing children's prior knowledge of chapter content.

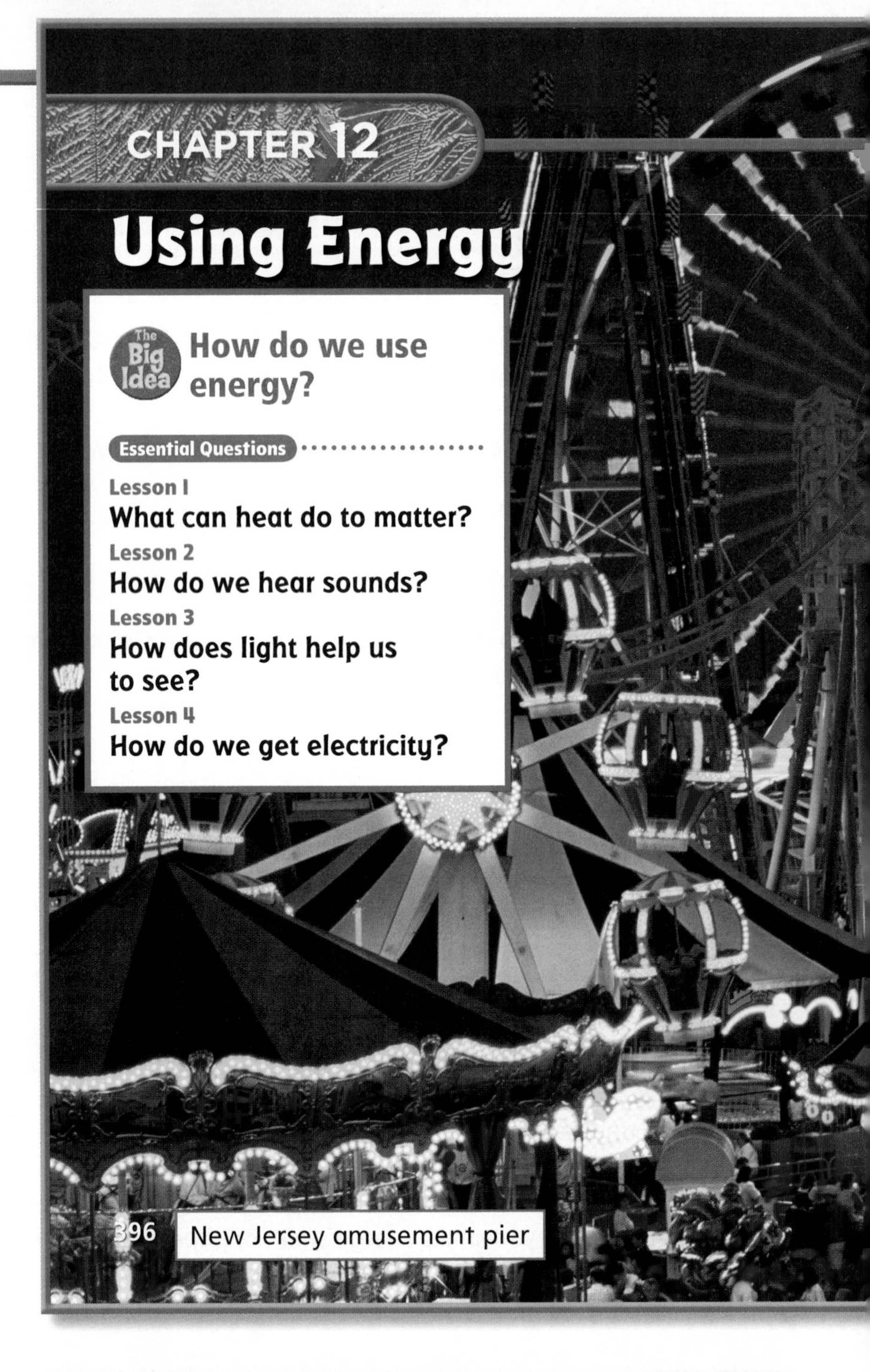

CHAPTER 12
Using Energy
How do we use energy?
Essential Questions
Lesson 1
What can heat do to matter?
Lesson 2
How do we hear sounds?
Lesson 3
How does light help us to see?
Lesson 4
How do we get electricity?
396


New Jersey amusement pier

### Differentiated Instruction

**Instructional Plan**

**Chapter Concept** different energy has different properties

**EXTRA SUPPORT** After covering heat, **Lesson 1,** pages 398–403, children who need to identify properties of sounds should do **Lesson 2,** pages 404–413.

**ON LEVEL** After covering heat, **Lesson 1,** pages 398–403, children who can describe properties of sounds can cover what sounds move through, **Lesson 2,** pages 410–411, before going to **Lesson 3,** pages 414–419.

**ENRICHMENT** For children who are ready to explore another form of energy, electricity, **Lesson 4,** pages 420–427, builds on the topic of electricity from Grade 1 by comparing static and current electricity.

### RESOURCES and TECHNOLOGY

- **School to Home Activities,** pp. 115–126
- **Reading and Writing,** pp. 224–246
- **Assessment,** pp. 146–159
- Classroom Presentation Toolkit CD-ROM
- PuzzleMaker CD-ROM

 www.macmillanmh.com

 e-Journal

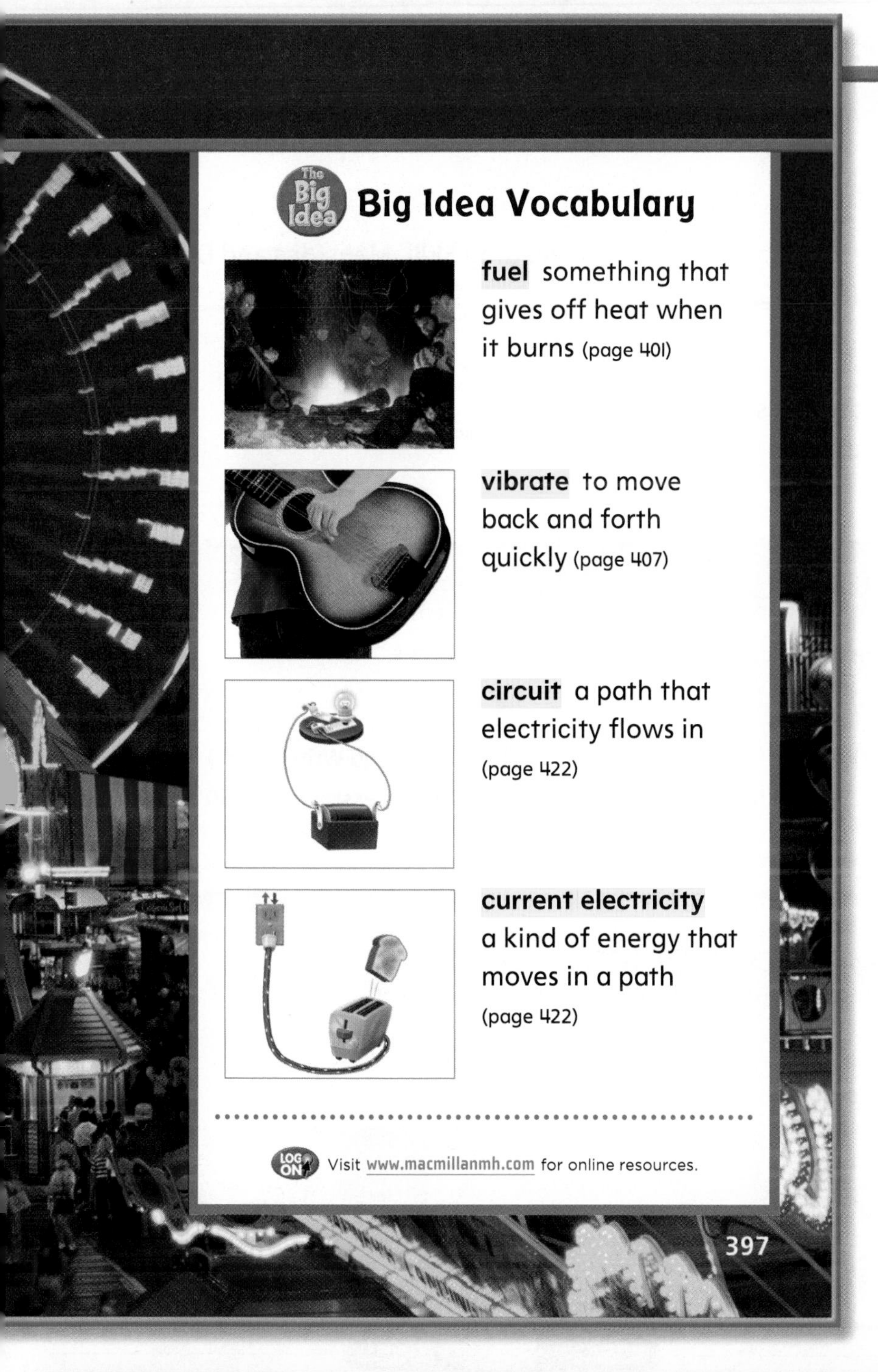

## Big Idea Vocabulary

**fuel** something that gives off heat when it burns (page 401)

**vibrate** to move back and forth quickly (page 407)

**circuit** a path that electricity flows in (page 422)

**current electricity** a kind of energy that moves in a path (page 422)

Visit www.macmillanmh.com for online resources.

397

# Big Idea Vocabulary

- Have a volunteer read the **Big Idea Vocabulary** words aloud to the class. Ask children to find one or two of the words in the chapter by using the given page references. Add these words and their definitions to a class Word Wall.
- Encourage children to use the illustrated glossary in the Student Edition's reference section. Guide children to explore the **e-Glossary**, which offers audio pronunciations, definitions, and sentences using the vocabulary words.

## Science Leveled Readers

ALSO ON AUDIO CD
Leveled Reader Library

APPROACHING

**Electricity** See how electricity is all around us and how people use it.

**ISBN:** 978-0-02-285859-9

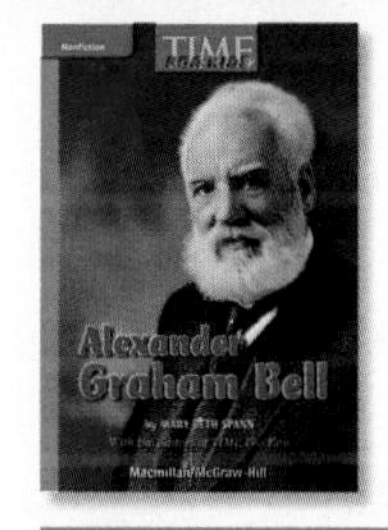


ON LEVEL

**Alexander Graham Bell** Read all about this brilliant man and his inventions.

**ISBN:** 978-0-02-285867-4

BEYOND

**The Camera's Eye** Find out how cameras work by studying the human eye.

**ISBN:** 978-0-02-286166-7

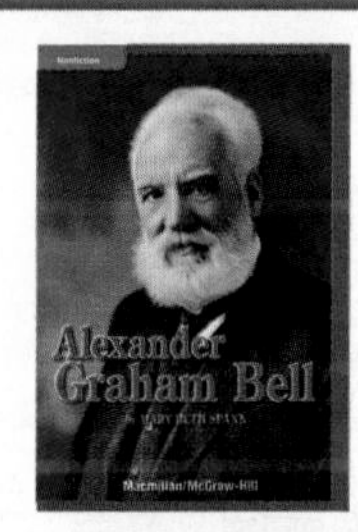


ELL

**Alexander Graham Bell** Uses sheltered language of On-Level Reader.

**ISBN:** 978-0-02-283428-9

See teaching strategies in the Leveled Reader Teacher's Guide. To order, call 1-800-442-9685.

 **Leveled Reader Database** Online Readers, searchable by topic, reading level, and keywords

# Stop Here to Plan Your Lesson

Lesson 1
Heat
Look and Wonder
This is a desert on a sunny day. How can you tell it is hot?


## Lesson 1 Heat

### Essential Question

What can heat do to matter?

### Objective

- Recognize that the Sun supplies heat and energy to Earth.

**Reading Skill** Main Idea and Details

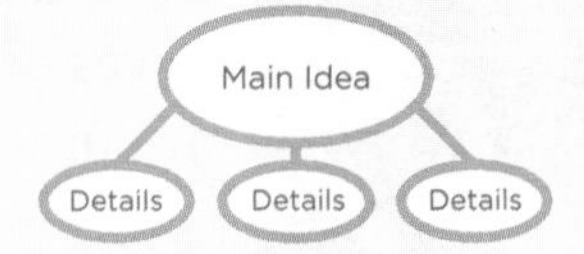


*Graphic Organizer 1*, p. TR3

## FAST TRACK

**Lesson Plan** When time is short, follow the Fast Track and use the essential resources.

**1 Introduce**
**Look and Wonder, p. 398**
Resource **Activity Lab Book, p. 183**

**2 Teach**
**Discuss the Main Idea, p. 400**
Resource **Visual Literacy, p. 36**

**3 Close**
**Think, Talk, and Write, p. 403**
Resource **Assessment, p. 150**

**Professional Development** Look for NSDL to find recommended Science Background resources from the National Science Digital Library.

## Reading and Writing

Name ______ Date ______ LESSON Outline

**Heat**

**Use your book to help you fill in the blanks.**

**What is heat?**

1. Energy makes ___matter___ move or change.
2. Heat is energy that can change the ___state___ of matter.
3. Heat can ___melt___ solids and turn liquids into gases.
4. The ___Sun___ gives Earth most of its heat.
5. We can also get heat ___energy___ from other things.
6. Something that gives off heat energy when it is burned is ___fuel___.
7. Heat energy can also come from ___motion___.

**Outline, pp. 225–226**
Also available as a student workbook

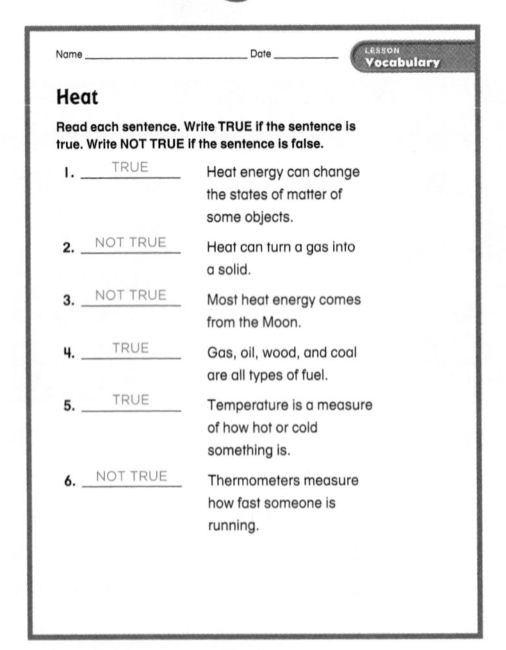

Name ______ Date ______ LESSON Vocabulary

**Heat**

**Read each sentence. Write TRUE if the sentence is true. Write NOT TRUE if the sentence is false.**

1. ___TRUE___ Heat energy can change the states of matter of some objects.
2. ___NOT TRUE___ Heat can turn a gas into a solid.
3. ___NOT TRUE___ Most heat energy comes from the Moon.
4. ___TRUE___ Gas, oil, wood, and coal are all types of fuel.
5. ___TRUE___ Temperature is a measure of how hot or cold something is.
6. ___NOT TRUE___ Thermometers measure how fast someone is running.

**Vocabulary, p. 227**
Also available as a student workbook

## Visual Literacy

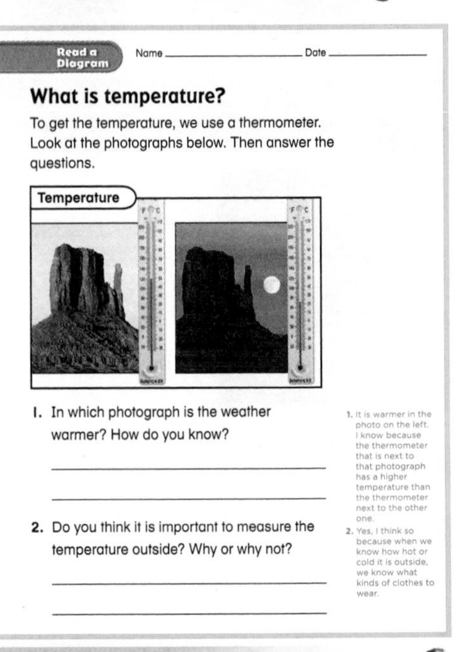

Read a Diagram Name ______ Date ______

**What is temperature?**

To get the temperature, we use a thermometer. Look at the photographs below. Then answer the questions.

Temperature

1. In which photograph is the weather warmer? How do you know?
2. Do you think it is important to measure the temperature outside? Why or why not?

1. It is warmer in the photo on the left. I know because the thermometer that is next to that photograph has a higher temperature than the thermometer next to the other one.
2. Yes, I think so because when we know how hot or cold it is outside, we know what kinds of clothes to wear.

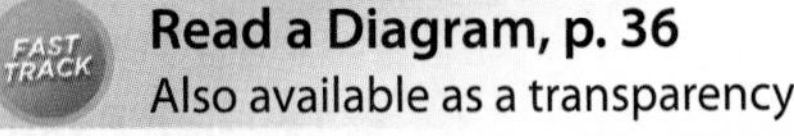


**Read a Diagram, p. 36**
Also available as a transparency

# Activity Lab Book

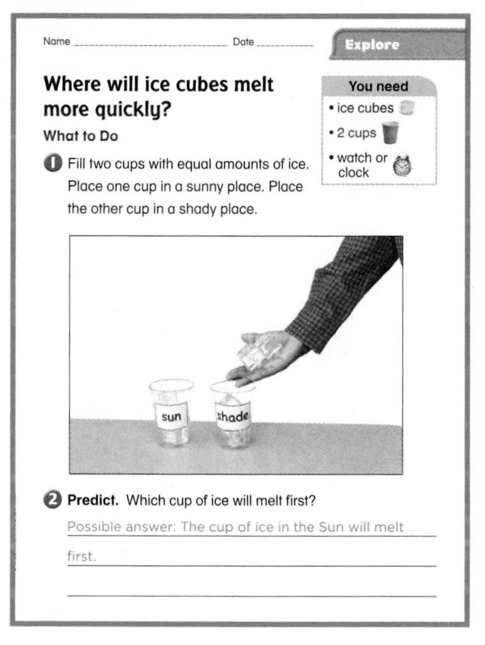

Name ______ Date ______ **Explore**

**Where will ice cubes melt more quickly?**

**What to Do**

**You need**
- ice cubes
- 2 cups
- watch or clock

1. Fill two cups with equal amounts of ice. Place one cup in a sunny place. Place the other cup in a shady place.



2. **Predict.** Which cup of ice will melt first?
   Possible answer: The cup of ice in the Sun will melt first.

**Explore, pp. 181–182**
Also available as a student workbook

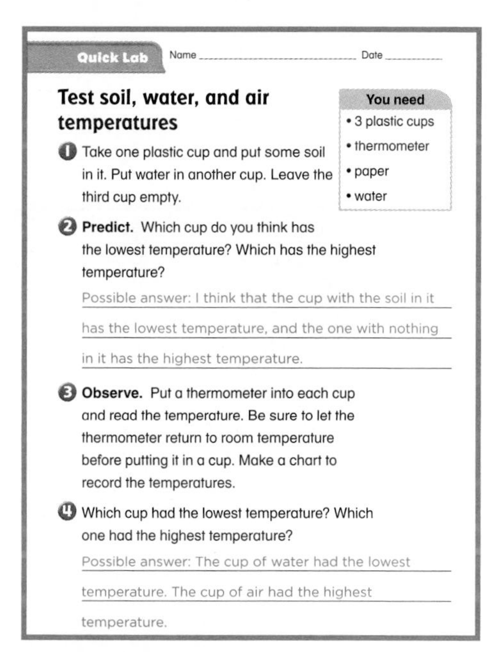

**Quick Lab** Name ______ Date ______

**Test soil, water, and air temperatures**

**You need**
- 3 plastic cups
- thermometer
- paper
- water

1. Take one plastic cup and put some soil in it. Put water in another cup. Leave the third cup empty.
2. **Predict.** Which cup do you think has the lowest temperature? Which has the highest temperature?
   Possible answer: I think that the cup with the soil in it has the lowest temperature, and the one with nothing in it has the highest temperature.
3. **Observe.** Put a thermometer into each cup and read the temperature. Be sure to let the thermometer return to room temperature before putting it in a cup. Make a chart to record the temperatures.
4. Which cup had the lowest temperature? Which one had the highest temperature?
   Possible answer: The cup of water had the lowest temperature. The cup of air had the highest temperature.

**Quick Lab, p. 184**
Also available as a student workbook

# Assessment

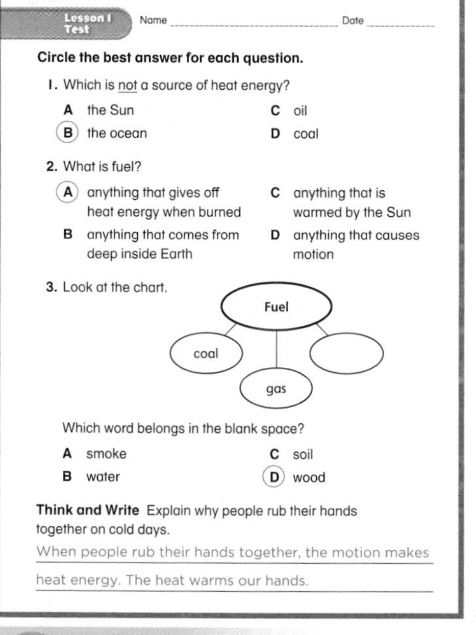

**Lesson 1 Test** Name ______ Date ______

**Circle the best answer for each question.**

1. Which is not a source of heat energy?
   - A the Sun
   - (B) the ocean
   - C oil
   - D coal
2. What is fuel?
   - (A) anything that gives off heat energy when burned
   - B anything that comes from deep inside Earth
   - C anything that is warmed by the Sun
   - D anything that causes motion
3. Look at the chart.



   Which word belongs in the blank space?
   - A smoke
   - B water
   - C soil
   - (D) wood

**Think and Write** Explain why people rub their hands together on cold days.
When people rub their hands together, the motion makes heat energy. The heat warms our hands.

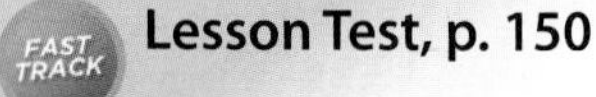

**Lesson Test, p. 150**

# ADDITIONAL RESOURCES

pp. 178–181

*Alexander Graham Bell*

pp. 55–56

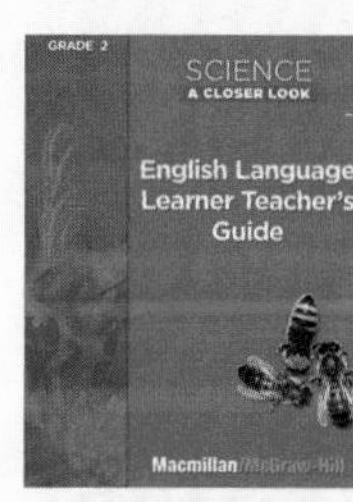


p. 120

p. 36

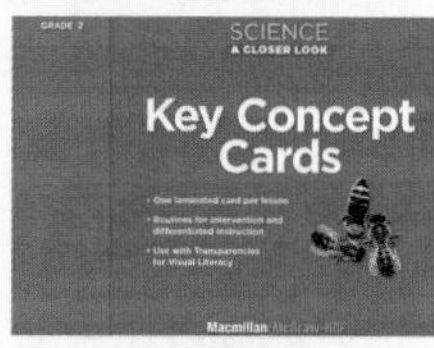


36

110–120

pp. 117–118

94–95

**Technology**

- Science Activity DVD
- TeacherWorks™ Plus CD-ROM
- Classroom Presentation Toolkit CD-ROM

-Review

# Lesson 1 Heat

**Objective**

- Recognize that the Sun supplies heat and energy to Earth.

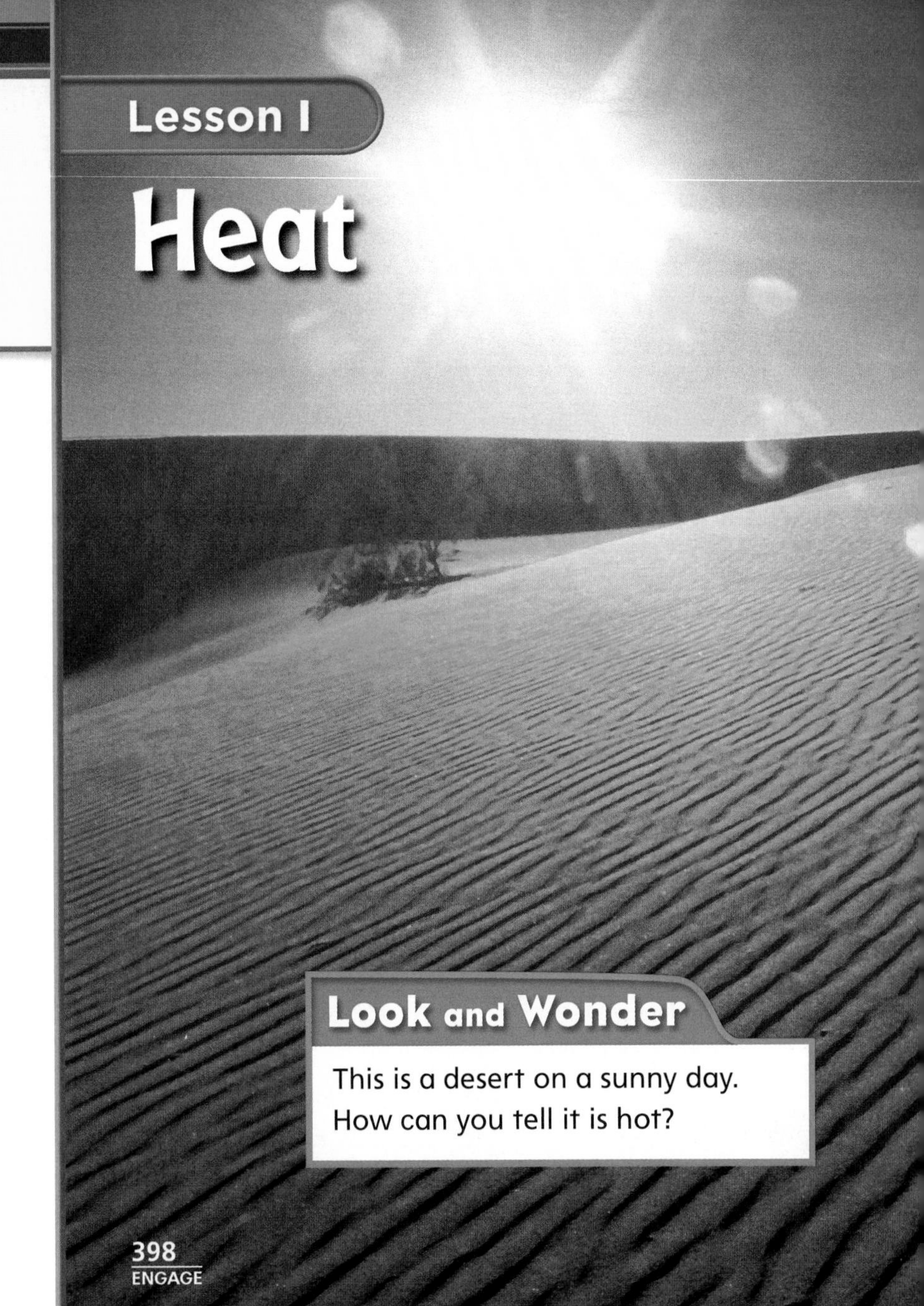

Lesson I
Heat
Look and Wonder
This is a desert on a sunny day. How can you tell it is hot?
398
ENGAGE


## 1 Introduce

### ▶ Assess Prior Knowledge

Have children share what they know about the Sun and heat energy. Ask:

- **What do you know about the Sun?**
- **Where are some places from which heat comes?**

Record children's answers in the What We Know column of the class **KWL** chart.

### Look and Wonder

Read the Look and Wonder statement and question about the temperature in the desert.

Invite children to share their responses. Encourage them to find clues in the picture to support their answers. Ask:

- **Which clues in the picture tell you it is hot?** Possible answers: sunlight beaming down on the land; land looks dry
- **Will temperature change in this desert during the night? Why or why not?** Possible answer: Yes, because when the Sun goes down, there is less heat.

### Warm Up

**Start with a Discussion**

Ask questions like these to prompt a discussion about the role of heat in people's lives:

- **What is the hottest day that you can remember?**

Have children describe what it felt like, what they did, and where they were. Ask:

- **What did you do to cool off?**

If a child repeats what another said, ask them to describe other ways they cooled off.

- **When it is not hot outside, how do people get the heat they need?** Possible answers: electricity, furnace, stove, fireplace

**RESOURCES and TECHNOLOGY**

- **Activity Lab Book,** pp. 181–183
- **Activity Flipchart,** p. 55
- Science Activity DVD

## Explore — Inquiry Activity

### Where will ice cubes melt more quickly?

**What to Do**

1. Fill two cups with equal amounts of ice. Place one cup in a sunny place. Place the other cup in a shady place.
2. **Predict.** Which cup of ice will melt first?
3. Record how long it takes for the ice in each cup to melt. Why did one cup of ice melt more quickly?

**Explore More**

4. **Predict.** Repeat the activity. Use equal amounts of water of the same temperature in two cups. How will each cup of water feel after one hour?

**You need**

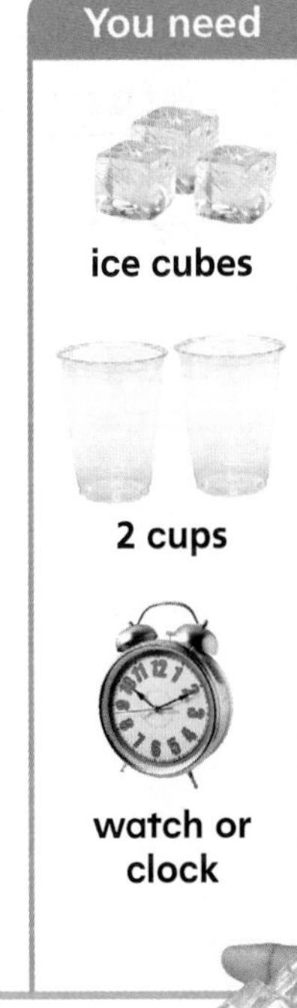

ice cubes

2 cups

watch or clock

Step 1



## Alternative Explore

### How can ice be melted quickly?

Have children **predict** which ice cube will melt more quickly: the one on the plate, or the one they pass around.

Help them understand how their body heat was transferred to the ice cube, which caused it to melt more quickly.

Name ________________ Date ________ Alternative Explore

**How can ice be melted quickly?**

In this activity, you will discover what happens when you add body heat to ice.

**You need**
- ice
- plate

**What to Do**

1. **Predict.** What do you think will happen if you and your classmates pass around an ice cube?
   Possible answer: It will slowly start to melt.
2. Using your hands, pass around an ice cube for at least five minutes. Place another ice cube on a plate. Leave it on the plate as you pass around the other ice cube.

**What Did You Find Out?**

3. Did the ice cube on the plate and the one you passed around melt at the same rate? Why or why not?
   Possible answer: No. The one that we passed around melted faster than the one on the plate. There was more heat energy in our hands than in the air. The ice cube we passed around melted faster because of the heat from our hands.

**Activity Lab Book, p. 183**

## Explore

 small groups  15 minutes

**Plan Ahead** If there is no source of sunlight in the classroom, use a bright lamp.

**Purpose** Children may have an informal understanding of how ice melts, but this activity will help develop their observational and recording skills.

**Structured Inquiry** **What to Do**

Explain to children that during this activity, they will measure and record how long it takes ice to melt.

1. Make sure children place the same number of cubes in each cup and label the cups.
2. **Predict** Accept all reasonable predictions. Ask children to explain what information they used to help make their predictions.
3. Have children record the time when they put the cups in the sunlight and shade, and the time when the ice melted completely in each cup. Help them do the math to find out the amount of time it took the ice in each cup to melt.

   Children should determine that the cup in the sunlight melted first because it got hotter more quickly than the cup in the shade.

**Guided Inquiry** **Explore More**

4. **Predict** Ask: **How will you use what we learned about how the ice melted to help predict how the water will feel?** Possible answer: The ice in the sunlight melted more quickly than the ice in the shade, so the water in the sunlight will feel warmer because it is heated by the Sun.

**Open Inquiry**

Encourage children to create further tests to explore the rate at which things melt. If they need help coming up with their own ideas, ask: **What else freezes? Will it melt at the same rate as ice? What might make ice melt more slowly? What might make ice melt more quickly?**

# 2 Teach

## Read Together and Learn

**Reading Skill** **Main Idea and Details** The main idea is the most important idea in the reading selection. Details give more information about the main idea.

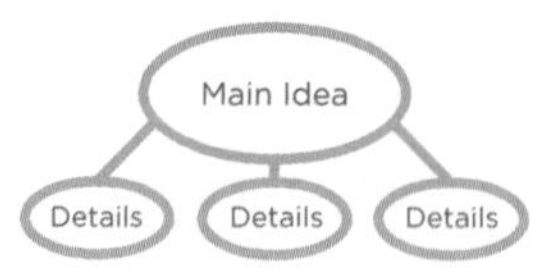


*Graphic Organizer 1*, p. TR3

## What is heat?

FAST TRACK

### Discuss the Main Idea

**Main Idea** Heat is a kind of energy that can change the state of matter.

Before reading, ask children to share what they know about heat.

After reading together, ask:

- **Where did you use or feel heat today?** Possible answers: walking in the sun; an adult heating food on the stove; using hot water in the shower
- **What are some other things that create heat?** Possible answers: clothes dryer, stove, furnace

**Read Together and Learn**

**Essential Question**
What can heat do to matter?

**Vocabulary**
heat
fuel

### What is heat?

Energy makes matter move or change. There are many kinds of energy. **Heat** is a kind of energy tha can change the state of matter. Heat can turn a solid into a liquid. Heat can turn a liquid into a gas.

We use heat every day. Most hea on Earth comes from the Sun. The Sun warms the air, the land, and th water on Earth.

On a hot day, the Sun warms the water and the land first. Then the air becomes warm.



## Science Background

**Heat** Heat is the flow of energy from one substance to another. Friction, caused when two things rub together, can cause heat. The Sun produces intense heat because of the nuclear reactions that occur in the Sun's core. Burning fuels, such as gas, oil, wood and coal, all produce heat.

See **Science Yellow Pages**, in the Teacher Resources section, for background information.

**Professional Development** For more Science Background and resources from NSDL visit http://nsdl.org/refreshers/science

## ELL Support

**Use Illustrations** Use the images on pages 400–401 to help children discuss how heat can change things.

**BEGINNING** Point to items and ask volunteers to identify them. If children cannot identify the items, name them. Have the group repeat the name of each item.

**INTERMEDIATE** Ask children to tell how heat is being used in each illustration. Ask them to describe ways that people use heat to cook or keep warm.

**ADVANCED** Have children explain what the people did in order to build a fire. Ask them to describe how the food might have felt before it was cooked.

Heat comes from other things
o. **Fuel** is something that gives
f heat when it burns. Gas, oil,
ood, and coal can be burned
fuel.

Heat can also come from
otion. Rub your hands together
ickly to make them warm.
w touch your hands to your
ce. Heat moved from your
nds to your face.

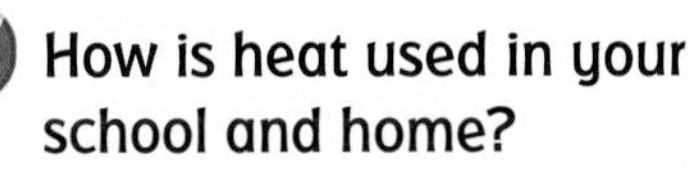

How is heat used in your school and home?

▲ People use fuel to keep warm.

ople use fuel to
k food. ▼

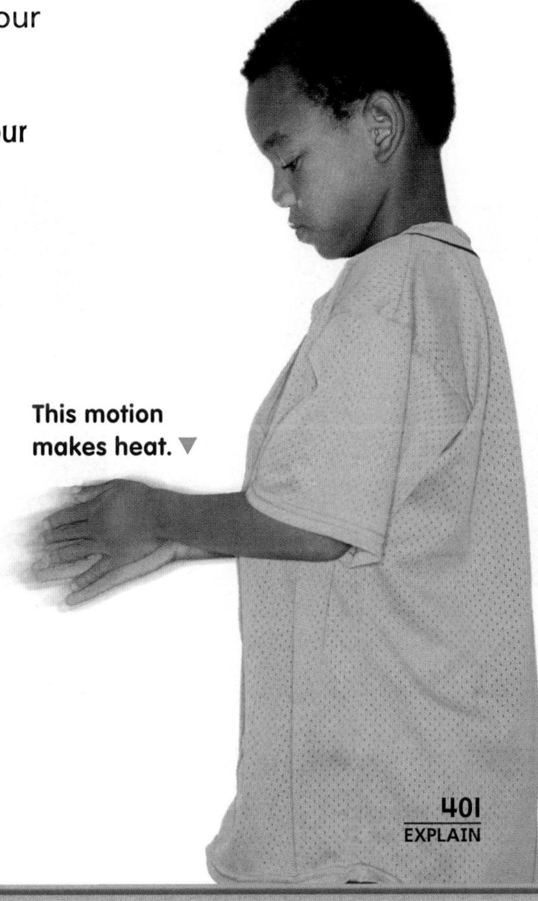

This motion makes heat. ▼





## ▶ Use the Visuals

Ask children the following:

- **What sources of heat are the people using in these pictures?** burning gas and wood, rubbing hands together
- **What are some other sources of heat energy that they could have used?** Possible answers: electric stove, heater, furnace

## ▶ Develop Vocabulary

**heat** Remind children that heat is a kind of energy. Have children use the word in a sentence in which they describe how heat can change the different states of water.

**fuel** Remind children that fuel is something people burn to get heat. Write *fuel* on the board and have children list things that can be burned as fuel.

## ▶ Explore the Main Idea

ACTIVITY Give children time to explore how it feels to rub their hands together, slowly at first, and then very fast. Explain that the increased motion creates more heat.

### Quick Check Answer

Possible answers: to cook food; to make the room warmer; to boil water

## Differentiated Instruction

### Leveled Activities

**EXTRA SUPPORT** Ask children to work with a partner to describe what a winter day without heat might be like. Encourage them to talk about how they would feel and what they could and could not do in the cold weather. Have them work together to write and illustrate their story.

**ENRICHMENT** Have children create a poster about heat energy. Ask them to find pictures in magazines or online of the different sources of heat, such as the Sun, fuel, and friction. Invite children to create collages of the different types of heat energy using the pictures. They should also describe the sources of energy and how they are used.

### RESOURCES and TECHNOLOGY

- **Reading and Writing,** pp. 225–227
- PuzzleMaker CD-ROM
- Classroom Presentation Toolkit CD-ROM

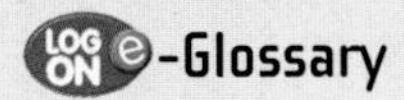

e-Glossary

## What is temperature?

### ▶ Discuss the Main Idea

**Main Idea** Temperature is how hot or cold something is. It is measured with a thermometer.

After reading, ask:

- **What causes something to become warmer?** the addition of heat
- **What causes something to become cooler?** the loss of heat

Children may have difficulty understanding that the cooling process is reflected by a lower temperature. Refer to this concept throughout the chapter to facilitate understanding.

**Read a Photo**

Ask a volunteer to tell what the temperature reading is in the desert during the day. Have the child explain how they got the answer.

If no one can accurately read the thermometer, explain that the top of the red line in the middle of the thermometer is next to a number. That number represents the temperature.

**Answer to Read a Photo** It is hotter during the day because the temperature is higher. The thermometer reads 122°F for the day, and 70°F for the night.

### ▶ Develop Vocabulary

Reinforce children's understanding of the lesson vocabulary with this word study activity.

Ask children to list words that describe *heat*, such as *hot, toasty,* or *steamy*. Ask volunteers to use each adjective in a sentence.

**RESOURCES and TECHNOLOGY**

- **School to Home Activities,** pp. 117–118
- **Reading and Writing,** p. 228
- **Activity Lab Book,** p. 184
- **Visual Literacy,** p. 36
- **e-Review** Narrated Summary and Quiz
- ***ExamView® Assessment Suite* CD-ROM**

## What is temperature?

Temperature is a measure of how hot or cold something is. We measure the temperature of air, water, and our bodies. To measure temperature, we use a tool called a thermometer. Some thermometers have a liquid inside. The liquid goes up or down with the temperature.

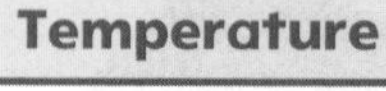

**Read a Photo**

Is it hotter during the day or night? How can you tell?



## Quick Lab

**Objective** Compare the temperature of soil, water, and air.

**You need** thermometers, three plastic cups per group, water, soil, crayons, paper

1. Have children predict whether the temperature of the soil, water, and air will be the same or different.
2. Ask children to use thermometers to measure and record the temperature of the soil, water, and air in the cups.
3. Have children compare the temperatures of each and discuss why the temperatures varied.

Temperature Chart

| air | soil | water |
| --- | --- | --- |
| 70°F | 62°F | 68°F |

## Quick Lab

Use a thermometer to compare the temperature of soil, water, and air.

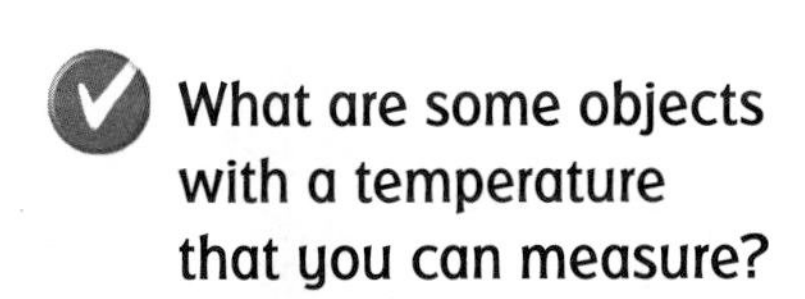

What are some objects with a temperature that you can measure?

## Think, Talk, and Write

1. **Main Idea and Details.** Where does most of our heat come from?
2. How do we measure temperature?
3. **Essential Question.** What can heat do to matter?

### Art Link

Look around your school or home for sources of heat. Draw them.

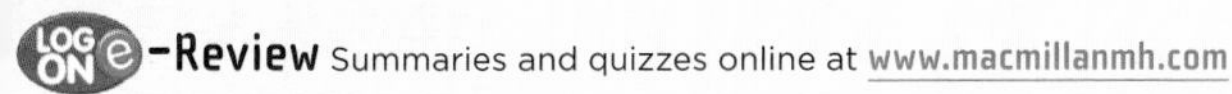

LOG ON e-Review Summaries and quizzes online at www.macmillanmh.com

403
EVALUATE

## Quick Check Answer

Possible answers: your body; outdoor and indoor air; water

# 3 Close

### ▶ Using the KWL Chart

Review with children what they have learned about how people use heat. Record their responses in the What We Learned column of the class **KWL** chart.

### ▶ Using the Reading Skill
#### Main Idea and Details

Use the reading skill graphic organizer to identify the main idea and details of the lesson.

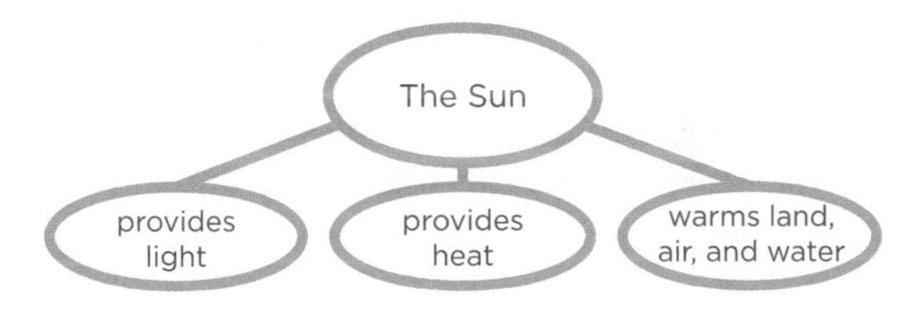


*Graphic Organizer 1*, p. TR3

## FAST TRACK Think, Talk, and Write

1. **Main Idea and Details** Most of Earth's heat comes from the Sun.
2. Thermometers are used to measure temperature.
3. **Essential Question** Possible answers: Heat can change the state of matter. Heat can change a solid into a liquid. Heat can change a liquid into a gas. Heat can change the temperature of matter.

### Art Link

Have children share their drawings with a partner and talk about the type of fuel each item uses.

## Formative Assessment

**What's Hot, What's Not!**

Divide children into *hot* and *cold* temperature groups. Have each child create a poster by listing items that belong in their temperature group. Ask children to illustrate each item on their list. Have them take turns acting out an item from their poster. The rest of the class can guess the item.

Key Concept Cards For student intervention, see the prescribed routine on **Key Concept Card 36.**

# Focus on Skills

small groups · 40 minutes

Use Your

Activity Flipchart, p. 56

### Objective

- Measure and compare temperatures by using thermometers.

**You need** three cups, three thermometers, tape, markers

**EXTEND** This inquiry skill activity will help children develop measurement skills as they record and compare temperatures of water and ice.

## Focus on Skills

### Inquiry Skill: Measure

You **measure** to find out about things around you. You can measure how long, how heavy, or how warm something is.

#### ▶ Learn It

A class wants to measure the temperature in different parts of their classroom. They measure the temperature by a sunny window. They measure the temperature in a shady place.

▲ a sunny window

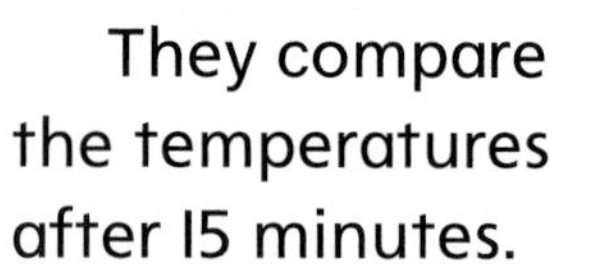
They compare the temperatures after 15 minutes.

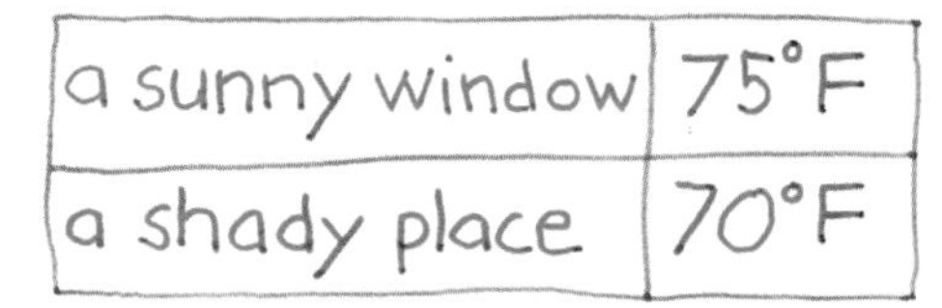

| a sunny window | 75°F |
|---|---|
| a shady place | 70°F |

▲ a shady place

## Inquiry Skill: Measure

### ▶ Learn It

Have a volunteer read aloud the Learn It section on page 56. Discuss how things are measured to get exact figures. For example, one bowl of water is warm and another bowl hot, but by using a thermometer the temperature of each bowl can be found and compared.

Discuss different tools that are used for measurement, such as rulers, scales, and clocks. Study the photos together. Ask:

- **Why are the temperature readings different?** Possible answer: Because the temperatures were taken in two different parts of the room.
- **Which place has the highest temperature? What is the difference between the two temperatures?** the sunny window; 5°F

## Integrate Math

### Measuring and Comparing Objects

Have children use rulers to measure the perimeter or distance around two different-sized books or other rectangular objects. Have children record the measurement of each side, and use those figures to write an addition sentence to determine the perimeter of each object. Ask:

- **How can you find the difference between the lengths of the two objects?** Subtract the shorter length from the longer length.
- **How can you find the difference between the widths of the two objects?** Subtract the shorter width from the longer width.

Have children write number sentences to show what they did.

Skill Builder

**Try It**

You can measure the temperature of ice, cold water, and warm water.

1. Fill cups with ice, cold water, and warm water.

2. **Predict.** What is the temperature in each cup? Record your predictions.
3. **Measure.** Put a thermometer in each cup for 5 minutes. Record each temperature.

4. **Compare.** Were your predictions close to your measurements?

| Measuring Temperature | | | |
|---|---|---|---|
| | ice | cold | warm |
| predict | | | |
| measure | | | |

Use with Chapter 12, Lesson 1 56

## ▶ Try It

Read the Try It section on page 56 with the class. If necessary, remind children how to read the temperatures on a thermometer.

1. Ask: **What do you think will happen to the temperature in each cup after an hour?**
2. **Predict** Children may make a chart like the one pictured on page 56 to record their predictions, or they may use graphic organizer 3 on page TR5 of this book. After children make their predictions, ask: **What information did you use to help you predict the temperature?**
3. **Measure** Make sure children place and remove the thermometers in each cup at the same time.
4. **Compare** As children compare their actual measurements of the three cups with their predictions, ask: **Which prediction was closest to the actual temperature?**

## ▶ Apply It

Have children continue observing the change in temperature in the cups in five-minute increments for another ten minutes.

Ask them to predict what the temperature might be after each five-minute period.

After every five minutes, have children measure and record the temperature for each cup. Encourage children to make a chart to compare their predictions with the actual temperatures.

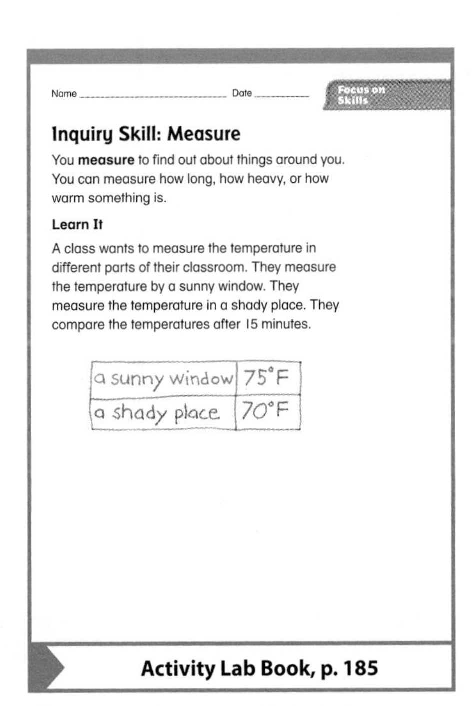
Name ______ Date ______ Focus on Skills

**Inquiry Skill: Measure**

You **measure** to find out about things around you. You can measure how long, how heavy, or how warm something is.

**Learn It**

A class wants to measure the temperature in different parts of their classroom. They measure the temperature by a sunny window. They measure the temperature in a shady place. They compare the temperatures after 15 minutes.

| a sunny window | 75°F |
|---|---|
| a shady place | 70°F |

**Activity Lab Book, p. 185**

**RESOURCES and TECHNOLOGY**

- **Activity Lab Book,** pp. 185–186
- **TeacherWorks™ Plus CD-ROM**

Stop Here to

# Plan Your Lesson

## Lesson 2 Sound

### Essential Question

How do we hear sounds?

### Objectives

- Discover how different sounds are produced.
- Describe the volume and pitch of sounds.

**Reading Skill** Problem and Solution

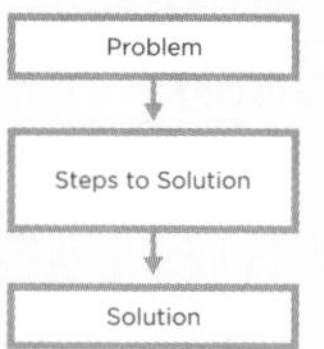


*Graphic Organizer 12*, p. TR14

## FAST TRACK

**Lesson Plan** When time is short, follow the Fast Track and use the essential resources.

**1 Introduce**

**Look and Wonder, p. 404**

Resource **Activity Lab Book, p. 189**

**2 Teach**

**Read a Diagram, p. 407**

**Discuss the Main Idea, p. 408**

Resource **Visual Literacy, p. 37**

**3 Close**

**Think, Talk, and Write, p. 411**

Resource **Assessment, p. 151**

**Professional Development** Look for NSDL to find recommended Science Background resources from the National Science Digital Library.

## Reading and Writing

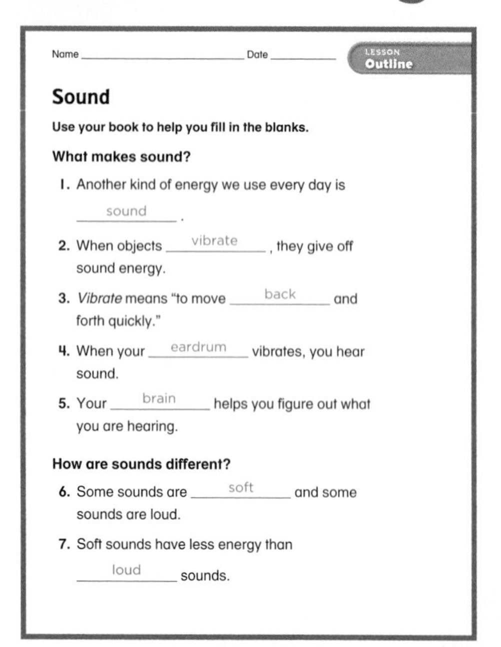

Name ____________ Date ________ LESSON Outline

**Sound**

**Use your book to help you fill in the blanks.**

**What makes sound?**

1. Another kind of energy we use every day is ___sound___.
2. When objects ___vibrate___, they give off sound energy.
3. *Vibrate* means "to move ___back___ and forth quickly."
4. When your ___eardrum___ vibrates, you hear sound.
5. Your ___brain___ helps you figure out what you are hearing.

**How are sounds different?**

6. Some sounds are ___soft___ and some sounds are loud.
7. Soft sounds have less energy than ___loud___ sounds.

**Outline, pp. 229–230**
Also available as a student workbook

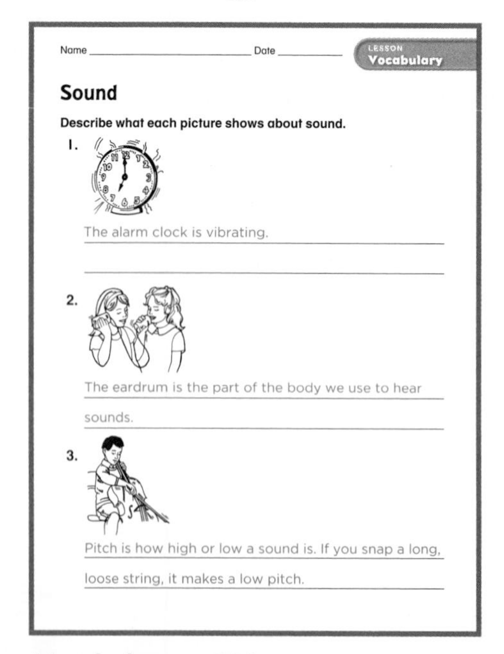

Name ____________ Date ________ LESSON Vocabulary

**Sound**

**Describe what each picture shows about sound.**

1. The alarm clock is vibrating.
2. The eardrum is the part of the body we use to hear sounds.
3. Pitch is how high or low a sound is. If you snap a long, loose string, it makes a low pitch.

**Vocabulary, p. 231**
Also available as a student workbook

## Visual Literacy

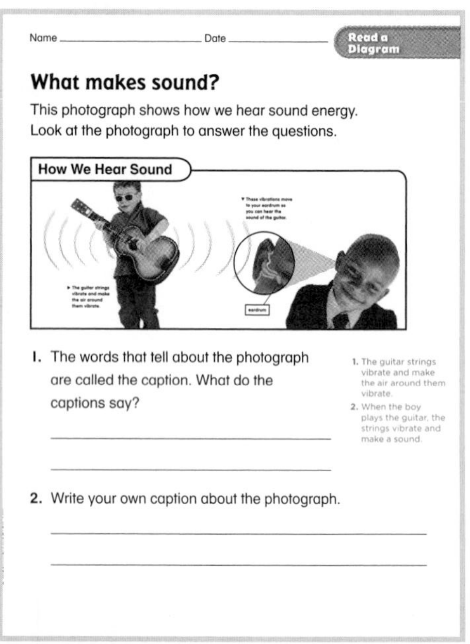

Name ____________ Date ________ Read a Diagram

**What makes sound?**

This photograph shows how we hear sound energy. Look at the photograph to answer the questions.

How We Hear Sound

1. The words that tell about the photograph are called the caption. What do the captions say?
2. Write your own caption about the photograph.

1. The guitar strings vibrate and make the air around them vibrate.
2. When the boy plays the guitar, the strings vibrate and make a sound.

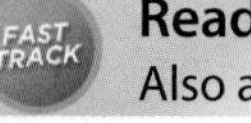

**Read a Diagram, p. 37**
Also available as a transparency

## Activity Lab Book

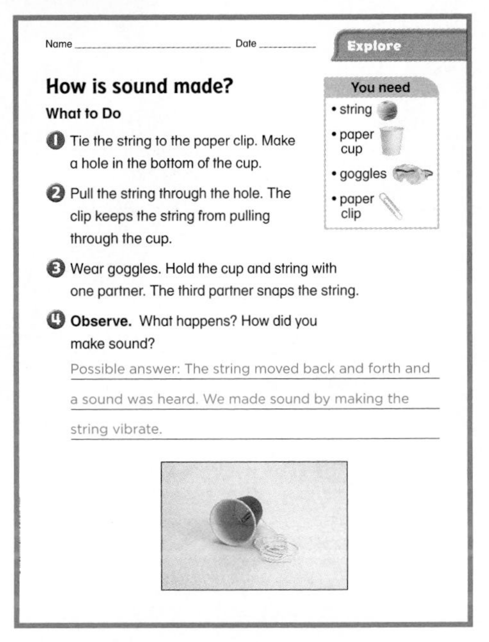

Name ______ Date ______ **Explore**

**How is sound made?**

**What to Do**

**You need**
- string
- paper cup
- goggles
- paper clip

1. Tie the string to the paper clip. Make a hole in the bottom of the cup.
2. Pull the string through the hole. The clip keeps the string from pulling through the cup.
3. Wear goggles. Hold the cup and string with one partner. The third partner snaps the string.
4. **Observe.** What happens? How did you make sound?

   Possible answer: The string moved back and forth and a sound was heard. We made sound by making the string vibrate.

**Explore, pp. 187–188**
Also available as a student workbook

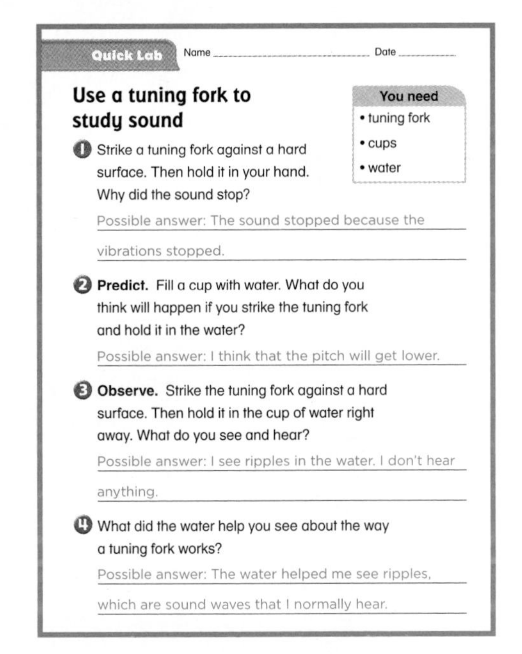

**Quick Lab** Name ______ Date ______

**Use a tuning fork to study sound**

**You need**
- tuning fork
- cups
- water

1. Strike a tuning fork against a hard surface. Then hold it in your hand. Why did the sound stop?

   Possible answer: The sound stopped because the vibrations stopped.
2. **Predict.** Fill a cup with water. What do you think will happen if you strike the tuning fork and hold it in the water?

   Possible answer: I think that the pitch will get lower.
3. **Observe.** Strike the tuning fork against a hard surface. Then hold it in the cup of water right away. What do you see and hear?

   Possible answer: I see ripples in the water. I don't hear anything.
4. What did the water help you see about the way a tuning fork works?

   Possible answer: The water helped me see ripples, which are sound waves that I normally hear.

**Quick Lab, p. 190**
Also available as a student workbook

## Assessment

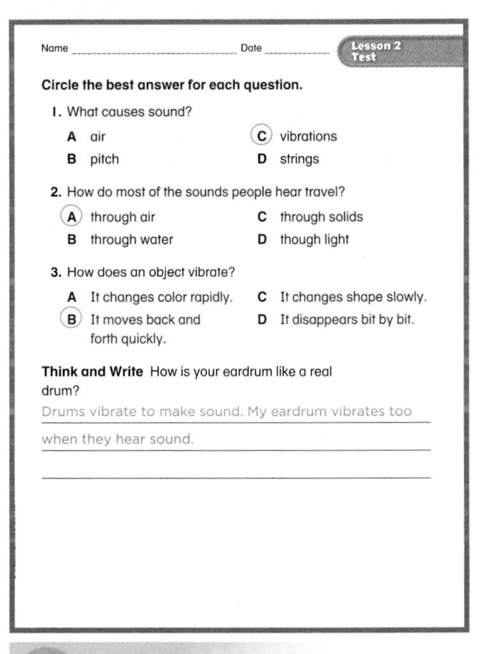

Name ______ Date ______ **Lesson 2 Test**

**Circle the best answer for each question.**

1. What causes sound?
   - A air
   - B pitch
   - (C) vibrations
   - D strings
2. How do most of the sounds people hear travel?
   - (A) through air
   - B through water
   - C through solids
   - D though light
3. How does an object vibrate?
   - A It changes color rapidly.
   - (B) It moves back and forth quickly.
   - C It changes shape slowly.
   - D It disappears bit by bit.

**Think and Write** How is your eardrum like a real drum?

Drums vibrate to make sound. My eardrum vibrates too when they hear sound.

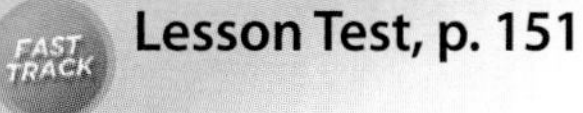

**Lesson Test, p. 151**

# ADDITIONAL RESOURCES

**pp. 182–187**

***Alexander Graham Bell***

**p. 57**

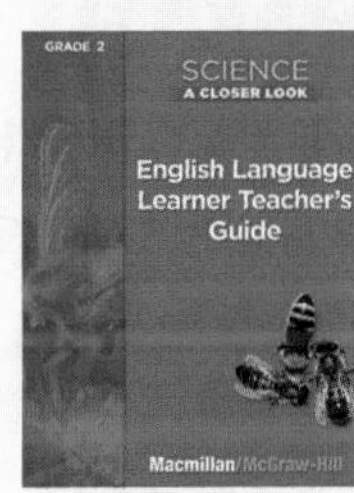


**p. 121**

**p. 37**

**37**

**110–120**

**pp. 119–120**

**96–98**

### Technology

- **Science Activity DVD**
- **TeacherWorks™ Plus CD-ROM**
- **Classroom Presentation Toolkit CD-ROM**
-  *Sound*
- **Science Songs CD** Track 14
- 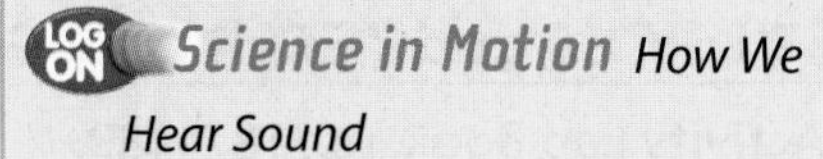 ***Science in Motion*** *How We Hear Sound*
-  **e-Review**
- 

# Lesson 2 Sound

### Objectives

- Discover how different sounds are produced.
- Describe the volume and pitch of sounds.

Lesson 2
Sound
Look and Wonder
Keep the noise down! How are sounds made? How can some sounds be different from others?
404
ENGAGE


## 1 Introduce

### ▶ Assess Prior Knowledge

Have children share what they know about sounds and how they are made.

- **What makes sounds?** Possible answers: people, animals, instruments
- **What are some words that describe sounds?** Possible answers: loud, soft, scary

Record children's answers in the What We Know column of the class **KWL** chart.

### FAST TRACK Look and Wonder

Read the Look and Wonder statement and questions about how sounds are made and how some sounds are different than others.

Invite children to share their responses. Ask:

- **What are the children in this picture doing to create sounds**? Possible answers: playing violins; drawing a bow across strings
- **What kind of sounds do you think they are making?** Possible answers: music, loud sounds, pretty sounds

### Warm Up

**Start With a Poem**

Find a poem that includes descriptions of different types of sound, for example, "Night Creature" in ***Sing a Song of Popcorn,*** by Lillian Moore (Scholastic, 1988). After reading the poem, discuss how night is described with quiet sounds, and day with loud sounds.

Ask children to describe sounds they might hear at night, and sounds that they might hear during the day. Record their answers on chart paper.

Have children write their own poem about night and day sounds. Invite children to share their poems with the class.

### RESOURCES and TECHNOLOGY

- **Activity Lab Book,** pp. 187–189
- **Activity Flipchart,** p. 57
- Science Activity DVD

## Explore — Inquiry Activity

### How is sound made?

**What to Do**

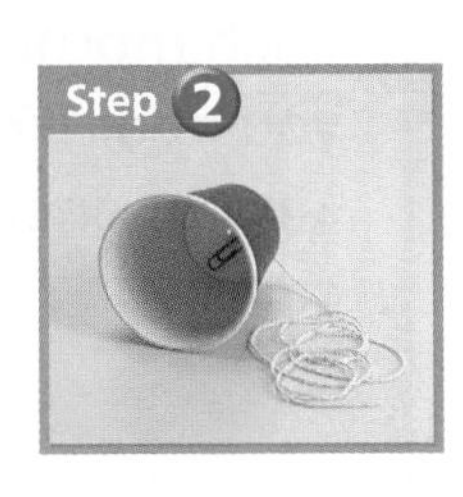
Step 2

1. Tie the string to the paper clip. Make a hole in the bottom of the cup.
2. Pull the string through the hole. The clip keeps the string from pulling through the cup.
3. Wear goggles. Hold the cup and string with a partner. The third partner snaps the string.
4. **Observe.** What happens? How did you make sound?

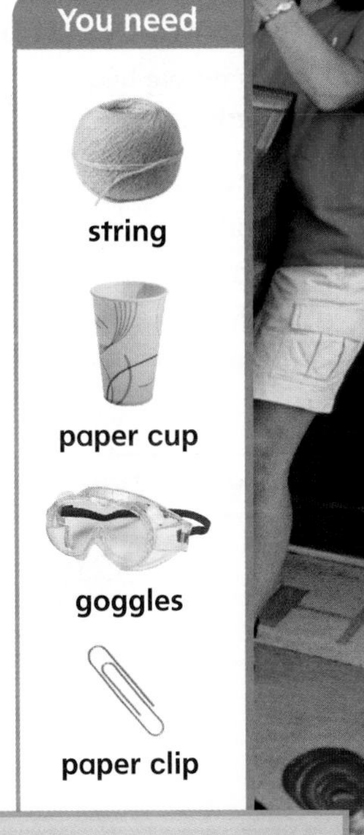
You need: string, paper cup, goggles, paper clip

**Explore More**

5. **Predict.** How will the sound be different if you change the length of the string? Try it.

Step 3

405 EXPLORE

## Alternative Explore

### How can a flute be made out of a straw?

Each child will need a plastic straw and a pair of scissors.

Help children cut one end of the straw into a V-shaped point and flatten it by pulling it between their fingers.

Have children blow into the end they cut. With practice, they should be able to make the straw vibrate and produce a loud sound.

Name ______ Date ______ Alternative Explore

**How can a flute be made out of a straw?**

In this activity, you will make a musical instrument out of a straw.

You need: • plastic straw • scissors

**What to Do**

1. Make a point at one end of the straw by making two small cuts. Flatten the cut end of the straw by pulling it between your fingers.
2. Describe what happens when you blow into the cut end of the straw.
   Possible answer: At first it seemed like nothing happened, but after a while I was able to make a sound.

**What Did You Find Out?**

3. What caused the straw to make a sound?
   Possible answer: When I blew through the small opening, the sides of the straw vibrated and made a sound.

Activity Lab Book, p. 189

## Explore

 small groups  30 minutes

**Plan Ahead** Collect enough paper cups, pre-cut string, and paper clips for every group. **Be Careful!** Children should wear goggles and take care when snapping the string.

**Purpose** This activity will enable children to see and feel what happens when a plucked string vibrates.

**Structured Inquiry** What to Do

Explain to children that they will investigate what happens when a string is plucked.

1. Some children may need help tying the string to the paper clips. Encourage children to help one another.
2. Have children pull the string tightly and snap it. Next, have them hold the string loosely and snap it.
3. Ask children to discuss with their partners the difference between the two sounds.
4. **Observe** Have children look closely at the string in order to see it move. Children should conclude that when the string vibrates (moves back and forth), sound is heard. Ask: **When the string stops vibrating, what happens to the sound?** It stops.

**Guided Inquiry** Explore More

5. **Predict** Ask children to think about what would happen if they changed the length of their string. Ask: **Would a longer piece of string make the same sound as a shorter piece?** Have children test their predictions by using different lengths of string. They may predict that a shorter string will make a higher sound and a longer string will make a lower sound.

**Open Inquiry**

Provide children with materials, such as thick and thin rubber bands, yarn, ribbon, thin metal wire, so they can investigate other factors that may play a role in the types of sound produced when something is plucked. Encourage children to share the results of their investigations with others.

# 2 Teach

## Read Together and Learn

**Reading Skill** Problem and Solution A problem is what needs to be done, found out, or changed. A solution fixes the problem.

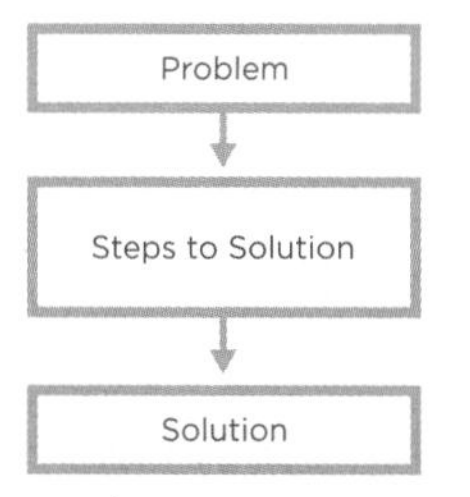


*Graphic Organizer 12*, p. TR14

## What makes sound?

### ▶ Discuss the Main Idea

**Main Idea** When objects vibrate, they create a type of energy called sound.

After reading together, ask:

- **What are some things that make sound?** Possible answers: strings, cymbals, bells, car engine, washing machine

Help children explore what parts of the things they mentioned might vibrate in order to create sound. Explain that when we talk or sing, the thin "strings" or vocal cords in our neck vibrate to create sound. Have children touch their neck while they hum or sing softly.

**Read Together and Learn**

▶ **Essential Question**
How do we hear sounds?

▶ **Vocabulary**
sound
vibrate
pitch

### What makes sound?

Ring! A loud alarm clock wakes you up each morning. How do you hear it? **Sound** is a kind of energy that we can hear.

▲ When the be on the alarm clock are hit they move b and forth qu

**How We Hear Sound**

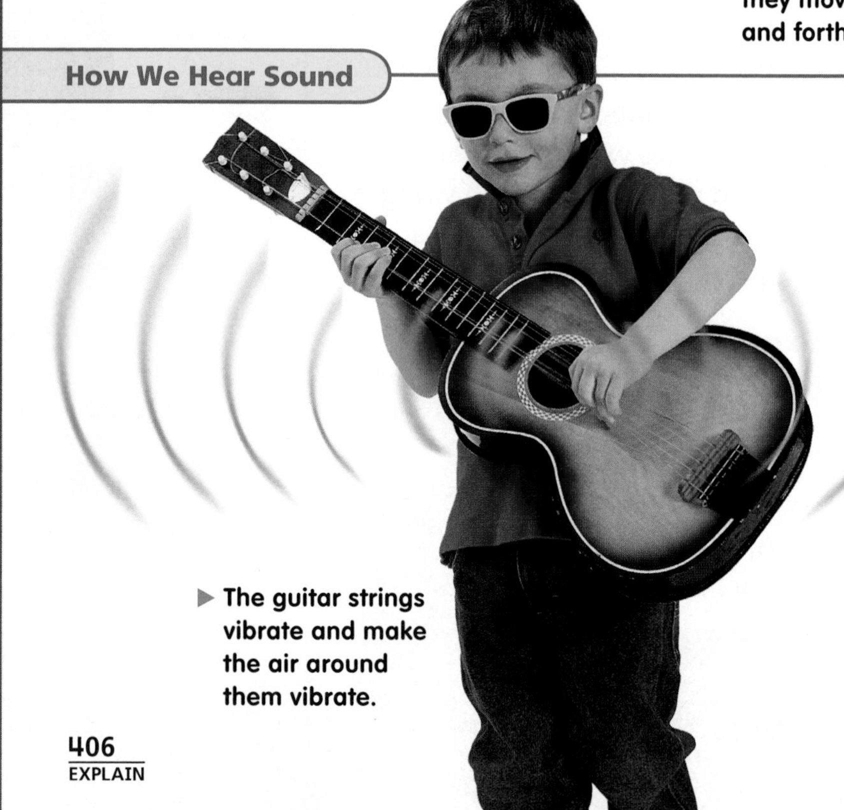

▶ The guitar strings vibrate and make the air around them vibrate.

406
EXPLAIN

## Science Background

**Sound** Sound is a type of energy that is made when an object vibrates. The energy of the vibrations moves through the air in the form of sound waves. When sound waves reach a listener's ears, they cause the eardrum to vibrate. These vibrations are transferred to the inner ear, which sends a message to the brain, which perceives the sound.

See **Science Yellow Pages**, in the Teacher Resources section, for background information.

**Professional Development** For more Science Background and resources from NSDL visit http://nsdl.org/refreshers/science

## ELL Support

**Use Realia** Collect objects that produce sounds and have children listen and then describe the sounds they hear.

**BEGINNING** Name the objects children use to make sounds, and have children identify whether the sound created is loud or soft.

**INTERMEDIATE** Have children describe the sounds using a simple sentence or phrase. For example: *The whistle made a loud sound*. If possible, make both loud and soft sounds with each object and have children distinguish between them.

**ADVANCED** Have children list adjectives that describe sounds, and use them in complete sentences to describe the sounds they hear.

Sound energy is made when ›jects vibrate. An object ›n **vibrate** when it moves ›ck and forth quickly. When ›mething vibrates, air around ›e object vibrates also.

The eardrum is the part our body we use to hear ›unds. Messages sent from ›ur ear to your brain tell you ›hat sound you heard.

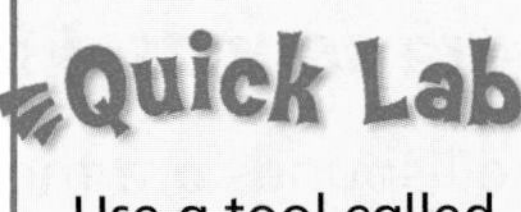

Use a tool called a tuning fork. **Observe** what happens when you strike it and place it in water.

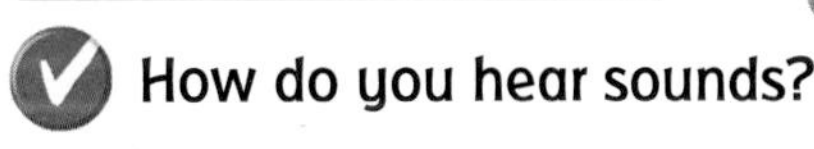

How do you hear sounds?

How did the sound travel from the guitar to the boy's ear?

Science in Motion Watch how sound travels at www.macmillanmh.com

These vibrations move to your eardrum so you can hear the sound of the guitar.

407
EXPLAIN

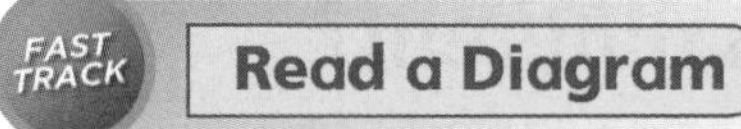

Help children discuss what they see in the diagram. Ask:

- **What shows how the sound travels?** the blue lines, which represent vibrating air

**Answer to Read a Diagram** The sound vibrations move from the guitar through the air to the boy's eardrum, so he can hear it.

 *Science in Motion How We Hear Sound*

## ▶ Develop Vocabulary

**sound** Explain to children that *sound* can be used as a noun meaning "energy made when objects vibrate," or a verb meaning "to pronounce or make an impression." Have children use *sound* in a sentence as a noun and as a verb. For example, noun: *I heard the sound*; verb: *The music sounds nice.*

**vibrate** Explain that vibrate means "to move back and forth quickly." Show children pictures of things that vibrate. Have them identify the item and tell whether it makes fast or slow vibrations. Use this sentence as a guide: *This ___ vibrates. It makes ____ vibrations.*

## ▶ Explore the Main Idea

ACTIVITY Have children sit quietly and listen for sounds for one minute. After the minute is up, have children write what sounds they heard on the board. Repeat this activity at different times of day. Compare the lists. Discuss why the time of day might have made a difference in the sounds the children heard.

### ✓ Quick Check Answer

When something vibrates, the air around it vibrates. The vibrating air enters the ear. Then the eardrum vibrates and sends messages to the brain.

## Quick Lab

 pairs  15 minutes

**Objective** Observe sound energy.

**You need** tuning fork, plastic cups, water

1. Have children strike a tuning fork against their desk and observe what happens.
2. Ask children to strike the tuning fork again, and quickly put their hand around it. Ask: **Why did the sound stop?** Because the vibrations stopped.
3. Have children strike the tuning fork again, and submerge it in water. Waves should appear.
4. Help children use what they learned to decide if sounds can be heard under water.

### RESOURCES and TECHNOLOGY

- **Reading and Writing,** pp. 229–231
- **Activity Lab Book,** p. 190
- **Visual Literacy,** p. 37
- **PuzzleMaker CD-ROM**
- **Classroom Presentation Toolkit CD-ROM**
-  *Sound*
- **Science Songs CD** Track 14
- 

# How are sounds different?

## Discuss the Main Idea

FAST TRACK

**Main Idea** Sounds can be loud or soft and have high or low pitches.

After reading together, ask:

- **What are some things that don't need much energy to make a sound?** Accept any answer that relates to a soft sound.
- **What are some things that need a great deal of energy to make a sound?** Accept any answer that relates to a loud sound.
- **What makes yelling different from whispering?** Possible answers: Yelling requires more energy than whispering. Yelling produces large vibrations, while whispering produces small vibrations.

## ▶ Use the Visuals

After reading, ask:

- **What are some other animals that make loud sounds?** Possible answers: elephant, monkey, tiger

Explain that a harp has strings that vary from short to long. Ask:

- **Which string on a harp would make the highest-pitched sound?** the shortest string
- **Which string on a harp would make the lowest-pitched sound?** the longest string

Explain that when the string of a violin or guitar is pressed down, it makes the area of the string that can vibrate shorter. When the string that is being held down is plucked, it produces a higher-pitched sound.

## How are sounds different?

Not all sounds are the same. You hear loud and soft sounds every day. You can make your voice loud or soft. A whisper has less energy than a shout. Try making loud and soft sounds.

▲ Small vibrations ma soft sounds. The me of a cat sounds soft.

▼ Big vibrations make loud sounds. The roar of a lion sounds loud.

408
EXPLAIN

## Differentiated Instruction

### Leveled Activities

**EXTRA SUPPORT** Give children musical instruments, such as triangles, bells, and small hand drums. Have them make sounds with the instruments and talk together about what part of each instrument vibrates in order to create sound.

**ENRICHMENT** Have children use books or the Internet to gather information on one particular instrument and how it produces vibrations to make sound.

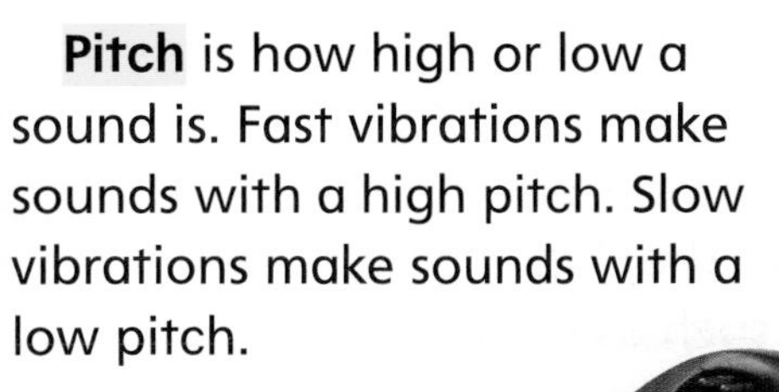

**Pitch** is how high or low a sound is. Fast vibrations make sounds with a high pitch. Slow vibrations make sounds with a low pitch.

▶ **If you snap a short or tight string, it makes a high pitch.**

▶ **If you snap a long or loose string, it makes a low pitch.**

harp

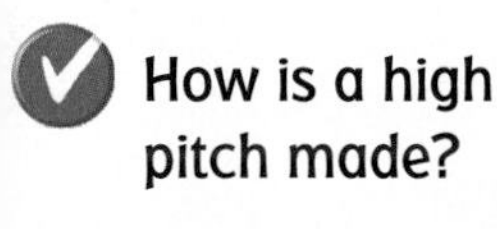

How is a high pitch made?

FACT Some sounds can not be heard by humans.



## ▶ Address Misconceptions

Children may think that humans can hear all sounds, but humans can only hear sound within a certain range of pitch. For example, humans can't hear the sound of a dog whistle, but dogs can hear that sound. Dogs can hear a wider range of sounds. Animals like bats and porpoises can hear very high-pitched sounds.

FACT **Some sounds can not be heard by humans.** Have children conduct research to identify sounds that can be heard by animals, but cannot be heard by humans.

## ▶ Develop Vocabulary

**pitch** ***Scientific vs. Common Use*** Explain that people often use the word *pitch* to describe throwing something, such as a baseball. In physical science, *pitch* means "how high or low a sound is." Slow vibrations make low-pitched sounds. Fast vibrations make high-pitched sounds.

## ▶ Explore the Main Idea

ACTIVITY Play a pitched instrument, such as a xylophone, and have children identify which sounds have a high pitch and which have a low pitch.

### ***Quick Check Answer***

Possible answers: when something vibrates quickly; when a small object or string vibrates

## Differentiated Instruction

### Leveled Questions

**EXTRA SUPPORT** Ask these questions to check children's understanding:

- **What type of vibrations make loud sounds?** big vibrations
- **What type of vibrations make high-pitched sounds?** fast vibrations

**ENRICHMENT** To develop children's higher-order thinking skills, ask:

- **Why do lions make louder sounds than cats?** Possible answer: A lion is stronger than a cat and can use more energy to make a louder sound.
- **What does a fire engine need to make a high-pitched sound?** Possible answer: A high pitch is needed to make other people aware that it's headed to an emergency.

### RESOURCES and TECHNOLOGY

Classroom Presentation Toolkit CD-ROM

*Sound*

# What do sounds move through?

## ▶ Discuss the Main Idea

**Main Idea** Sounds can move through solids, liquids, and gases.

After reading together, ask:

- **What are some other examples of solids that sound can travel through?** Possible answers: walls, doors, glass windows

## ▶ Explore the Main Idea

ACTIVITY Have a child stand outside the classroom. Shut the door and ask the class to say the child's name three times, once very soft, once at normal level, and once very loud. Have the child come back into the classroom and tell how many times she heard her name. Discuss why only the louder sounds traveled through the walls and door of the classroom.

## ▶ Develop Vocabulary

Reinforce children's understanding of *pitch* with this word study activity. Show children various pictures of things that make sound. Have them identify whether the item makes a low- or high-pitched sound. Once they've identified the pitch, have them use the item in the picture and the word *pitch* in a sentence.

### What do sounds move through?

Place your ear against your desk. Now gently tap the desk with your pencil. You hear vibrations through the desk. Sound moves through solids, such as wood or plastic.

Sound moves through liquids also. Have you ever heard sound under water? When a sound is made, the water vibrates and you hear the sound.

▼ **Dolphins and other animals make sounds under water to communicate with each other.**

410
EXPLAIN

### RESOURCES and TECHNOLOGY

- **School to Home Activities,** pp. 119–120
- **Reading and Writing,** p. 232
- SCIENCE QUEST *Sound*
- LOG ON *Science in Motion* *How We Hear Sound*
- LOG ON e-**Review** Narrated Summary and Quiz
- **ExamView® *Assessment Suite* CD-ROM**

### Classroom Equity

Children benefit from exposure to "real-life" role models to which they can relate. Such models can help children envision themselves in similar careers someday.

Use this unit as an opportunity to invite a female or minority scientist or engineer from your community to your classroom. Have him or her discuss his or her career and work. Have children create their own questions to interview him or her about their career.

Most sounds you hear
ve through air. Air is
de of gases. The closer
u are to a sound, the
der it sounds. The
ther you are from a
und, the softer it sounds.

▲ How can you tell when a fire engine is close by or far away?

What can sounds move through?

## Think, Talk, and Write

1. **Problem and Solution.** How would you get a guitar string to make a sound with a high pitch?
2. Why do your hands make a sound when you clap them together?
3. **Essential Question.** How do we hear sounds?

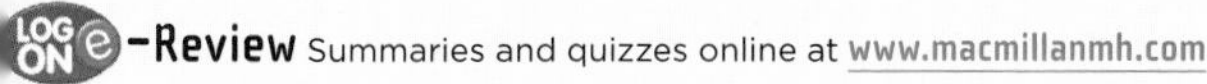

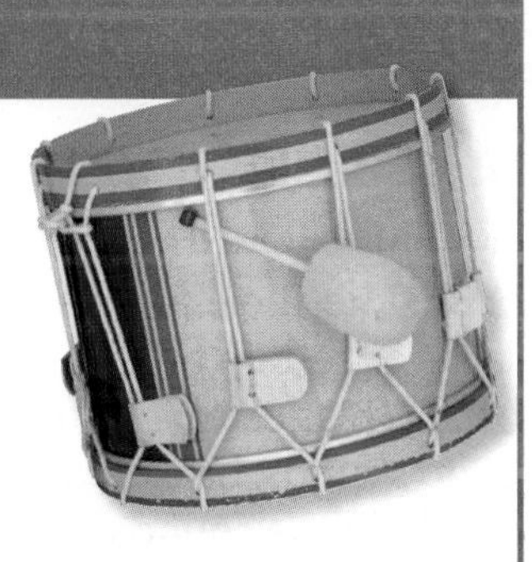

## Music Link

Make your own musical instruments. Stretch rubber bands over the open end of a plastic cup. Vibrate the rubber bands to make different pitches.

LOG ON e-Review Summaries and quizzes online at www.macmillanmh.com

411
EVALUATE

## Formative Assessment

### Making Music

Show children a picture of an instrument that the class discussed during the lesson. Have them write what people must do in order for the instrument to produce a sound. Ask children to also describe the kinds of sounds the instrument can make.

**Key Concept Cards** For student intervention, see the prescribed routine on **Key Concept Card 37.**

### Quick Check Answer

Sounds can move through solids, liquids, and gases.

Read the caption under the fire engine photo. Possible answer: If the sound is loud, the fire engine is near. If the sound is soft, it is far away.

# 3 Close

### ▶ Using the KWL Chart

Review the **KWL** chart and ask children to describe what they now know about how sound is produced and moves. Record their responses in the What We Learned column of the class **KWL** chart.

### ▶ Using the Reading Skill
**Problem and Solution**

Use the reading skill graphic organizer to identify problems and solutions in the lesson.

You fell down and needed help.

↓

You took a deep breath and called out with all the energy you had.

↓

The sound made by your vibrating vocal cords traveled through the air and into the ears of the people far away who came and helped you.

*Graphic Organizer 12,* p. TR14

### FAST TRACK Think, Talk, and Write

1. **Problem and Solution** Possible answers: pluck a short string; hold a string down against the handle to make the part that vibrates shorter
2. When your hands hit together, they make the air around them vibrate. The vibrating air moves to your ears, so a sound is heard.
3. **Essential Question** Possible answer: When an object vibrates, the vibrations move through the air to our eardrum. Our eardrum vibrates and sends messages about the sound to our brain.

## Music Link

Give children thick and thin rubber bands. Have children wrap rubber bands over the open end of a plastic cup and then pluck the strings.

# Writing in Science

## Objective

- Identify details about sounds in a description.

## Sound Off!

### Talk About It

Read the top paragraph aloud with children. Discuss what the pictured items may sound like. Ask:

- **How can we describe sounds?** Possible answers: loud or soft in volume; high or low in pitch
- **Why are we able to hear sounds?** Possible answer: because sounds make vibrations that travel to our eardrums, so we can hear the sounds

### Learn About It

Ask children to make a list of sounds that are heard every day and choose one sound from the list. Have children describe the volume, pitch, and how the sound makes them feel. As a class, write a sample paragraph about the sound on chart paper. Ask:

- **How do we use this sound?** Possible answer: People use alarm clocks to wake up.
- **Why are sounds important?** Possible answer: They help people get information.

Read the Remember box aloud. Show children how the main idea and details have been organized in the class's sample paragraph. Read aloud the paragraph.

### Write About It

Before writing, make sure that children can see both the sound list and the sample paragraph. Invite volunteers to read their paragraphs to the class.

**RESOURCES and TECHNOLOGY**

- **Reading and Writing,** pp. 233–234

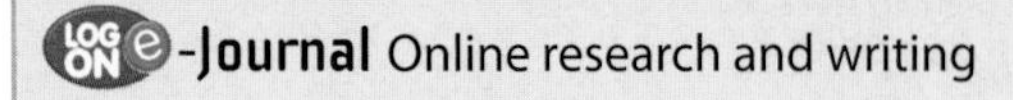

**-Journal** Online research and writing

---

## Writing in Science

# Sound Off!

Think about the sounds you hear every day. Some sounds are loud and others are soft. Some sounds are high and others are low.

### Write About It

Describe the pitch and volume of a sound you hear every day. How do we use sounds? Why are sounds important?

**Remember**

When you describe something, you give details.

**-Journal** Write about it online at www.macmillanmh.com

412
EXTEND

---

## Integrate Writing

### Sound Books

Have children write books about sounds they hear at different places or certain times of the year, such as camping sounds, spring sounds, basketball game sounds, home sounds, school sounds, or playground sounds.

Ask children to draw a picture of the person or thing that makes the sound and write one or two sentences that describe the sound on each page.

Invite volunteers to share their books with the class.

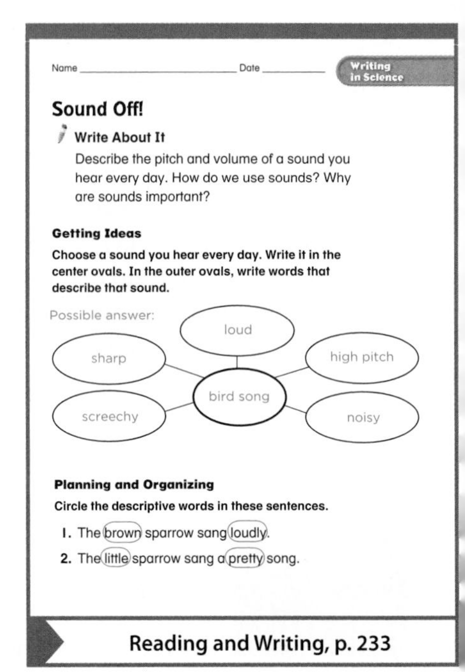

Name ______________ Date ________ Writing in Science

**Sound Off!**

**Write About It**

Describe the pitch and volume of a sound you hear every day. How do we use sounds? Why are sounds important?

**Getting Ideas**

**Choose a sound you hear every day. Write it in the center ovals. In the outer ovals, write words that describe that sound.**

Possible answer: bird song — loud, sharp, high pitch, screechy, noisy

**Planning and Organizing**

**Circle the descriptive words in these sentences.**

1. The (brown) sparrow sang (loudly).
2. The (little) sparrow sang a (pretty) song.

**Reading and Writing, p. 233**

ath in Science

# rum Fun

Miss Lee sells four different drums in her store. e first drum is 10 centimeters wide. The second um is 20 centimeters wide. The third drum is 30 ntimeters wide.

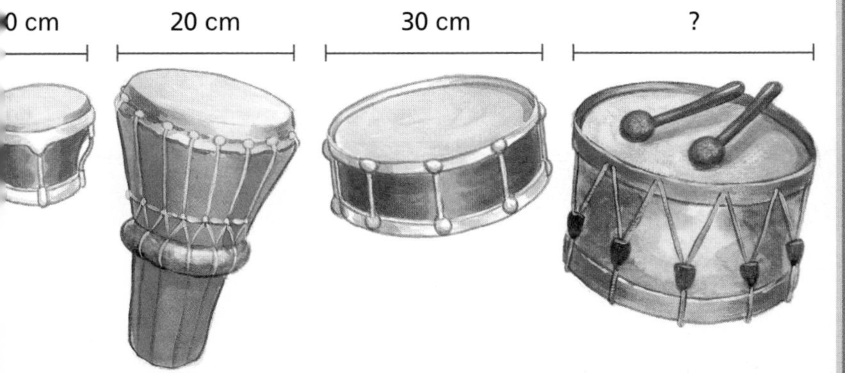


### llow the Pattern

w wide is the fourth drum?
llow this number pattern:
+ 10 = 20
+ 10 = 30
+ ? = ?

ss Lee knows the smallest drum s the highest pitch. Which drum s the lowest pitch?

**Remember**
You can use a pattern to help you solve problems.

413
EXTEND

# Math in Science

**Objective**

- Apply a pattern to solve a drum width problem.

## Drum Fun

### Talk About It

Read the first paragraph on page 413 with children. Draw pictures of the 10, 20, and 30 centimeter drums on the board and label them. Ask:

- **What do you notice about the numbers?** They get bigger.
- **What is the amount the size increases?** 10 cm

### Learn About It

As a class, make a list of four numbers that show a clear pattern in how they change. Ask:

- **What is this pattern?**

Work together with children to write a story problem about the pattern.

### Try It

Read aloud the Remember box together. Then read the math problem with children and invite them to solve it. Ask:

- **How did you find the answer?**

Encourage children to share the different ways they solved the problem. Model for children how they could use number sentences, such as those shown on page 413, to show how they solved the problem.

Ask:

- **Which drum would have the lowest pitch?** the largest drum
- **Which drum would have the highest pitch?** the smallest drum

### Integrate Math

**Sound Patterns**

Demonstrate a sound pattern, such as: 1 clap, 2 claps, 3 claps, 4 claps.

Encourage children to copy the pattern. Ask:

- **How does the pattern change?** It increases by 1 clap each time.

Explain that the pattern can be described as a plus 1 pattern.

Invite children to make and describe new clapping patterns.

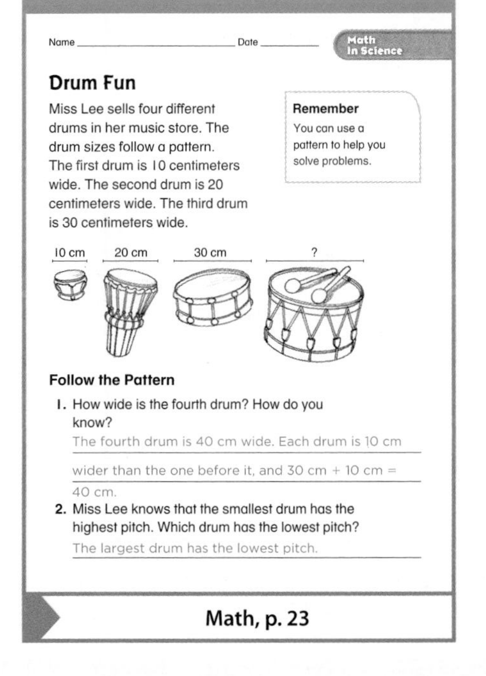

Name ____________ Date ________ Math in Science

**Drum Fun**

Miss Lee sells four different drums in her music store. The drum sizes follow a pattern. The first drum is 10 centimeters wide. The second drum is 20 centimeters wide. The third drum is 30 centimeters wide.

**Remember**
You can use a pattern to help you solve problems.

10 cm 20 cm 30 cm ?

**Follow the Pattern**

1. How wide is the fourth drum? How do you know?
The fourth drum is 40 cm wide. Each drum is 10 cm wider than the one before it, and 30 cm + 10 cm = 40 cm.
2. Miss Lee knows that the smallest drum has the highest pitch. Which drum has the lowest pitch?
The largest drum has the lowest pitch.

Math, p. 23

**RESOURCES and TECHNOLOGY**

**Math,** pp. 23–24

# Plan Your Lesson

## Lesson 3 Light

### Essential Question

How does light help us to see?

### Objective

- Identify the composition and properties of light.

**Reading Skill** Sequence

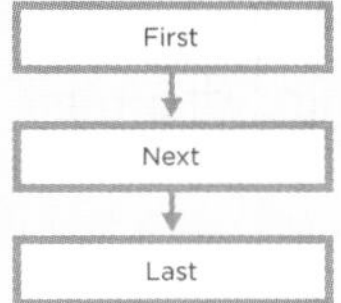


*Graphic Organizer 7,* p. TR9

## FAST TRACK

**Lesson Plan** When time is short, follow the Fast Track and use the essential resources.

**1 Introduce**
Look and Wonder, p. 414
Resource **Activity Lab Book, p. 193**

**2 Teach**
Discuss the Main Idea, p. 416
Resource **Visual Literacy, p. 38**

**3 Close**
Think, Talk, and Write, p. 419
Resource **Assessment, p. 152**

**Professional Development** Look for NSDL to find recommended Science Background resources from the National Science Digital Library.

## Reading and Writing

Name ____________ Date ______ LESSON Outline

**Light**

**Use your book to help you fill in the blanks.**

**What is light?**

1. Did you know that ___light___ energy helps you see things?
2. Some light comes from ___lightbulbs___ and flashlights.
3. Most light on Earth comes from the ___Sun___.
4. Light ___reflects___ off of objects and goes into our eyes to help us see.
5. The dark area made when something is blocking light is called a ___shadow___.
6. Some ___solid___ objects can block light and make shadows.

**How do we see color?**

7. White light is really a mix of different ___color___s of light.

**Outline, pp. 235–236**
Also available as a student workbook

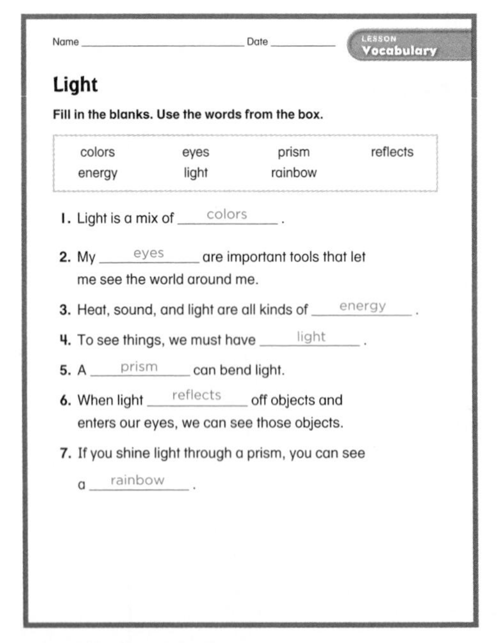

Name ____________ Date ______ LESSON Vocabulary

**Light**

**Fill in the blanks. Use the words from the box.**

| colors | eyes | prism | reflects |
|---|---|---|---|
| energy | light | rainbow | |

1. Light is a mix of ___colors___.
2. My ___eyes___ are important tools that let me see the world around me.
3. Heat, sound, and light are all kinds of ___energy___.
4. To see things, we must have ___light___.
5. A ___prism___ can bend light.
6. When light ___reflects___ off objects and enters our eyes, we can see those objects.
7. If you shine light through a prism, you can see a ___rainbow___.

**Vocabulary, p. 237**
Also available as a student workbook

## Visual Literacy

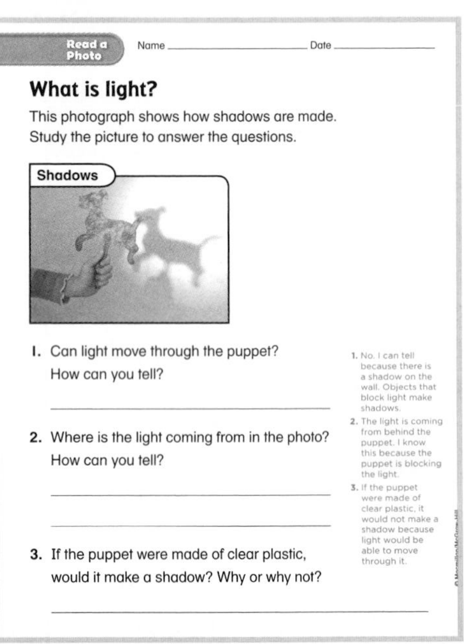

Read a Photo Name ____________ Date ______

**What is light?**

This photograph shows how shadows are made. Study the picture to answer the questions.

Shadows

1. Can light move through the puppet? How can you tell?
2. Where is the light coming from in the photo? How can you tell?
3. If the puppet were made of clear plastic, would it make a shadow? Why or why not?

1. No. I can tell because there is a shadow on the wall. Objects that block light make shadows.
2. The light is coming from behind the puppet. I know this because the puppet is blocking the light.
3. If the puppet were made of clear plastic, it would not make a shadow because light would be able to move through it.

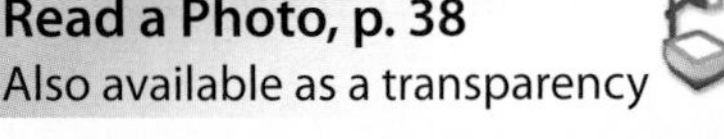

**Read a Photo, p. 38**
Also available as a transparency

# Activity Lab Book

Name ________ Date ________ **Explore**

**What does light pass through?**

**What to Do**

| You need |
|---|
| • flashlight |
| • cardboard |
| • plastic wrap |
| • various items |

1. **Predict.** Which materials will light pass through? Which will block the light?
   Possible answer: I predict that light will pass through clear objects and that dark objects will block the light.
2. Work with a partner. Hold up the cardboard. Hold plastic wrap three inches in front of the board. Your partner shines the flashlight on the object.
3. **Observe.** Did the plastic wrap block the light or did the light pass through it?
   The light passed through it.
4. **Compare.** Which objects block the light and which let light pass through?
   Possible answer: Cardboard and other dark objects block the light. Plastic wrap and other light objects let light pass through.

**Explore, pp. 191–192**
Also available as a student workbook

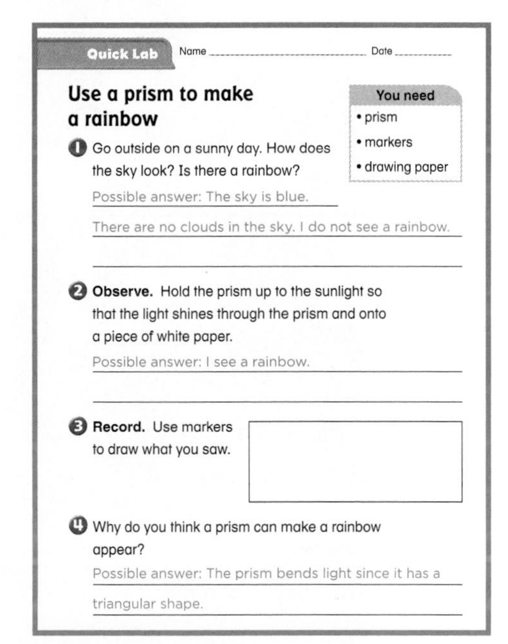

**Quick Lab** Name ________ Date ________

**Use a prism to make a rainbow**

| You need |
|---|
| • prism |
| • markers |
| • drawing paper |

1. Go outside on a sunny day. How does the sky look? Is there a rainbow?
   Possible answer: The sky is blue. There are no clouds in the sky. I do not see a rainbow.
2. **Observe.** Hold the prism up to the sunlight so that the light shines through the prism and onto a piece of white paper.
   Possible answer: I see a rainbow.
3. **Record.** Use markers to draw what you saw.
4. Why do you think a prism can make a rainbow appear?
   Possible answer: The prism bends light since it has a triangular shape.

**Quick Lab, p. 194**
Also available as a student workbook

# Assessment

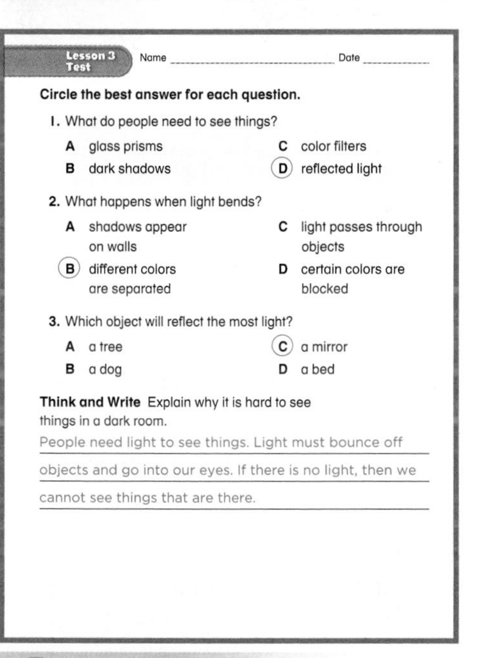

**Lesson 3 Test** Name ________ Date ________

**Circle the best answer for each question.**

1. What do people need to see things?
   A glass prisms C color filters
   B dark shadows (D) reflected light
2. What happens when light bends?
   A shadows appear on walls C light passes through objects
   (B) different colors are separated D certain colors are blocked
3. Which object will reflect the most light?
   A a tree (C) a mirror
   B a dog D a bed

**Think and Write** Explain why it is hard to see things in a dark room.
People need light to see things. Light must bounce off objects and go into our eyes. If there is no light, then we cannot see things that are there.

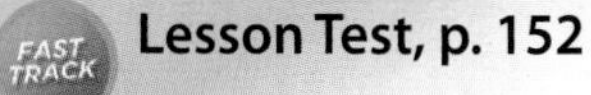


**Lesson Test, p. 152**

## ADDITIONAL RESOURCES

pp. 188–191

*Alexander Graham Bell*

pp. 58–59

p. 122

Transparencies for Visual Literacy

p. 38

38

110–120

pp. 121–122

99–100

### Technology

- Science Activity DVD
- TeacherWorks™ Plus CD-ROM
- Classroom Presentation Toolkit CD-ROM

-Review

NSDL

# Lesson 3 Light

## Objective

- Identify the composition and properties of light.

# 1 Introduce

## Assess Prior Knowledge

Have children share what they know about light. Ask:

- **Where does light come from?** the Sun, light bulbs, fire, fireflies
- **What kinds of things does light pass through?** Possible answers: clear liquids, windows

Record children's answers in the What We Know Column of the class **KWL** chart.

## Look and Wonder

Read the Look and Wonder questions. Invite children to share their responses to the questions.

Ask:

- **How do you know that light travels?** Possible answer: If you turn on a flashlight, the light travels from the flashlight to the object.

Lesson 3
Light
Look and Wonder
Where is this light coming from?
What is blocking some of the light?
414
ENGAGE


## Warm Up

### Start with a Demonstration

Shine a flashlight onto a piece of colored construction paper. Ask children to describe what they see.

Shine the flashlight at a small mirror, so the light is reflected onto a wall or desk. Have children describe what they see.

Ask:

- **What is happening to the light?** Possible answer: The light is bouncing off the mirror.

Shine the flashlight through a piece of colored plastic wrap. Ask children to describe what they think is happening to the light.

## RESOURCES and TECHNOLOGY

- **Activity Lab Book,** pp. 191–193
- **Activity Flipchart,** p. 58
- Science Activity DVD

## Explore Inquiry Activity

### What does light pass through?

#### What to Do

1. **Predict.** Which materials will light pass through? Which will block the light?
2. Work with a partner. Hold up the cardboard. Hold plastic wrap three inches in front of the board. Your partner shines the flashlight on the object.
3. **Observe.** Did the plastic wrap block the light, or did the light pass through it?
4. **Compare.** Which objects block the light and which let light pass through?

#### Explore More

5. **Predict.** What might happen with other classroom items? Try it.

You need

Step 2

## Alternative Explore

### How much light passes through paper?

Distribute flashlights and paper to pairs. Ask children to predict whether light will go through each kind of paper.

Have one child hold the flashlight, while another child holds the paper. Have a board behind the child holding the paper and darken the room.

Have children observe and compare how much light different papers allowed through.

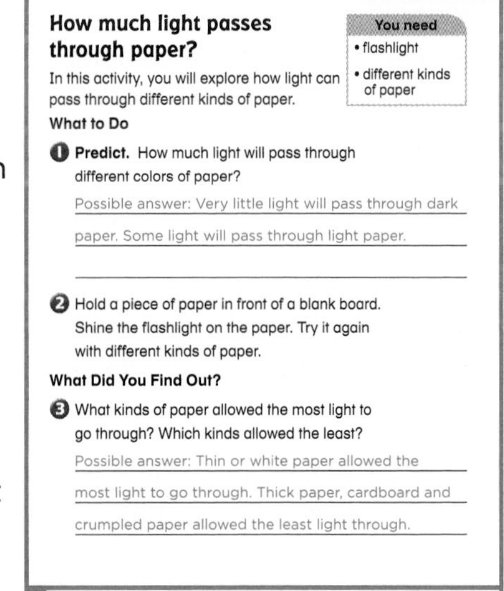

Name ______ Date ______ Alternative Explore

**How much light passes through paper?**

You need
- flashlight
- different kinds of paper

In this activity, you will explore how light can pass through different kinds of paper.

**What to Do**

1. **Predict.** How much light will pass through different colors of paper?
   Possible answer: Very little light will pass through dark paper. Some light will pass through light paper.
2. Hold a piece of paper in front of a blank board. Shine the flashlight on the paper. Try it again with different kinds of paper.

**What Did You Find Out?**

3. What kinds of paper allowed the most light to go through? Which kinds allowed the least?
   Possible answer: Thin or white paper allowed the most light to go through. Thick paper, cardboard and crumpled paper allowed the least light through.

Activity Lab Book, p. 193

## Explore

 pairs  15 minutes

**Plan Ahead** Provide enough flashlights and materials for pairs. Check flashlights to make sure each emits a strong beam.

**Purpose** Children will observe that different amounts of light pass through different materials.

**Structured Inquiry** What to Do

Have children suggest a way to test whether or not different materials allow light to pass through them. Explain to children that in this activity, they will hold an object in front of the cardboard to see whether or not light can shine through it.

1. **Predict** Have children explain their predictions. Ask: **Does the object's color affect the amount of light that will pass through it?**
2. **Be Careful!** Remind children not to direct light into people's eyes. Ask: **What happens to the beam of light if it is not aimed directly on the object being tested?**
3. **Observe** Ask: **How can you tell that light passed through the plastic wrap?** Have children observe what would happen if there was more than one layer of plastic wrap.
4. **Compare** Ask: **How are the objects that block light and the objects that let light through the same? How are they different?**

**Guided Inquiry** Explore More

5. **Predict** Have children discuss their predictions with their partners.

**Open Inquiry**

Encourage children to explore further by asking: **What are some other questions you have about how light is reflected?** Help children find materials and resources they will need to answer some of their questions.

# 2 Teach

## Read Together and Learn

**Reading Skill** Sequence The order in which things happen.

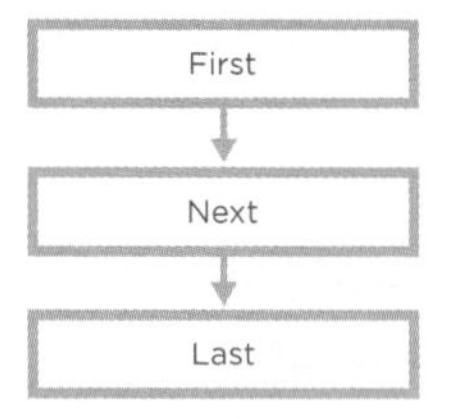


*Graphic Organizer 7,* p. TR9

### What is light?

FAST TRACK **Discuss the Main Idea**

**Main Idea** Light is a kind of energy that can only pass through some objects.

After reading together, ask:

- **Why does your body make a shadow when you are outside on a sunny day?** My body blocks light.
- **What does light do to help people see things?** Light bounces off things.
- **Will the teddy bear's arms reflect more or less light than the mirrors on its vest?** less

**Read Together and Learn**

**Essential Question**
How does light help us to see?

**Vocabulary**
light
reflect

### What is light?

You need light to see things. **Ligh** is a kind of energy. You see things because light will **reflect**, or bounc off things around you. Light that reflects off objects enters your eye Then you can see the objects.

Some sources of light are the Su light bulbs, and flashlights. Most light on Earth comes from the Sun.

Smooth, shiny objects, such as mirrors, reflect a lot of light.

416
EXPLAIN

## Science Background

**Light** Different colors of light have different wavelengths. White light is a mixture of all colors in the visible part of the electromagnetic spectrum. Red has the longest wavelength and violet has the shortest. Infrared and ultraviolet light are outside of the spectrum of visible light. Infrared light has a wavelength that's too long for the human eye to detect. Ultraviolet light has a wavelength that's too short for the human eye to detect.

See **Science Yellow Pages**, in the Teacher Resources section, for background information.

 **Professional Development** For more Science Background and resources from NSDL visit http://nsdl.org/refreshers/science

## ELL Support

**Use Realia/Demonstrate** In a darkened room, shine a flashlight onto objects, such as a mirror, prism, hand lens, and book. A blackboard should be behind the objects. Have children say the words *shadow* or *light,* depending on what appears.

**BEGINNING** Have children sort the objects used in the activity into groups of shadow and light. Have children repeat the names of the objects.

**INTERMEDIATE** Have children use short sentences, such as, *The pencil makes a shadow,* to describe what they see on the board.

**ADVANCED** Have children explain why they see light on the board or a shadow for each object used in the activity.

Have you ever made a shadow on a
all? A shadow is a dark area where light
oes not reach.

Different objects let different amounts
f light through. A book is a solid object.
can block light and make a shadow.
lass is clear. It does not make a shadow
ecause light passes through it.

**Shadows**

**Read a Photo**

How is this shadow made?

What are some sources of light?

417
EXPLAIN

## Read a Photo

Look at the photo on page 417 and read the Read a Photo question. Ask:

- **What in the picture gives a clue that there is a shadow?** The shadow looks like the dog puppet.
- **What do you think would happen if the light source was turned on and off?** The shadow would appear and disappear.

**Answer to Read a Photo** The puppet is blocking the light.

## Develop Vocabulary

**light** Explain that the word *light* is a homophone and that homophones are words which sound the same, but have different meanings. Ask children if they know what the homophone for *light* means. Explain that *light* can also mean something that does not weigh a lot. Ask children to think of other homophones and discuss their meanings. buy/by; flour/flower; sail/sale

**reflect** Write the word *reflect* on the board. Ask children to describe what they think it means. Explain that light reflects or "bounces off a surface." Underline the prefix *re-*, and explain that it means "again." Have children try to think of other words with *re-* as a prefix, such as: *redo, repeat, recall, remake*.

## *Quick Check Answer*

Possible answers: the Sun, light bulbs, flashlights

## Differentiated Instruction

### Leveled Activities

**EXTRA SUPPORT** Have children shine a flashlight onto an object with a white piece of paper set underneath. Use objects that allow different amounts of light to pass through. Explain why sometimes they see shadows and sometimes they do not.

**ENRICHMENT** Ask children to bring in or collect different objects around the classroom that reflect light or let light pass through, such as foil, spoons, drinking glass, or water. Divide children into groups. Have them use a flashlight to test whether light bounces off the objects or passes through them. Children can write a report about their findings. They should note similarities among the group of items that let light pass through and explain how these features help light shine through. Have children predict other materials that would not block light.

## RESOURCES and TECHNOLOGY

- **Reading and Writing,** pp. 235–237
- **Visual Literacy,** p. 38
- **PuzzleMaker CD-ROM**
- **Classroom Presentation Toolkit CD-ROM**
-  **e-Glossary**

# How do we see color?

## ▶ Discuss the Main Idea

**Main Idea** White light is made up of all colors and can be separated into the different colors when it is bent.

After reading together, ask:

- **What happens when light bends?** It separates and allows people to see all the different colors.
- **How can you create a purple light?** Place thin purple paper or plastic over a light.

## ▶ Use the Visuals

Look at the pictures on pages 418 and 419 and read the captions. Ask:

- **What happens to light when it hits raindrops?** It bends and becomes separated into colors.
- **Why do we see a green traffic light?** A filter in front of the light bulb only allows green light to pass through.

## ▶ Develop Vocabulary

To reinforce children's understanding of vocabulary used in the lesson, use this word study activity. Tell children that the word *light* can also be used as a verb meaning "to give out light." Ask children to use the verb in a sentence. Possible answers: It's time to light the candles. The Sun lights Earth.

### RESOURCES and TECHNOLOGY

- **School to Home Activities,** pp. 121–122
- **Reading and Writing,** p. 238
- **Activity Lab Book,** p. 194
- LOG ON **e-Review** Narrated Summary and Quiz
- **ExamView® *Assessment Suite* CD-ROM**

## How do we see color?

Did you know light can bend? Light is a mix of all colors. When white light bends, it separates into different colors. Then we can see the colors of the rainbow.

A prism is an object that can make light bend.

### Quick Lab

Use a prism and sunlight to see the colors of the rainbow. **Observe** and draw what you see.

Raindrops can act like prisms and separate light to make rainbows.

418
EXPLAIN

## Quick Lab

 small groups  15 minutes

**Objective** Observe what happens to light as it passes through a prism.

**You need** prisms, a light source, colored pencils, paper

1. Give each group a prism.
2. Ask children to shine light through a prism onto a white sheet of paper.
3. Have children observe and discuss what is happening to the light. Encourage children to record results by drawing a picture using colored pencils.

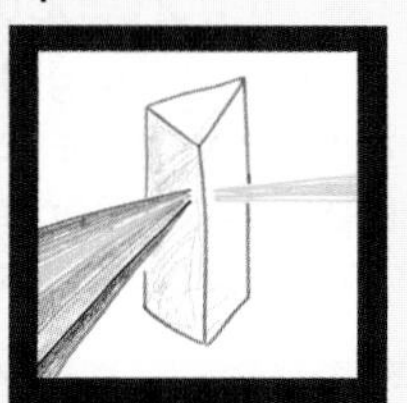

Have you ever seen colored ghts? A filter is a tool that ets only certain colors of light ass through it.

Some filters let only one olor pass through. A red ilter blocks all colors except ed. You see only red light vith a red filter.

Colored glass makes a white light look red, green, or yellow.

What color is most light we see?

## Think, Talk, and Write

1. **Sequence.** What happens when we see objects?
2. What kind of objects make shadows?
3. **Essential Question.** How does light help us to see?

### Art Link

Make a filter. Cover a flashlight with colored plastic wrap. Then make shadow puppets!

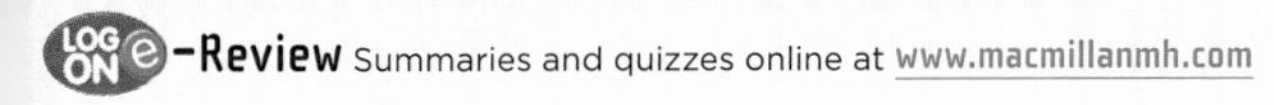
**e-Review** Summaries and quizzes online at www.macmillanmh.com

## Formative Assessment

**Rainbow Connection**
Have children draw diagrams to show how raindrops can act as prisms and describe what is needed to create a rainbow. Remind them to show the light source in their drawings.

Key Concept Cards

**Key Concept Cards** For student intervention, see the prescribed routine on **Key Concept Card 38.**

### Quick Check Answer

Most light we see is white light.

# 3 Close

## Using the KWL Chart

Review with children what they have learned about light. Record their responses in the What We Learned column of the class **KWL** chart.

## Using the Reading Skill

### Sequence

Use the reading skill graphic organizer to show the sequence of how light is seen.

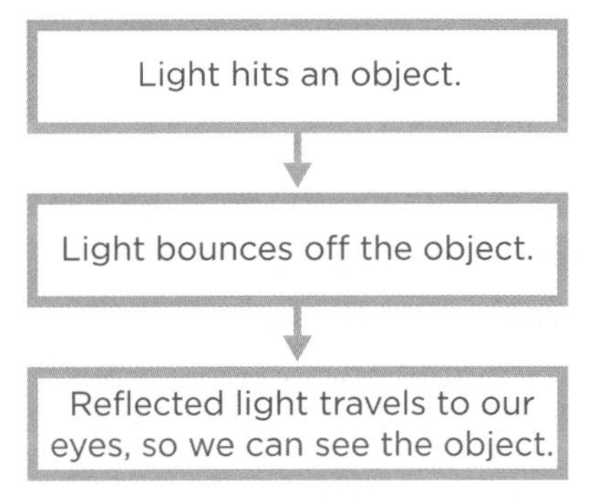
Light hits an object.
Light bounces off the object.
Reflected light travels to our eyes, so we can see the object.

*Graphic Organizer 7*, p. TR9

## FAST TRACK Think, Talk, and Write

1. **Sequence** First, light hits an object. Next, light reflects off the object. Last, the reflected light travels to our eyes, so we can see the object.
2. Objects that do not let light pass through them make shadows.
3. **Essential Question** Possible answers: When light reflects off an object, it enters our eyes and we see the object. We can also see things that make their own light, such as a flashlight or the Sun.

## Art Link

Encourage children to use a variety of colored, transparent papers to make filters. Discuss which colors work best to create vivid shadows on the wall.

# Be a Scientist

small groups · 30 minutes

Use Your

Activity Flipchart, p. 59

**Skills** compare, predict, record data

## Objective

- Compare how sunlight affects the temperature of light and dark objects.

**You need** one black cloth, one white cloth, two thermometers, clock

**Plan Ahead** Be sure to have the same type of cloth in black and white, so that thickness will not affect the outcome of the experiment. Do the activity at a time of day when the Sun is bright enough to warm the cloth quickly. This activity should be repeated at least three times as part of the scientific method.

**Be Careful!** Remind children to handle thermometers carefully because they can be broken.

**EXTEND** Children will measure, record, and compare the effects of sunlight on different colors.

# Be a Scientist

**You need**

black cloth

white cloth

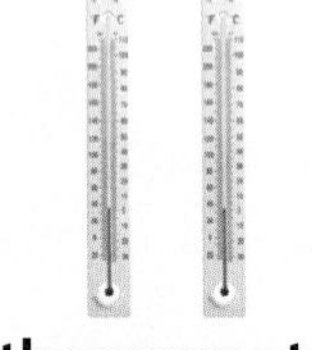

2 thermometers

clock

## How does sunlight affect the temperature of light and dark objects?

### What to Do

1. Record the temperature of each thermometer on a chart. Wrap one thermometer in black cloth as show[n]. Wrap the other in white cloth.

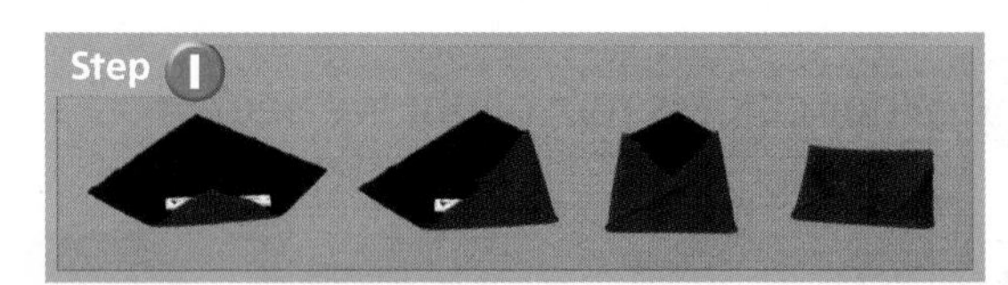


2. Place the wrapped thermometers o[n a] sunny windowsill. Wait 15 minutes.

**Structured Inquiry** What to Do

## How does sunlight affect the temperature of light and dark objects?

1. Have children feel two different-colored cloths that have been stored together in the shade. Ask: **What can you tell me about the temperatures of the two cloths?** They are the same temperature. Children may use graphic organizer 11 with the titles *Black Cloth* and *White Cloth* at the top to record temperatures. Encourage them to draw a line to make two rows on the chart. Ask them to record the initial temperatures in the first row.

### RESOURCES and TECHNOLOGY

- **Activity Lab Book,** pp. 195–196
- **TeacherWorks™ Plus CD-ROM**

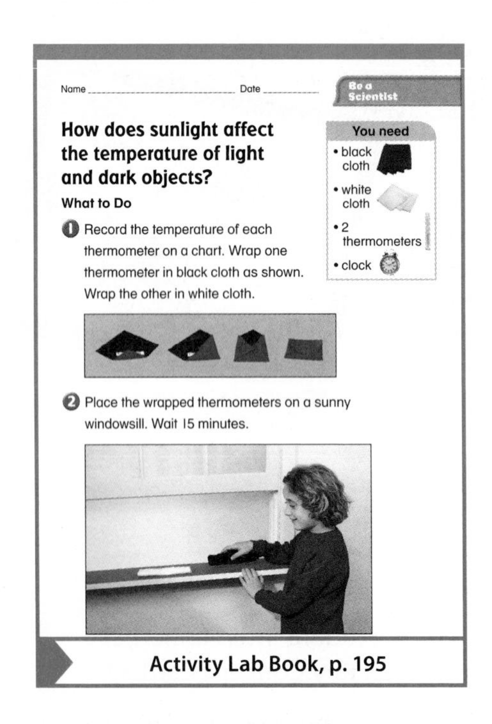

Name ______ Date ______

Be a Scientist

**How does sunlight affect the temperature of light and dark objects?**

**What to Do**

1. Record the temperature of each thermometer on a chart. Wrap one thermometer in black cloth as shown. Wrap the other in white cloth.

2. Place the wrapped thermometers on a sunny windowsill. Wait 15 minutes.

**You need**
- black cloth
- white cloth
- 2 thermometers
- clock

Activity Lab Book, p. 195

## Inquiry Investigation

Compare. Feel each cloth with your hands after 15 minutes. Which color cloth feels warmer?

Predict. Which color will have the higher temperature? Why do you think so?

Record Data. Unwrap each cloth and record each temperature on the chart.

Compare the temperatures. What happened to the temperature of each cloth? Was your prediction correct?

### nvestigate More

ompare. What other dark colors and light olors can you test? Make a plan and test it.

Use with Chapter 12, Lesson 3 59

## Integrate Math

### Comparing Heights

Provide pairs of children with rulers and yardsticks.

Ask children how they would measure each other's height. Once they develop a strategy, tell them how many inches tall you are. Using your height as a basis, have children make estimates of their own height.

Ask children to measure and record their height in inches, and make a class height graph with the results.

2. Encourage children to handle the thermometers with care.
3. **Compare** Discuss with children that people's senses are not as accurate as a thermometer.
4. **Predict** Have children explain the reasons behind their predictions. Try to relate the predictions to actual experiences, such as what it feels like to wear dark clothing on a hot day.
5. **Record Data** Remind children to use the same unit of temperature that they used in step 1.
6. Have children discuss their predictions and the results of the activity. Ask: **Would the two different cloths have different temperatures in a dark room?** No, they would both be room temperature. Explain that the black cloth absorbed more heat. Note that the more direct sunlight the object receives, the greater the temperature difference.

### Guided Inquiry Investigate More

**Compare** Have children repeat the activity using a wider range of colors. Ask them to make new predictions for temperatures of the different colors. Ask: **Will different light sources heat the cloth differently?** Use different light sources to find out. **Be Careful!** Some light sources are very hot.

### Open Inquiry

Have children generate additional questions about heat absorption and color, such as: **Do dark colors cool more quickly or more slowly than light colors?** Children may use a cooler or refrigerator to find out the answer.

Children may also investigate whether colored water absorbs heat differently. Ask: **Does different-colored water heat differently in the sun?** Children may use food coloring and water to find out.

Have children make a plan to find out the answers.

# Plan Your Lesson

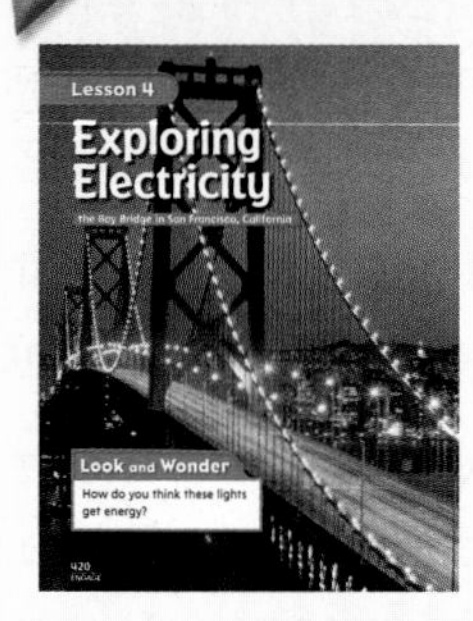

Lesson 4
Exploring Electricity
Look and Wonder
How do you think these lights get energy?


## Lesson 4 Exploring Electricity

### Essential Question

How do we get electricity?

### Objective

- Identify forms of electricity and their uses.

**Reading Skill** Cause and Effect

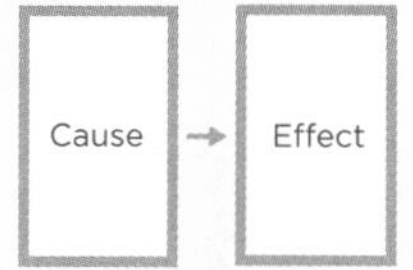


*Graphic Organizer 8,* p. TR10

## FAST TRACK

**Lesson Plan** When time is short, follow the Fast Track and use the essential resources.

**1 Introduce**
**Look and Wonder, p. 420**
Resource **Activity Lab Book, p. 199**

**2 Teach**
**Discuss the Main Idea, p. 422**
Resource **Visual Literacy, p. 39**

**3 Close**
**Think, Talk, and Write, p. 425**
Resource **Assessment, p. 153**

**Professional Development** Look for NSDL to find recommended Science Background resources from the National Science Digital Library.

## Reading and Writing

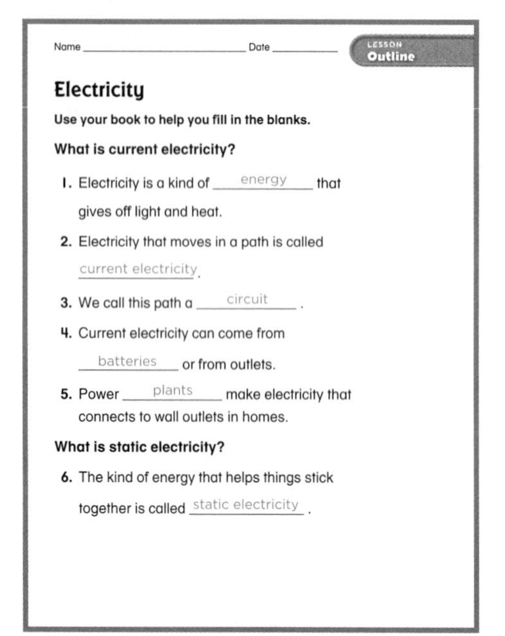

Name ____________ Date ________ LESSON Outline

**Electricity**

**Use your book to help you fill in the blanks.**

**What is current electricity?**

1. Electricity is a kind of ___energy___ that gives off light and heat.
2. Electricity that moves in a path is called ___current electricity___.
3. We call this path a ___circuit___.
4. Current electricity can come from ___batteries___ or from outlets.
5. Power ___plants___ make electricity that connects to wall outlets in homes.

**What is static electricity?**

6. The kind of energy that helps things stick together is called ___static electricity___.

**Outline, pp. 239–240**
Also available as a student workbook

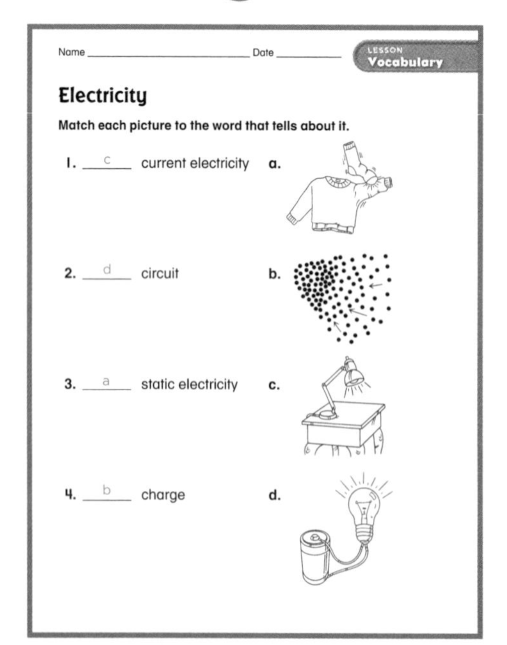

Name ____________ Date ________ LESSON Vocabulary

**Electricity**

**Match each picture to the word that tells about it.**

1. ___c___ current electricity  a.
2. ___d___ circuit  b.
3. ___a___ static electricity  c.
4. ___b___ charge  d.

**Vocabulary, p. 241**
Also available as a student workbook

## Visual Literacy

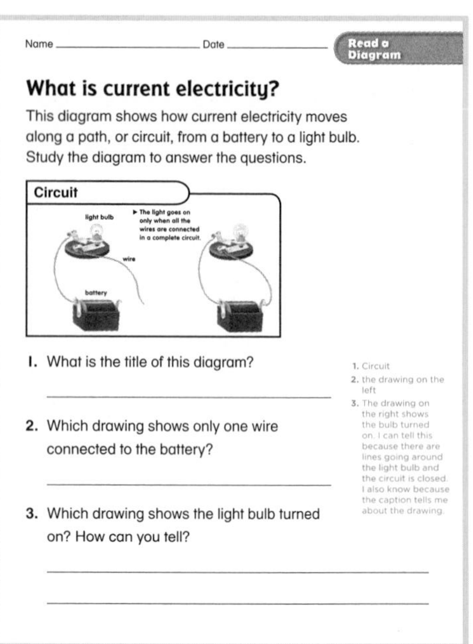

Name ____________ Date ________ Read a Diagram

**What is current electricity?**

This diagram shows how current electricity moves along a path, or circuit, from a battery to a light bulb. Study the diagram to answer the questions.



1. What is the title of this diagram?
2. Which drawing shows only one wire connected to the battery?
3. Which drawing shows the light bulb turned on? How can you tell?

1. Circuit
2. the drawing on the left
3. The drawing on the right shows the bulb turned on. I can tell this because there are lines going around the light bulb and the circuit is closed. I also know because the caption tells me about the drawing.

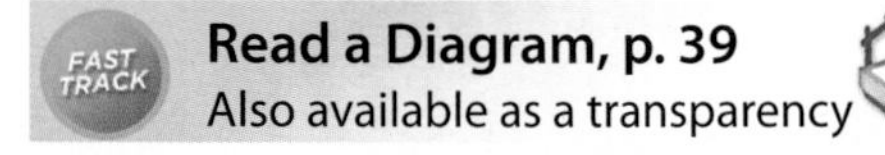

**Read a Diagram, p. 39**
Also available as a transparency

# Activity Lab Book

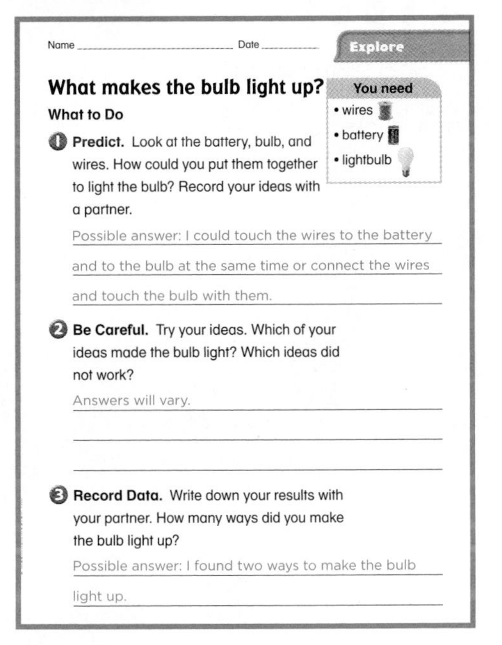
Name ____ Date ____ **Explore**

### What makes the bulb light up?

**You need**
- wires
- battery
- lightbulb

**What to Do**

1. **Predict.** Look at the battery, bulb, and wires. How could you put them together to light the bulb? Record your ideas with a partner.
   Possible answer: I could touch the wires to the battery and to the bulb at the same time or connect the wires and touch the bulb with them.
2. **Be Careful.** Try your ideas. Which of your ideas made the bulb light? Which ideas did not work?
   Answers will vary.
3. **Record Data.** Write down your results with your partner. How many ways did you make the bulb light up?
   Possible answer: I found two ways to make the bulb light up.

**Explore, pp. 197–198**
Also available as a student workbook

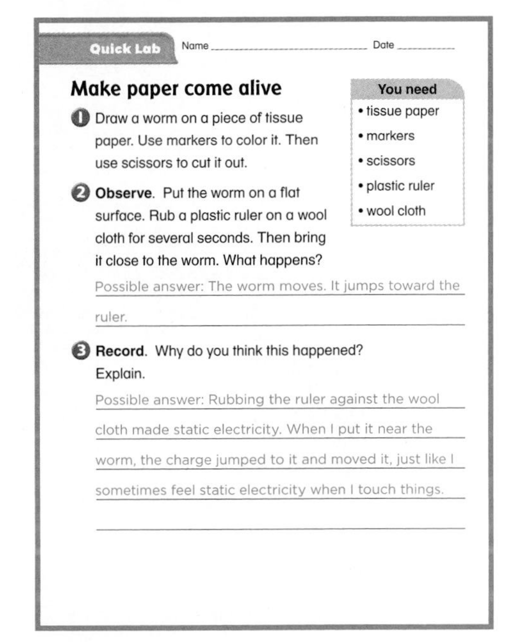
**Quick Lab** Name ____ Date ____

### Make paper come alive

**You need**
- tissue paper
- markers
- scissors
- plastic ruler
- wool cloth

1. Draw a worm on a piece of tissue paper. Use markers to color it. Then use scissors to cut it out.
2. **Observe.** Put the worm on a flat surface. Rub a plastic ruler on a wool cloth for several seconds. Then bring it close to the worm. What happens?
   Possible answer: The worm moves. It jumps toward the ruler.
3. **Record.** Why do you think this happened? Explain.
   Possible answer: Rubbing the ruler against the wool cloth made static electricity. When I put it near the worm, the charge jumped to it and moved it, just like I sometimes feel static electricity when I touch things.

**Quick Lab, p. 200**
Also available as a student workbook

# Assessment

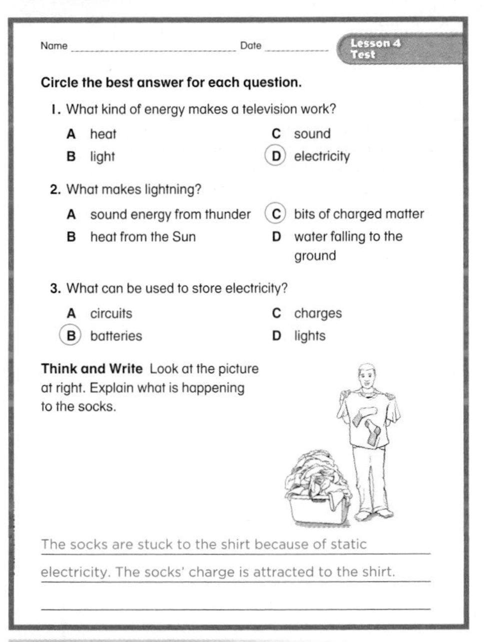
Name ____ Date ____ **Lesson 4 Test**

**Circle the best answer for each question.**

1. What kind of energy makes a television work?
   A heat  C sound
   B light  (D) electricity
2. What makes lightning?
   A sound energy from thunder  (C) bits of charged matter
   B heat from the Sun  D water falling to the ground
3. What can be used to store electricity?
   A circuits  C charges
   (B) batteries  D lights

**Think and Write** Look at the picture at right. Explain what is happening to the socks.

The socks are stuck to the shirt because of static electricity. The socks' charge is attracted to the shirt.

**Lesson Test, p. 153**

## ADDITIONAL RESOURCES

**pp. 192–195**

***Alexander Graham Bell***

**p. 60**

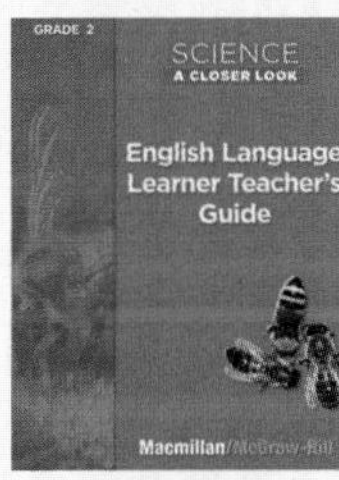

**p. 123**

**p. 39**

**39**

**110–120**

**pp. 123–124**

**101–103**

### Technology

- Science Activity DVD
- TeacherWorks™ Plus CD-ROM
- Classroom Presentation Toolkit CD-ROM
-  e-Review
- 

# Lesson 4 Exploring Electricity

**Objective**

- Identify forms of electricity and their uses.

## 1 Introduce

### ▶ Assess Prior Knowledge

Have children share what they know about electricity. Ask:

- **What is electricity?** Accept all reasonable answers.
- **From where does electricity come?** Possible answers: batteries, outlets, lightning, electrical wires
- **What are some things that use electricity?** Accept all reasonable answers.

Record children's answers in the What We Know column of the class **KWL** chart.

### Look and Wonder

Read the Look and Wonder question. Invite children to share their responses. Then ask:

- **What else in this picture uses electricity?** lights in houses and buildings
- **How do lights get the electricity they need?** through electrical wires

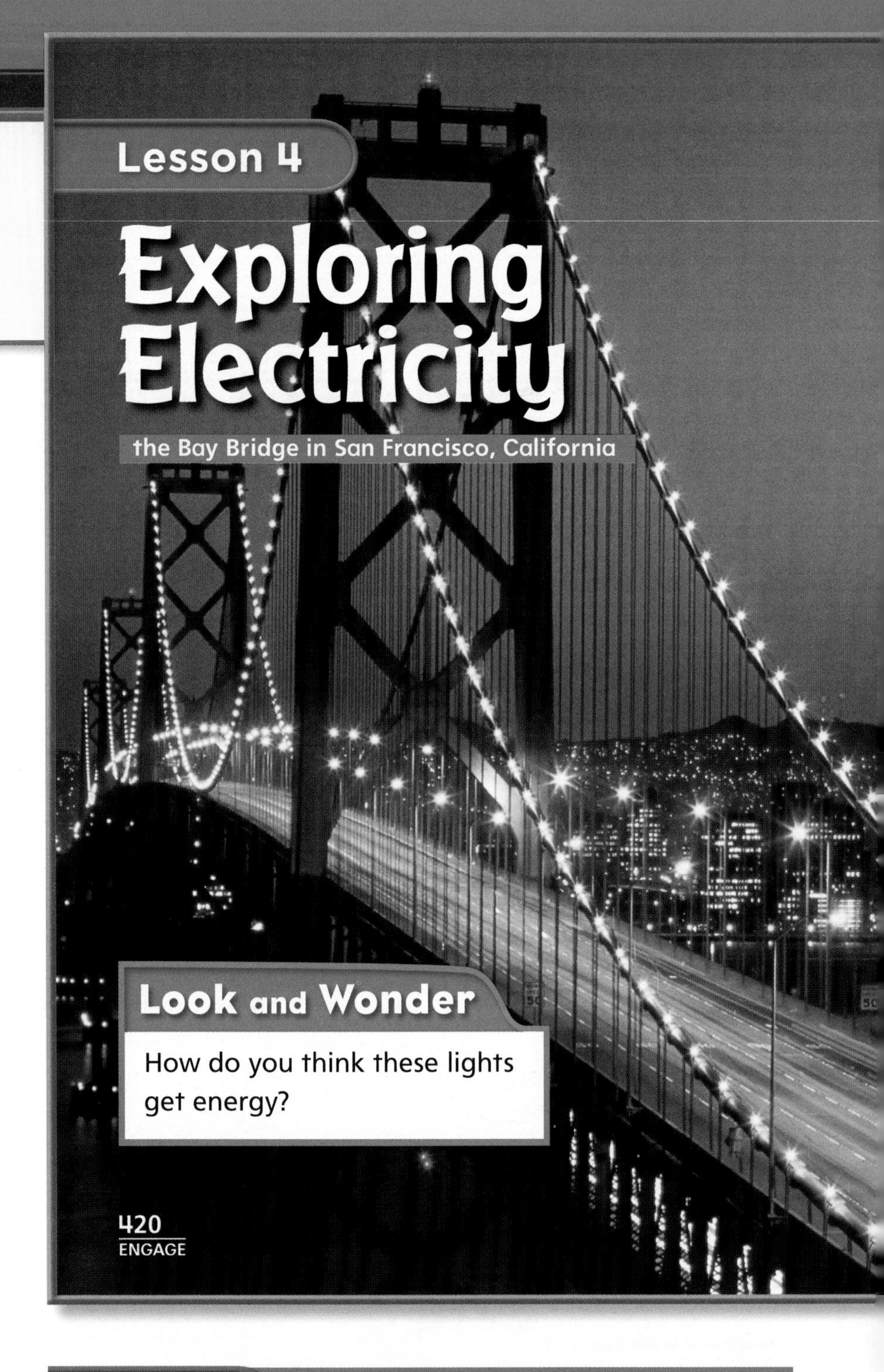

Lesson 4
Exploring Electricity
the Bay Bridge in San Francisco, California
Look and Wonder
How do you think these lights get energy?
420
ENGAGE


### Warm Up

**Start With a Book**

Before reading aloud ***Switch On, Switch Off,*** by Melven Berger (HarperCollins, 2001), show children the cover and ask:

- **What do you notice about the illustration?** Possible answers: The boy closes his eyes when the light is on; the switch is up in one picture and down in another.

While reading, encourage children to listen for facts. Ask:

- **Which shape is most like a closed circuit? Why?** a circle; The electricity goes around the circuit in a loop.
- **How does a switch affect a light bulb?** Possible answer: It controls whether it is on or off.

After reading, ask children to share the facts they learned. Encourage them to discuss parts of the book that they didn't understand. Record their responses in the class **KWL** chart.

**RESOURCES and TECHNOLOGY**

- **Activity Lab Book,** pp. 197–199
- **Activity Flipchart,** p. 60
- Science Activity DVD

## Explore — Inquiry Activity

### What makes the bulb light up?

#### What to Do

1. **Predict.** Look at the battery, bulb, and wires. How could you put them together to light the bulb? Record your ideas with a partner.
2. △ **Be Careful.** Try your ideas. Which of your ideas made the bulb light? Which ideas did not work?
3. **Record Data.** Write down your results with your partner. How many ways did you make the bulb light up?

#### Explore More

4. **Predict.** How could you make a second bulb light up? What else would you need?

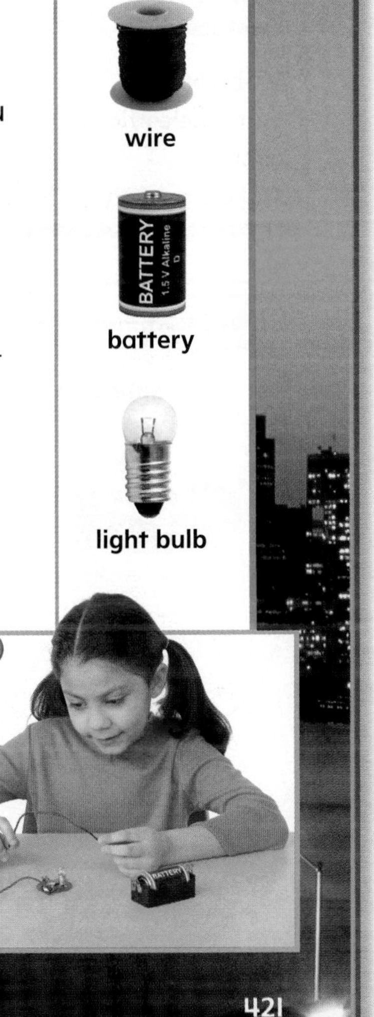




## Explore

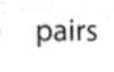

**Plan Ahead** Plan how to group children to perform this investigation. Consider pairing stronger children with less-able partners.

If a spool of wire is used for this activity, it is necessary to cut the wires and strip about half an inch of the plastic coating off the wires, so that the bare wires can make contact with the battery.

**Be Careful!** Warn children not to have wet hands when working with wires and electricity. Children should handle the bulb gently, so it doesn't break.

**Purpose** This activity will help children discover on their own how to create a closed circuit to make a bulb light up.

**Structured Inquiry** What to Do

1. **Predict** Encourage children to explain their ideas to one another. Have them attach a red-dot sticker to one of the wires, and a blue-dot sticker to the other, to make it easier for them to record what they will try.
2. Have children test their ideas and encourage them to try other things that they may not have thought of in step one.
3. **Record Data** Have children draw diagrams to show what worked, and share their results with others. Help children discuss why some things worked and others did not.

**Guided Inquiry** Explore More

4. **Predict** Give children an additional wire and bulb and help them connect the second bulb to the circuit. Have them observe what happens. Ask: **Why do you think the bulbs do not shine as brightly?** They are sharing the electricity.

**Open Inquiry**

Encourage children to think about what other questions they have about how electricity works. Review the **KWL** chart and see if they can use what they know now to generate additional questions. For example: *What would happen if we attached ten light bulbs to one battery?*

## Alternative Explore

### What does a light bulb need?

Draw a diagram on the board showing the path of electricity through a battery and light bulb. The diagram on page 422 may be used as an example.

Explain that the flow of electricity to and from the battery must not be interrupted if the bulb is to light up.

Have children draw a diagram that shows a bulb that will not light up—one in which the flow is broken.

Name ______ Date ______ Alternative Explore

**What does a light bulb need?**

What to Do

You need
- pencil
- drawing paper

1. **Observe.** Work with a partner to draw the path of the electric current that travels from the battery to the bulb and back again.
2. **Predict.** What do you think will happen if the electric current cannot make a circuit between the light bulb and the battery?
   Possible answer: I think that the bulb will not light up.
3. Draw a diagram that shows an incomplete circuit.

What Did You Find Out?

4. What can keep a light bulb from lighting up?
   Possible answers: If the flow of electricity is stopped somewhere, it could keep a bulb from lighting up.

Activity Lab Book, p. 199

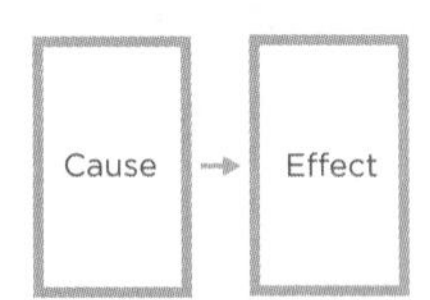

# 2 Teach

## Read Together and Learn

**Reading Skill** Cause and Effect A cause is why an event happens. An effect is the event that happens.

| Cause | → | Effect |
|---|---|---|

*Graphic Organizer 8*, p. TR10

## What is current electricity?

### Discuss the Main Idea

FAST TRACK

**Main Idea** Current electricity moves in a path called a circuit.

Before reading the text together, ask children to list things that need electricity in order to work.

After reading together, ask:

- **Why do you need to plug things into an outlet?** Possible answers: to create a complete circuit; to allow electricity to get to the appliance and make it work
- **Why do things have an on/off switch?** Possible answers: to get the electricity to flow to the object; to shut it off; to open and close the circuit

**Read Together and Learn**

**Essential Question**
How do we get electricity?

**Vocabulary**
current electricity
circuit
static electricity

## What is current electricity?

Do batteries make some of your toys work? Batteries make a kind o electricity. **Current electricity** is a kind of energy that moves in a path The electricity moves along a path called a **circuit**. The circuit needs to be completely connected for the electricity to move.

**Circuit**

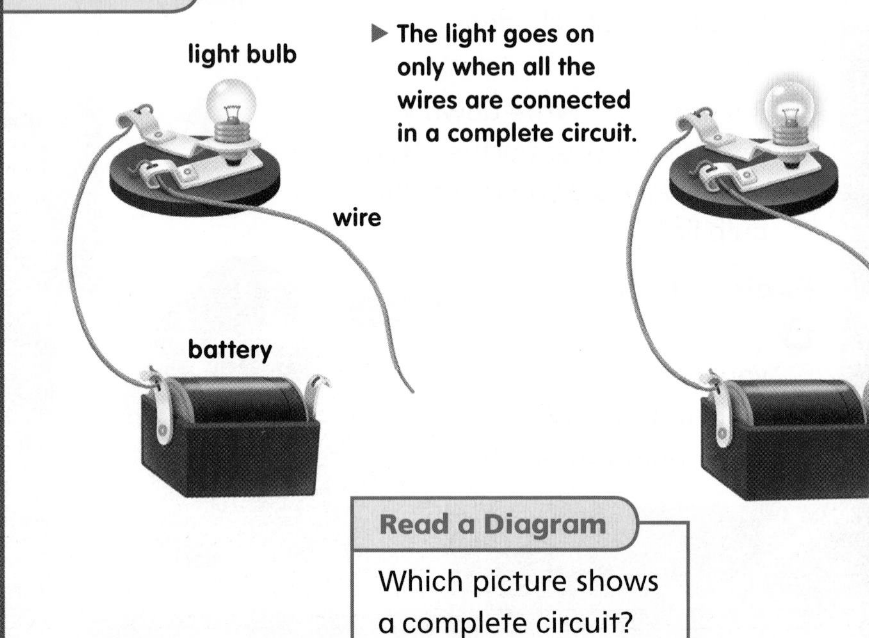

▶ The light goes on only when all the wires are connected in a complete circuit.

**Read a Diagram**

Which picture shows a complete circuit?



## Science Background

**Electricity** Electricity is the flow of electrical energy, commonly defined as the movement of electrons through wires and batteries. Unlike heat or light energy, electricity is a secondary energy source. It's usually made by the conversion of some other source, such as the burning of coal or the use of wind or water power. Electricity rarely occurs naturally, except in the form of lightning.

See **Science Yellow Pages**, in the Teacher Resources section, for background information.

**Professional Development** For more Science Background and resources from NSDL visit http://nsdl.org/refreshers/science

## ELL Support

**Use Realia** Collect things with on/off switches and give children the opportunity to use them.

**BEGINNING** Name each object and ask a child to turn it on. Say: *(Child's Name) turned on the* ____. When the child turns it off, have the group repeat the sentence changing the word *on* to *off*. Repeat until everyone has had a turn.

**INTERMEDIATE** Have children take turns turning things on and off. They should describe what they did.

**ADVANCED** Have children describe, in their own words, what causes the objects to turn on and off. Help them use the terms *current electricity* and *circuit* in their descriptions.

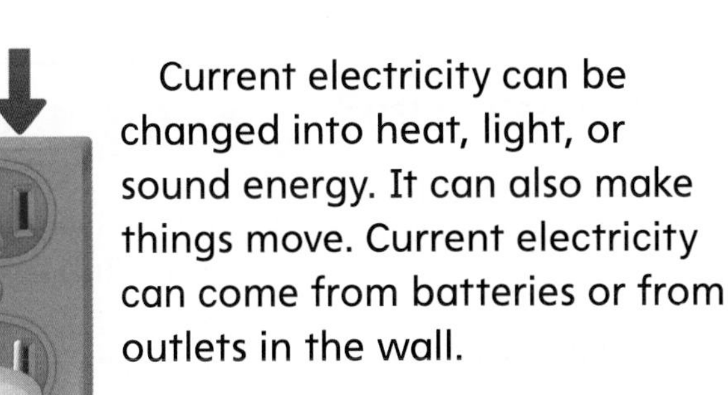

Current electricity can be changed into heat, light, or sound energy. It can also make things move. Current electricity can come from batteries or from outlets in the wall.

Buildings called power plants change other kinds of energy into electricity. The electricity runs through wires into your house and into the outlets. When you plug in your toaster and turn it on, you complete the circuit with the power plant.

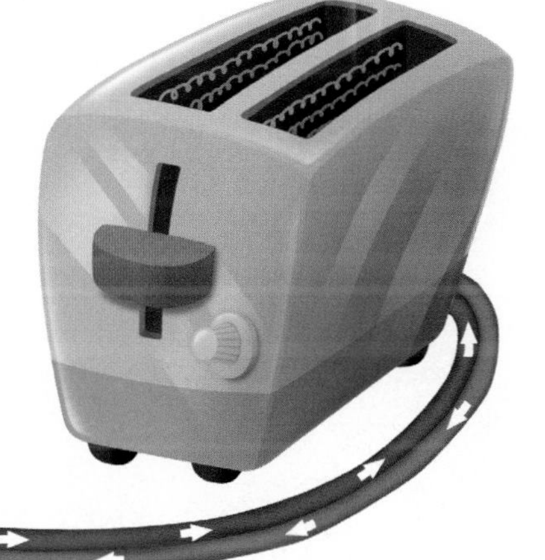

▶ **Electricity can move through the circuit when the toaster is plugged in.**

**How do you use current electricity every day?**

**FACT** Electricity comes through wires from power plants, not from your wall.



## ▶ Address Misconceptions

Children may think that electricity comes from the wall. Explain that electricity starts at power plants, flows through power lines and stations, and ends up in buildings, such as schools and houses.

**FACT** **Electricity comes through wires from power plants, not from your wall.** If possible, show children electrical wires connecting the school to power lines or point out power supply boxes.

### Read a Diagram

Direct children to the diagram on page 422. Ask:

- **What is the same about both circuits?** They both have a light bulb, battery, and wires.
- **What could you do to make the light on the left light up?** Connect the loose wire to the battery.

**Answer to Read a Diagram** the picture on the right; Because all wires are connected and the bulb is lit.

## ▶ Develop Vocabulary

**current electricity** Explain to children that the word *current* means "a steady flow of water or air in a particular direction." Using this definition, ask children to explain why we call electricity that moves through an electric circuit, *current electricity*.

**circuit** Write *circuit* on the board. Ask: **What is the name of a shape that sounds like *circuit*?** circle Write *circle* below the word *circuit* and ask a volunteer to underline the base *circ*. Tell children that a *circuit* means "a route or path that finishes at the point it began," just like a circle does. Have children create a circular flowchart showing how a light bulb, battery, and wires create a circuit.

## Quick Check Answer

Accept all reasonable answers.

### RESOURCES and TECHNOLOGY

- **Reading and Writing,** pp. 239–241
- **Visual Literacy,** p. 39
- **PuzzleMaker CD-ROM**
- **Classroom Presentation Toolkit CD-ROM**
- 

## Differentiated Instruction

### Leveled Activities

**EXTRA SUPPORT** Help children list things that turn electricity into other forms of energy. Have children use a sentence frame to write complete sentences about the items on the list. For example: *A* ____ *changes current electricity into* ____ *energy.*

*A* toaster *changes current electricity into* heat *energy. An* alarm clock *changes current electricity into* sound *energy.*

**ENRICHMENT** Discuss with children different sources of electrical energy, such as solar energy, hydroelectric energy, or wind energy. Have children use books or the Internet to research one of the ways electricity is made. Have them write a report or create an illustrated poster to share what they learned with others.

## What is static electricity?

### ▶ Discuss the Main Idea

**Main Idea** Static electricity is a kind of electricity made by tiny pieces of matter that attract and repel each other.

After reading pages 424 and 425 together, have the children describe a time they felt static electricity. Have them describe what they were doing and how it felt.

### ▶ Use the Visuals

Read and discuss the caption for the lightning picture on page 424. Ask:

- **What happens when charges build up in clouds during a storm?** lightning; static electricity
- **What ways could you stay safe in a lightning storm?** Possible answers: go inside, leave the pool, don't step in puddles (because electric currents can travel through water, just as sound can)

### ▶ Develop Vocabulary

**static electricity** Explain that the word *static* means "something that does not move." Have children talk to a partner about why they think people call the electricity that is stored in clouds, a balloon, or in cat's fur *static electricity*. Encourage children to reinforce the meaning of the term by comparing *static electricity,* which jumps from one object to another when objects build up charges, with *current electricity*, which flows in a path. Have the pairs share with the whole group.

**RESOURCES and TECHNOLOGY**

- **School to Home Activities,** pp. 123–124
- **Reading and Writing,** p. 242
- **Activity Lab Book,** p. 200
- LOG ON e-**Review** Narrated Summary and Quiz
- **ExamView® *Assessment Suite* CD-ROM**

## What is static electricity?

You take your clothes out of the dryer. They are stuck together! This happens because of static electricity.

**Static electricity** is a kind of energy made by tiny pieces of matter. You can not see these pieces of matter, but they are everywhere.

Like magnets, some of these pieces of matter attract or repel each other.

**Quick Lab**

Make a tissue paper worm. Rub a ruler to charge it. Observe how the ruler moves the worm.

Lightning is static electricity. Charges made in a storm jump between the clouds and the ground.

424
EXPLAIN

**Quick Lab**

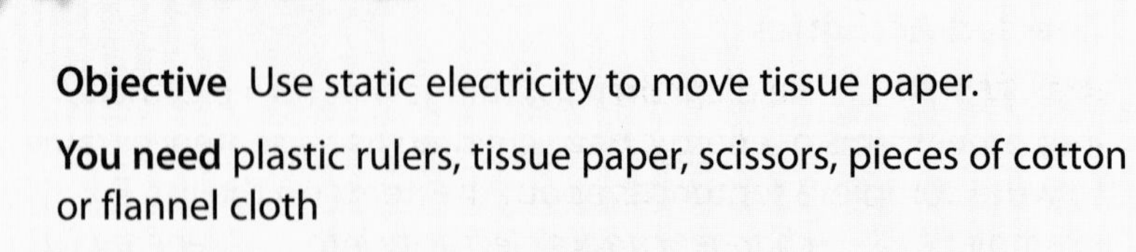


**Objective** Use static electricity to move tissue paper.

**You need** plastic rulers, tissue paper, scissors, pieces of cotton or flannel cloth

1. Have children cut a worm shape out of the tissue paper.
2. Tell children to rub the plastic ruler with cotton, flannel, or through their hair, and then place the ruler near their tissue-paper worm.
3. Have children observe what happens. The paper should be attracted or repelled by the charged ruler.

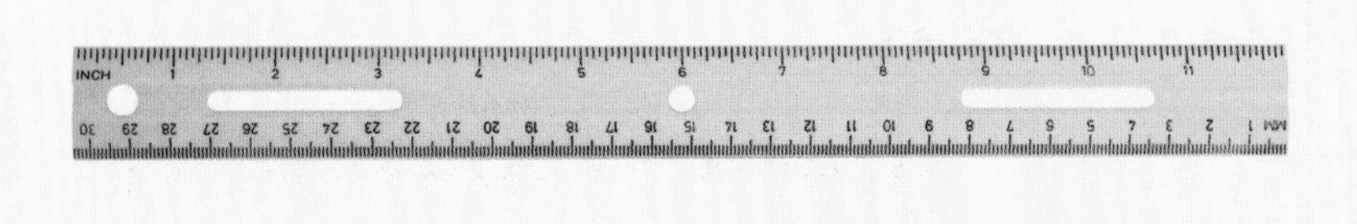

Vhen the tiny pieces
matter attract or repel
ch other, they have a
tic charge.

A static charge can build
on one object and jump
another. Sometimes you
n see or hear a static
arge move from one
ject to another.

What are some examples of static electricity?

The cat's fur is attracted to the charged balloon, so it sticks up.

## Think, Talk, and Write

1. **Cause and Effect.** How does a battery make your toy work?
2. What kind of energy causes your socks to stick together?
3. **Essential Question.** How do we get electricity?

### Social Studies Link

Research and write about how people use electricity.

**e-Review** Summaries and quizzes online at www.macmillanmh.com



## Formative Assessment

**Draw a Diagram**

Have children draw a diagram showing how an electric current starts at a power plant, travels to homes, flows through a device to make it work, and returns to the power plant.

**Key Concept Cards** For student intervention, see the prescribed routine on **Key Concept Card 39.**

### Quick Check Answer

lightning; clothes sticking together in a dryer; a charged balloon sticking to the wall; getting a shock when you touch a doorknob

# 3 Close

## ▶ Using the KWL Chart

Review with children what they have learned about electricity and how it works. Record their responses in the What We Learned column of the class **KWL** chart.

## ▶ Using the Reading Skill

### Cause and Effect

Use the Reading Skill graphic organizer to identify causes and effects in the lesson. Ask: **How does turning on a flashlight change the flashlight?**

| Cause | | Effect |
|---|---|---|
| Turn the flashlight switch on. | → | The circuit is completed. Electricity flows through the flashlight, creating light energy. The light bulb lights. |

*Graphic Organizer 8,* p. TR10

## FAST TRACK Think, Talk, and Write

1. **Cause and Effect** The battery makes current electricity to make the toy work.
2. Static electricity makes socks stick together.
3. **Essential Question** Possible answers: We can get electricity from batteries. We can also get electricity from a power plant. The electricity travels through wires into the outlets in our homes.

## Social Studies Link

Ask children to think about common uses for electricity, such as running lights or computers. Encourage them to also consider the many tools health care workers use that are powered by electrical energy, such as X-ray machines, special thermometers, microscopes, and machines that monitor heart rate.

# Reading in Science

**Objective**

- Identify the effect of electricity on the machines people use.

## It's Electric!

**Genre: Nonfiction** Stories or books about real people and events.

Have children read the title and look at the illustration. Ask:

- **What does this illustration show you?** Possible answer: It shows how electricity is made.

### Before Reading

Make a list with children of the things they use that need electricity to work. Have them think about how each thing gets electricity. Then ask:

- **Why do some machines need electricity?** Possible answer: The electricity gives the machines power to work.
- **Where does electricity you use in your home come from?** Possible answer: a power plant

### During Reading

Read the text together. Have children look for the words *generator* and *power plant* as they read. Ask:

- **Which part of the illustration shows where energy is formed?** the power plant
- **What is the name of the machine that creates electricity?** generator
- **Where does the electricity you use in your home begin?** at a power plant

Talk about other ways electricity that's generated at power plants can be used by people.

**RESOURCES and TECHNOLOGY**

- **Reading and Writing,** pp. 243–244
- **Technology: A Closer Look,** Lesson 4

Real World

## Reading in Science

# It's Electric

You can flip a switch to turn on a light, a computer, or a dishwasher. They all use electricity.

The electricity starts at a power plant. At the plant, energy turns a large wheel called a turbine. The energy might come from burning coal or oil, flowing water, wind, or nuclear reactions.

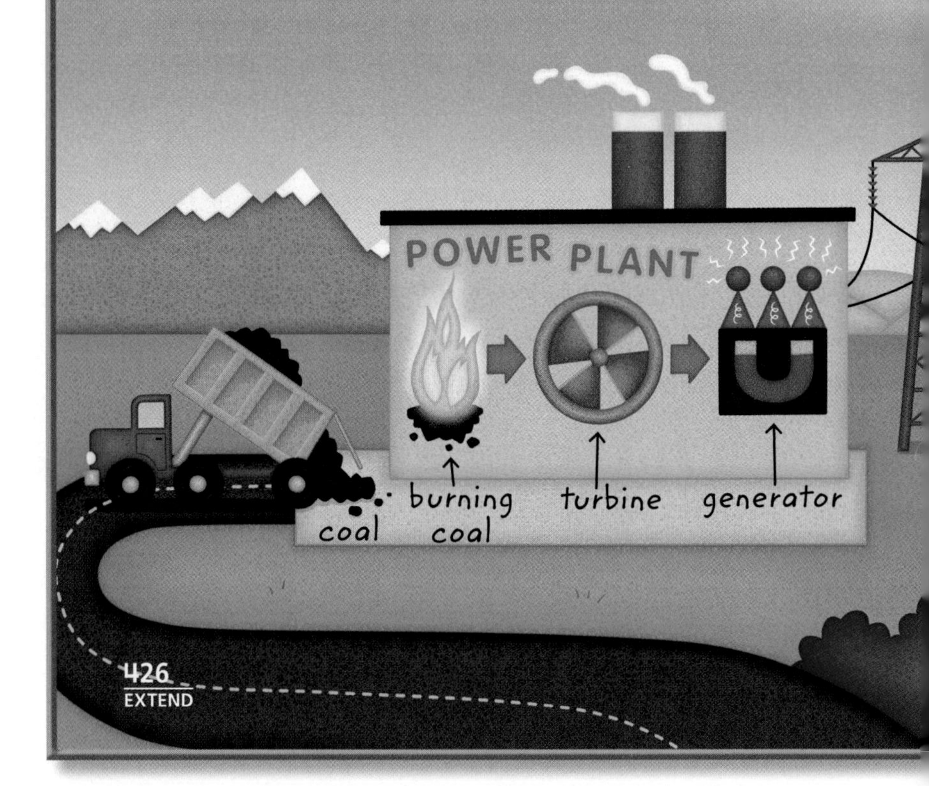


### ELL Support

**Use Illustrations** Ask children to point to or name things that use electricity in the classroom. Have them look at the illustration on pages 426 and 427. Point to the boy with the lamp and ask: *Where does electricity come from?* Walk children through the illustration as they trace the path of electricity with their fingers. Help them match the captions to ideas in the article.

**BEGINNING** Have children answer yes or no questions about the illustration, such as: *Does electricity travel through power lines?*

**INTERMEDIATE** Have children say a phrase or sentence to describe the illustration.

**ADVANCED** Children can answer detailed questions about the illustration, such as: *What happens to the turbine?*

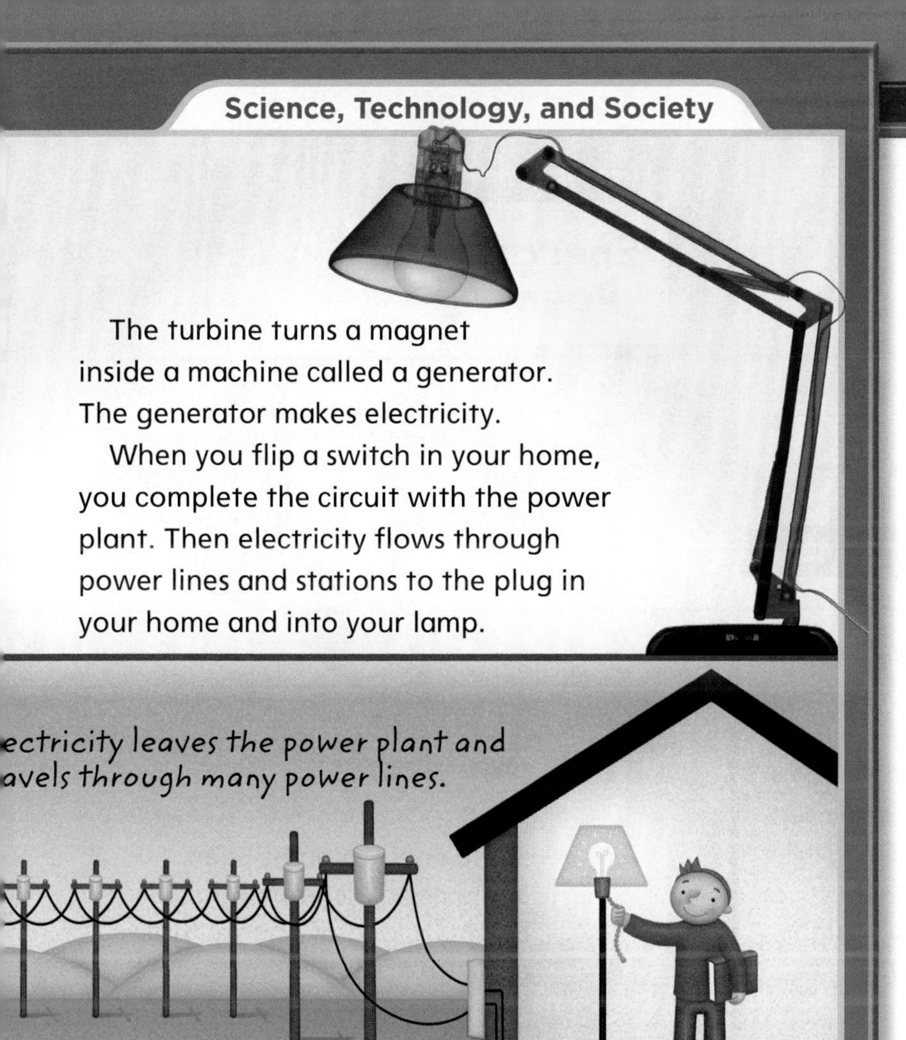

The turbine turns a magnet inside a machine called a generator. The generator makes electricity.

When you flip a switch in your home, you complete the circuit with the power plant. Then electricity flows through power lines and stations to the plug in your home and into your lamp.



**lk About It**

**use and Effect.** What kes the light go on in ur home?

Connect to AMERICAN MUSEUM & NATURAL HISTORY at www.macmillanmh.com

427
EXTEND

## ▶ Address Misconceptions

Children may believe that electricity only comes from power plants or batteries, but there are many forms of electricity. Walking across a rug produces static electricity, a kind of energy made by tiny pieces of matter that attract and repel each other. Static electricity causes sparks, crackling, or the attraction of dust and air. Lightning is also static electricity. During a thunderstorm, charges jump between the clouds and the ground.

## After Reading

Have children talk about the different uses of electricity. Ask:

- **What would life be like without electricity?**

Explain that without electricity, people could not operate machines like lights, computers, or machines to wash clothes.

Display Graphic Organizer 8. Explain to children that a cause is why an event happens, and an effect is the event that happens. Ask:

- **How does electricity affect people's lives?**

Fill in the left and right boxes with children's responses.

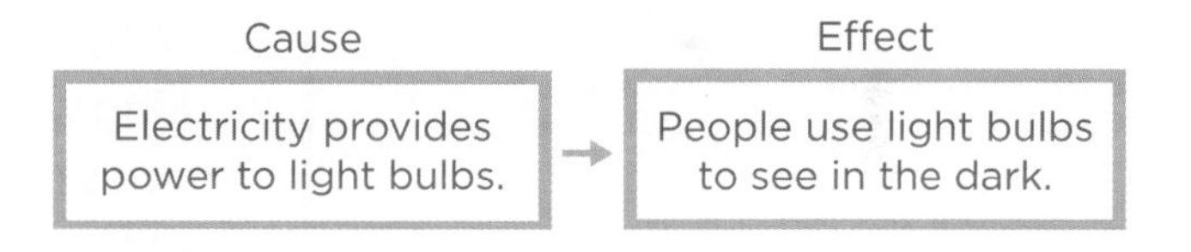

Cause
Effect
Electricity provides power to light bulbs.
People use light bulbs to see in the dark.

*Graphic Organizer 8,* p. TR10

## Talk About It

Possible answers: The electricity starts at the power plant. Then it goes through power lines. Then it comes to my home. I pull the cord or flip the light switch. The light goes on.

If children have difficulty answering the question, direct their attention to the illustration. Ask them to point to the part of the illustration that shows electricity in the home. Work backward to explain the parts of the process that create and deliver electricity.

Once children understand the process, ask them to identify the parts of the picture that show the cause and effect of electricity.

## Extended Reading

### Visit the Library

Read aloud ***Flick a Switch: How Electricity Gets to Your Home,*** by Barbara Seuling (Holiday House, 2003).

Review how electricity gets to people's homes. Ask:

- **What generates electricity for a big city?**
- **How does electricity travel from a power plant to your home?**

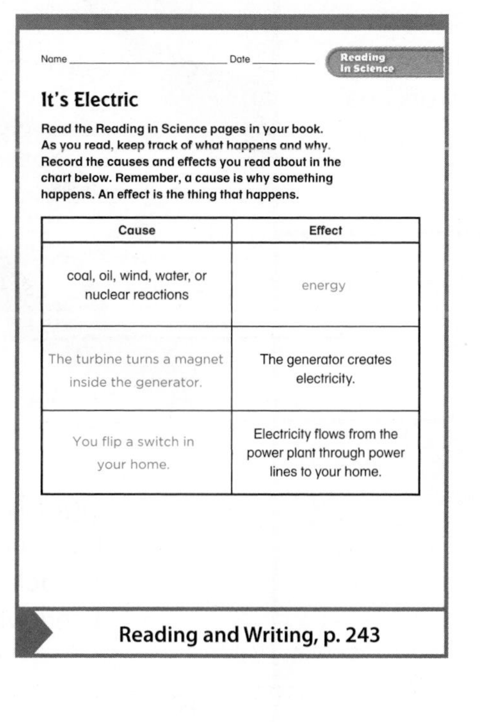

Name ____________ Date ________ Reading in Science

**It's Electric**

**Read the Reading in Science pages in your book. As you read, keep track of what happens and why. Record the causes and effects you read about in the chart below. Remember, a cause is why something happens. An effect is the thing that happens.**

| Cause | Effect |
|---|---|
| coal, oil, wind, water, or nuclear reactions | energy |
| The turbine turns a magnet inside the generator. | The generator creates electricity. |
| You flip a switch in your home. | Electricity flows from the power plant through power lines to your home. |

**Reading and Writing, p. 243**

## I Read to Review

### Objective

- Review types of energy with independent reading.

### Buddy Reading

- During small group instructional time, have children work together in pairs to read the selection to one another.
- Pair weaker readers with stronger partners to ensure success during work period.

### Independent Reading

- Give children copies of School to Home Activities, pages 125 and 126. Have them assemble pages to make their own book.
- Encourage children to take their book home and read it aloud to a family member.

PAGES **428–429** ▶

Ask: **What are some different things that make heat?**
Possible answers: the Sun, oven, fire, car engines, toasters, microwave ovens

Ask: **What type of vibrations make a high-pitched sound?**
fast

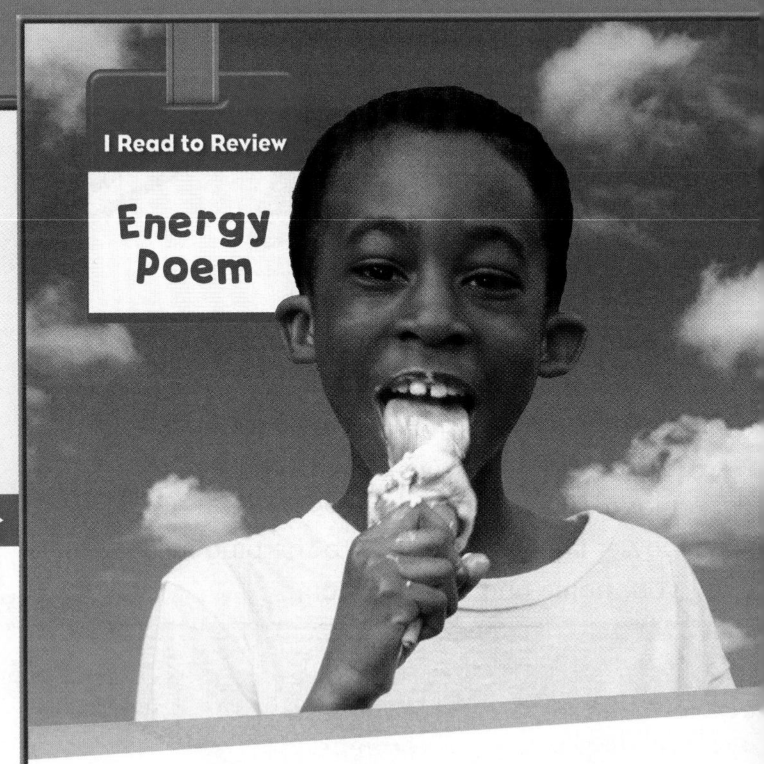


Some energy can be seen.
Some energy can be felt.
Heat is a kind of energy that
can make solids melt.

428

PAGES **430–431** ▶

Ask: **Why do toys stop working when the batteries in them get too old?**
The toy stops working because there is not enough electricity. To make the toy work again, you need to put in new batteries.

Light is a kind of energy.
Light can bounce and bend.
I block light to make shadows.
I see white when all colors blend.

430

**RESOURCES and TECHNOLOGY**

- **School to Home Activities,** pp. 125–126
- **Assessment,** pp. 158–159
- **TeacherWorks™ Plus CD-ROM**

Sound is a kind of energy.
I hear pitches high and low.
I hear because of vibrations,
small, big, fast, and slow.

Electricity is a kind of energy.
It makes many things run.
Without electricity, my toy
would be no fun!





## Make an Energy Book

 individual  25 minutes

**Materials** drawing paper, colored pencils

1. Tell children that they are going to write a book to show what they know about different types of energy.
2. Have children make a book by folding two pieces of paper and stapling on the fold. When they have finished, ask a volunteer to count the number of pages in the book.
3. To help children organize their writing, encourage them to list different types of energy and jot down details about each. Ask: **How do you use energy? Where does energy come from?**
4. Before they begin writing, explain that the inside pages should have text and illustrations to explain how people use energy. Ask children to include all different kinds of energy in their writing. Remind children that they have eight pages on which they can write. Suggest that they complete the inside pages before the front cover, which should illustrate some form of energy.

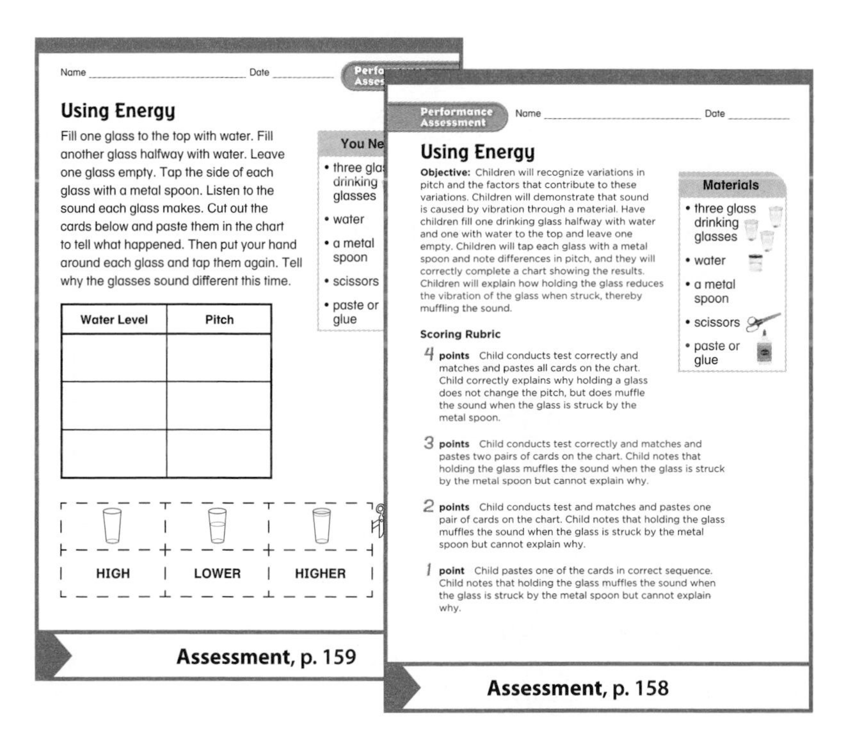

Name ________ Date ________

### Using Energy

Fill one glass to the top with water. Fill another glass halfway with water. Leave one glass empty. Tap the side of each glass with a metal spoon. Listen to the sound each glass makes. Cut out the cards below and paste them in the chart to tell what happened. Then put your hand around each glass and tap them again. Tell why the glasses sound different this time.

| Water Level | Pitch |
|---|---|
| | |
| | |
| | |

HIGH | LOWER | HIGHER

Assessment, p. 159

Performance Assessment Name ________ Date ________

### Using Energy

**Objective:** Children will recognize variations in pitch and the factors that contribute to these variations. Children will demonstrate that sound is caused by vibration through a material. Have children fill one drinking glass halfway with water and one with water to the top and leave one empty. Children will tap each glass with a metal spoon and note differences in pitch, and they will correctly complete a chart showing the results. Children will explain how holding the glass reduces the vibration of the glass when struck, thereby muffling the sound.

**Materials**
- three glass drinking glasses
- water
- a metal spoon
- scissors
- paste or glue

**Scoring Rubric**

4 **points** Child conducts test correctly and matches and pastes all cards on the chart. Child correctly explains why holding a glass does not change the pitch, but does muffle the sound when the glass is struck by the metal spoon.

3 **points** Child conducts test correctly and matches and pastes two pairs of cards on the chart. Child notes that holding the glass muffles the sound when the glass is struck by the metal spoon but cannot explain why.

2 **points** Child conducts test and matches and pastes one pair of cards on the chart. Child notes that holding the glass muffles the sound when the glass is struck by the metal spoon but cannot explain why.

1 **point** Child pastes one of the cards in correct sequence. Child notes that holding the glass muffles the sound when the glass is struck by the metal spoon but cannot explain why.

Assessment, p. 158

# CHAPTER 12 Review

For information on Depth of Knowledge levels, see page 433B.

## ▶ Use the KWL

Review the **KWL** chart that the class made at the beginning of the chapter. Help children compare what they know about energy now with what they knew then. Add any additional information to the What We Learned column of the **KWL** chart.

## ▶ Make a FOLDABLES Study Guide

Create a four-door Foldables. Give the Lesson 1 group index cards labeled *Heat, Energy,* and *Temperature.* On each card, have children define the term, provide examples, and illustrate the examples. Glue these under the first door.

Distribute index cards to members of the Lesson 2 group. On each card, ask them to write and illustrate responses to the following questions: *What is sound? How is sound made? How do we hear sound? What are different kinds of sound? What things do sounds move through?* Paste the cards under the second door.

Follow a similar procedure for the Lesson 3 group. Ask them to respond to questions about light energy. Attach the cards under the third door. Ask the Lesson 4 group to compare current and static electricity in a Venn diagram. Glue the diagram under the fourth door.

See page TR38 in the back of the Teacher's Edition for more information on Foldables.

**Vocabulary**

1. vibrate
2. current electricity
3. static electricity
4. reflect
5. circuit

**RESOURCES and TECHNOLOGY**

- **Reading and Writing,** pp. 245–246
- **Assessment,** pp. 146–149, 154–157
- *ExamView® Assessment Suite* CD-ROM
- PuzzleMaker CD-ROM
-  Vocabulary Games

## CHAPTER 12 Review

**Vocabulary** DOK 1

**Use each word once for items 1–5.**

| circuit |
|---|
| current electricity |
| reflect |
| static electricity |
| vibrate |

1. Sound is made when objects ________.
2. Energy that moves through wires is called ________.
3. Energy that jumps from object to object is called ________.
4. We can see objects because of the light they ________.
5. This picture shows a complete ________.

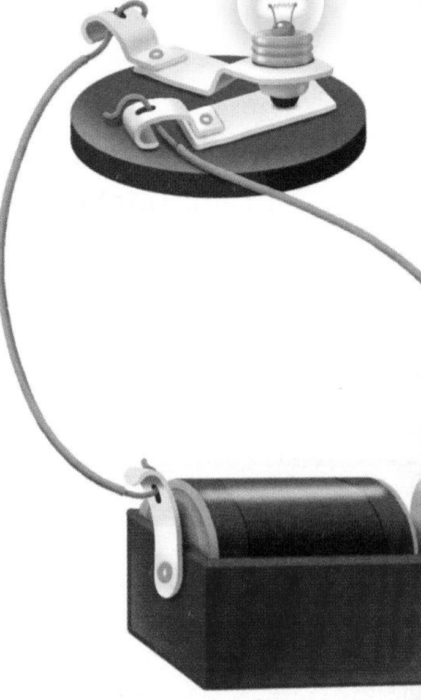

  -Glossary Words and definitions online at www.macmillanmh.com

Name ______________ Date ________ CHAPTER Vocabulary

**Using Energy**

**Match the vocabulary word on the left with the letter of the phrase that describes it.**

| | | |
|---|---|---|
| 1. d | current electricity | a. to move backward and forward quickly |
| 2. a | vibrate | b. a path for electricity |
| 3. b | circuit | c. energy that can change the state of matter |
| 4. e | charge | d. energy that can be changed to heat, light, or sound energy |
| 5. c | heat | e. a force that makes tiny pieces of matter sometimes attract or repel each other |

**Reading and Writing, p. 245**

Chapter Test A Name ______________ Date ________

**Using Energy**

**Write the word or words that best complete each sentence in the spaces below. Words may be used only once.**

| | | | |
|---|---|---|---|
| charge | fuel | pitch | vibrate |
| circuit | heat energy | sound | |
| eardrum | light | temperature | |

1. A kind of energy we can hear is called sound.
2. Tiny pieces of matter that attract or repel each other have a charge.
3. One way to change the state of matter is to use heat energy.
4. The kind of energy that helps us see is called light.
5. When an object moves back and forth quickly, it vibrates.
6. Something that gives off heat when it is burned is called fuel.
7. A sound's pitch tells how high or low it is.
8. The path on which electricity travels is called a circuit.
9. A thermometer measures temperature.
10. The part of our body that hears sound is the eardrum.

**Assessment, p. 146**

## cience Skills and Ideas

DOK 2

**swer the questions below.**

What happens to a sound when it moves away from you?

**Measure.** How many degrees Celsius is the temperature on this thermometer?

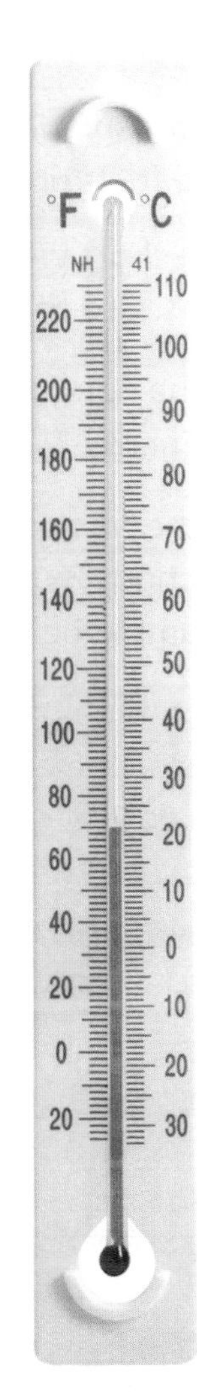


What can heat do?

**Main Idea and Details.** Why can you see a rainbow with a prism?

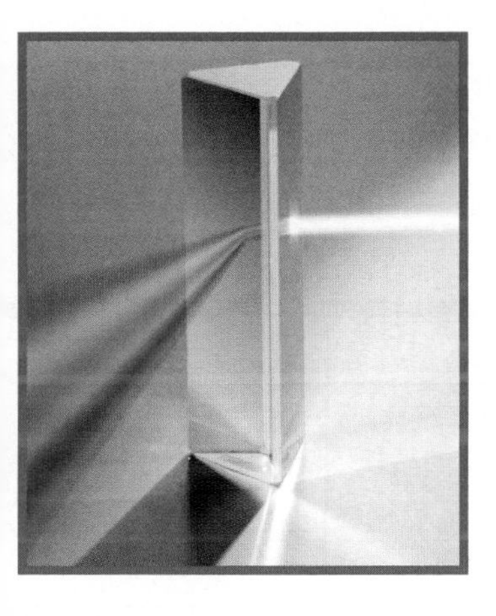

**10.** How do we use energy?

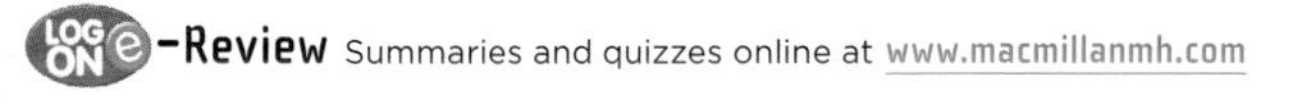


Summaries and quizzes online at www.macmillanmh.com

433

## Science Skills and Ideas

**6.** It gets softer.

**7. Measure** 21°C

**8.** Heat can turn solids into liquids, turn liquids into gases, melt things, and warm things.

**9. Main Idea and Details** Encourage children to complete the main idea and details graphic organizer. A prism bends white light to separate it into different colors.

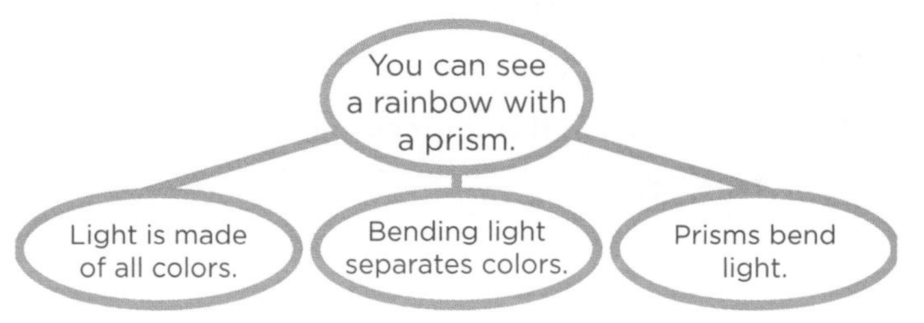


*Graphic Organizer 1,* p. TR3

**10.** Accept all reasonable answers. Children should address concepts taught in each lesson, such as: using heat energy to stay warm; hearing sound energy; using light energy to see things; and how we get and use electricity.

## Summative Assessment and Intervention

**Assessment** provides a summative test for Chapter 12.

**Leveled Readers** may be used to reteach lesson content in an alternative format. Leveled readers deliver chapter content at different readability levels. The back cover of each reader provides comprehension building activities specific to the book content (see page 397).

**Key Concept Cards 36–39** contain prescribed routines for student intervention.

# Performance Assessment

## Energize!

**Materials** paper, pencils

### ▶ Teaching Tips

To save time, prepare blank charts for children in advance.

1. Review with children the different types of energy, their sources, and uses.
2. Help children create a chart with five rows and three columns. The headings should be as follows: *Energy Type; Sources; Uses.* The energy types are *Heat, Sound, Light,* and *Electricity.*
3. Have children fill in the sources and uses for each type of energy.

**Performance Assessment** DOK 2

## Energize!

Make a chart showing different types of energy and facts about the types of energy.

- ▶ Make a chart with five rows and three columns. Use the example to help you.
- ▶ Label the columns with the following titles: *Energy Type, Sources, Uses.*
- ▶ Write the names of four types of energy in the chart. Put one in each row like the example below.
- ▶ For each type of energy, fill in the sources of the energy and ways you use it each day.

| Energy Type | Sources | Uses |
|---|---|---|
| Heat | the Sun, rubbing hands, fuel | to keep warm, cooking |
| Sound | | |
| Light | | |
| Electricity | | |

433A

### Scoring Rubric

**4 Points** Child accurately fills in information for all four types of energy.

**3 Points** Child correctly fills in information for three types of energy.

**2 Points** Child correctly fills in information for two types of energy.

**1 Point** Child accurately fills in information for one type of energy.

At the train station, the sound of the train gets louder and louder.

What do you know?

- **A** The train is moving toward you.
- **B** The train is moving away from you.
- **C** The train is slowing down.
- **D** The train is speeding up.

DOK 2

What happens when white light bends?

- **A** It speeds up.
- **B** It separates into different colors.
- **C** It makes a shadow.
- **D** It prevents us from seeing.

DOK 1

Look at the picture.

Which part of this circuit makes electricity?

- **A** the light bulb
- **B** the switch
- **C** the wires
- **D** the battery

DOK 1

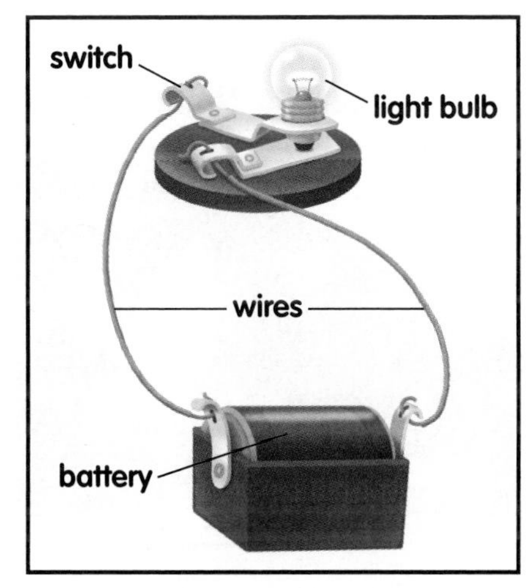




# Test Preparation

1. A: The train is moving toward you. As objects move farther away, the noise they make sounds softer. We cannot tell the speed of an object from the sound we hear it make.
2. B: It separates into different colors. When white light bends, how fast it travels does not change. It will not cause a shadow or change how it is reflected either.
3. D: the battery. Electricity only travels through the wires, the switch, and the light bulb.

## Depth of Knowledge

**Level 1 Recall** Level 1 requires memory of a fact, a definition, or a procedure. At this level, there is only one correct answer.

**Level 2 Skill/Concept** Level 2 requires an explanation or the ability to apply a skill. At this level, the answer reflects a deep understanding of the topic.

**Level 3 Strategic Reasoning** Level 3 requires the use of reasoning and analysis, including the use of evidence or supporting information. At this level, there may be more than one correct answer.

**Level 4 Extended Reasoning** Level 4 requires the completion of multiple steps and requires synthesis of information from multiple sources or disciplines. At this level, the answer demonstrates careful planning and complex reasoning.

# Careers in Science

**Objective**

- Describe how crash testers help protect people.

## Crash Tester

**Genre: Nonfiction** Stories or books about real people and events.

Discuss with children what they know about titles of books and articles. Remind them that a title helps them know what the story will be about. Ask:

- **What does the title tell you about this article?** Possible answer: It is about testing something by crashing it.

## Talk About It

Read the text together. Ask:

- **What parts of a car help make it safer?** Possible answers: seat belts, air bags, bumpers, strong doors
- **Why would someone crash cars to make them safer?** Possible answer: They need to know what happens in a real crash.

## Learn About It

Explain that mechanics and car designers also need to know about science. Ask:

- **What forces do a car mechanic and a car designer need to think about?** Possible answer: They need to know about friction to make the car stop.

## Write About It

Have children write five facts that they learned about crash testers. Ask:

- **Why are crash testers important?**

**RESOURCES and TECHNOLOGY**

# Careers in Science

## Crash Tester

If you like to learn about cars and safety, you could become a crash tester. Crash testers learn how to make cars safer by setting up crashes!

These workers explore what happens to dummies, or big dolls, in a car crash. Then the crash testers decide how to make the cars safer. Crash testers study air bags and seat belts to make help protect people better.

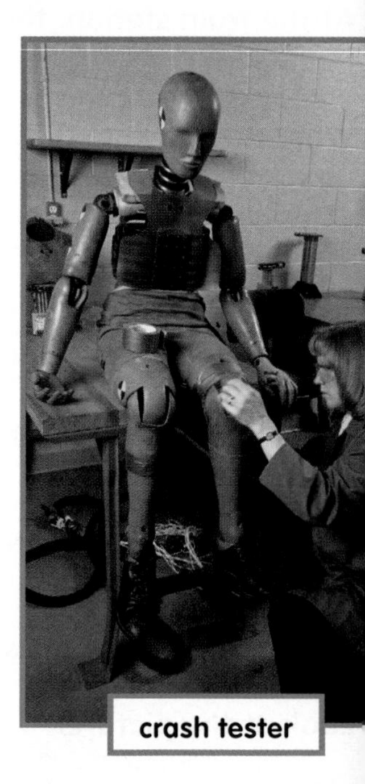

crash tester

### More Careers to Think About

mechanic

car designer

## Integrate Writing

**Writing About Pictures**

Have children look closely at the pictures in their books. Tell them they will write detailed captions for each picture. To help them organize their writing, ask:

- **What is each person doing in each picture?**
- **What kinds of problems might they have to solve?**
- **What kinds of simple machines does a crash tester, mechanic, or car designer use?**

Next, invite children to write a few sentences about the crash tester, auto mechanic, or car designer. They may extend their writing by discussing how science plays a role in the career or including imaginary quotes that reveal the point of view of the worker.

# Reference

## Science Handbook



## Health Handbook

# Measurements

### Objectives

- Measure length in nonstandard units.
- Use the standard units of centimeters and inches to measure length.

### ▶ Assess Prior Knowledge

Display a ruler and a pencil of different lengths. Ask:

- **Which is longer, the pencil or the ruler?**
- **How did you draw that conclusion?** Possible answer: I put them side by side and compared them.

### ▶ Use the Visuals

Have children study the pictures on page R2. Ask:

- **What object is used to measure the first string?** paper clips
- **What is used to measure the second string?** hands

Read the captions under the pictures, and have children count the paper clips and hands to verify the lengths. Ask:

- **What do you notice about the eight clips and the two hands?** Possible answer: All the clips are the same length and the two hands are the same length.

### ▶ Discuss the Main Idea

Point out to children that many kinds of objects can be used as measuring tools, but they should be of the same type and size. The objects should also be placed in the same direction (end to end). Explain to children that this way of measuring objects is called *nonstandard measurement*.

### ▶ Explore the Main Idea

ACTIVITY Provide children with paper clips of uniform size to measure their science books. Have children do the Try It activity using different kinds of nonstandard measuring tools.

## Measurements

### Nonstandard

You can use objects to measure the length of some solids. Line up objects and count them. Use objects that are alike. They must be the same size.

▲ **This string is about 8 paper clips long.**

▲ **This string is about 2 hands long.**

### Try It

**Measure a solid in your classroom. Tell how you did it.**

R2
SCIENCE HANDBOOK

## ELL Support

**Use Language Pattern/Labels** Give children objects to use as measuring tools. Ask each child to select an object in the room to measure. Help them make a chart to show what they measured.

**BEGINNING** Help children name the object and identify the measurement.

**INTERMEDIATE** Ask each child to draw and label the measurements of the object. Help them complete a sentence frame that describes the length, width, or height of the object they measured.

**ADVANCED** Have each child explain to the class how they measured at least two attributes of the object.

## ndard

ı can also use a ruler to measure length of some solids. You can asure in a unit called **centimeters**.

◀ **This toy is about 8 centimeters long. This is written as 8 cm.**

ı can also use a ruler to measure ı unit called **inches**. One n is longer than I centimeter.

◀ **This toy is about 3 inches long. This is written as 3 in.**

### It

**mate the length of this toy . Then find its exact length.**

## Differentiated Instruction

### Leveled Activities

**EXTRA SUPPORT** Distribute 30 cm (12 in.) lengths of string to children. Ask them to use an object in the classroom to measure the string. Have children write the length of the string in terms of the object they used. Then ask them to use a ruler to measure the string and record the length.

**ENRICHMENT** Invite children to compare the measurements of two objects in the classroom. Ask children to use an object in the classroom as a measuring tool. Have them repeat the activity using a ruler as a measuring tool. Encourage children to make a chart to record their findings.

## ▶ Assess Prior Knowledge

Find out what children know about rulers. Ask:

- **What units are marked on a ruler and used for measurement?** Possible answers: centimeters and inches

## ▶ Use the Visuals

Explain that *standard measurement* means "the most common way to measure things," and that people usually measure things with a ruler. Have children compare the two pictures shown on page R3. Ask:

- **What is alike and different about the alligator in the pictures?** The alligator is the same length in both pictures, but it is measured in centimeters in the top picture and in inches in the bottom picture.

## ▶ Discuss the Main Idea

Explain how the distance between each number represents one standard measuring unit, with centimeters on the top ruler and inches on the bottom ruler. Model holding a ruler along a solid object, and counting the number of units from one end of the object to the other.

## ▶ Explore the Main Idea

**ACTIVITY** Have children practice measuring by looking at the toy alligators. Ask:

- **Where does the tail of the alligator line up on the top ruler?** with the 0
- **Where does the nose of the alligator line up on the top ruler?** with the 8

Repeat the questions using the bottom ruler.

Before doing the Try It activity, explain to children that an estimate is not an exact number. Have children estimate by looking at the pictures of the rulers above. Then provide them with rulers to get precise measurements. Have children estimate and measure the lengths of other solids.

# Measurements

## Objectives

- Use a measuring cup to measure volume.
- Use a balance to measure mass.
- Use a clock to measure the passage of time.
- Use a thermometer to compare and read temperatures.

### ▶ Assess Prior Knowledge

Hold up a measuring cup and ask:

- **Why do people use measuring cups like this when they cook or bake?** Possible answer: to make sure they use the right amounts of ingredients

Display a scale and a balance. Ask:

- **How is a scale different from a balance?** Possible answer: A scale measures the weight of an object; a balance compares the mass of two objects.

### ▶ Use the Visuals

Have children look at the photographs on page R4. Ask:

- **How much liquid is in the measuring cup?** Possible answers: about 1 cup; 8 ounces
- **What will happen when something is put in the yellow pan of the balance?** Possible answers: It will be lower; the arrow will move to the right of the line.

### ▶ Discuss the Main Idea

Explain that liquids can be measured in different kinds of standard units. The measuring cup shows cups on the left and ounces on the right. Point out to children that the mass of objects can be measured using a balance.

### ▶ Explore the Main Idea

ACTIVITY Provide children with a pair of objects and a balance to do the Try It activity. Remind children to have the arrow point to the line of the balance before objects are placed in the buckets.

## Measurements

### Volume

You can measure the volume of a liquid with a **measuring cup**. Volume is the amount of space a liquid takes up.

▲ This measuring cup has 1 cup of liquid.

### Mass

You can measure mass with a **balance**. The side that has the object with more mass will go down.

▲ Before you compare the mass of two objects, be sur[e] the arrow points to the line

### Try It

**Place two objects on a balance. Which has more mass?**

R4
SCIENCE HANDBOOK

## ELL Support

**Make Comparisons** Ask children to make labels to reinforce the meanings of words. Have children take turns using the pan balance to compare the mass of two objects, and practice using language to describe what they discover.

BEGINNING Help children in label the objects they are comparing. Ask children to point to and name the object with more mass.

INTERMEDIATE Ask children to use a sentence to describe which object has more mass.

ADVANCED Have children compare the masses of multiple objects and describe in their own words what they have discovered about the mass of each object.

## me

u can measure time with a **clock**. :lock measures in units called hours, nutes, and seconds. There are 60 nutes in 1 hour.

## mperature

u can measure nperature with **hermometer**. ermometers measure units called degrees.

Degrees Fahrenheit

Degrees Celsius

◀ The temperature is 29 degrees Celsius.

## y It

: a thermometer to find the nperature outside today.



## Differentiated Instruction

### Leveled Activities

EXTRA SUPPORT Provide pairs of children with clock faces marked in minutes. Ask:

- **How many minutes are there up to the 3?** 15
- **How many minutes are there up to the 6?** 30

ENRICHMENT Provide pairs of children with two clock faces that have movable hands. Tell them it is 9:15. Have one partner show this time on a clock face. Have the other partner show what the time would be 10 minutes later on the second clock face. Ask:

- **What time is it now?** 9:25
- Repeat the activity so that each partner has a chance to show a start time and a finish time at least twice.

## ▶ Assess Prior Knowledge

Have children describe clocks and watches they have seen and tell how they are used. Then ask them to describe a thermometer they have seen and how it was used.

- **Why is it helpful to know the temperature?** Possible answers: It can help me decide what clothing to wear or what activities to do.

## ▶ Discuss the Main Idea

Discuss with children different kinds of clocks, such as digital clocks and standard clocks with hands. Explain that an hour, a minute, and a second are standard units of time.

Display a clock with a second hand. Ask a volunteer to use their finger to trace the path the minute hand would move to complete a full hour. Have children point to the minute lines. Ask:

- **How many minutes are there in half an hour?** 30

Explain that a degree is a standard unit of temperature. Ask:

- **What are the two types of degrees that are found on most thermometers?** Celsius and Fahrenheit

## ▶ Use the Visuals

Have children look at the clock and thermometer pictures and read the labels on page R5. Ask:

- **What time is it on the pictured clock?** 10 o'clock
- **How can you tell?** The hour hand is on 10 and the minute hand is on 12.
- **What is the temperature in degrees Celsius on the pictured thermometer?** about 30°C

## ▶ Explore the Main Idea

ACTIVITY Have children decide where to place the thermometer outside. Then have them complete the Try It activity. Encourage them to draw a chart showing the hours they are in class. Ask volunteers to record the temperature hourly on the chart.

# Science Tools

## Objectives

- Recognize that a computer can be used to gather information.
- Use a hand lens to magnify objects.

### ▶ Assess Prior Knowledge

Have children identify tools that scientists use, such as computers, telescopes, and microscopes. Discuss any experience children may have had with computers and the Internet. Ask:

- **What can you do on a computer?** Accept all reasonable answers.
- **What tools make it easier to see very small objects?** Possible answers: hand lens, a microscope

### ▶ Use the Visuals

Have children look at the pictures on page R6 and read the labels. Ask volunteers to describe what each part of the computer labeled in the picture does. Ask:

- **How can you tell the computer what to do?** Possible answers: Type on the keyboard; tap or click the mouse.

Have children compare the images of the ladybugs or beetles. Ask:

- **Using the hand lens, what details are easier to see on the beetle?** Possible answer: the legs

### ▶ Discuss the Main Idea

Explain to children that not all information found on the Internet may be accurate. Point out that Web sites from reliable sources, such as national science or educational institutions, are the best to use for research. Read the text about computers together.

### ▶ Explore the Main Idea

ACTIVITY Before children do the Try It activity, show them how to move the hand lens to get a clear focus of the object. Encourage them to draw what the object looks like as seen with the unaided eye, then through the hand lens.

## Science Tools

### Computer

A computer is a tool that can help you get information. You can use the Internet to connect to other computers around the world.

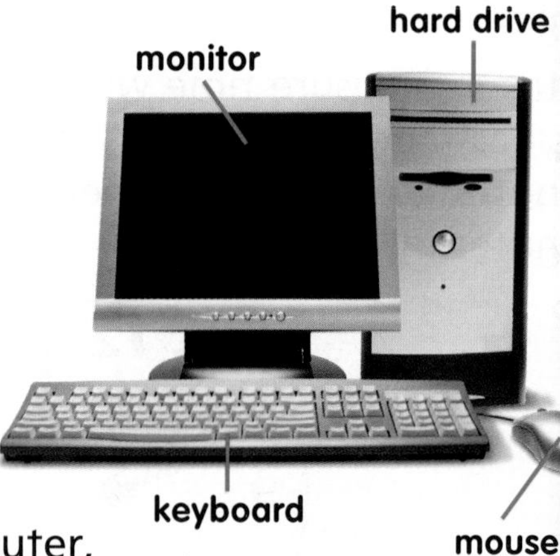


When you use a computer, make sure an adult knows what you are working on.

### Hand Lens

A hand lens is another tool that can help you get information. A hand lens makes objects seem larger.

### Try It

**Use a hand lens to look at an object. Draw what you see.**

R6
SCIENCE HANDBOOK

## ELL Support

**Draw Pictures** Reinforce comprehension by having children draw pictures of vocabulary words.

**BEGINNING** Ask children to draw, label, and name aloud the parts of a computer.

**INTERMEDIATE** Have children draw the parts of a computer and use short phrases to describe the function of each part.

**ADVANCED** Encourage children to draw a computer and use complete sentences to write simple instructions for how to use it.

## raphs

### ır Graphs

r graphs organize data. The title the graph tells you what the data ıbout. The shaded bars tell you w much of each thing there is.

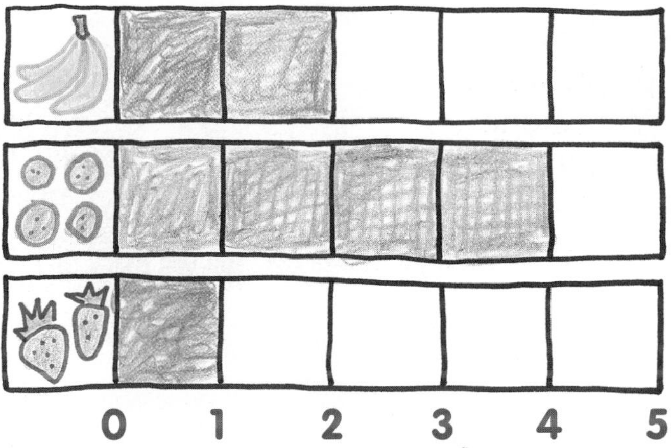


### y It

ıke a bar graph that ɔws your classmates' ɾorite fruits.

## Differentiated Instruction

### Leveled Questions

EXTRA SUPPORT Ask questions such as these to check children's understanding of the material.

- **What is the title of the bar graph?** Favorite Fruits
- **What do the colored-in squares tell you?** how many children like that fruit best

ENRICHMENT Use these types of questions to develop children's higher-order thinking skills.

- **If three children liked apples, how could this information be added to the graph?** Possible answer: Add another row with a picture of apples on the left, and color three boxes next to the apples.
- **What are some other things a bar graph can show?** Accept all reasonable answers.

# Graphs

## Objective

- Explain how to read and make a bar graph.

### ▶ Assess Prior Knowledge

Have children describe what they know about how bar graphs can organize data. Ask:

- **Why are bar graphs useful?** Possible answers: They make information easy to understand; they can be used to compare things.

### ▶ Use the Visuals

Have children look at the bar graph on page R7. Ask:

- **What does this graph show?** the types of fruit people like best
- **How many people liked bananas?** two
- **What was the most popular fruit?** oranges

### ▶ Discuss the Main Idea

Display other bar graphs to the class. Point out the different information that bar graphs can show. Explain the information shown on the vertical and horizontal axes. Ask volunteers to read the graphs to the class.

### ▶ Explore the Main Idea

ACTIVITY Make a list on the board of the fruits the class likes best, using tally marks to count the number how many children like each type of fruit. Have children use the list to complete the Try It activity.

Instruct children to use the graph shown on page R7 as a model. Display the graphs so children can review the results as a class.

# Your Body

## Objectives

- Explain that the skeletal system gives the body its shape.
- Describe how the muscular system helps the body move.
- Recognize how the nervous system tells the body what to do.

### ▶ Assess Prior Knowledge

Have children describe what they know about skeletons. Ask:

- **Where are there bones in your body?** Possible answers: in my arms and legs; in my hands and feet

Invite children to share any experiences they have had with broken bones.

### ▶ Use the Visuals

Review the diagram and labels on page R8 with children. Ask:

- **What bones are labeled on this boy's body?** skull, spine, arm bones, leg bones

Have children locate these bones in their own bodies.

### ▶ Discuss the Main Idea

Read the text together. Have children think about what they would look like if they had no bones. Ask:

- **What might your body feel like without bones?** Possible answers: squishy, floppy
- **Do you think you could walk without leg bones? Why or why not?** Possible answer: No, because my legs wouldn't hold me up.
- **Why are bones important?** Possible answer: They give a body its shape.

### ▶ Explore the Main Idea

ACTIVITY Have children do the Try It activity. Encourage them to try and count the number of bones they can feel in their arms. Have them make a drawing to show what they felt.

# Your Body

## Skeletal System

Your body has many parts. All your parts work together to help you live.

Bones are hard body parts inside your body. They help you stand straight. Your bones give your body its shape.

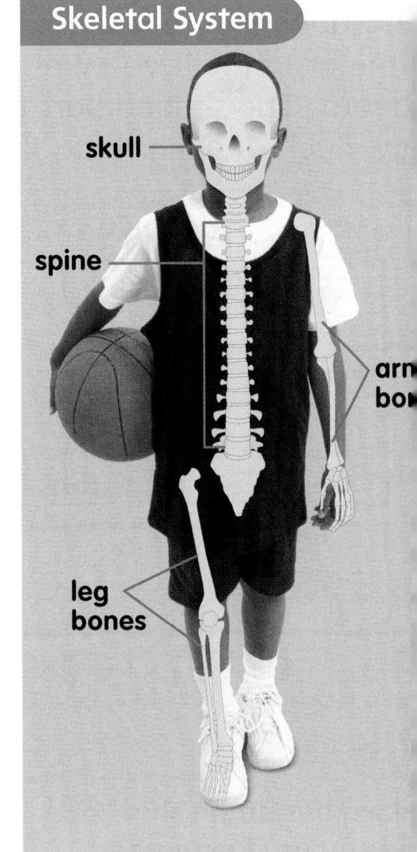


### Try It

**How many bones do you think there are in one arm? Count them.**

R8
HEALTH HANDBOOK

## ELL Support

**Use Descriptive Words** Have children talk about what their bones feel like and what they do.

**BEGINNING** Encourage children to use one- or two-word responses to questions about their arm bone's length and how it feels. Repeat with one of the bones in their hand.

**INTERMEDIATE** Have children use short sentences to describe one of their arm bones. Ask them describe a bone in their hand, and compare the two bones.

**ADVANCED** Have children feel their spine and use the illustration on page R8 to describe what the bones in the spine look and feel like. Help children discuss how the structure of the spine allows people to move their back.

## ...uscular System

...uscles are body parts ...at help you move. ...ey are inside your body.

...uscles get stronger when ...u exercise them.

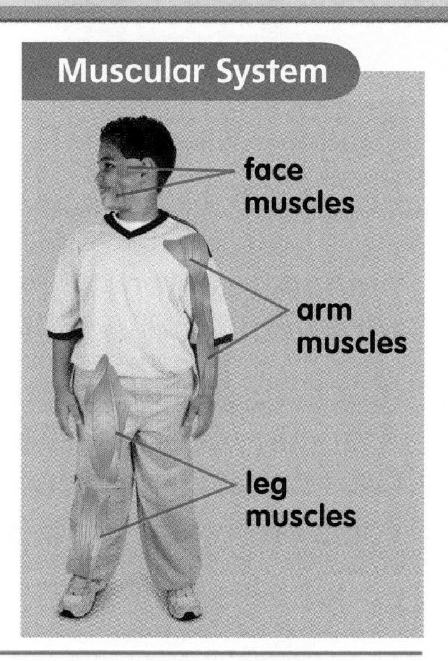


## ...rvous System

...ur brain sends messages ... through your body. The ...essages travel along tiny ...dy parts called nerves.

...ese messages tell your ...dy parts to move. They ...n also alert you of danger.

### ...y It

**...mp up and down in place. ...ich muscles did you use?**

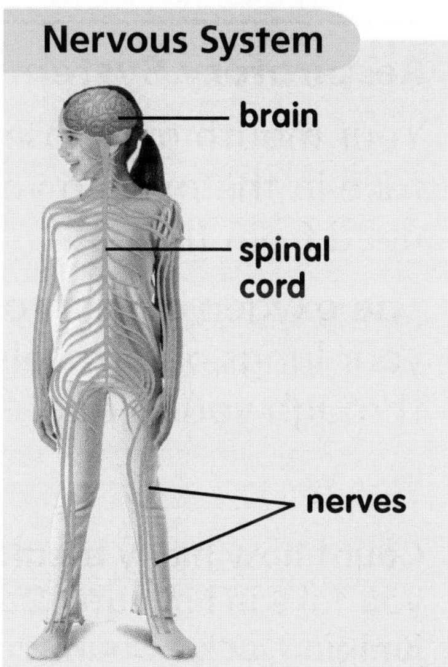




## Integrate Writing

### Describe a Movement

Have children work in pairs to observe and describe body movements. Invite them to choose a movement to observe. Then, ask one child to do the movement while the other observes. Have children switch roles.

Encourage children to draw and write about the body parts that were needed to complete the movement. Ask children to include body parts from the skeletal, muscular, and nervous systems.

## ▶ Assess Prior Knowledge

Invite children to lift their hands over their head and slowly move them back down to their sides. Ask:

- **What helps your body move?** muscles
- **How can you make your muscle show?** Possible answers: by bending or flexing my arm or leg

## ▶ Use the Visuals

Review the diagram of the muscular system on page R9 with children. Ask:

- **What muscles are in the diagram?** face muscles, arm muscles, leg muscles
- **What do face muscles help you do?** Possible answers: eat, talk, laugh, yawn

Look at the diagram of the nervous system on page R9 with children. Ask:

- **According to the diagram, where are the nerves in your body?** Possible answers: in your arms; in your legs; all over your body
- **What are the different parts of the nervous system?** brain, spinal cord, nerves

## ▶ Discuss the Main Idea

After reading the text, explain to children that without muscles, people would not be able to move. Without the spinal cord and nerves, messages about how to move would not be able to get from the brain to the muscles.

## ▶ Explore the Main Idea

ACTIVITY Have children do the Try It activity. Invite them to list each body part whose muscles helped them jump.

Then encourage them to try and jump without moving their arms at all to see if using their arm muscles makes it easier to jump.

**RESOURCES and TECHNOLOGY**

- **Human Body,** Lessons 1–2

# Your Body

## Objectives

- Explain that the circulatory system moves blood through the body.
- Recognize that the respiratory system moves oxygen through the body.
- Explain that the digestive system breaks down food in the body.

### ▶ Assess Prior Knowledge

Invite children to hold their hands over their chest and take in a couple of deep, slow breaths. Ask:

- **What do you feel?** Possible answers: I feel my heart beating. I feel my chest move in and out.

### ▶ Use the Visuals

Review the diagram of the circulatory system on page R10 with children. Ask:

- **What are the main parts of the circulatory system?** heart, arteries, veins

Have children look at the diagram of the respiratory system on page R10. Ask:

- **What is the body part between the mouth and the lungs?** the trachea
- **What is the body part below the lungs?** the diaphragm

### ▶ Discuss the Main Idea

Explain how the respiratory and circulatory systems work together. Bodies need oxygen to live. Air is taken in through the nose and mouth, and travels down to the lungs.

### ▶ Explore the Main Idea

ACTIVITY Have stopwatches available for the Try It activity. Explain that the body needs more oxygen during exercising because the heart pumps blood faster to move the oxygen more quickly through the body.

**RESOURCES and TECHNOLOGY**

**Human Body,** Lessons 1, 3–4

## Your Body

### Circulatory System

Blood travels through your body. Your heart pumps this blood through blood vessels.

Blood vessels are tubes that carry blood inside your body. Arteries and veins are blood vessels.

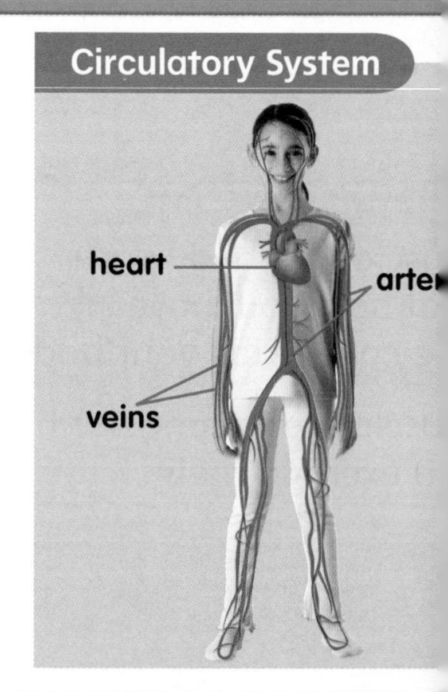


### Respiratory System

Your mouth and nose take in the oxygen you need from the air.

The oxygen goes into your lungs and travels through your blood.

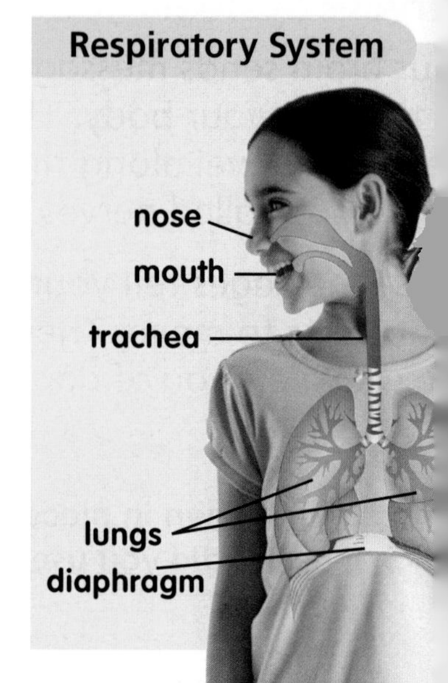


### Try It

**Count how many breaths you take in I minute. Do ten jumping jacks. Count again.**

R10
HEALTH HANDBOOK

## ELL Support

**Use Diagrams** Have children label or identify the circulatory and respiratory systems in diagrams or in pictures of people.

BEGINNING Have children make labels like the ones on the diagram on page R10. Display pictures of people from magazines or books. Guide children in reading each label as they apply the labels to a picture of a person in a book or magazine.

INTERMEDIATE Ask children to use complete sentences to identify the various body parts labeled in the two diagrams on page R10.

ADVANCED Have children discuss with a partner the two diagrams on page R10 and explain in their own words how each system works and why it is important.

igestive System

hen you eat, your body uses food
r energy. Food enters your body
rough your mouth. Your stomach
d intestines help you get nutrients
om the food in your body.

Digestive System

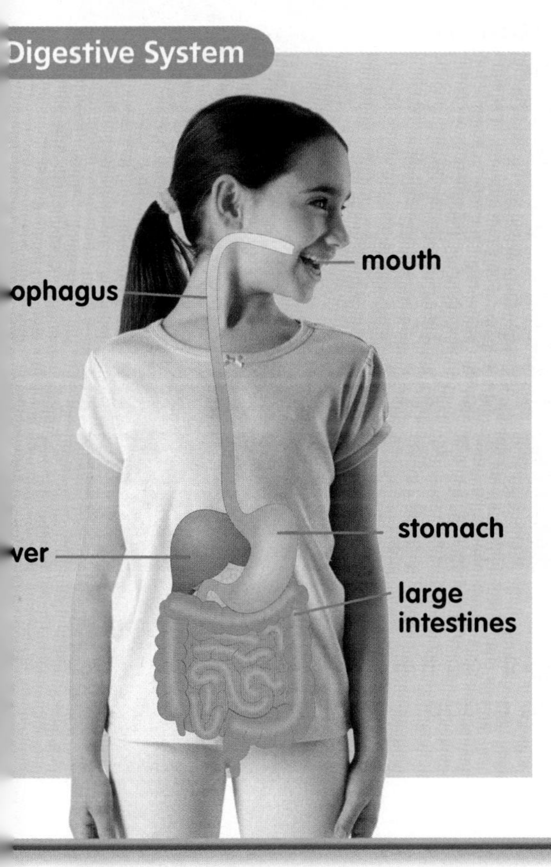


**Try It**

**Write a story about why your body needs food.**

RII
HEALTH HANDBOOK

## Integrate Writing

**The Journey of an Apple**

Have children write a story about being an apple and getting eaten by a person. Allow children to choose another food if they wish. Ask children to describe the apple's journey through the human body.

Model how to write in the first person, for example: *The girl took a bite out of me. I felt her teeth chew me to bits. Before I knew it, I slid down her throat.*

Remind children to include the different parts of the digestive system in their stories. Encourage children to be creative and use descriptive words. Suggest that children draw illustrations with their writing. Have children share their stories with classmates.

## ▶ Assess Prior Knowledge

Have children describe how they eat food. Ask:

- **What happens to the food after you chew it?** Possible answer: I swallow the food, and it goes down my throat into my stomach.
- **Why do we need food?** Possible answers: to give us energy; to get the vitamins our bodies need

## ▶ Use the Visuals

Review the diagram of the digestive system on page R11 with children. Ask:

- **What parts of the digestive system are labeled in the picture?** mouth, esophagus, liver, stomach, large intestines, small intestines
- **What body part is connected to the esophagus?** the stomach

## ▶ Discuss the Main Idea

Explain that the different parts of the digestive system break down food so the body can use it. Ask:

- **What happens to the food when you put it in your mouth?** Possible answers: I chew the food into small pieces; it dissolves and slides down the esophagus; it mixes with the saliva in my mouth.

Explain that the stomach and intestines continue to break food into simple compounds that can be transferred from the intestines into the nutrients, vitamins, and minerals that bodies use for energy and good health.

## ▶ Explore the Main Idea

ACTIVITY Give each child a cracker. Encourage them to chew the cracker for a long time and feel how the cracker changes texture, shape, and flavor. After they have swallowed, invite volunteers to describe what happened to the cracker as they chewed.

For the Try It activity, ask children to explain why eating three meals a day is essential to a healthful diet. Encourage them to compare how they feel before and after eating a meal.

# Healthful Foods

### Objectives

- Recognize the importance of eating healthful foods.
- Comprehend that nutrients in foods help store energy and help bodies grow.

## ▶ Assess Prior Knowledge

Discuss with children what they think is a healthful breakfast or lunch. Ask:

- **What foods are good for you?**
- **What foods should not be eaten very often?**

## ▶ Use the Visuals

Invite children to study the food pyramid on page R12. Ask them to name the kinds of foods they see in each of the colored sections in the pyramid. Have them compare the sizes of the labels at the bottom of each section. Ask:

- **Which section of the pyramid is the largest?**
  Possible answers: the orange one; Grains
- **Which section of the pyramid is the smallest?**
  Possible answers: the purple one; Meats & Beans

Have children choose a food they like to eat and place it in the correct category.

## ▶ Discuss the Main Idea

Discuss what it means to have a healthy diet. Explain that this food pyramid is meant to help children choose foods that are good for their bodies. Point out that the human body needs nutrients from each of these groups to be healthy.

## ▶ Explore the Main Idea

ACTIVITY Have children select a breakfast, lunch, or dinner meal for the Try It activity. Encourage them to make a menu for their healthful meal. Ask them to place the foods in the order that they would be served.

Remind children that not all the foods for each group is shown in the pyramid. For example, apples, peaches, pineapples, and melons are fruits, too. Have children tell which food groups their meals came from.

Have children tell which food groups the foods in their meal come from.

# Healthful Foods

### MyPyramid

MyPyramid is a guide for healthful eating. A healthful meal contains foods from the five food groups. A food group is a group of foods that are alike.

Eat more foods from the largest slice of the pyramid. Eat less from the smallest slice.

Grains | Vegetables | Fruits | Milk | Me… Be…

### Try It

**Plan a healthful meal. Include one food from each group.**

R12
HEALTH HANDBOOK

## ELL Support

**Matching Games** Make a set of food cards containing two cards for each food group. The name of the food group should be written on each card. Place the cards face down. Have children take turns turning over two cards. If they match, they keep the cards. If not, they turn them face down again, and the next child takes a turn.

**BEGINNING** Help children read the names of the food group aloud as they play.

**INTERMEDIATE** Ask children to read the name of the food group and find it on the pyramid.

**ADVANCED** Encourage children to describe the food group and name foods that are in that group.

## ealthful Foods

chickpeas

utrients are materials in foods that ake you healthy. Nutrients called rbohydrates store energy in your ody. Proteins help your body grow.

ople around the world get utrients from different foods.

Foods Around the World

| Food | Part of the world | Nutrient |
|---|---|---|
| Rice | Asia | carbohydrate |
| Tortilla | Central America | carbohydrate |
| Millet | Africa | carbohydrate |
| chickpeas | Middle East | protein |
| olives | Europe | oils |



### y It

t your favorite foods. Find out at nutrients are in the food.



## Differentiated Instruction

### Leveled Activities

**EXTRA SUPPORT** Have children fold a piece of paper in half and write *Protein* at the top of one half and *Carbohydrate* at the top of the other. Invite children to draw and label pictures of foods that contain these nutrients on the correct half of the paper.

**ENRICHMENT** Tell children that most foods contain more than one kind of nutrient. Have children use the Internet to research a particular food and how many different nutrients it contains. Invite children to share their research with the class.

## ▶ Assess Prior Knowledge

Invite children to share what they know about nutrients in foods. Ask:

- **What kinds of things are in food that your body needs?** Possible answers: nutrients; vitamins
- **Which foods have less of the things your body needs?** Possible answers: candy, cake, soda
- **Which foods have more of the things your body needs?** Possible answers: eggs, chicken, nuts, beans

## ▶ Use the Visuals

Invite children to study the chart called *Foods Around the World* on page R13. Review the column titles. Ask:

- **Which carbohydrates are listed on the chart?** rice, millet, tortilla
- **Which protein food is listed on the chart?** chickpeas
- **What is the oil listed on the chart?** olive oil

## ▶ Discuss the Main Idea

Read the text on page R13 together. Explain that people around the world might eat different foods to stay healthy. Discuss how many kinds of foods can provide the body with the nutrients that it needs. Ask:

- **What nutrient stores energy in the body?** carbohydrates
- **What nutrient helps the body grow?** proteins

## ▶ Explore the Main Idea

**ACTIVITY** Have children do the Try It activity. If their favorite meal includes many kinds of food put together, such as a cheese, tomato, and onion omelet, encourage them to write down each kind of food separately. To help children research nutrients in their favorite foods, give them resources such as ingredient and nutrition labels from different foods.

**RESOURCES and TECHNOLOGY**

▶ **Human Body,** Lesson 7

# Healthy Living

## Objective

- Identify ways to take care of your body and stay healthy.

### ▶ Assess Prior Knowledge

Ask children to describe things they do to take good care of their bodies. Ask:

- **What do you do every day to stay healthy?** Possible answers: brush my teeth; eat vegetables; exercise

### ▶ Use the Visuals

Review the photographs and read the captions on page R14 with children. Ask:

- **What are the children in the left picture doing?** Possible answers: karate, judo, kung fu, martial arts
- **How does exercise help people stay healthy?** Possible answers: It helps their lungs and heart work better; it makes their muscles stronger.
- **Why do you think the boy is at the doctor's office?** Possible answers: He might have a sore throat; he might be getting a check-up.

### ▶ Discuss the Main Idea

Have children describe what it feels like to be sick. Then ask them what it feels like to be healthy. Explain that people have more energy and are happier when they are healthy and active.

### ▶ Explore the Main Idea

ACTIVITY Have children do the Try It activity. Ask them to review the bar graph style shown on page R7. Discuss the kinds of labels they might use for the horizontal and vertical axes in their graphs.

Point out to children that there are some healthful activities that they may not think of as exercise, such as walking, raking leaves, and doing chores.

Invite children to make a bar graph and to color in a box each time they exercise.

**RESOURCES and TECHNOLOGY**

**Human Body,** Lesson 8

## Healthy Living

### Stay Healthy

Be active every day. Exercise keeps your heart and lungs healthy.

Doctors and dentists can help you stay healthy as you grow.

▲ **Exercise is important for a healthy body.**

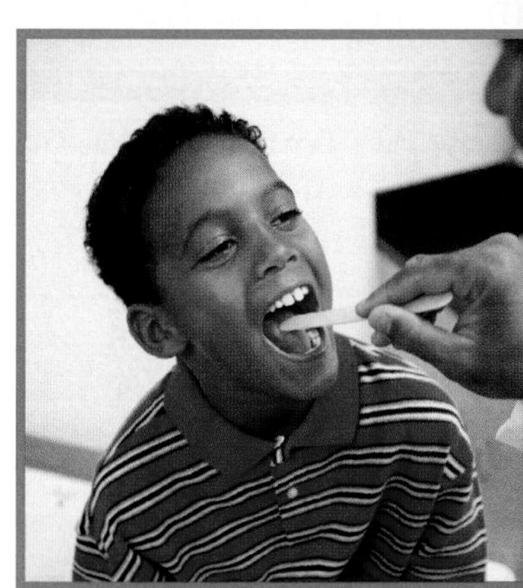

▲ **Get checkups from a docto and a dentist every year.**

### Try It

**Record how many times you exercise in one week.**



## ELL Support

**Role Play** Have children demonstrate ways that they are physically active and describe how it makes their bodies feel when they do a particular activity.

**BEGINNING** Help children name physical activities that members of their group are demonstrating.

**INTERMEDIATE** Have children use short phrases to describe physical activities that members of their group are demonstrating, and name the body parts being used.

**ADVANCED** Encourage children to use full sentences to describe how they feel when they do the activity they are acting out. Ask them to tell which parts of their body that activity exercises.

## ake Care of Your Body

obacco and alcohol harm you. obacco smoke can make it hard ɔ breathe. Alcohol slows down our mind and body.

▲ Only take medicines that your parent or doctor gives you.

Here are some ways to take care of your body. ▼

### Try It

Make a poster about being drug free. Share it with your school.



## Extended Reading

### Visit the Library

Read the book, ***Oh the Things You Can Do That Are Good for You!***, by Dr. Seuss (Random House, 2001). Discuss as a class what is happening in the illustrations. Ask:

- **What kinds of exercises are the characters doing?** Possible answers: running, jogging, biking
- **What do the characters do to keep healthy?** Possible answers: bathe, exercise

After reading, flip through the pages and have children identify whether the characters are doing healthy or unhealthy things.

## ▶ Assess Prior Knowledge

Have children name ways they take care of their body every day. Ask:

- **Why is it important to take care of your body?** so that I stay healthy

## ▶ Use the Visuals

Study the chart on page R15 together. Ask:

- **Why is taking a shower or a bath important?** Possible answers: because it gets rid of dirt; because it kills the germs that could make me sick
- **Why is it important to brush and floss your teeth?** Possible answers: It gets food out of your teeth; keeps teeth strong; it prevents cavities.

Have children think of other items that could be added to the list, such as washing hands, being active, or eating healthful foods.

## ▶ Discuss the Main Idea

Read the text together. Discuss with children things that do not help keep their bodies healthy. Ask:

- **Why is smoking bad for your health?** Possible answer: It can make it difficult to breathe.

Invite children to share information they have learned about other things that can damage their health, such as unhealthy foods, poison, pollution, or germs.

## ▶ Explore the Main Idea

ACTIVITY Ask children to create a chart like the one on page R15. Have them include three things they do every day to take care of their body. Guide them to use at least one activity that is not in the chart on page R15.

Have children do the Try It activity. Provide them with poster board, markers, crayons, and other materials for the posters. Encourage them to explain in writing why drugs are bad for their health, and include it in their poster.

Obtain permission to display the posters outside of the classroom

# Safety

**Objective**

- Identify everyday dangers and how to stay safe indoors and outdoors.

## ▶ Assess Prior Knowledge

Ask children to name things at home that are dangerous. Ask:

- **What should you do if you see something that looks dangerous?** Possible answers: Move away from it; tell an adult about it right away.

## ▶ Use the Visuals

Have children identify the items in the circular pictures on page R16 and tell why they are dangerous. Ask:

- **Why are these things dangerous?** Possible answers: They could burn me if I touch them; they could start a fire.

## ▶ Discuss the Main Idea

Read the text together. Show children a warning symbol on a household product, such as on a bottle of bleach. Remind children that they should never touch any item that has a warning symbol.

Explain to children that fire can be used safely by an adult, but it can be very dangerous if used carelessly or incorrectly.

Remind children that they can call 911 from any telephone if they ever have an emergency. Be sure to inform them that it is against the law to call 911 when there is no emergency.

## ▶ Explore the Main Idea

ACTIVITY Have each child find one thing in the classroom that could be dangerous or make them sick if ingested or used improperly. Children should be able to identify simple items, such as soap, chalk, staplers, or scissors.

Discuss ways to make the classroom safer.

Have children do the Try It activity.

**RESOURCES and TECHNOLOGY**

▶ **Human Body,** Lesson 8

# Safety

### Safety Indoors

To stay safe indoors, do not touch dangerous things. Tell an adult about them right away. Never taste anything without permission.

In case of a fire, get out fast. If your clothes catch fire, remember to stop, drop, and roll.

▲ **Do not touch these things.**

### Try It

**Practice stop, drop, and roll. Teach it to a friend.**

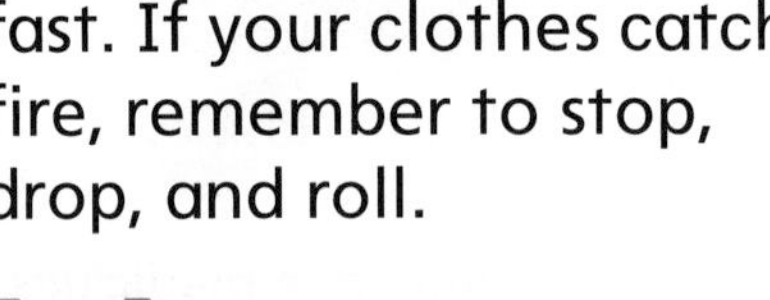

## Extended Reading

**Visit the Library**

Read together ***Clifford The Firehouse Dog,*** by Norman Bridwell (Cartwheel, 1994). After reading, walk children through the book again. Ask:

- **What things on this page are warning people of danger?** Possible answers: sirens, smoke, firefighters
- **What are the people doing to stay safe or get to a safe place?**

Review as a class the school's fire safety instructions, such as those near alarms and fire exits.

## afety Outdoors

e safe outdoors. Follow these rules.

▲ Wear a helmet.

▲ Cross at a crosswalk.

▲ Wear your seat belt.

▲ Follow game rules.

### ·y It

hoose one of the rules. Make poster showing the safety rule.



## Differentiated Instruction

### Differentiated Questions

EXTRA SUPPORT Ask questions such as these to check children's understanding of the material.

- **How can you stay safe in a car?** a seat belt
- **What are some things in the house that can be dangerous?** Possible answers: poison, ovens, electrical outlets, knives

ENRICHMENT Use these types of questions to develop children's higher-order thinking skills.

- **Why shouldn't you cross in the middle of the street?** Possible answer: because someone driving a car might not be able to see you
- **Why is the stove dangerous?** Possible answer: because it can get hot and burn you or start a fire

## ▶ Assess Prior Knowledge

Invite children to list outdoor places or activities that could be dangerous, such as parking lots, construction sites, playing in the street, or climbing a tree.

## ▶ Use the Visuals

Direct children's attention to the pictures on page R17. Ask:

- **What is each of these children doing?** Possible answers: riding a bicycle; crossing the street; riding in a car; and playing soccer
- **What safety rules are these children following?** Possible answers: wearing a helmet; crossing at the crosswalk; crossing with a crossing guard; wearing a seat belt; kicking the ball properly; wearing shin guards

## ▶ Discuss the Main Idea

Discuss why it is important to follow safety rules. Ask:

- **Why is it important to wear a helmet?** Possible answer: A helmet will protect my head if I fall.
- **Why is it important to cross a street at the crosswalk?** Possible answers: People in cars will be able to see you better; a crossing guard can help you cross; the lights can tell you when it is safe to cross.
- **What are some ways to stay safe on the playground?** Possible answers: Watch out for other children, so you don't bump into them; play fair so that no one feels like they have to fight; stay out of the way of bats and balls.

## ▶ Explore the Main Idea

ACTIVITY Give each child a sticker and instruct them to give the sticker to a classmate who is being safe during recess time or gym class.

Provide children with paper, paint, markers, and collage materials for the Try It activity. Request permission to hang the posters where everyone can be reminded about safety. For example, if someone makes a poster about not jumping on the stairs, ask for permission to hang it near a stairway.

# Safety

## Objective

- Explain how to use a wind scale to determine whether it is safe to be outside.

### ▶ Assess Prior Knowledge

Invite children to describe storms that they have seen. Ask:

- **What did you see during the storm?** Possible answers: lots of rain; snow; wind; dark clouds
- **What did it look like just before the storm?** Possible answers: the wind picked up; the sky got dark

### ▶ Use the Visuals

Have children study the chart on page R18. Ask:

- **What do the pictures on the left show?** They show what each level of wind on the wind scale looks like.
- **How fast is the wind at a scale of 3?** 8–12 miles per hour
- **Which numbers on the wind scale might be dangerous?** Possible answers: numbers 9 or 12

### ▶ Discuss the Main Idea

Read the text on page R18 together. Explain that wind is natural and can feel good on a hot day. Strong winds can be dangerous and might mean a storm is coming. A wind scale can tell you if it is dangerous to go outside. Ask:

- **What kind of storm happens when the wind blows over 73 miles per hour?** a hurricane

### ▶ Explore the Main Idea

Have children do the Try It activity. Invite children to draw what they observe and write a sentence about what the wind is like outside.

## Safety

### Wind Scale

Very windy weather can be dangerous. Do not play outdoors before a storm. Scientists use a scale like this one to tell how hard the wind is blowing.

▼ This is a Beaufort scale.

**Wind Scale**

| Number | What You Can See | Wind Speeds |
|---|---|---|
| 0 | calm | less than 1 mile per hour |
| 3 | gentle breeze | 8–12 miles per hour |
| 6 | strong breeze | 25–30 miles per hour |
| 9 | very strong wind | 47–54 miles per hour |
| 12 | hurricane | more than 73 miles per hour |

### Try It

**Look out a window. What kind of wind do you see? Use the wind scale to help.**

R18
HEALTH HANDBOOK

## ELL Support

**Relate to Personal Experience** Have children describe their experiences with different levels of wind. Use a fan to demonstrate what different winds may feel like. Start with the fan's lowest setting. Ask: **How does the wind feel? How does the wind affect you?** Continue to raise the setting, repeating the questions each time. Turn off the fan. Have children describe how they feel with no wind.

BEGINNING Have children use or repeat descriptive words to explain how the wind feels.

INTERMEDIATE Encourage children to use phrases and short sentences to describe how the wind feels at each setting.

ADVANCED Ask children to describe in complete sentences how the wind changes.

# Glossary

**adaptation** a body part or a way an animal acts that helps it stay alive (page 70)
**The anteater's long snout is an adaptation.**

**amphibian** an animal that lives part of its life in water and part on land (page 57)
**A salamander is an amphibian.**

**anemometer** a tool that measures the speed of wind (page 227) **The stronger the wind is, the faster the anemometer spins.**

**Arctic** a very cold place near the North Pole (page 132) **Animals in the Arctic have layers of fat to keep them warm.**

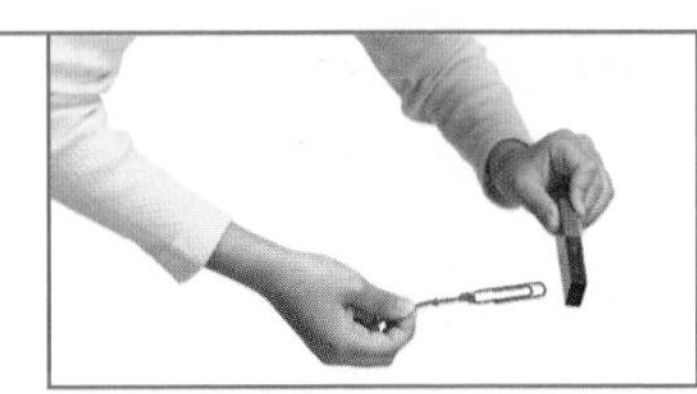

**attract** to pull toward something (page 386) **A magnet can attract some objects.**

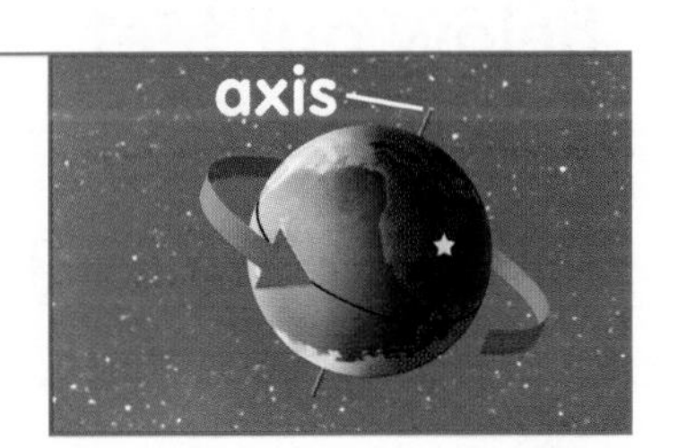


**axis** a center line that an object spins around (page 255) **Earth spins on its axis.**

-Glossary Words and definitions online at www.macmillanmh.com

# C

**camouflage** a way that animals blend into their surroundings (page 71) **Animals use camouflage to stay safe.**

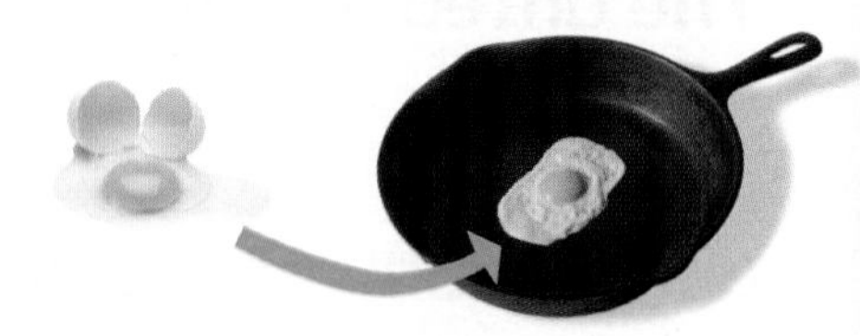

**chemical change** when matter changes into different matter (page 328) **Cooking an egg makes a chemical change.**

**circuit** a path that electricity flows in (page 422) **A bulb will light when connected with wires in a circuit.**

**cirrus** thin, wispy clouds high in the sky (page 239) **The wind blows cirrus clouds into wispy streams.**

**condense** to change from a gas to a liquid (pages 233, 334) **Water vapor can condense on a cold glass.**

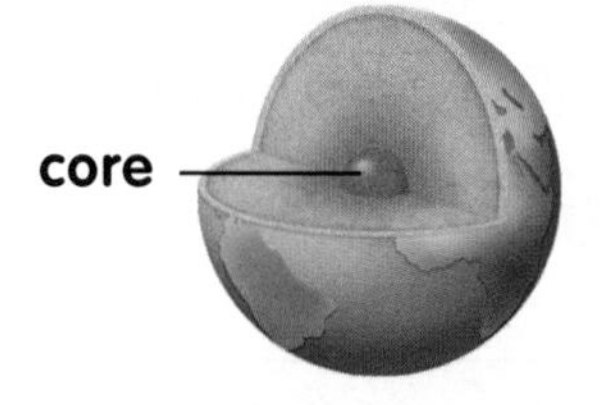


**core** Earth's deepest and hottest layer (page 160) **The core is thousands of miles below our feet.**

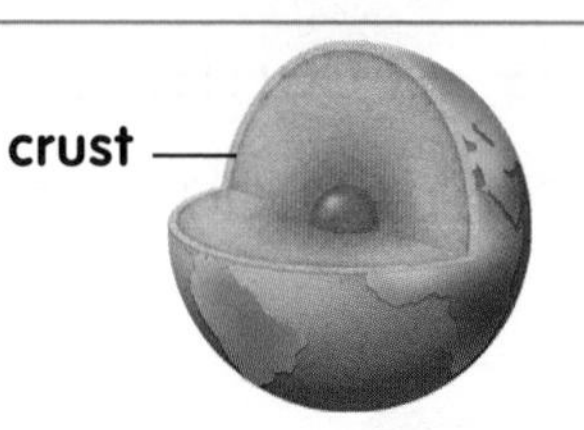


**crust** Earth's outer layer (page 160) **We live on Earth's crust.**

**cumulus** white, puffy clouds (page 238) **Small cumulus clouds appear in good weather.**

**current electricity** a kind of energy that moves in a path (page 422) **When you use a toaster, you use current electricity.**

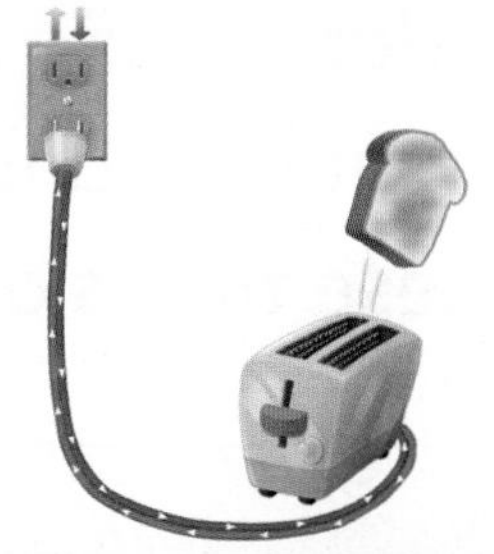

## D

**decompose** when plant and animal parts rot or break down (page 198) **This log will decompose over time.**

**desert** a dry habitat that gets very little rain (page 130) **A desert is hot and dry.**

**dissolve** to mix evenly with a liquid and form a solution (page 343) **Sugar will dissolve when it is mixed with water.**

**drought** a long period of time with little or no rain (page 104) **Plants can die in a drought.**

-Glossary Words and definitions online at www.macmillanmh.com

## E

**earthquake** a shake in Earth's crust (page 174) **An earthquake damaged this road.**

**endangered** when many of one kind of animal die and only a few are left (page 106) **These tigers are endangered.**

**evaporate** to change from a liquid to a gas (pages 232, 333) **Water can evaporate from oceans, rivers, lakes, or land.**

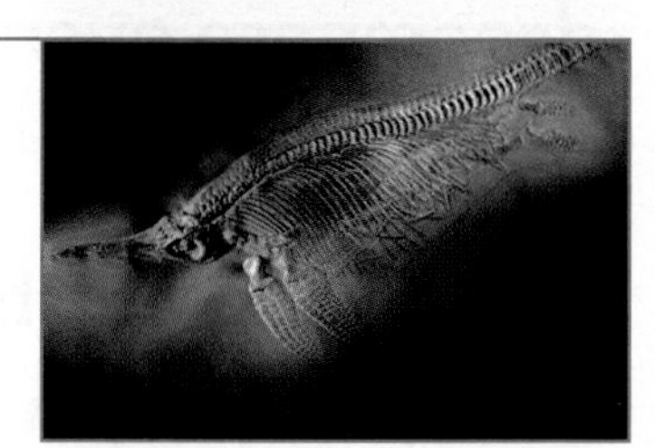

**extinct** when a living thing dies out and no more of its kind live on Earth (page 109) **Dinosaurs are extinct.**

## F

**flood** water that flows over land and can not easily soak into the ground (page 175) **This man is walking in a flood.**

**flower** the plant part that makes seeds (page 30) **Some flowers can grow into fruit.**

food chain a model of the order in which living things get the food they need (page 96) **A food chain begins with the Sun.**

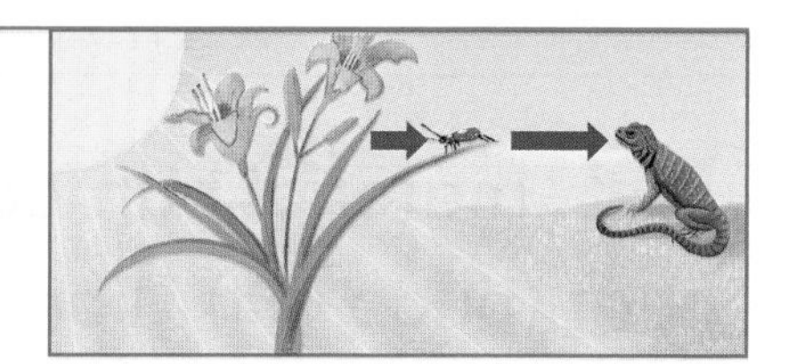

food web two or more food chains that are connected (page 99) **This picture shows a desert food web.**

force a push or pull on an object (page 369) **When you kick a ball, you are using a kind of force.**

fossil what is left of a living thing from the past (page 108) **This fish fossil was found in the desert.**

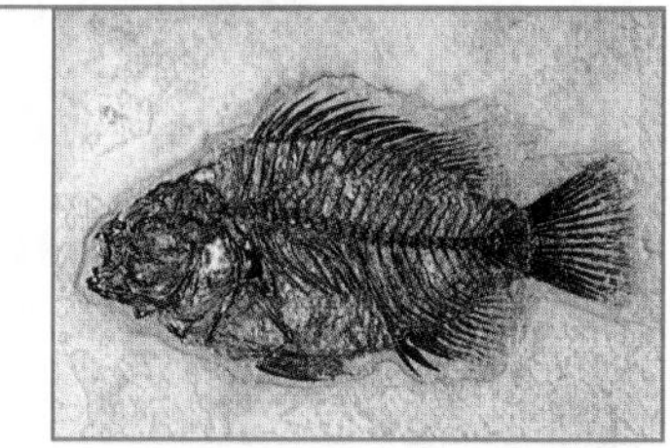

fresh water water that is not salty (page 165) **Fresh water is found in lakes, ponds, rivers, and streams.**

friction a force that slows down moving things (page 371) **A skate makes friction when the stopper rubs against the ground.**

LOG ON e-Glossary Words and definitions online at www.macmillanmh.com

**fuel** material burned to make power or heat (page 401) **Wood is fuel for fire.**

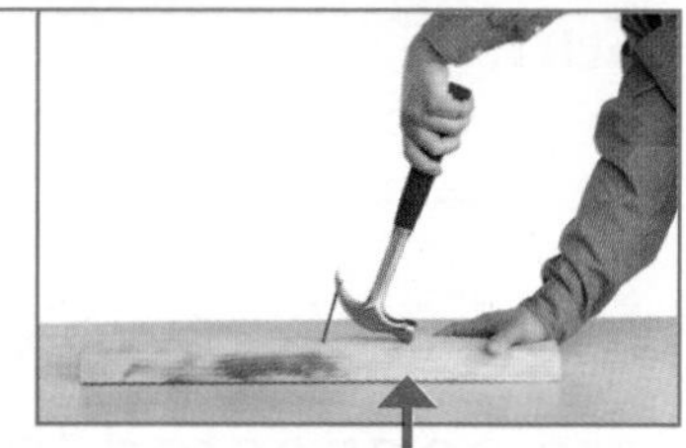


**fulcrum** the point that a lever moves against (page 378) **This piece of wood can act as a fulcrum.**

## G

**gas** matter that spreads to fill the space it is in (page 312) **The tube is filled with gas.**

**gravity** a kind of force that pulls down on everything on Earth (page 370) **Gravity is the force that pulls a ball to the ground.**

## H

**habitat** a place where plants and animals live (page 90) **A habitat can be wet, dry, windy, or cold.**

**heat** a kind of energy that makes objects warmer (page 400) **The Sun gives us heat.**

## I

**insect** an animal with six legs, antennae, and a hard outer shell (page 58) **An ant is an insect.**

## L

**landform** one of the different shapes of Earth's land (page 156) **This landform is called a valley.**

**landslide** sudden movement of soil from higher to lower ground (page 175) **Buildings can be damaged in a landslide.**

**larva** stage in the life cycle of some animals after they hatch from an egg (page 64) **A caterpillar is a larva.**

**lever** a simple machine made of a bar that turns around a point (page 378) **A lever can help you move or lift objects.**

**life cycle** how a living thing grows, lives, has young, and dies (pages 34, 62) **The pictures show the life cycle of a chicken.**

LOG ON e-Glossary Words and definitions online at www.macmillanmh.com

**light** a kind of energy that lets us see (page 416) **We get light from the Sun.**

**liquid** matter that takes the shape of the container it is in (page 310) **Water is a liquid.**

## M

**mammal** an animal with hair or fur that feeds milk to its young (page 56) **A lion is a mammal.**

**mantle** the very hot layer below Earth's crust (page 160) **The mantle is too hot for living things.**

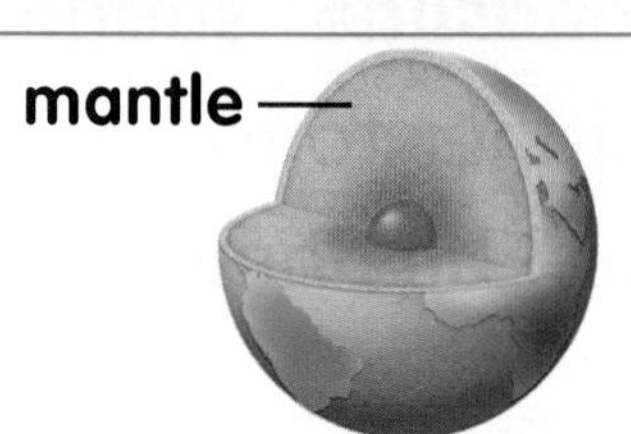


**mass** the amount of matter in an object (page 296) **The larger boot has more mass.**

**matter** anything that takes up space and has mass (page 296) **Everything around us is made of matter.**

**minerals** bits of rock and soil that help plants and animals grow (pages 26, 190) **Plants use minerals in the ground to grow.**

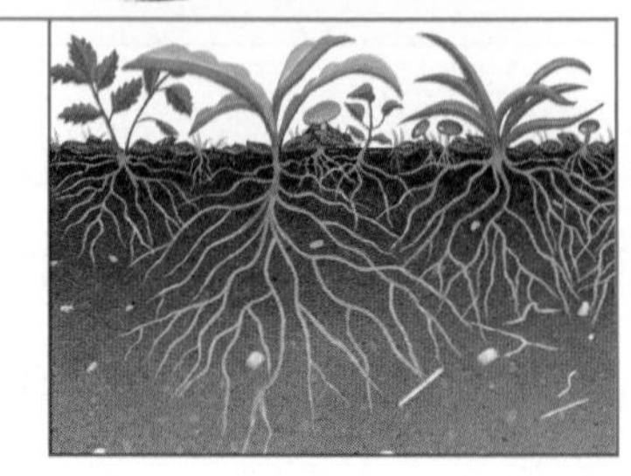

**mixture** two or more things that keep their own properties when mixed together (page 340) **This snack food is a mixture.**

**motion** a change in the position of an object (page 363) **This dog is in motion.**

## N

**natural resource** a material from Earth that people use in daily life (page 188) **Rocks are a natural resource.**

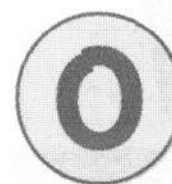

## O

**ocean** a large body of salt water (pages 136, 166) **Oceans cover most of Earth**.

**orbit** the path Earth takes around the Sun (page 262) **Earth orbits the Sun each year.**

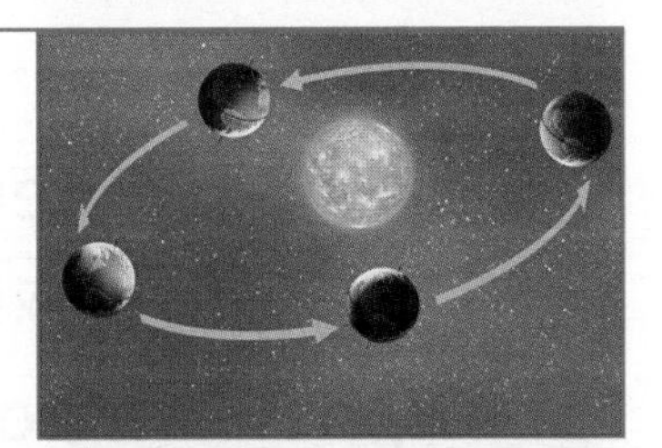

**oxygen** a gas found in the air we breathe (page 27) **Living things need oxygen.**

-Glossary Words and definitions online at www.macmillanmh.com

# P

**phase** the Moon's shape as we see it from Earth (page 271) **The Moon's phase will change each night.**

**physical change** a change in the size or shape of matter (page 326) **When you fold matter, you make a physical change.**

**pitch** how high or low a sound is (page 409) **Short, tight strings make a high pitch.**

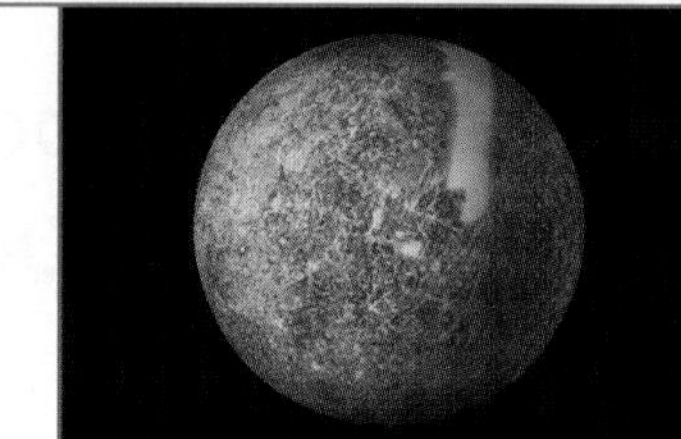

**planet** a huge object that travels around the Sun (page 276) **Mercury is the planet closest to the Sun.**

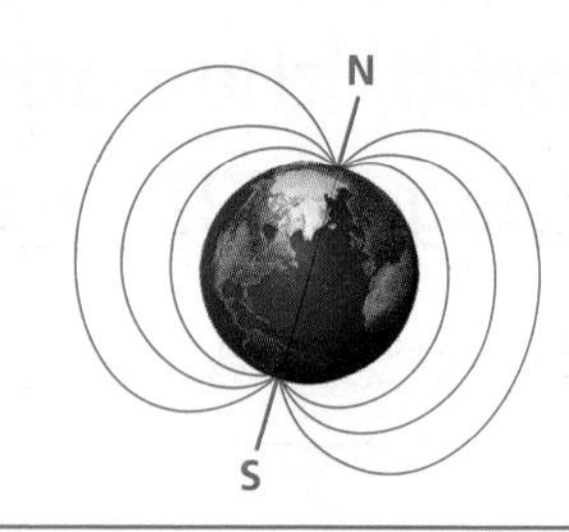


**poles** the two ends of a magnet, or either end of Earth's axis (page 388) **Earth has two poles, a north pole and a south pole.**

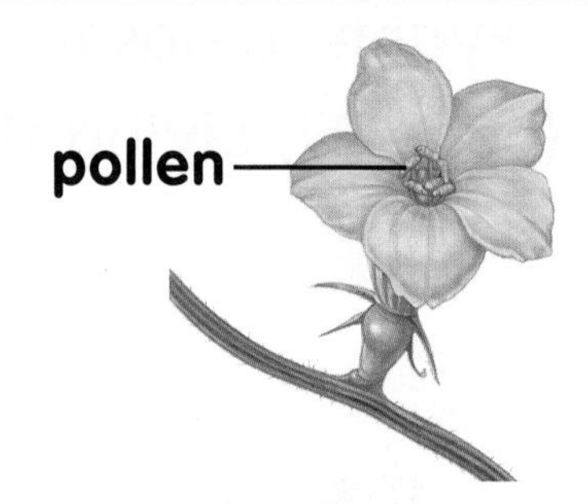


**pollen** a sticky powder inside a flower that helps make seeds (page 30) **Pollen can move from flower to flower.**

**pollution** anything that makes air, land, or water dirty (page 204) **Garbage is one kind of pollution.**

**pond** a small body of fresh water (page 138) **A pond is home to plants and animals.**

**position** the place where something is (page 362) **The position of the dog is above the cat.**

**precipitation** water falling from the sky as rain, snow, or hail (page 225) **Rain is one kind of precipitation.**

**predator** an animal that hunts other animals for food (page 97) **A predator must be fast to catch its food.**

**prey** **the** animals that are eaten by predators (page 97) **The bird catches prey in its beak.**

LOG ON e-Glossary Words and definitions online at www.macmillanmh.com

**property** the look, feel, smell, sound, or taste of a thing (page 298) **One property of this toy toucan is that it is soft.**

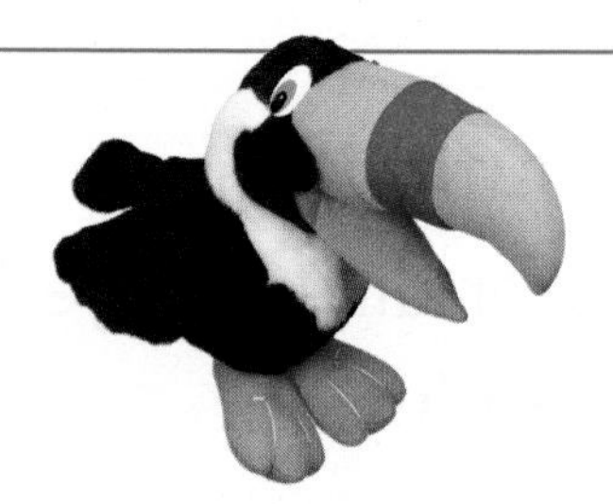

**pupa** the stage in a butterfly's life cycle when it makes a hard case around itself (page 64) **The pupa hangs from a branch.**

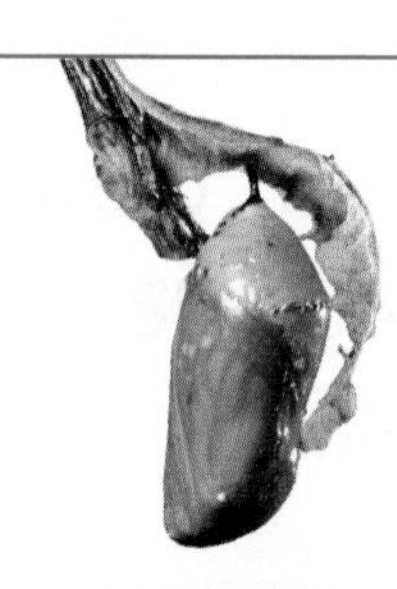

## R

**rain forest** A habitat where it rains almost every day (page 124) **Many kinds of plants and animals live in a rain forest.**

**ramp** A simple machine with a flat, slanted surface (page 379) **A ramp can be used to move an object from one level to another.**

**recycle** To make new items out of old items (page 206) **You can recycle paper.**

**reduce** to cut back on how much you use something (page 206) **We should reduce the amount of water we use.**

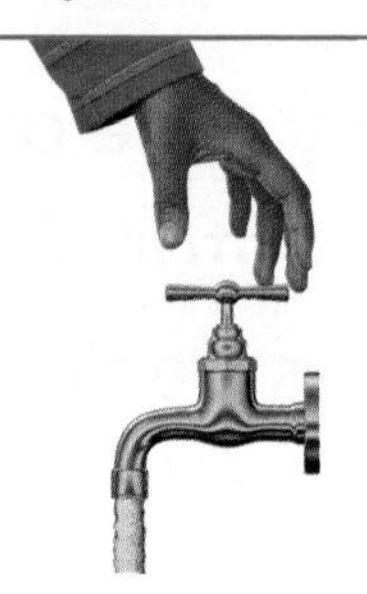

**reflect** to bounce off something (page 416) **Light can reflect better off shiny objects.**

**repel** to push away or apart (page 388) **The two south poles of a magnet repel each other.**

**reptile** an animal with rough, scaly skin (page 57) **An alligator is a reptile.**

**reuse** to use something again (page 206) **We can reuse items to cut down on waste.**

**rock** a hard, nonliving part of Earth (page 188) **A rock like this can be used as a tool.**

**rotation** a turn or spin (page 254) **Earth makes one rotation in 24 hours.**

LOG ON e-Glossary Words and definitions online at www.macmillanmh.com

## S

**seed** a plant part that can grow into a new plant (page 30) **A seed can grow with water, warmth, and air.**

**seedling** a young plant (page 34) **A seedling will grow into an adult plant.**

**simple machine** a tool that can change the size or direction of a force (page 378) **This simple machine is called a ramp.**

**soil** a mix of tiny rocks and bits of dead plants and animals (page 196) **Most plants need soil to grow.**

**solar system** the Sun, eight planets, and their moons (page 276) **Planets in our solar system orbit the Sun.**

**solid** matter that has a shape of its own (page 302) **This chair is a solid.**

**solution** a kind of mixture with parts that do not easily separate (page 343) **Water and this drink mix make a solution.**

**sound** a type of energy that is heard when objects vibrate (page 406) **An alarm clock makes a loud sound.**

**speed** how far something moves in a certain amount of time (page 364) **A cheetah has a fast running speed.**

**star** an object in space made of hot, glowing gases (page 272) **The Sun is a star that we see during the day.**

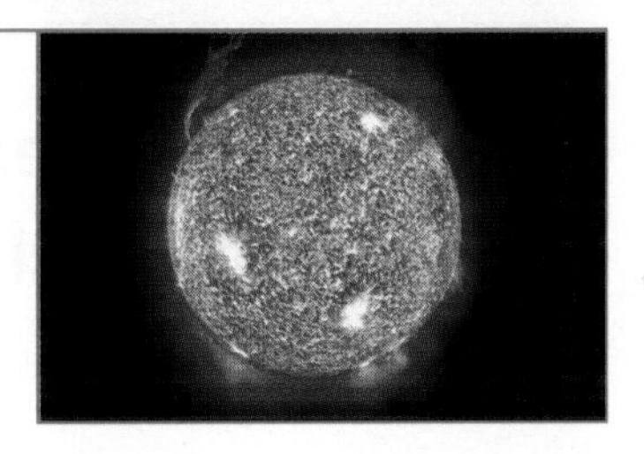

**static electricity** a kind of energy made by tiny pieces of matter that attract and repel each other (page 424) **Static electricity makes the cat's hair stick to the balloon.**

**stratus** thin clouds that form into layers like sheets (page 239) **Stratus clouds can cover the whole sky.**

LOG ON **e-Glossary** Words and definitions online at www.macmillanmh.com

**temperature** a measurement of how hot or cold something is (page 224) **A low temperature means something is cold.**

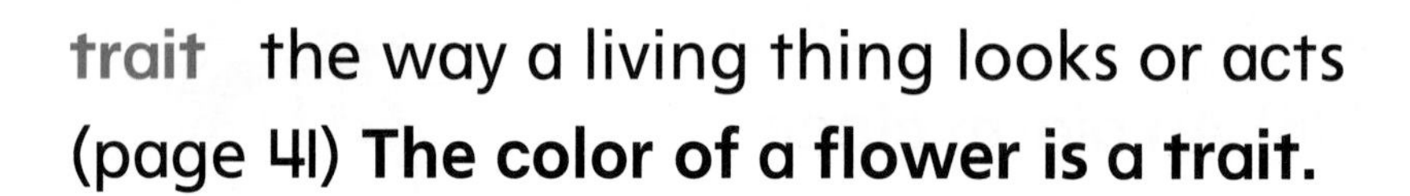

**trait** the way a living thing looks or acts (page 41) **The color of a flower is a trait.**

**vibrate** to move back and forth quickly (page 407) **Strings vibrate to make sound.**

**volcano** an opening in Earth's crust (page 174) **A volcano can change the land quickly.**

**volume** the amount of space something takes up (page 311) **You can measure the volume of a liquid with measuring cups.**

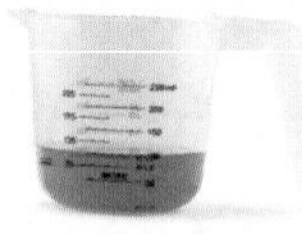

**woodland forest** a habitat that gets enough rain and sunlight for trees to grow well (page 122) **Many deer live in a woodland forest.**

# Science Skills

**classify** to group things by how they are alike (page 5) **You can classify animals by how many legs they have.**

**communicate** to write, draw, or tell your ideas (page 9) **You can communicate the ways you can change a piece of clay.**

Changing Clay
1. I roll the clay.
2. I pinched the clay.
3. I squeezed the clay.
4. I poked the clay.

**compare** to observe how things are alike or different (page 5) **You can compare how a cat and a dog are alike and different.**

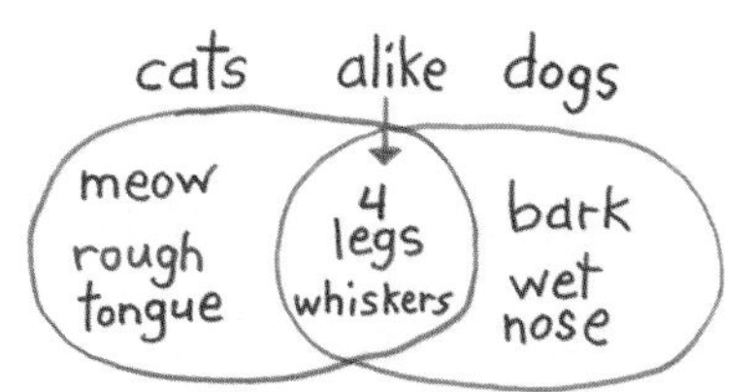


**draw conclusions** to use what you observe to explain what happens (page 9) **You can draw conclusions about why the stick will make a shadow.**

**infer** to use what you know to figure something out (page 7) **From these tracks, you can infer what animal was here.**

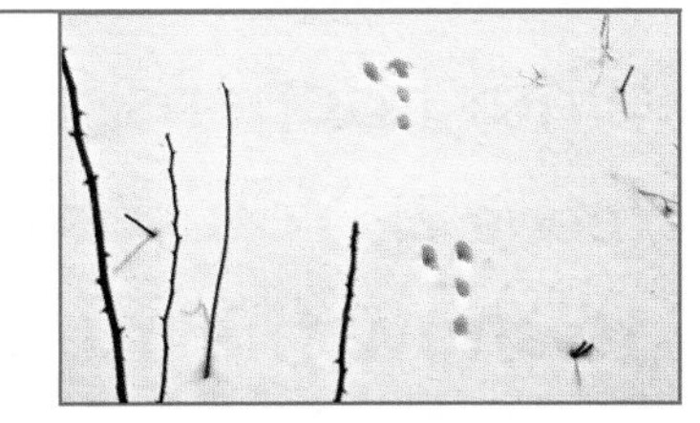

**investigate** to make a plan and try it out (page 8) **You can investigate how long it takes the car to stop rolling.**

**make a model** to make something to show how something looks (page 4) **You can make a model of a mountain in the ocean.**

**measure** to find out how far something moves, or how long, how much, or how warm something is (page 6) **You can measure temperature with a thermometer.**

**observe** to see, hear, taste, touch, or smell (page 4) **You can observe how the flower looks, smells, and feels.**

**predict** to use what you know to tell what you think will happen (page 8) **You can predict what the weather will be like today.**

**put things in order** to tell or show what happens first, next, or last (page 7) **You can put things in order to show the life cycle of a plant.**

**record data** to write down what you observe (page 6) **You can record data about what your class had for lunch.**

# Credits

**Abbreviation key:** MMH=Macmillan/McGraw-Hill

**Cover Photography Credits:** Steven M. Rollman/Natural Selection Stock Photography, **Back** Gerry Ellis/Digital Vision, Ltd./Getty Images, **Inside Back** Gerry Ellis/Digital Vision, Ltd./Getty Images, **Inside Front** Gerry Ellis/Digital Vision, Ltd./Getty Images, **Spine** Gerry Ellis/Digital Vision, Ltd./Getty Images.

**Literature Credits:** "The Seed" from *Always Wondering* by Aileen Fisher. Copyright © 1991 by Aileen Fisher. HarperCollins Publishers. Used by permission of the author, who controls all rights.
"In Payment" from *Out in the Dark and Daylight* by Aileen Fisher. Copyright © 1980 by Aileen Fisher. Published by Harper & Row Publishers. All rights reserved.
"Sunflakes" from *Country Pie* by Frank Asch. Copyright © 1979 by Frank Asch. Published by Greenwillow Books. All rights reserved.
"Popcorn Hop" from *Kindergaten Kids* by Stephanie Calmenson. Text Copyright © 2005 by Stephanie Calmenson. Published by HarperCollins Publishers. All rights reserved.

**Photography Credits:** **0–i** Gerry Ellis/Digital Vision, Ltd./Getty Images; **iv** (tl)George Grall/National Geographic Image Collection, (tl)Ken Cavanagh/MMH, (bl)MMH, (br)Siede Preis/Photodisc Green/Getty Images; **vi** (b)Chris Mattison/NHPA; **vii** (b)Annie Griffiths Belt/National Geographic Image Collection; **viii** (bl)E. R. Degginger/Photo Researchers, Inc., (bc)Valerie Giles/Photo Researchers, Inc., (br)Geoff Dann/DK Images, (bkgd)Clive Streeter/DK Images; **ix** (l)Digital Vision/PunchStock, (c)Space Telescope Science Institute/NASA/Photo Researchers, Inc., (r)StockTrek/Getty Images; **x** (b)Ken Cavanagh/MMH; **xi** Roger Ressmeyer/CORBIS; **xii** (l)EyeWire/Getty Images, (r)Photodisc/Getty Images, (others)The McGraw–Hill Companies; **xiii** (t)Photodisc/Getty Images (others)The McGraw–Hill Companies; **xiv** (t)Dynamic Graphics Group/IT Stock Free/Alamy, (bl)MMH, (bc)Peter Anderson/DK Images, (br)Tom Bean/CORBIS; **xv** (t)Studio Photogram/Alamy, (cl bc)Colin Keates/Courtesy of the Natural History Museum, London/DK Images, (c)Visuals Unlimited/Getty Images, (bl)Ken Cavanagh/MMH, (br)Arco Images/Alamy; **xvi** (t)C Squared Studios/Getty Images, (b)Ken Karp/MMH; **1** (bkgd)Don Paulson Photography/SuperStock; **2** (inset)George McCarthy/naturepl.com; **2–3** (bkgd)Hans Schouten/Foto Natura/Minden Pictures; **3** (b)Ken Karp/MMH, (cw from top)Photodisc/Getty Images, (others)MMH; **4** (tl)Brand X Pictures/PunchStock, (tr)Michael Orton/Getty Images, (b)Michael P. Gadomski/Photo Researchers, Inc.; **5** (l)Martin Ruegner/Getty Images, (r)Marie Read/Animals Animals/Earth Scenes; **6** (tl)Ken Cavanagh/MMH, (tr)Klaus Uhlenhut/Animals Animals/Earth Scenes, (bl)Brand X Pictures/PunchStock, (br)Siede Preis/Photodisc/Getty Images, (inset)George Grall/National Geographic Image Collection; **7** (tl)Siede Preis/Getty Images, (br)Norbert Rosing/National Geographic Image Collection, (others)Photodisc/Getty Images; **8** (l)David Boag/PictureQuest/Jupiterimages, (c)Hans Pfletschinger/Peter Arnold, Inc.; **8–9** Michael Durham/Minden Pictures; **9** Photodisc/Getty Images; **10–11** (l)Bob Elsdale/Getty Images; **11** (t)Kim Taylor/Bruce Coleman USA, (tc)D. Parer & E. Parer–Cook/Auscape/Minden Pictures, (b)Stephen Dalton/Minden Pictures, (bc)Jeff Foott/Getty Images, (inset)Photodisc/Getty Images; **12** (l)Geoff Brightling/DK Images, (r)Dan Suzio/Photo Researchers, Inc.; **12–13** Ken Karp/MMH; **13** (r)James Balog/Getty Images; **14** (l)James Balog/Getty Images, (t)Siede Preis/Photodisc/Getty Images; **15** (t)Photodisc/Getty Images, (bl)Geoff Brightling/DK Images, (br)Dan Suzio/Photo Researchers, Inc.; **16** (tl)Michael Groen/MMH, (others)Ken Cavanagh/MMH; **17** Daniel Cox/Jupiterimages; **20–21** Frank Krahmer/Masterfile; **21** (t)Papilio/Alamy, (c)Photo Courtesy of Pdphoto.org, (b)Tom Bean/CORBIS; **22** CORBIS; **22–23** Peter Finger/CORBIS; **23** (t)Ken Karp/MMH, (others)MMH, **24** (t)Meul/ARCO/naturepl.com, (b)John Cancalosi/Peter Arnold, Inc.; **25** (tl)Steve Taylor/SPL/Photo Researchers, Inc., (tc)Papilio/Alamy, (tr)du Sud/Photolibrary.com, (b)CORBIS; **26** (t)Neil Fletcher/DK Images, (b)Burke/Getty Images, (bkgd)CORBIS; **27** (t)CORBIS, (b)Siede Preis/Getty Images; **28–29** Interfoto Pressebildagentur/Alamy; **29** (t)Michael Scott/MMH, (others)MMH; **30** Claudius Thiriet/Grandeur Nature/Hoa–Qui/Imagestate; **31** (tl)Karen Tweedy–Holmes/CORBIS, (tr)Ingram Publishing/age fotostock, (br)Malcolm Case–Green/Alamy; **32** (l)Blickwinkel/Alamy, (cl)David Sieren/Visuals Unlimited, (cr)Peter Gardner/DK Images, (r)Derek Hall/DK Images; **33** (tr)DAJ/Getty Images, (others)Ken Cavanagh/MMH; **34** (tl)Doug Sokell/Visuals Unlimited, (tc)britishcolumbiaphotos.com/Alamy, (cr)Peter Anderson/DK Images, (bl)David R. Frazier Photolibrary, Inc./Alamy, (br)Tom Bean/CORBIS; **35** (l)Peter Arnold Inc./Alamy, (r)Clark W. Biedel, MD; **36** imagestate/Alamy; **37** (l)Herbert Kehrer/zefa/CORBIS, (c)DIOMEDIA/Alamy, (r)PhotoAlto/PunchStock; **38–39** Tim Laman/Getty Images; **39** (l)Ken Karp/MMH, (others)MMH; **40** (tr)Brand X Pictures/PunchStock, (cl)Siede Preis/Getty Images, (cr)Papilio/Alamy, (b)Gary Crabbe/Alamy; **41** (t)Brand X Pictures/PunchStock, (b)Russell Illig/Getty Images; **42** (l)David Noton/Getty Images, (r)Dave G. Houser/CORBIS; **42–43** Fabrice Bettex/Alamy; **45** (l)Edward Parker/Alamy, (r)Dominique Halleux/Peter Arnold, Inc.; **46** (inset)Peter Fakler/Alamy, (bkgd)Arco Images/Alamy; **47** Wilson Goodrich/Index Stock Imagery; **48** (inset)Nico Tondini/age fotostock, (bkgd)Masterfile; **49** (inset)Jerome Wexler/Visuals Unlimited, Danita Delimont/Alamy; **50** (l)MMH, (r)Photo Courtesy of Pdphoto.org; **51** (tl)Jim Allan/Alamy, (tr)John Henwood/Alamy, (b)George H. H. Huey/CORBIS; **51A** MMH; **52–53** Dwight Kuhn Photography; **53** (t)franzfoto.com/Alamy, (tc)DK Images, (tc)Jane Burton/DK Images, (b)Juergen & Christine Sohns/Animals Animals/Earth Scenes, (bc)John P. Marechal/Bruce Coleman USA; **54–55** Tui De Roy/Minden Pictures; **55** (t)Kevin Schafer/zefa/CORBIS, (tcl)Michael & Patricia Fogden/CORBIS, (tcr)Werner H. Mueller/zefa/CORBIS, (b)Gary Bell/oceanwideimages.com, (bcl)Fritz Rauschenbach/zefa/CORBIS, (bc)David A. Northcott/CORBIS; **56** franzfoto.com/Alamy; **57** (tl)Herve Berthoule/Jacana/Photo Researchers, Inc., (tr)Danita Delimont/Alamy, (cr)Bruce Coleman, Inc./Alamy, (b)Wendell Metzen/Bruce Coleman USA; **58** (l)Xavier Desmier/Rapho/Imagestate, (r)CORBIS; **59** (t)Bob Elsdale/Getty Images, (c)E. R. Degginger/Animals Animals/Earth Scenes, (b)Tim Ridley/Getty Images; **60–61** Brand X Pictures/PunchStock; **61** Nancy Ney/Digital Vision/Getty Images; **62** (tl tc) Reuters/CORBIS, (tr)Garrison Ron/CORBIS, (bl)DK Images, (others) Jane Burton/DK Images; **63** (t)Daniel A. Bedell/Animals Animals/Earth Scenes, (b)Jane Burton/DK Images; **64** (l)Papilio/Alamy, (c)John P. Marechal/Bruce Coleman USA, (r)John T. Fowler/Alamy; **65** (l)PhotoTake/Alamy, (r)DIOMEDIA/Alamy; **66** Courtesy of Nancy Simmons; **66–67** (bkgd)Sundell Larsen/Photodisc/Getty Images; **67** (t)Jane Burton/Bruce Coleman USA, (b)Wildlife Pictures/Peter Arnold, Inc.; **68–69** Chris Mattison/NHPA; **69** (b)Ken Karp/MMH, (others) MMH; **70** (l)Comstock/PunchStock, (tr)Michael Dick/Animals Animals/Earth Scenes, (b)Dynamic Graphics Group/IT Stock Free / Alamy, (br)Juergen & Christine Sohns/Animals Animals/Earth Scenes; **71** (tl)Richard Moran/AlaskaStock.com, (tc)David Fritts/Animals Animals/Earth Scenes, (tr)Steve Maslowski/Visuals Unlimited, (b)David A. Northcott/CORBIS; **72** (tl)Dietmar Nill/Foto Natura/Minden Pictures, (tr)Arthur Morris/CORBIS, (b)Georgette Douwma/Getty Images; **73** (tl)Hemera/age fotostock, (tr)Creatas/PunchStock, (b)PhotoDisc/Getty Images; **74** (l)David Shale/naturepl.com, (r)Steve Bloom Images/Alamy; **75** Jane Burton/DK Images; **76–77** Richard Du Toit/naturepl.com; **77** CORBIS; **78** (l)Anup Shah/Getty Images; **78–79** (bkgd)John Conrad/CORBIS; **80** (b)David A. Northcott/CORBIS; **81** (tl)Arthur Morris/CORBIS, (tr)CORBIS, (cr)Photodisc/Getty Images, (bl)John P. Marechal/Bruce Coleman USA, (bcl)DIOMEDIA/Alamy, (br)John T. Fowler/Alamy; **81A** (tl)CORBIS, (tr)David Shale/naturepl.com, (bl)Eric Crichton/DK Images, (br)Dana Hursey/Masterfile; **82** (t)John and Karen Hollingsworth/U.S. Fish and Wildlife Service, (bl)Chuck Place/Alamy, (br)John Daniels/Ardea.com; **83** Ted Mead/Getty Images; **86–87** Victoria McCormick/Animals Animals/Earth Scenes; **87** (t)Annie Griffiths Belt/National Geographic Image Collection, (tc)Brand X/CORBIS, (b)Ken Lucas/Visuals Unlimited, (bc)Thinkstock/Jupiterimages; **88–89** Georgette Douwma/Getty Images; **89** (b)Stephen Krasemen/NHPA, (others) MMH; **90** (t)Annie Griffiths Belt/National Geographic Image Collection, (b)Michael Quinton/Minden Pictures; **90–91** Klein/Peter Arnold, Inc.; **91** (r)Karl Switakx/NHPA; **93** DK Limited/CORBIS; **94–95** Brand X/CORBIS; **95** (bl)Gary Meszaros/Visuals Unlimited, (bc)Guenter Rossenbach/zefa/CORBIS, (br)Arthur Morris/CORBIS, (others) MMH; **99** Tom Vezo/Minden Pictures; **101** (tl)George Grall/Getty Images, (tr)Creatas/Punchstock, (cl b)Frank Greenaway/DK Images, (others)Dynamic Graphics Group/IT Stock Free/Alamy; **102–103** Bob Krist/CORBIS; **103** (br)Ken Karp/MMH, (others)MMH; **104** (l)Thinkstock/Jupiterimages; **104–105** BRUCE COLEMAN INC./Alamy; **105** (inset)Catherine Ledner/ Getty Images, Jim Zuckerman/CORBIS; **106** (t)Schafer & Hill/Getty Images; **106–107** Arthur Morris/Visuals Unlimited; **107** Fred Bavendam/Minden Pictures; **108** (inset)Ken Lucas/Visuals Unlimited, James L. Amos/National Geographic Image Collection; **109** Jonathan Blair/CORBIS; **110** Dennis Finnin/American Museum of Natural History; **110–111** (bkgd)CORBIS; **111** (t b)Courtesy American Museum of Natural History; **112–113** C Squared Studios/Getty Images; **113** (t)CORBIS, (b)Photodisc/Getty Images; **114** Jane Burton/DK Images; **114–115** C Squared Studios/Getty Images; **115** (l)Jane Burton/DK Images, (r)PhotoDisc/Getty Images; **116** Tom Vezo/Minden Pictures; **117** (l)CORBIS, (r)ThinkStock/Jupiterimages; **118–119** David Pike/naturepl.com; **119** (t)Richard Hamilton Smith/CORBIS, (tc)Steven Kazlowski/Peter Arnold, Inc., (b)Tim Krieger/Alamy, (bc)Konrad Wothe/Minden Pictures; **120–121** Perry Mastrovito/Jupiterimages; **121** (l)Ken Karp/MMH, (others)MMH; **122** (inset)Markus Botzek/zefa/CORBIS; **122–123** (b)Mickey Gibson/Animals Animals/Earth Scenes; **123** (cl)NPS Photo by Jim Peaco, (t to b)Gordon & Cathy Illg/Animals Animals/Earth Scenes, (2)Myron Jay Dorf/CORBIS, (3)Getty Images, (4)Punchstock, (5)Gary W. Carter/CORBIS, (6)Alan & Sandy Carey/zefa/CORBIS; **124** (t)Dynamic Graphics Group/IT Stock Free/Alamy, (c)Theo Allofs/CORBIS, (b)Lothar Lenz/zefa/CORBIS; **124–125** Richard Hamilton Smith/CORBIS; **125** (t)Eric Crichton/DK Images, (b)Jim Battles/Dembinsky Photo Associates; **126** (tl)Claus Meyer/Minden Pictures, (tr)Courtesy of Rebecka Eriksson/American Museum of Natural History, (b)Pete Oxford/naturepl.com; **126–127** Ron Giling/Peter Arnold, Inc.; **127** Jordi Bas Casas/NHPA; **128–129** Mick Roessler/Index Stock Imagery; **129** (l)Ken Karp/MMH, (others)MMH; **131** (tl)Steve Cooper/Photo Researchers, Inc., (tr)Jim Merli/Visuals Unlimited, (bl)Joe McDonald/Animals Animals/Earth Scenes, (br)Karen McGougan/Bruce Coleman USA; **132** (t)Steven Kazlowski/Peter Arnold, Inc., (b)Phil A. Dotson/Photo Researchers, Inc.; **132–133** Yva Momatiuk/John Eastcott/Minden Pictures;

**133** (r)Tui De Roy/Minden Pictures; **134–135** Konrad Wothe/Minden Pictures; **135** (l)Ken Karp/MMH, (others)MMH; **137** (t)Fred Bavendam/Minden Pictures, (tr)Dave King/DK Images, (c)Kelvin Aitken/Peter Arnold, Inc., (bl)Millard H. Sharp/Photo Researchers, Inc., (br)Eric Haucke/Greg Ochocki Productions/Photo Researchers, Inc.; **138** (l)Suzanne L. and Joseph T. Collins/Photo Researchers, Inc., (t)Winfried Wisniewski/Foto Natura/Minden Pictures, (r)Edward Kinsman/Photo Researchers, Inc., (b)Alan and Sandy Carey/Photo Researchers, Inc.; **138–139** Tim Krieger/Alamy; **139** (t)Patrice Ceisel/Visuals Unlimited, (bl)Photodisc/Getty Images; **140** Ariel Skelley/CORBIS; **141** (l)E. R. Degginger/Photo Researchers, Inc., (t)Michael Redmer/Visuals Unlimited, (r)Geoff Brightling/DK Images, (b)Stockdisc/PunchStock; **142** (inset)G.K. & Vikki Hart/Getty Images, (bkgd)Matthia Breiter/Minden Pictures; **143** (inset)Patrick Endres/AlaskaStock.com, (bkgd)Creatas/PunchStock; **144** (inset)Photodisc/Getty Images, (bkgd)altrendo nature/Getty Images; **145** (t)David Muench/CORBIS, (cr)Rene Krekels/Foto Natura/Minden Pictures, (b)Photonica/Getty Images, (inset)Stockdisc/PunchStock; **146** (t)Mick Roessler/Index Stock Imagery, (cl)Steven Kazlowski/Peter Arnold, Inc., (cr)Mickey Gibson/Animals Animals/Earth Scenes, (bl)Konrad Wothe/Minden Pictures, (br)Karen Kasmauski/CORBIS; **147** Tim Krieger/Alamy; **148** (t)Douglas Faulkner/Photo Researchers, Inc., (bl)Angela Rowlings/AP Images, (br)Casey & Astrid Witte Mahaney/Lonely Planet Images; **149** Nicholas Pitt/Alamy; **150** (bl)E. R. Degginger/Photo Researchers, Inc.; **150–151** (bkgd)Clive Streeter/DK Images; **151** (t)Ingram Publishing/Alamy, (bl)Geoff Dann/DK Images, (bl)Valerie Giles/Photo Researchers, Inc.; **152–153** Ian Grant/Alamy; **153** (t)Tibor Bognar/CORBIS, (tc)ABC Basin Ajansi/CORBIS, (b)Carlo Allegri/Getty Images, (bc)Weatherstock/Peter Arnold, Inc.; **154–155** Francesc Muntada/CORBIS; **155** (t)Darrell Gulin/Photodisc/Getty Images, (tcl)Rainer Hackenberg/zefa/CORBIS, (tcr)Bananastock/PictureQuest, (bl)Tibor Bognar/CORBIS, (bc)Eastcott Momatiuk/Getty Images, (br)Bruce Coleman, Inc./Alamy; **156** Tibor Bognar/CORBIS; **157** (t)Masterfile, (b)Mark A. Johnson/CORBIS; **158** Krys Bailey/Marmotta PhotoArt; **161** Ken Karp/MMH; **162–163** Wayne Levin/Getty Images; **164** (t)Frank Lukasseck/zefa/CORBIS, (b)Mitsuaki Iwago/Minden Pictures; **165** Oscar Garcia Bayeri/age fotostock; **166** Studio Photogram/Alamy; **166–167** (b)A.N.T. Photo Library/NHPA; **167** Steve Dunwell/Getty Images; **168** (tl)Frank Greenaway/DK Images, (tr)Owaki/Kulla/CORBIS, (b)S. Michael Bisceglie/Animals Animals/Earth Scenes; **169** (t)PhotoDisc/Getty Images, (c)James Steinberg/Photo Researchers, Inc., (b)Jules Frazier/Getty Images; **170–171** JupiterMedia/Alamy; **171** (b)Michael Scott/MMH, (others) MMH; **172** (t)Ken Samuelsen/Photodisc/Getty Images, (b)Tony Wilson–Bligh/Papilio/CORBIS; **173** (l)Richard Hamilton Smith/CORBIS, (c)Joseph Van Os/Getty Images, (r)Susan Rayfield/Photo Researchers, Inc.; **174** ABC Basin Ajansi/CORBIS; **174–175** Weatherstock/Peter Arnold, Inc.; **175** ABC Basin Ajansi/CORBIS; **176** John Elk III/Lonely Planet Images; **176–177** (t)C Squared Studios/Getty Images, (b)David Greedy/Lonely Planet Images; **177** (l)Michael S. Yamashita/CORBIS, (r)Michael S. Yamashita/CORBIS; **178** Studio Photogram/Alamy; **178–179** Galen Rowell/CORBIS; **179** SW Productions/Getty Images; **180–181** Jeremy Woodhouse/Masterfile; **181** Carlo Allegri/Getty Images; **183** (t)Ken Karp/MMH, (b)Tibor Bognar/CORBIS; **184–185** Digital Vision/Getty Images; **185** (t)Profimedia.CZ s.r.o/Alamy, (tc)Arco Images/Alamy, (b)Jules Frazier/Photodisc/Getty Images, (bc)Clive Streeter/DK Images; **186–187** John A. Karachewski/MMH; **187** (b)Michael Scott/MMH, (others)MMH; **188** (t)Ken Karp/MMH; **188–189** (b)Profimedia.CZ s.r.o./Alamy; **189** (t)Jules Frazier/Photodisc/Getty Images, (b)Glow Images/Alamy; **190** (t bcl)Colin Keates/Courtesy of the Natural History Museum, London/DK Images, (tcl)Visuals Unlimited/Getty Images, (tcr)PhotoLink/Getty Images, (cr)Jacques Cornell/MMH, (bl)Arco Images/Alamy, (bcr)CORBIS, (br)Demetrio Carrasco/Courtesy of the Hubbell Trading Post National Historic Site, Arizona/DK Images, (others)Ken Cavanagh/MMH; **190–191** GC Minerals/Alamy; **191** Colin Keates/DK Images; **192** Isaac Newman/Alamy; **193** (l to r)Brand x Pictures/Punchstock, (2)Ken Karp/MMH, (b)CORBIS; **194** (inset)Ingram Publishing/Alamy; **194 & 195** (inset)Phillipe Dumar/Alamy; **194–195** (bc)Per Karlsson, BKWine.com/Alamy, Holt Studios International Ltd/Alamy; **195** (l)Ken Karp/MMH, (bl bc)Susan E. Degginger/Alamy, (inset)G.K. & Vikki Hart/Getty Images, (others)MMH; **196** (l)Franck Jeannin/Alamy, (t)Ken Karp/MMH, (b)Image Source/Punchstock; **196–197** CORBIS; **197** (b)Image Source/PunchStock, (bkgd)Michael P. Gadomski/Photo Researchers, Inc., (others)Ken Karp/MMH; **198** (b)Phillipe Dumar/Alamy; **198–199** (t)CORBIS; **199** (l)David Muench/CORBIS, (tr)PhotoAlto/PunchStock, (br)Photodisc/Getty Images, (inset)Digital Vision/PunchStock; **200–201** Robert McGouey/Alamy; **201** Ken Karp/MMH; **202** (l)Klaus Hackenberg/zefa/CORBIS, (r)Jonathan Nourok/PhotoEdit, (b)Terry W. Eggers/CORBIS; **203** (tl)Ryan McVay/Photodisc/Getty Images, (tr)Siede Preis/Getty Images, (bl)CORBIS, (bc)Andy Crawford/DK Images, (br)C Squared Studios/Getty Images; **204–205** (t)Ken Karp/MMH, (b)Wilmar Photography.com/Alamy; **205** (t)David Young–Wolff/PhotoEdit, (c)Jacques Cornell/MMH; **206** (cl)Tom Schierlitz/Getty Images, (others)Ken Karp/MMH; **207** (l)Digital Vision; **208–209** (b)Julie Houck/CORBIS; **209** (tl)Worldwide Picture Library/Alamy, (tr)Jeremy Horner/CORBIS; **210** (inset)Victoria Snowber/Getty Images; **210–211** (t)John M. Roberts/CORBIS, (b)Ken Cavanagh/MMH; **212** Stockdisc/PunchStock; **212–213** Jeremy Woodhouse/Photodisc/Getty Images; **213** Ken Karp/MMH; **214** (br)Comstock Images; **215** (tr)Arco Images/Alamy, (bl)CORBIS, (bc)Michael P. Gadomski/Photo Researchers, Inc., (br)Franck Jeannin/Alamy, (others)Colin Keates/Courtesy of the Natural History Museum, London/DK Images; **216** (t)Nicholas DeVore/Bruce Coleman USA, (bl)Philippe Psaila/Photo Researchers, Inc., (br)David R. Frazier/Photo Researchers, Inc.; **217** Galen Rowell/CORBIS; **220–221** CORBIS; **221** (t)MMH, (tc)Steve Satushek, (b)Warren Faidley/Weatherstock, (bc)Robert Glusic/Getty Images; **222–223** Steve Satushek; **223** (c)MMH, (b)Stockbyte/Getty Images; **224** (cl)Stockbyte/Getty Images, (r)Stuart Pearce/Pixtal/age fotostock, (others)MMH; **225** (tr)Layne Kennedy/CORBIS, (bl)Steve Satushek/Getty Images, (br)Tony Freeman/PhotoEdit; **226** (l)Andrew Holt/Alamy, (r)Steven Puetzer/Masterfile; **227** (t)Markus Dlouhy/Peter Arnold, Inc., (b)Tony Freeman/PhotoEdit; **228** Rudi Von Briel/PhotoEdit; **229** (l)Dennis MacDonald/PhotoEdit, (r)MMH; **230–231** Frans Lemmens/Getty Images; **231** (b)Ken Karp/MMH, (others)MMH; **232–233** CORBIS; **233** CORBIS; **234** (cl)Digital Archive Japan/Alamy; **234–235** (inset)David Middleton/Photo Researchers, Inc., (bkgd)Alison Wright/CORBIS; **235** (r)Alan Kearney/Getty Images; **236–237** Richard Wong/Alamy; **238** Robert Glusic/Photodisc/Getty Images; **239** (t)Gerald & Buff Corsi/Visuals Unlimited, (b)Warren Faidley/Weatherstock; **240** (l)Stockbyte/PunchStock, (tcr)Courtesy of USDA Natural Resources Conservation Service, (tr)Tina Manley/Alamy, (bcr)CORBIS, (br)Steven Puetzer/Masterfile; **241** (t)Jim Reed/Getty Images, (b)Comstock/PunchStock; **242–243** (t)Peter Menzel/SPL/Photo Researchers, Inc., (b)Kent Wood/Photo Researchers, Inc.; **243** (r)David R. Frazier/Photo Researchers, Inc., (inset)NOAA/AP Images; **244** Gallo Images/CORBIS; **244–245** Macduff Everton/CORBIS; **246–247** Bertram G. Murray/Animals Animals/Earth Scenes; **247** imagebroker/Alamy; **248** (l)Mimotito/Digital Vision/Getty Images, (r)Andre Jenny/Alamy; **249** (inset)PhotoLink/Getty Images, Adrian Sherratt/Alamy; **249A** (t)MMH, (b)The McGraw–Hill Companies; **250–251** Brand X Pictures/PunchStock; **251** (t)Shigemi Numazawa/Atlas Photo Bank/Photo Researchers, Inc., (b)Brand X Pictures/Punchstock; **252–253** Walter Bibikow Index Stock Imagery; **253** (t)MMH, (b)Ken Karp/MMH; **257** Jon Hicks/CORBIS; **258–259** Andre Jenny/Alamy; **259** (5)Ken Karp/MMH, (others)MMH; **260** (l)Bill Frymire/Masterfile, (r)Richard Hutchings/PhotoEdit; **261** (l)Photodisc/Getty Images, (tr)Tony Anderson/Getty Images, (br)Stockdisc/PunchStock; **264** (l)CORBIS, (r)Shoot Pty. Ltd./Index Stock Imagery; **266–267** Bill Brooks/Masterfile; **267** (b)Ken Karp/MMH, (others)MMH; **269** Stockbyte/PunchStock; **270** Larry Landolfi/Photo Researchers, Inc.; **270–271** NASA/Photo Researchers, Inc.; **271** Larry Landolfi/Photo Researchers, Inc.; **272** Gerard Lodriguss/Photo Researchers, Inc.; **273** Picto's/Jupiterimages; **274** (1)World Perspectives/Getty Images, (2)Kevin Anthony Horgan/Getty Images, (3)Digital Vision/PunchStock, (4)StockTrek/Getty Images, (others)USGS/Photo Researchers, Inc.; **274–275** (bkgd)Otto Rogge/CORBIS, Digital Vision/PunchStock; **275** (bl)Space Telescope Science Institute/NASA/Photo Researchers, Inc., (bc)StockTrek/Getty Images, (br)Ken Karp/MMH, (others)MMH; **278** (tl)USGS/Photo Researchers, Inc., (tcl)World Perspectives/Getty Images, (tcr)Kevin Anthony Horgan/Getty Images, (tr)Digital Vision/PunchStock, (bkgd)Josh Westrich/zefa/CORBIS, StockTrek/Getty Images; **278–279** (bkgd)Josh Westrich/zefa/CORBIS, Digital Vision/PunchStock; **279** (l)Space Telescope Science Institute/NASA/Photo Researchers, Inc., (r)StockTrek/Getty Images; **280** The Granger Collection, NY; **280–281** Joshua Ets–Aiken/Getty Images; **281** Gerard Lodriguss/Photo Researchers, Inc.; **282** Robert Spoenlein/zefa/CORBIS; **283** Werner H.Mueller/zefa/CORBIS; **284** Stock Connection/Alamy; **285** Stock Connection/Alamy; **286** (l)Picto's/Jupiterimages, (r)Brand X Pictures/PunchStock; **287** (l)Gordon & Cathy Illg/Animals Animals/Earth Scenes, (b)Stockbyte/PunchStock; **287A** C Squared Studios/Getty Images; **288** (t)image100/age fotostock, (bl)Mauro Fermariello/Photo Researchers, Inc., (br)Syracuse Newspapers/David Lassman/The Image Works; **289** Matthias Kulka/CORBIS; **292–293** Andre Jenny/Alamy; **293** (tc)C Squared Studios/Getty Images, (bc)Michel Tcherevkoff/Getty Images, (others)Ken Karp/MMH; **294–295** Paul Chesley/Getty Images; **295** MMH; **296** Ariel Skelley/CORBIS; **297** Ken Karp/MMH; **298** (l)Kevin Schafer/zefa/CORBIS, (r)Ken Cavanagh/MMH; **298–299** Ken Karp/MMH; **299** (l)Ken Karp/MMH, (r)Digital Zoo/Getty Images; **300–301** Ken Karp/MMH; **301** (b)Ken Karp/MMH, (others)MMH; **302** (l)Colin Keates/Courtesy of the Natural History Museum, London/DK Images, (c)Ken Karp/MMH, (r)Nikreates/Alamy; **303** Ken Karp/MMH; **304** (tc)Brand X Pictures/PictureQuest, (tr)Photodisc/Getty Images, (c)Ken Karp/MMH, (b)MMH; **305** Ken Karp/MMH; **306** (c)Gary Braasch/CORBIS, (c)Ryan McVay/CORBIS, (others)C Squared Studios/Getty Images; **307** (t)V&A Images/Alamy, (c)Louie Psihoyos/Getty Images, (bkgd)CORBIS; **308–309** Ken Karp/MMH; **309** (b)Ken Karp/MMH, (others)MMH; **310** (l)Digital Vision/Punchstock, (tr)Michel Tcherevkoff/Getty Images, (br)Mark Viker/Getty Images; **311** The McGraw–Hill Companies; **312** (l)Buzz Pictures/Alamy, (c)Ken Karp/MMH, (r)Heka agence photo/Alamy; **313** (cl)Keate/Masterfile, (cr)Stock Food/SuperStock, (others)Ken Karp/MMH; **314** Sharon L. Jonz/Photis/PictureQuest; **315** Ken Karp/MMH; **316** (l)Steven May/Alamy, (inset)Ken Karp/MMH; **317** Ken Karp/MMH; **318** Jeff Greenberg/Alamy; **318–319** (br)Ken Karp/MMH; **319** Gerald Lopez/DK Images; **320** MMH; **321** (t)Ken Karp/MMH, (b)C Squared Studios/Getty Images; **321A** MMH; **322–323** David Chasey/Getty Images; **323** (tc)F. Schussler/Getty Images, (others)Ken Karp/MMH; **324–325** Ken Karp/MMH; **325** (b)Ken Karp/MMH, (others)MMH; **326** (tc)CORBIS, (others)Ken Karp/MMH; **327** (t)David M. Dennis/Animals Animals/Earth Scenes, (bl)imagebroker/Alamy, (br)Mike McClure/Index Stock Imagery; **328** (t)Steve Schott/DK Images, (tcl tcr)Bryan Mullennix/Getty Images, (bcl)Siede Preis/Getty Images, (bcr)Siede Preis/Photodisc/Getty Images, (others)Steve Gorton/DK Images; **329** (l)Photodisc/Getty Images, (r)F. Schussler/Getty Images; **330–331** Douglas Peebles Photography /Alamy; **331** (b)Ken Karp/MMH, (others)MMH; **332** (l)PhotoDisc/Getty Images, (r)Charles O'Rear/CORBIS; **332–333** John A. Rizzo/Getty Images; **334** (t)Nora Good/Masterfile, (b)Arno Gasteiger/Stock Central/age fotostock; **335** Ken Karp/MMH; **336** (tl)William Thomas Cain/Getty Images, (others)Photo provided by and reproduced with the permission of Crayola LLC; **336–337** (b)Stockdisc/Punchstock; **337** (tl)Steve Rasmussen/AP Images, (tr)William Thomas Cain/Getty Images, (b)Stockdisc/Punchstock, (br)Dave Mager/Index Stock Imagery; **338–339** Garry Gay/Alamy; **339** (b)Ken Karp/MMH, (others)MMH; **340–341** Ken Karp/MMH; **342** (tcl)CORBIS, (cl cr)Ken Karp/MMH, (bl)Stockbyte/PunchStock,(br)CORBIS, (others)Stockdisc/PunchStock; **343** (t)Ken Karp/MMH, (b)DIOMEDIA/Alamy; **344** (l)Ken Karp/MMH, (r)Charles D. Winters/Photo Researchers, Inc.; **346** (t)Ken Karp/MMH, (cl)Enigma/Alamy Images, (c)Greg Kuchik/Getty Images, (bl)Stockdisc/Punchstock, (br)CORBIS; **347** (tc)Robert Hicks/Alamy, (b)Ken Cavanagh/MMH, (others)MMH; **348** (t)Stockdisc/PunchStock, (b)chris fotoman smith/Alamy; **349** (t)Ken Karp/MMH, (b)D. Hurst/Alamy; **350** Brooke Slezak/Getty Images; **351** Deborah Jaffe/Photodisc/Getty Images; **352** Graham Charles/age fotostock; **353** (l)imagebroker/Alamy, (r)Dorling Kindersley/Getty Images; **353A** (tc)CORBIS, (bl)Photodisc/Getty Images, (br)F. Schussler/Getty Images, (others)Ken Karp/MMH; **354** (t)Ben Weaver/Getty Images, (bl)Jose Luis Pelaez, Inc./CORBIS, (br)Jeff Greenberg/PhotoEdit; **355** Lester Lefkowitz/Getty Images; **356** (t)Jeff Lepore/Photo Researchers, Inc., (b)Edward Kinsman/Photo Researchers, Inc.; **357** (t)Michael Durham/Minden Pictures, (b)Dr. Merlin D. Tuttle/Photo Researchers, Inc.; **358–359** Jean Brooks/Getty Images; **359** (t)Daniel Dempster/Bruce Coleman USA, (tc)Richard Hutchings/PhotoEdit, (others)Michael Scott/MMH; **360** (inset)Pat Doyle/CORBIS; **360–361** Adam Alberti/911 Pictures; **361** (t)Mark Raycroft/Minden Pictures; **362** (tl)CORBIS, (bl)Stockbyte/PictureQuest, (br)Pat Doyle/CORBIS, (others)C Squared Studios/Getty Images; **363** (tr)Tim Davis/Getty Images, (c)Daniel Dempster/Bruce Coleman USA, (bl)Karen & Ian Stewart/Bruce Coleman USA, (br)Ian & Karen Stewart/Bruce Coleman USA; **364** (t)G.K. & Vikki Hart/Getty Images, (b)Michael Scott/MMH; **364–365** Andy Rouse/Getty Images; **365** (t)Anup Shah/Photodisc/Getty Images, (c)Gerard Lacz/Animals Animals/Earth Scenes, (b)Tracey Rich/naturepl.com; **366–367** David Madison/Getty Images; **367** (b)Michael Scott/MMH, (others)MMH; **368** (tl)Thomas Barwick/Photodisc/Getty Images, (tr)Photodisc/Getty Images, (b)Creatas/PunchStock; **369** (t)Norbert Schaefer/CORBIS, (b)Rolf Bruderer/CORBIS; **370** (t)Photodisc/Getty Images, (c)Dan Burn–Forti /Getty Images; **370–371** (t)Photodisc/Getty Images, (b)Creatas/PunchStock; **371** (tl)Richard Hutchings/PhotoEdit, (tr)David Young–Wolff/PhotoEdit; **372** (l)Peter Arnold, Inc./Alamy, (r)Robert W. Ginn/PhotoEdit; **373** Scott T. Baxter/Getty Images; **374** Denis Finnin/American Museum of Natural History; **374–375** Roger Ressmeyer/CORBIS; **375** Jean–Charles Cuillandre/CFHT/Photo Researchers, Inc.; **376–377** Romy Ragan/Pixtal/age fotostock; **377** (b)Ken Karp/MMH, (others)MMH; **378** Ken Karp/MMH; **379** Steve Cole/Photodisc/Getty Images; **380** Kim Karpeles/Alamy; **381** (t)DK Images, (c)Image Source/PunchStock, (b)C Squared Studios/Getty Images; **382** Doug Allen/naturepl.com; **384–385** Ken Karp/MMH; **385** (l)Michael Scott/MMH, (others)MMH; **386** (br)Charles Bowman/Alamy, (others)Ken Karp/MMH; **387** (t)Comstock Images/Alamy, (bl)PhotoLink/Getty Images, (bc)Ken Cavanagh/MMH, (br)Ken Karp/MMH, (others)MMH; **388** (b)photolibrary.com pty. ltd./Index Stock Imagery, (others)Michael Scott/MMH; **389** (l)Ingram Publishing/Alamy, (r)Planetary Visions Ltd/Photo Researchers, Inc.; **394** (t)David Young–Wolff/PhotoEdit, (b)Ken Karp/MMH; **395** (t)Ken Karp/MMH, (b)First Light/CORBIS; **396–397** Kelly–Mooney Photography/CORBIS; **397** (t)Jim West/Alamy, (b)Comstock/Punchstock; **398–399** Designpics/Inmagine Corporation LLC; **399** (b)Ken Karp/MMH, (others)MMH; **400–401** Darryl Leniuk/Masterfile; **401** (t)Jim West/Alamy, (c)David Young–Wolff/PhotoEdit, (b)CORBIS; **402** (l)Marvin E. Newman/Getty Images, (cr)Chad Ehlers/Alamy, (others)Ken Karp/MMH; **403** (t)Ken Karp/MMH, (b)Stock Connection Distribution/Alamy; **404–405** Dennis MacDonald/PhotoEdit; **405** (tl b)Michael Scott/MMH, (others)MMH; **406** (t)Image Source/Alamy, (b)Comstock/PunchStock; **407** (t)C Squared Studios/Getty Images, (b)Comstock/PunchStock; **408** (t)Yann Arthus–Bertrand/CORBIS, (b)PCL/Alamy; **409** C Squared Studios/Getty Images; **410** image100/PunchStock; **411** (t)Philip Rostron/Masterfile, (b)C Squared Studios/Getty Images; **412** (tl)Robert Slade/Alamy, (tr)Gary W. Carter/CORBIS, (bl)Robert W. Ginn/PhotoEdit, (br)Craig Orsini/Index Stock Imagery; **414–415** Sanford/Agliolo/CORBIS; **415** (b)Ken Karp/MMH, (others)MMH; **416 417** Ken Karp/MMH; **418** (t)ThinkStock LLC/Index Stock Imagery, (b)Lee Cohen/CORBIS; **419** (t)Chris Collins/CORBIS, (b)Comstock/PunchStock; **420–421** Larry Fisher/Masterfile; **421** (b)Ken Karp/MMH, (others)MMH; **424–425** Carson Ganci/PictureQuest; **425** Roger Ressmeyer/CORBIS; **427** Nicholas Veasey/Getty Images; **428** (t)Creatas Images/Punchstock, (bkgd)Masterfile; **429** Gavriel Jecan/CORBIS; **430** Image Source/Punchstock; **431** Ken Karp/MMH; **433** (l)ThinkStock LLC/Index Stock Imagery, (r)Ken Karp/MMH; **434** (t)Peter Yates/CORBIS, (bl)Tony Freeman/PhotoEdit, (br)Michael Rosenfeld/Getty Images; **R01** (tr)MMH, (bc)Jules Frazier/Getty Images; **R03** MMH; **R04** (t)MMH, (b)Ken Karp/MMH; **R05** MMH; **R06** (t)Ryan McVay/Getty Images, (b)Photodisc/Getty Images, (br)MMH; **R07** (br)Jules Frazier/Getty Images; **R08** ImageState/PunchStock; **R09 R10 R11** MMH; **R13** (t)David Loftus/Getty Images, (others)C Squared Studios/Getty Images; **R14** (l)Purestock/PunchStock, (r)Photodisc Collection/Getty Images; **R15** Stockdisc/PunchStock; **R16** (tr)MMH, (others)Richard Hutchings/PhotoEdit; **R17** (tl)Jack Hollingsworth/Getty Images, (tr)David Buffington/Getty Images, (bl)Photodisc/PunchStock, (br)D. Berry/PhotoLink/Getty Images; **R18** (t)Brand X Pictures/PunchStock, (t)PhotoLink/Getty Images, (c)Richard Chung/Reuters/CORBIS, (b)StockTrek/Getty Images, (bc)Photo by Master Sgt. Michael A. Kaplan/U.S. Air Force; **R19** (t to b)Juergen & Christine Sohns/Animals Animals/Earth Scenes, (1)Bruce Coleman, Inc./Alamy, (2)Tony Freeman/PhotoEdit, (3)Steven Kazlowski/Peter Arnold, Inc., (4)Ken Karp/MMH; **R20** (1)Photodisc/Getty Images, (2)F. Schussler/PhotoLink/Getty Images, (3)Gerald & Buff Corsi/Visuals Unlimited, (4)Nora Good/Masterfile, (others)David A. Northcott/CORBIS; **R21** (1)Digital Vision/PunchStock, (2)David Muench/CORBIS, (3)Mick Roessler/Index Stock Imagery, (4)Ken Karp/MMH, (5)Thinkstock/Jupiterimages, (others)Robert Glusic/Photodisc/Getty Images; **R22** (1)Schafer & Hill/Getty Images, (2)CORBIS, (3)Jonathan Blair/CORBIS, (4)Carlo Allegri/Getty Images, (others)ABC Basin Ajansi/CORBIS; **R23** (2)Photodisc/Getty Images, (3)Ken Lucas/Visuals Unlimited, (4)age fotostock/SuperStock, (5)Richard Hutchings/PhotoEdit, (others)Thomas Barwick/Photodisc/Getty Images; **R24** (t to b)Jim West/Alamy, (3)Photodisc/Getty Images, (4)Creatas/PunchStock, (5)Annie Griffiths Belt/National Geographic Image Collection, (6)Darryl Leniuk/Masterfile, (others)Ken Karp/MMH; **R25** (1)Tibor Bognar/CORBIS, (2)ABC Basin Ajansi/CORBIS, (3)John P. Marechal/Bruce Coleman USA, (4)Michael Scott/MMH, (5)DK Images, (6 7)Jane Burton/DK Images, (others)Burke/Triolo Productions/Brand X Pictures/Getty Images; **R26** (t to b)Sanford/Agliolo/CORBIS, (1)Michel Tcherevkoff/Getty Images, (2)Mark Viker/Getty Images, (3)franzfoto.com/Alamy, (4)Ken Karp/MMH, (5)Ariel Skelley/CORBIS; **R27** (t to b)Ken Karp/MMH, (2)Daniel Dempster/Bruce Coleman USA, (3)Profimedia. CZ s.r.o/Alamy, (4)Konrad Wothe/Minden Pictures, (5)CORBIS; **R28** (t to b)Shigemi Numazawa/Atlas Photo Bank/Photo Researchers, Inc., (3)C Squared Studios/Getty Images, (4)USGS/Photo Researchers, Inc., (5)Planetary Visions – Ltd./Photo Researchers, Inc., (others)Ken Karp/MMH; **R29** (t to b)Wilmar Photography.com/Alamy, (1)Tim Krieger/Alamy, (2)CORBIS, (3)C Squared Studios/Getty Images, (4)Pat Doyle/CORBIS, (5)Steve Satushek/Getty Images, (7)Tom Vezo/Minden Pictures, Brand X/CORBIS; **R30** (t to b)Ken Cavanagh/MMH, (1)John T. Fowler/Alamy, (2)Richard Hamilton Smith/CORBIS, (3)Steve Cole/Photodisc/Getty Images, (6)Matthias Kulka/CORBIS, (others)Ken Karp/MMH; **R31** (1)Michael Scott/MMH, (2)Wendell Metzen/Bruce Coleman USA, (4)Jules Frazier/Photodisc/Getty Images, (others)Ken Karp/MMH; **R32** (1)Tom Bean/CORBIS, (2)Steve Cole/Photodisc/Getty Images, (3)Clive Streeter/DK Images, (4)Geoff Dann/DK Images, (5)E. R. Degginger/Photo Researchers, Inc., (6)Valerie Giles/Photo Researchers, Inc., (7)Jonathan Nourok/PhotoEdit, (others)Papilio/Alamy; **R33** (t to b)Ken Karp/MMH, (1)Image Source/Alamy, (2)Andy Rouse/Getty Images, (3)Brand X Pictures/Punchstock, (5)Warren Faidley/Weatherstock; **R34** (t to b)MMH, (1)Brand X Pictures/Punchstock, (2)Comstock/Punchstock, (3)Weatherstock/Peter Arnold, Inc., (4)Ken Karp/MMH, (5)Mickey Gibson/Animals Animals/Earth Scenes, The McGraw–Hill Companies; **R35** (1)Stockdisc/Punchstock, (2)GK & Vikki Hart/Getty Images, (3)Ken Karp/MMH, (4)Walter H. Hodge/Peter Arnold, Inc., (5)Michael Scott/MMH, (6)MMH, (others)ITStock/Punchstock; **R36** (2)Photodisc/PunchStock, (3)Peter Anderson/DK Images, (4)Image Farm, Inc., (5)Robert Brenner/PhotoEdit, (6 8)Papilio/Alamy, (7)Holt Studios International Ltd./Alamy, (others)Ken Karp/MMH.

# Teacher Resources

## Teacher Resources



## Science Yellow Pages



## Scope & Sequence



## Index & Correlations

## Additional Teacher Resources

The **Teacher's Desk Reference** offers information tailored to today's science teacher.

It includes practical, hands-on support in several areas of *Professional Development,* including: Managing Inquiry Based Science and Scaffolding Inquiry Instruction, the 5 *E* Instructional Model, Differentiated Instruction, Gender Equity, Why Kindergarten Science Is Important, and Assessment in the Science Classroom.

This book also traces how science is taught across the grades. Beginning with First Grade, the *Science Across the Grades* section compares the key concepts of each chapter and lesson with the concepts in the corresponding chapters and lessons of the previous and subsequent grades. Teachers can, therefore, place the material they will be teaching in the greater context of the students' overall learning path from first to sixth grade.

Name ______________________ Date ______________________

Graphic Organizer 1

# Main Idea and Details

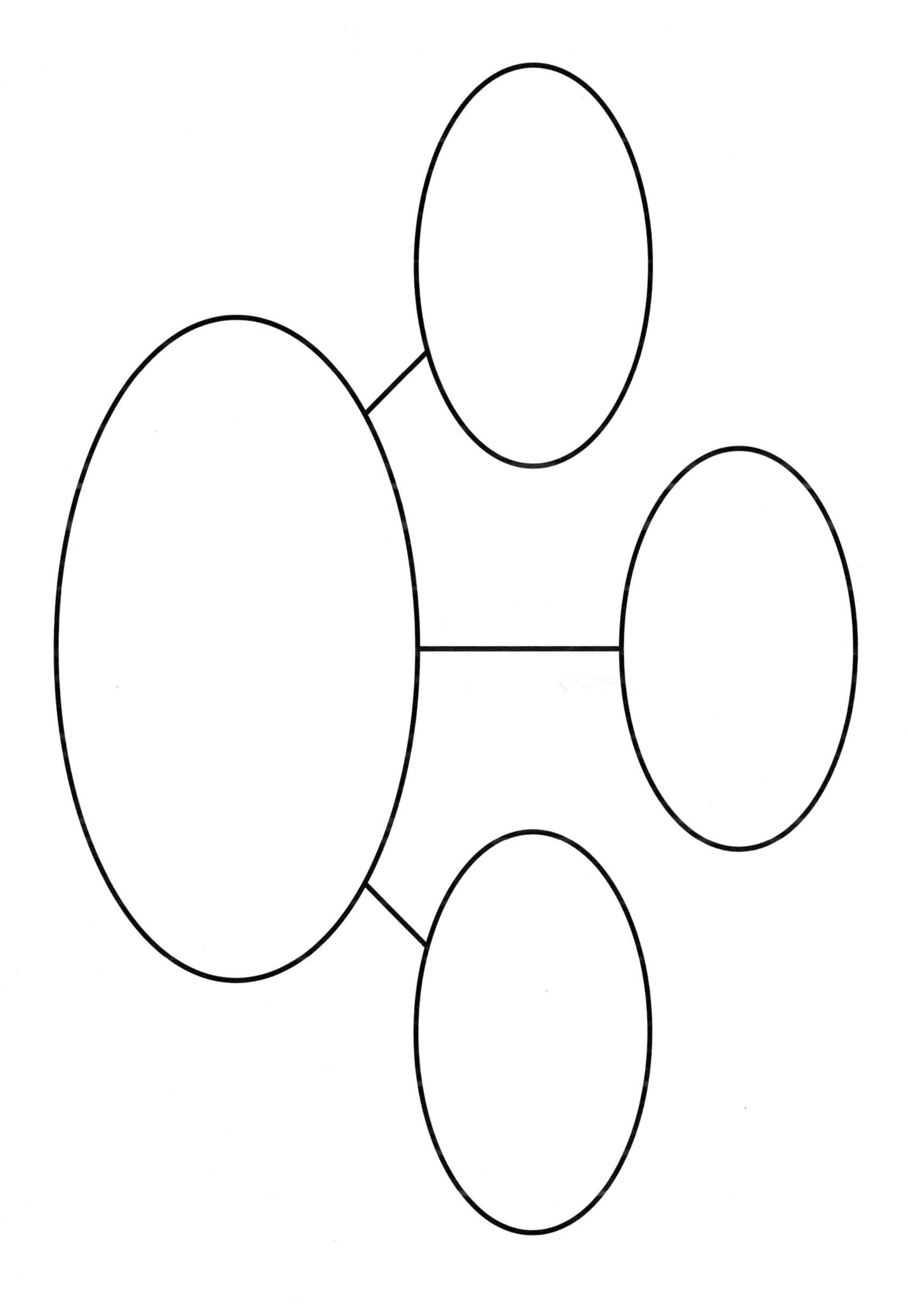

Name ______________________ Date ______________________

Graphic Organizer 2

## Main Idea and Details

| Details | | | | | | |
|---|---|---|---|---|---|---|
| Main Idea | | | | | | |

**Graphic Organizer 3**

## Predict

| What I Predict | What Happens |
| --- | --- |
| | |
| | |

Graphic Organizer 4

## Predict

| My Prediction | |
|---|---|
| What Happens | |

Name ______________________ Date ______________________

Graphic Organizer 5

## Summarize

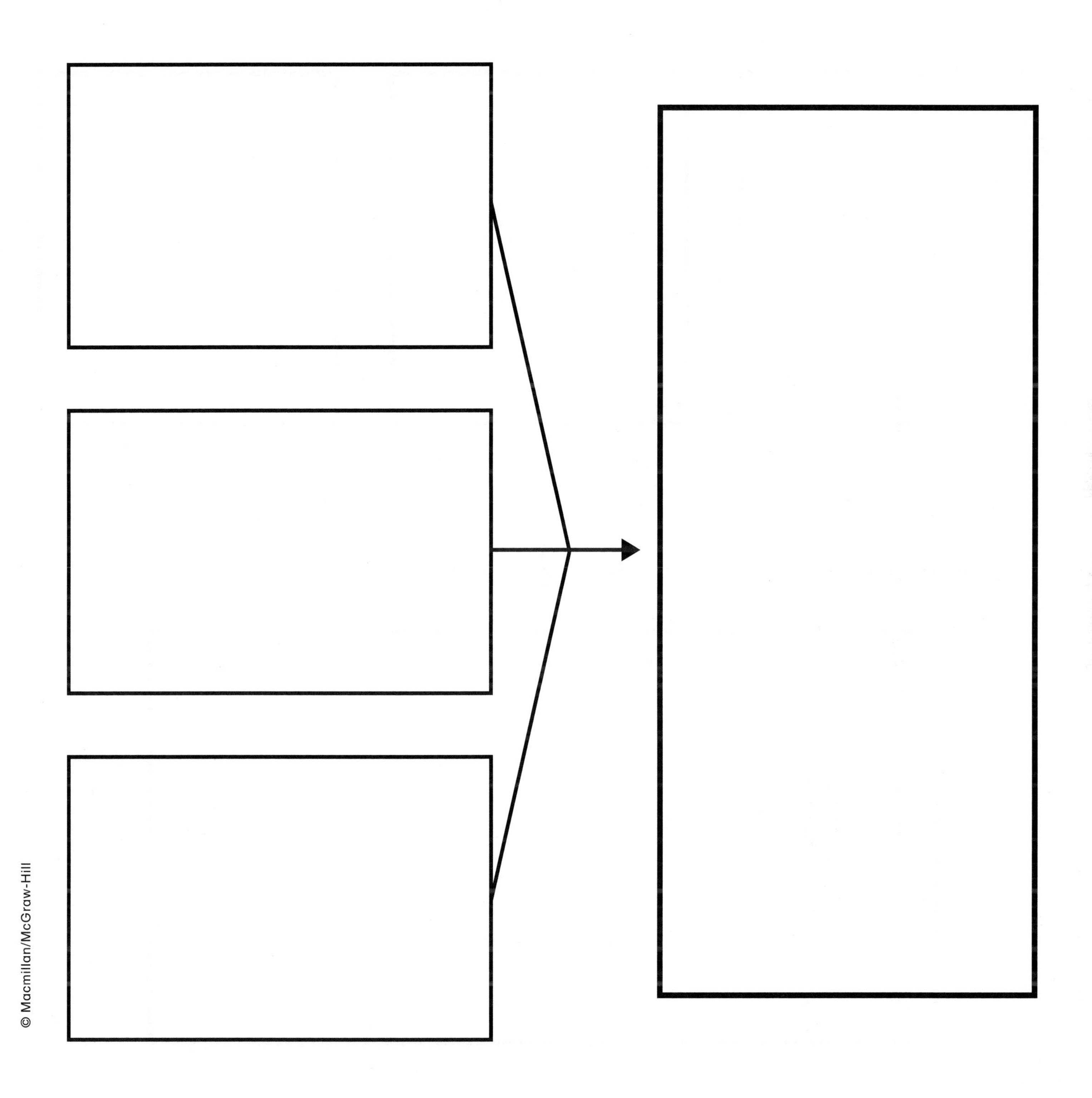

Name ______________________ Date ______________

## Graphic Organizer 6

# Summarize

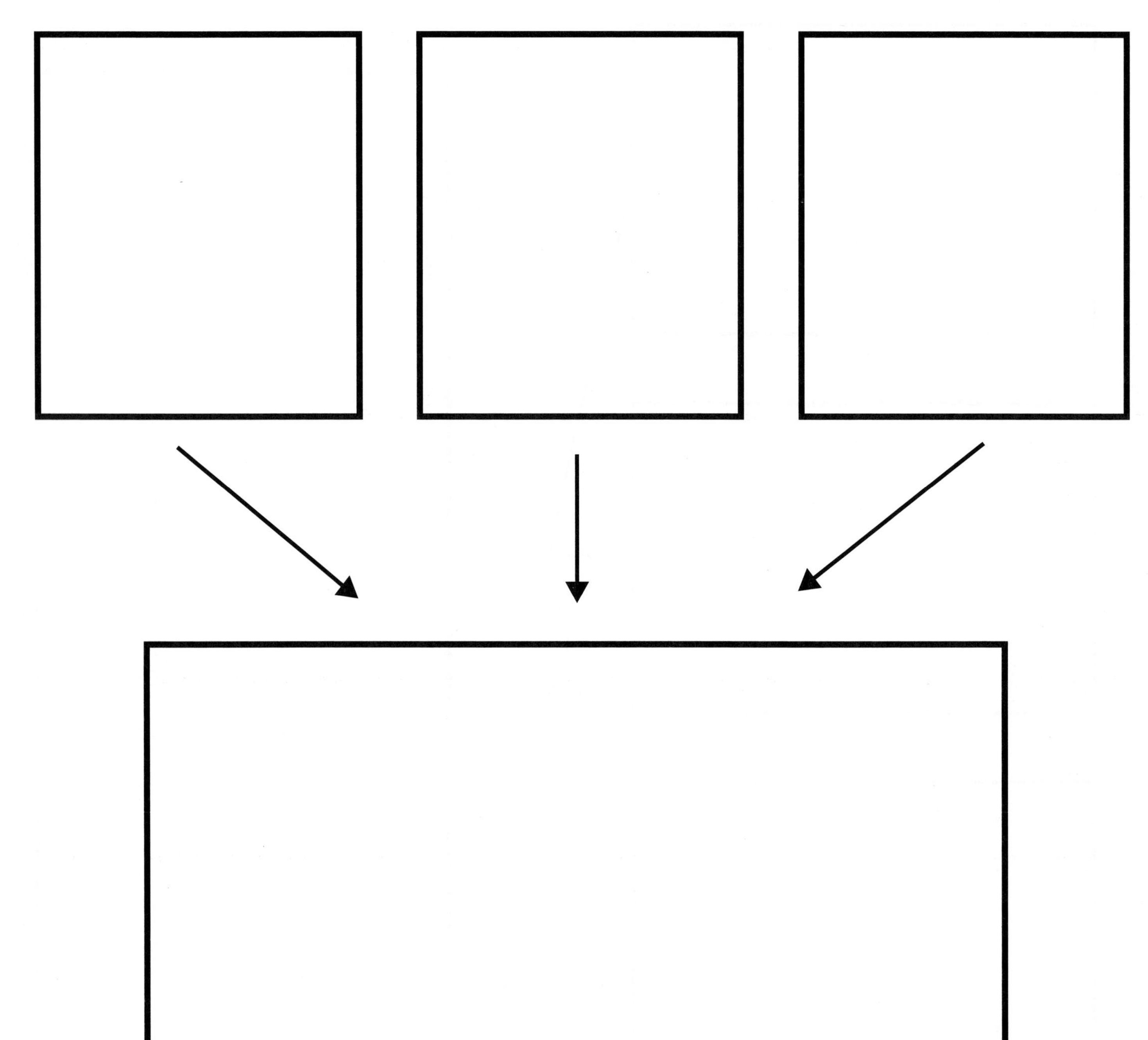

Name ______________________ Date ______________________

Graphic Organizer 7

## Sequence

First

Next

Last

Name ______________________ Date ______________________

Graphic Organizer 8

## Cause and Effect

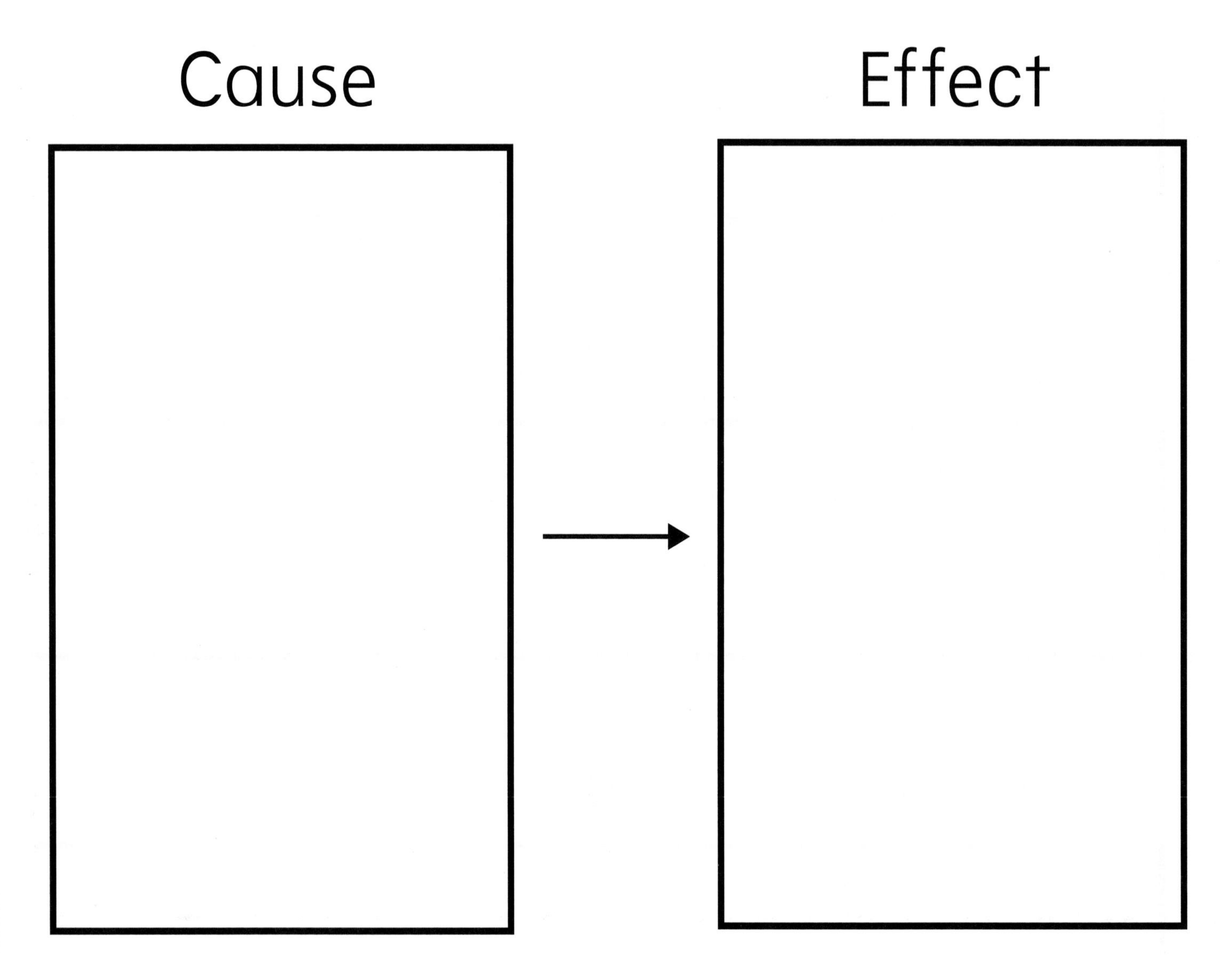

Name ______________________ Date ______________________

Graphic Organizer 9

## Cause and Effect

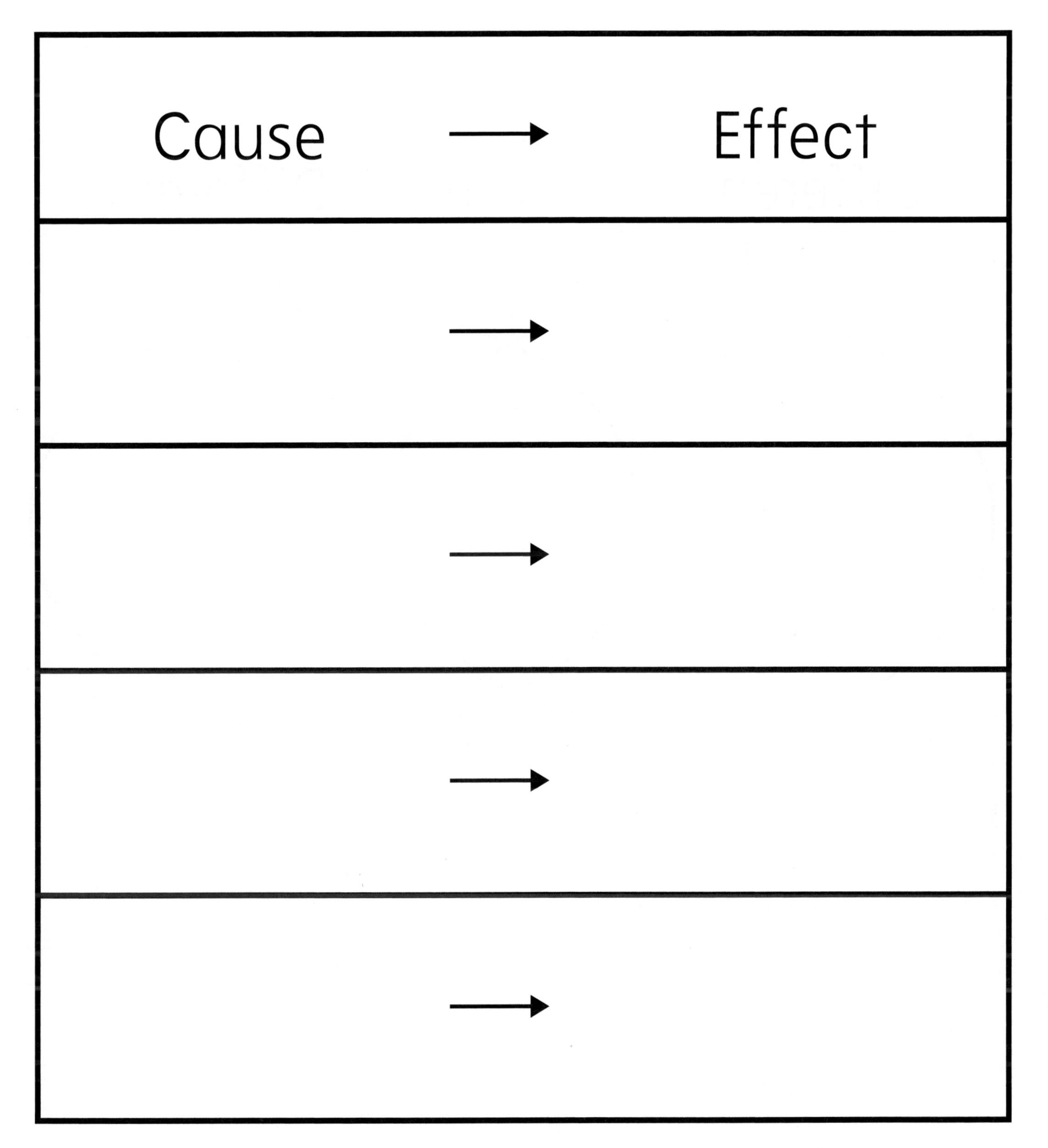

Name ______________________ Date ______________________

Graphic Organizer 10

## Compare and Contrast

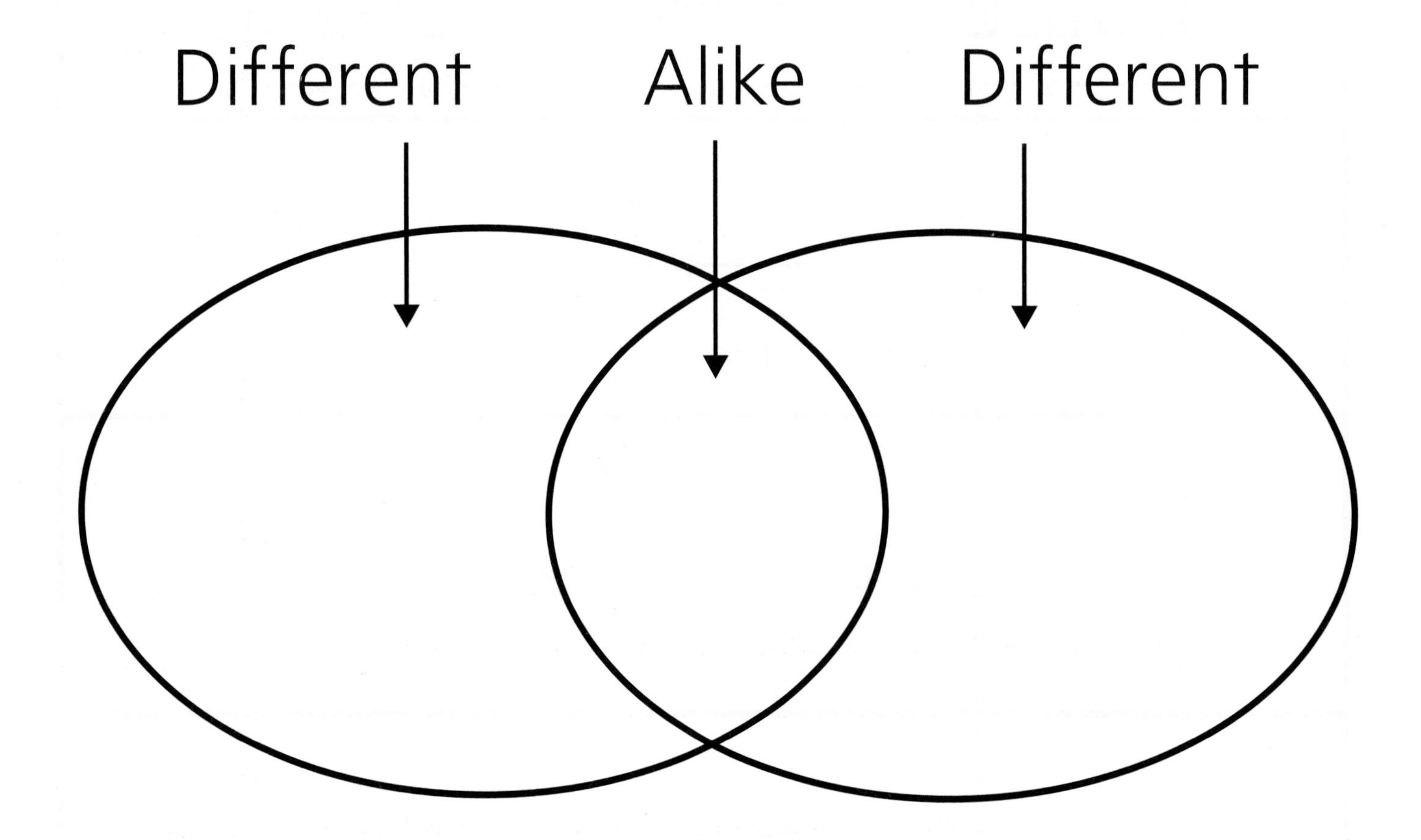

Name ______________________ Date ______________________

**Graphic Organizer 11**

## Classify

| | |
|---|---|
| | |
| | |

Name ______________________ Date ______________________

Graphic Organizer 12

## Problem and Solution

Problem

Steps to Solution

Solution

Name ______________________ Date ______________

**Graphic Organizer 13**

## Draw Conclusions

| Text Clues | Conclusions |
| --- | --- |
| | |
| | |

Name ______________________ Date ______________

Graphic Organizer 14

## Infer

| Clues | What I Know | What I Infer |
| --- | --- | --- |
| | | |
| | | |

Name ______________________ Date ______________________

Graphic Organizer 15

## Fact and Opinion

| Opinion | |
|---|---|
| Fact | |

Name ___________________________ Date ___________________

# United States Map

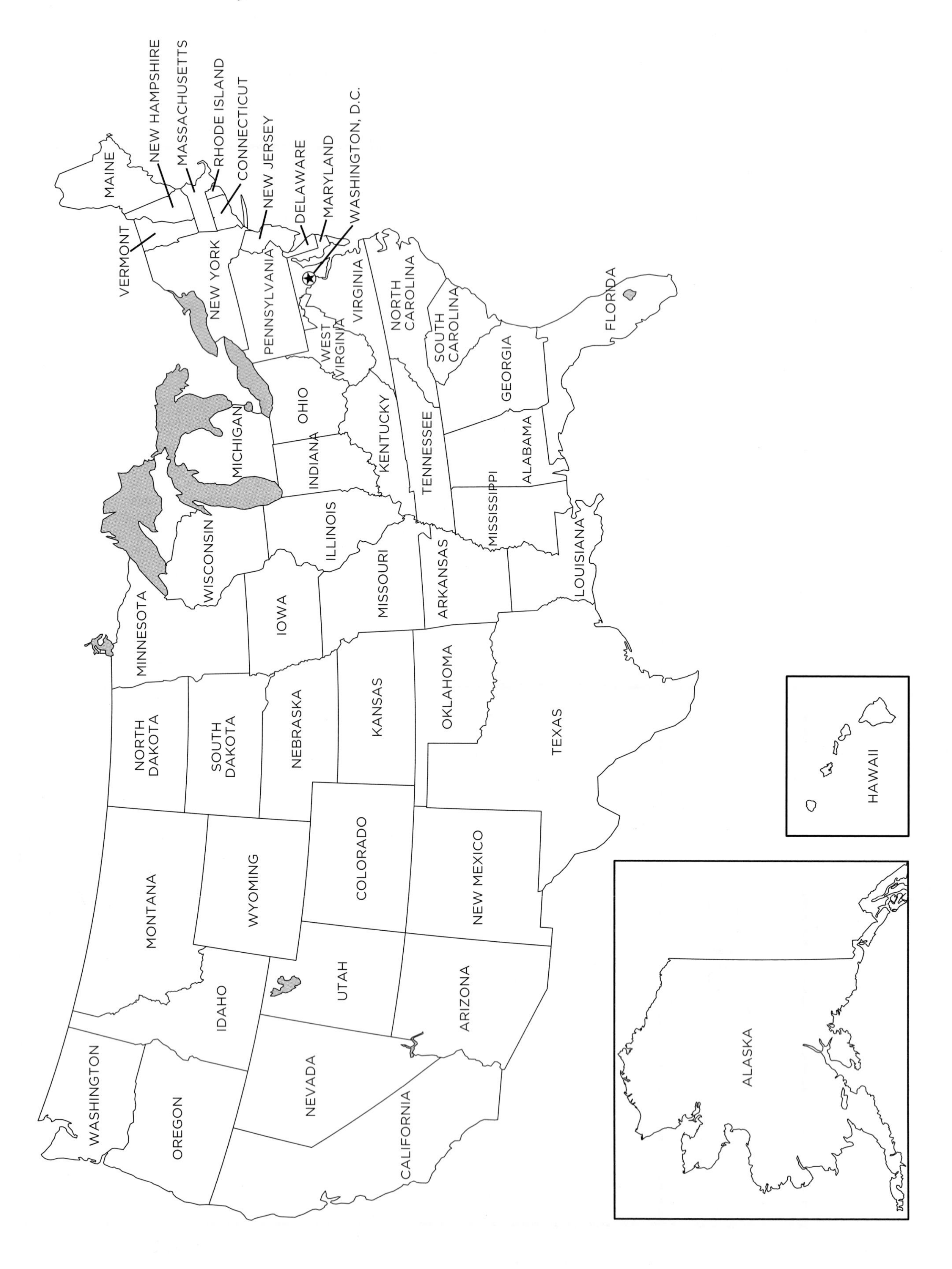

Name ______________________ Date ______________

## World Map

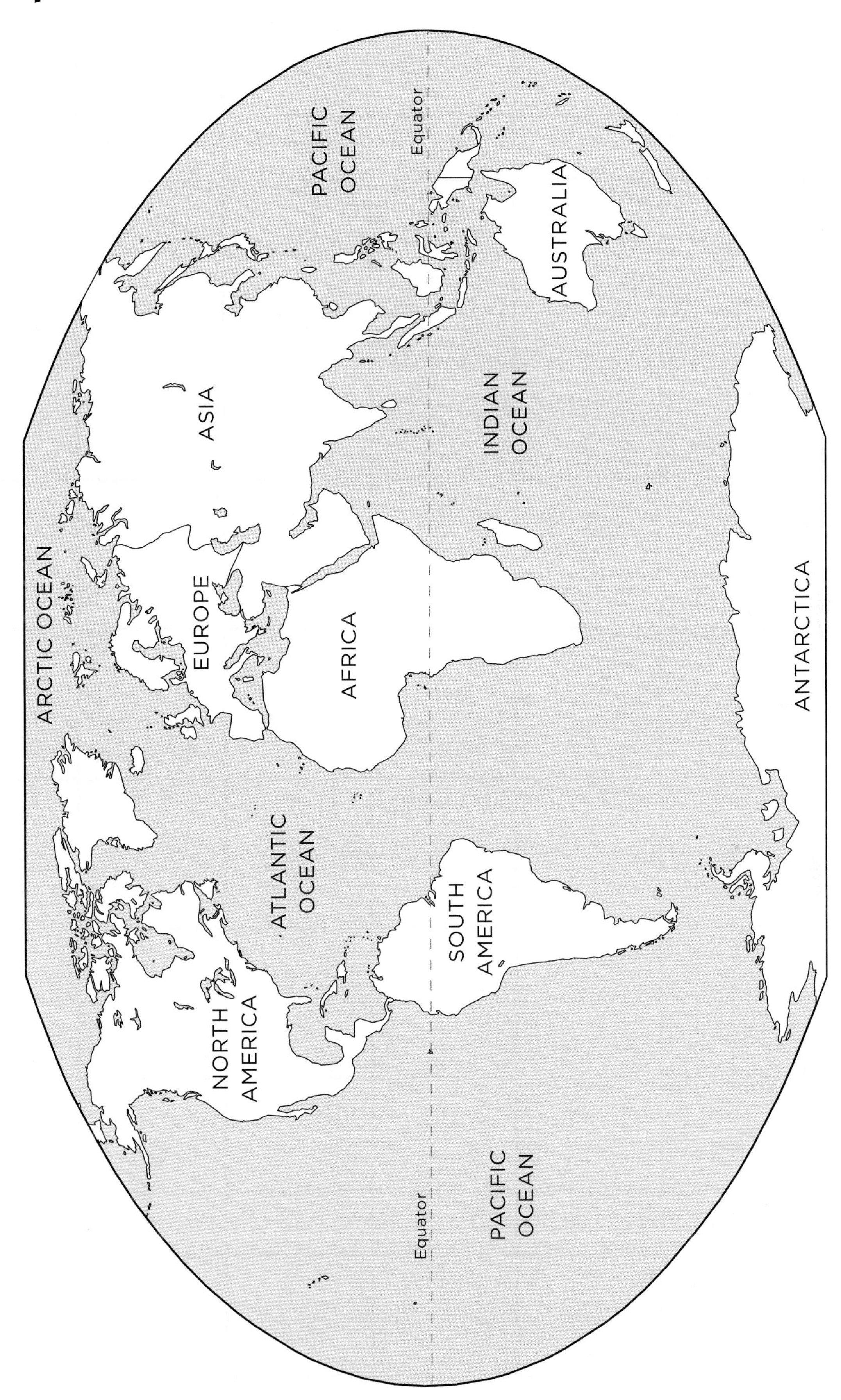

Name ______________________ Date ______________________

## Graph Paper

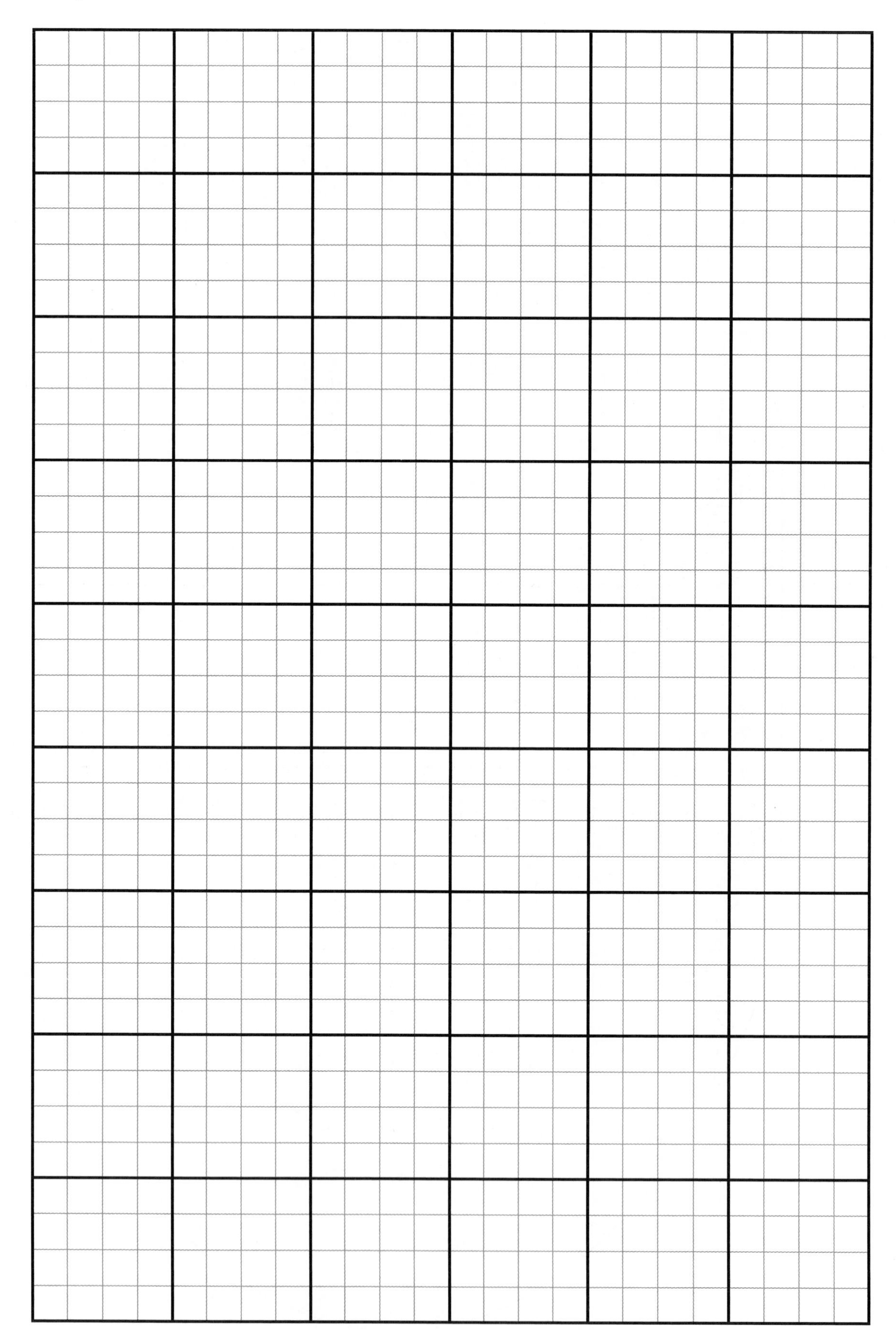

Name ______________________ Date ______________

## Calendar

| Sunday | Monday | Tuesday | Wednesday | Thursday | Friday | Saturday |
| --- | --- | --- | --- | --- | --- | --- |
| | | | | | | |
| | | | | | | |
| | | | | | | |
| | | | | | | |
| | | | | | | |

Name ______________________ Date ______________________

# Rulers

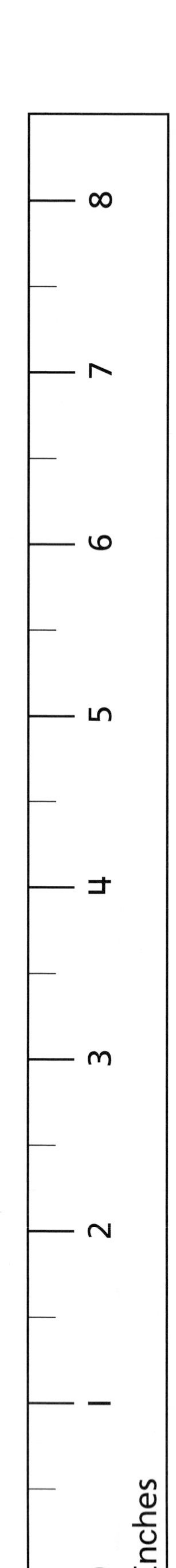


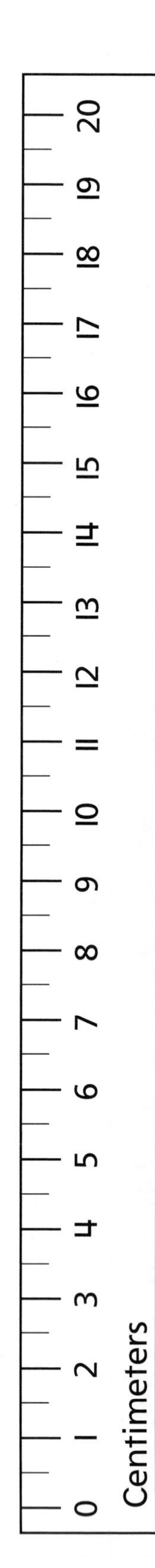

Name ____________________ Date ____________________

# Thermometer

| | | | | |
|---|---|---|---|---|
| Hot | Hot | Hot | Hot | Hot |
| Warm | Warm | Warm | Warm | Warm |
| Cool | Cool | Cool | Cool | Cool |
| Cold | Cold | Cold | Cold | Cold |

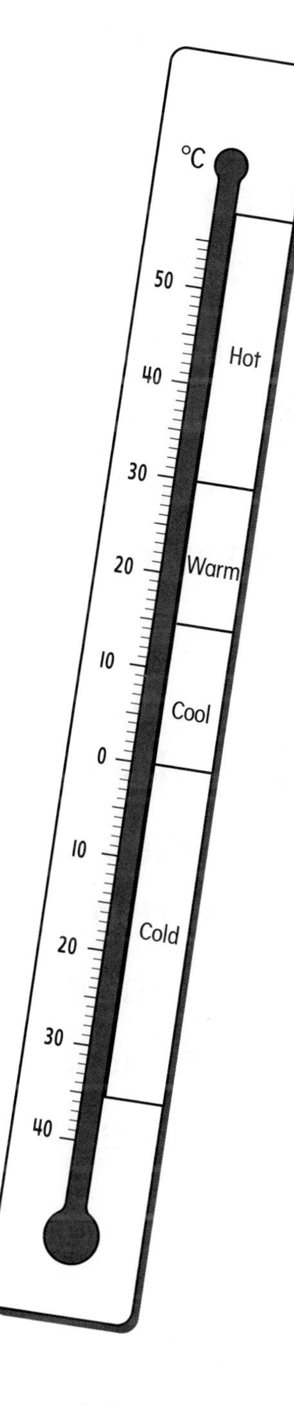

**To Teacher: Directions for making a primary thermometer:**

1. Cut out the five strips.
2. Color each *cold* section blue.
3. Color each *cool* section green.
4. Color each *warm* section yellow.
5. Color each *hot* section orange.
6. Align the strip on the thermometer as follows:
   Cold (below 0°C)
   Cool (1°C–15°C)
   Warm (15°C–30°C)
   Hot (above 30°C)
7. Once aligned, tape the entire strip to the *°F* side of the thermometer using clear tape. This will help to make it waterproof.

# The Fruit and the Flower



Do you know why plants grow flowers and fruits?
Did you know that's how they reproduce?
The fruits have seeds inside.
And the little seeds provide
Their own food to help them grow shoots and roots.
For protection little seeds have a coat.
On the wind, oceans, and rivers they can float.
Fruits once were flowers.
Fruits and seeds can be devoured
By a bird, monkey, rabbit, or a goat!

Plants need light, food, and water to grow tall;
Doesn't matter if they're big or small.
Creatures eat the fruit they find
Leaving lots of seeds behind
To grow in the soil wherever they may fall.
The pistil takes in pollen to make seeds
Found in fruit on different kinds of plants and trees.
From the stamen to the pistil
On each daffodil and thistle
Pollen's transferred by the birds and the bees.



Science Songs CD Track 1

# The Cycle of Life



Plants start out as tiny little seeds.
An oak makes small acorns that grow into big trees.
A seed sprouts and grows into a baby plant.
Seeds can be as big as a volleyball or smaller than an ant.

A life cycle shows how something lives and dies.
Watch a seedling grow up high into the sky.
Some plants live weeks and some live many years
From seed to plant to making seeds
Before they disappear.

Animals have babies that share their traits, it's true.
Baby plants look and act like their parents, too.
Sunflower seeds can be a yummy treat.
Plant those seeds and see a sunflower grow
Over seven feet!



Science Songs CD Track 2

# From Caterpillar to Butterfly

Unlike puppies who look like little dogs
Tadpoles don't look much like grown-up frogs.
Same for butterflies who go through stages,
Four different changes at different ages.

Baby butterflies hatch from eggs
As little caterpillars with lots of legs.
Also known as larva, they eat tons of leaves
To get all the nourishment that they need.
They're born on a leaf but it tastes so good
They eat it right up 'cause their home is food.

Three weeks later they must find a spot.
A leaf or a branch is the best they've got.
The caterpillar builds itself a good strong shell.
It's called a pupa now and protects itself well.
From each pupa comes a great surprise
When days later out comes a butterfly!



Science Songs CD Track 3

# What Animals Wear



Populations
Vary lots.
They are a group
Of creatures
All one kind
Though you may find
They may have
Different features.

A big black bear
Has fur not hair
That keeps her warm
In winter.
Depending on
Where she comes from
Her fur is dark or lighter.

Giraffes eat leaves
High in the trees.
Their necks help them
Reach taller.
Their fur has spots
That blend in lots
With all the local colors.

Snakes have scales
From heads to tails.
Be careful when you
See one
Because it might
Pack a poisonous bite
Like skins of certain
Frogs can.

Peacocks might
Have feathers bright
But only if they're
Male birds.
They wave their wings
And loudly sing
To make sure they're
Seen and heard.



Science Songs CD Track 4

# Rocks Are Amazing Things



Minerals are what make up rocks like shale or chalk.
Granite's made of minerals too 'cause it's a rock.
It has feldspar, quartz, and mica, too.
Minerals make rocks black, pink, or blue
And so many colors.
Rocks are amazing things.

Rocks are hard or soft depending on what kind.
Hardest is the diamond and it's hard to find.
Talc is soft and did you know
Pencil's lead's made of a mineral
And it's known as graphite.
Rocks are amazing things.

Geologists study rocks and all their properties.
A property tells you more about the thing you see.
Luster is how a rock shines in light.
It might shine like gold but it's just pyrite.
Sometimes rocks can fool you.
Rocks are amazing things!



Science Songs CD Track 5

# Down in the Dirt

Soil can be so many colors.
Look down where you tread.
Soil with lots of clay is pretty
'Cause iron makes the clay red.
Other soil is light and sandy.
Topsoil's black or brown.
Pebbles keep the soil from flooding
So little plants won't drown.
Minerals in rocks,
Minerals in rocks,
Minerals give soil its color.
Weather turns big rocks
Into little rocks
That make the soil their home.
Plants get their minerals from soil.
People eat plants to get minerals.
Gophers, worms, and ants
Like to dig their own tunnels
And they mix air and water
Deep into the topsoil.
Plants grow better
Thanks to their trouble and toil.



Science Songs CD Track 6

# Fossil Bones



Plants and animals long ago
Left behind their footprints so
We could find them,
Study them,
Learn about what they did.
Bones and teeth, prints and eggs,
Things with wings and long tall legs
Stuck in things like
Amber, rock,
Tar, ice, sand, and mud.

Amber comes from the sap of a tree.
When it's fresh it's so sticky.
Over time it gets
Fossilized —
Hard and golden brown.
Scientists learn about the Earth
From the present back to its birth.
They use fossils to
Get a clue
How things used to be.



Science Songs CD Track 7

# New Views, Old Clues



Fossils help scientists know
About things that lived so long ago.
They're clues that show scientists how
Things were different from how they are now.
A paleontologist
Is a fossil specialist.
He studies bones one by one,
And a full set is a skeleton.

Some cold places now are hot
And some warm places now are not.
There were creatures great and small.
Some were over 100 feet tall!
Extinction is when things die out,
Resulting from changes like drought
Or possibly a disease.
Imagine a dinosaur sneeze!
Imprints are shapes found in stones,
Left by things such as leaves, shells, and worms
Preserved in the rock, mud, and sand
Giving clues about how life began.



Science Songs CD Track 8

## Nature's Gifts



Things we use for living
Earth keeps giving.
They're called
Natural resources, it's true!
Water is a resource
Natural, of course;
So are the rocks
And plants and soil.
Wind's a great
Power source —
What a resource!
So are water and crude oil.
Water is for drinking,
Cooking, cleaning.
Plants and people need it
To grow tall.
Minerals in rocks
Are building blocks.
We use them to make
Jewelry, glass, and salt.

We use different plants
For food or pants
'Cause cotton is a plant
Used to make clothes.
Rubber from a tree
Amazingly
Makes rubber bands,
Erasers, or a rubber hose!
Paper's made of wood.
For desks it's good.
It also makes our pencils
As we know.
Pencil lead is graphite.
Classroom chalk's white.
Minerals in the classroom
Help us so!



Science Songs CD Track 9

# What Makes Us Go



Fuel is what we need to give us energy.
People get their fuel from all the food they eat.
Fruits and vegetables and sometimes meat
Give us the fuel we all need.

Lots of people love to eat their chicken fried.
Other folks like grilled beef with beans on the side.
Maybe you'll have fish and if you can't decide
It could be time to eat a pizza pie.

Plants aren't only used for yummy food, you know.
Wood is used for building everywhere you go.
When you come inside from playing in the snow
A wood fire will heat your home and warm your toes.

Clothing can be made from animals or plants.
Maybe you are wearing comfy cotton pants.
Shiny leather shoes can make it fun to dance.
Give Mother Nature's fashion a chance.

Oil and coal are fuels that heat up many homes.
They're made from plants and animals from long ago.
Earth's pressure squeezed them and don't you know
Millions of years later they make our cars go.



Science Songs CD Track 10

# You're Moving Now!

Science Songs

Objects move around.
They go up and down
Or back and forth,
To the south or the north.
Forces make you go
Either fast or slow.
No matter what the speed
As long as you proceed
You're moving now.

You could travel by train
Or fly a plane.
Take a ride in a car,
Drive near or far.
Some people run
'Cause it's more fun.
And if walking is too slow
A bike's the way to go.
You're moving now.

The cheetah moves fast.
See him zooming past,
While the turtle goes slow.
It's like watching
Grass grow.
When something moves,
Time and distance prove
Just how fast
An object goes.
From your fingers
To your toes
You're moving now.



Science Songs CD Track 11

# A Force Called Friction



Friction is a force that slows things down.
Treads on tires grip the ground.
Without friction, tires would go round and round,
Never stopping.
On a rough surface things move slow.
When it's smooth watch things flow.
That's why your sneakers are rough below
To keep you from slipping.
Watch how a ballet dancer glides.
See her feet, how they slide
'Cause slippers are smooth on the underside.
See how she's turning.
Changing direction uses force —
Speeding up, too, of course.
Now can you tell me, what's the source?
What gets things stopping? ... Friction!



Science Songs CD Track 12

# What Brings Me Down



Gravity is what pulls things down.
You feel it each time you fall.
Gravity keeps your feet on the ground
Though you can't see it at all.
Earth's gravity is a powerful force
The more it pulls, the more you weigh;
Unless you're in outer space, of course,
Where you might just float away.
We measure all objects' weights with a scale
With ounce, gram, pound, even ton.
You'll find Earth's gravitational pull
Affects each and every one.



Science Songs CD Track 13

# Sound Moves



Sound is energy that you hear.
Yell from far, whisper near,
But never shout in someone's ear.
Sound moves air.
When objects vibrate they make sound
Like the way jackhammers pound
Creating noise by vibrating the ground.
Sound moves air.
Volume measures from quiet to loud.
You'd better speak up in a crowd.
When teacher is talking no whispering's allowed.
Sound moves air.
Pitch is a sound quality
Measuring its frequency —
Low tuba to a soprano's high C.
Sound moves air.
Vibrations are fast when pitch is high
When cats meow or babies cry
Or the sound of an airplane racing by.
Sound moves air.
When your ear hears a sound and the pitch is low
Air vibrations are slow.
That's how you hear a sound, you know.
Sound moves air.



Science Songs CD Track 14

by Dinah Zike

## Folding Instructions

The following pages offer step-by-step instructions to make the Foldables study guides.

### Half-Book

1. Fold a sheet of paper ($8\frac{1}{2}$" x 11") in half.
2. This book can be folded vertically like a hot dog or ...
3. ... it can be folded horizontally like a hamburger.

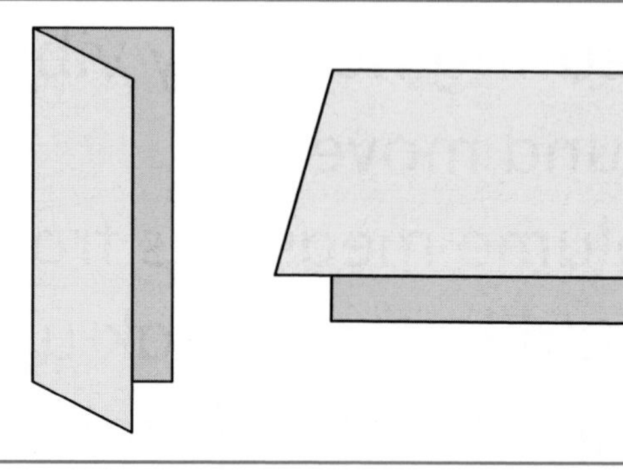

### Folded Book

1. Make a half-book.
2. Fold in half again like a hamburger. This makes a ready-made cover and two small pages inside for recording information.

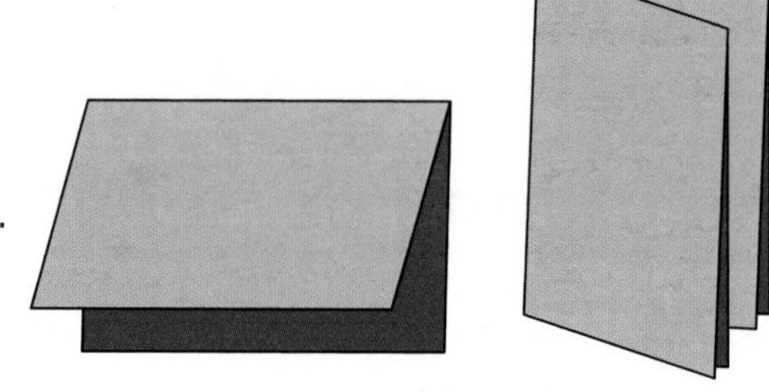

### Pocket Book

1. Fold a sheet of paper ($8\frac{1}{2}$" x 11") in half like a hamburger.
2. Open the folded paper and fold one of the long sides up two inches to form a pocket. Refold along the hamburger fold so that the newly formed pockets are on the inside.
3. Glue the outer edges of the two-inch fold with a small amount of glue.

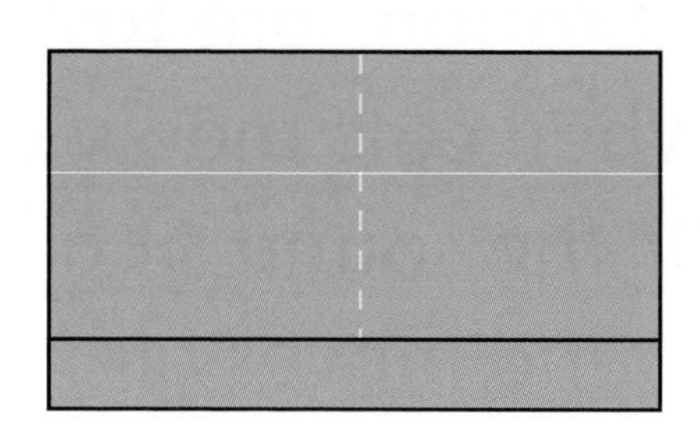

### Shutter Fold

1. Begin as if you were going to make a hamburger, but instead of creasing the paper, pinch it to show the midpoint.
2. Fold the outer edges of the paper to meet at the pinch, or midpoint, forming a shutter fold.

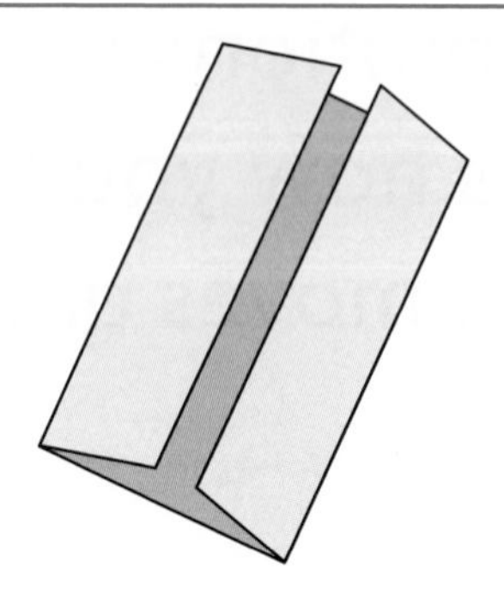

## Trifold Book

1. Fold a sheet of paper ($8\frac{1}{2}$" x 11") into thirds.
2. Use this book as is, or cut into shapes.

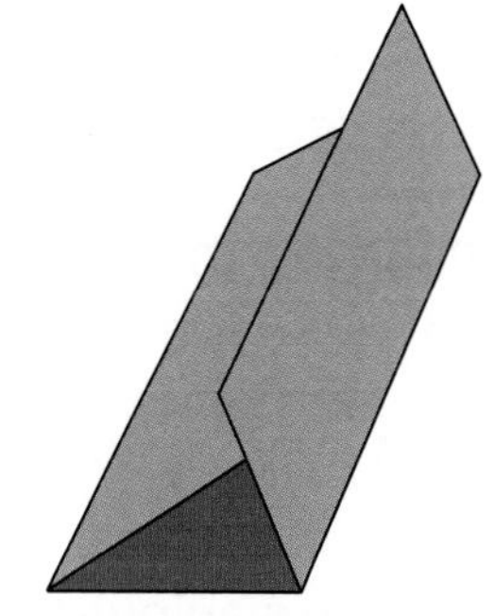

## Three-Tab Book

1. Fold a sheet of paper like a hot dog.
2. With the paper horizontal and the fold of the hot dog up, fold the right side toward the center, trying to cover one half of the paper.
3. Fold the left side over the right side to make a book with three folds.
4. Open the folded book. Place one hand between the two thicknesses of paper and cut up the two valleys on one side only. This will create three tabs.

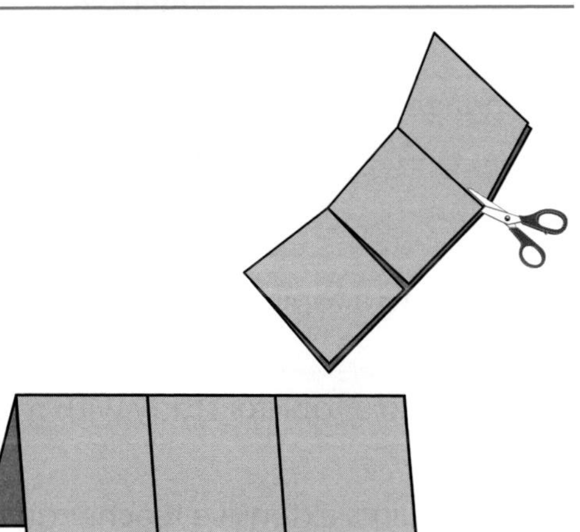

## Layered-Look Book

1. Stack two sheets of paper ($8\frac{1}{2}$" x 11") so that the back sheet is one inch higher than the front sheet.
2. Bring the bottoms of both sheets upward and align the edges so that all of the layers or tabs are the same distance apart.
3. When all the tabs are an equal distance apart, fold the papers and crease well.
4. Open the papers and glue them together along the valley, or inner center fold, or staple them along the mountain.

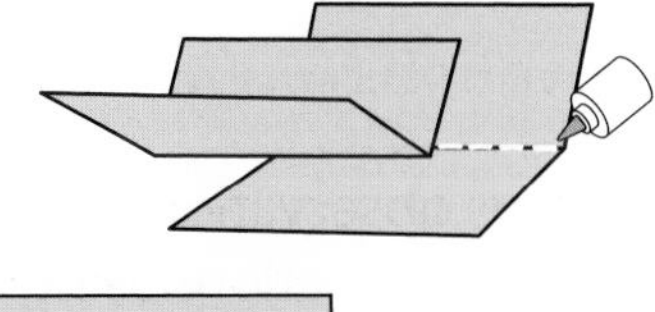

## Folded Table or Chart

1. Fold the number of vertical columns needed to make the table or chart.
2. Fold the horizontal rows needed to make the table or chart.
3. Label the rows and columns.

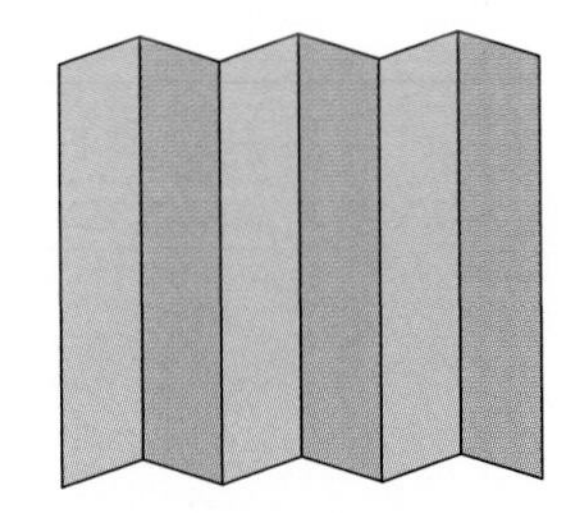

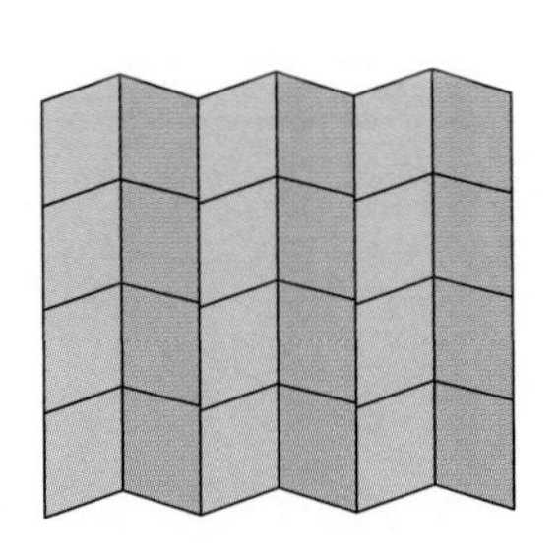

# 4-Point Activity Rubrics

Assessing Abilities Necessary to Do Scientific Inquiry

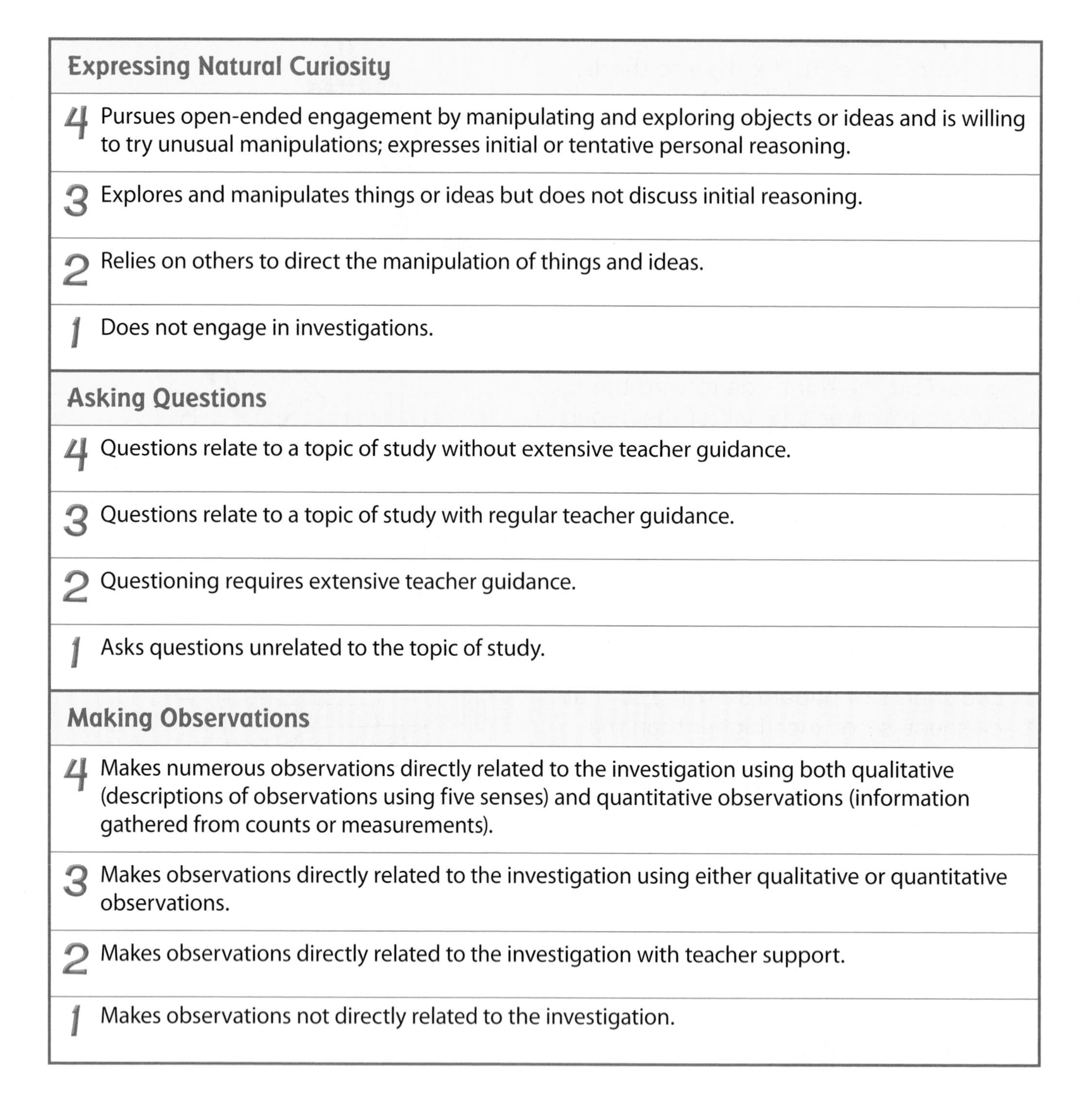

| Score | Expressing Natural Curiosity |
|---|---|
| 4 | Pursues open-ended engagement by manipulating and exploring objects or ideas and is willing to try unusual manipulations; expresses initial or tentative personal reasoning. |
| 3 | Explores and manipulates things or ideas but does not discuss initial reasoning. |
| 2 | Relies on others to direct the manipulation of things and ideas. |
| 1 | Does not engage in investigations. |
| | **Asking Questions** |
| 4 | Questions relate to a topic of study without extensive teacher guidance. |
| 3 | Questions relate to a topic of study with regular teacher guidance. |
| 2 | Questioning requires extensive teacher guidance. |
| 1 | Asks questions unrelated to the topic of study. |
| | **Making Observations** |
| 4 | Makes numerous observations directly related to the investigation using both qualitative (descriptions of observations using five senses) and quantitative observations (information gathered from counts or measurements). |
| 3 | Makes observations directly related to the investigation using either qualitative or quantitative observations. |
| 2 | Makes observations directly related to the investigation with teacher support. |
| 1 | Makes observations not directly related to the investigation. |

| Using Tools to Extend the Senses | |
|---|---|
| 4 | Selects and correctly uses appropriate tools independently. |
| 3 | Selects and uses tools appropriately with minimal assistance. |
| 2 | Selects appropriate tools with some assistance but may make mistakes in their use. |
| 1 | Unable to correctly select and use tools. |

| Answering Questions as a Result of the Investigations | |
|---|---|
| 4 | Describes and compares findings in terms of characteristics (number, shape, texture, size, weight, color), relative order (before/middle/after, top/bottom), and motion (fast/slow, straight/curved); draws pictures that correctly portray most features of the thing being investigated. Accurately describes the process of the inquiry and what information was gained from the inquiry. |
| 3 | Describes and compares findings in terms of characteristics, relative order, or motion; draws pictures that correctly portray some features of the thing being investigated. Offers some information about the process of inquiry. |
| 2 | Describes things in terms of characteristics; draws pictures that illustrate some features of the thing being described. |
| 1 | Describes things in terms of characteristics only. |

## Writing Links

*Writing Links* provide opportunities for teachers to integrate writing into the science curriculum and further prepare children for the writing assessment tests they will be taking. The writing logo is used in the Student Edition to identify writing tasks, and can be found in the following places:

- At the end of every lesson there is a writing question in the Think, Talk, and Write lesson review.
- A *Writing in Science* feature is in every chapter. A *Write About It* prompt is found in each feature. Also look for the on this page for online writing opportunities for children.
- Look for *Integrate Writing* boxes in the *Be a Scientist*, *Focus on Skills*, and *Writing in Science* features in the Teacher's Edition for more effective ways to include writing throughout each lesson.

### Connecting the Rubrics to the Writing Modes

The *Writing Links* four-point rubric features six modes of writing. These writing modes are personal narrative, descriptive writing, writing a story, explanatory (or "how-to") writing, writing that compares and expository writing. A *7-Trait Writing in Science* rubric is provided to assess each mode of writing.

Each of the six writing modes is designed to build writing skills that are essential to good writing in general and to science in particular, such as developing a clearly organized central (or main) idea with supporting facts and details and using varied sentence structure. These and additional writing skills are stressed on writing assessment tests that children will be taking and in the *7-Trait Writing in Science* rubrics provided.

### Macmillan/McGraw-Hill Writing Modes

- **Writing About Yourself** mode found in the text will help children craft a detailed true story about a personal experience within the framework of a clearly organized sequence of events. Most writing assessment tests call for the writing of narrative text that is organized in a clear and logical way.
- **Writing That Describes** tasks will help children learn to include vivid, sensory details in their writing, while enabling them to choose colorful or specific vocabulary. These skills will be useful to them in writing observation reports, and in both their narrative and expository writing.
- **Writing a Story** found in the text will help children craft fictional narrative—for example, a piece of science fiction—with vivid details and a thoughtfully planned story line that organizes events from beginning to end. Most writing assessment tests require the writing of a narrative, whether it would be a personal narrative based on a real-life event, or a fictional story.
- **Writing That Explains** tasks ask children to explain how to complete a task or a process, such as a science experiment. Children's ability to organize their writing into a step-by-step format is an essential tool for writing in science. Giving clear details and organizing events into a sequence is also required for all good writing.
- **Writing That Compares** focuses on skills necessary to write an essay or a report that compares or contrasts two items or findings. This objective mode of writing is often used when writing about science.
- **Writing That Gives Information** focuses on skills necessary to write a summary, an informational or research report, or an essay. This objective mode of writing is the one used most often when writing about science. It is also consistent with the writing mode tested most frequently on writing assessment tests.

## Using the Scoring Rubrics

Use the 4-Point Writing Rubrics to score children's responses to the writing modes and prompts featured in the Writing Links.

### 4-Point Writing Rubric

**To determine the appropriate score:**

- Find the description of the mode of writing featured in the Writing Link. These six modes are Writing About Yourself, Writing That Describes, Writing a Story, Writing That Explains, Writing That Compares, and Writing That Gives Information (such as a report).
- Then find the description of the writing trait that best expresses the quality of the child's writing in that mode. Assess the child's writing as 4 Excellent, 3 Good, 2 Fair, or 1 Unsatisfactory.
- Consider how well the response achieves the writer's purpose. Be sure the response addresses the features of **7-Trait Writing in Science:**
  - Ideas & Content
  - Organization
  - Voice
  - Word Choice
  - Sentence Fluency
  - Conventions
  - Presentation
- Assign a single score of 1–4 based on how well the child's writing corresponds to the descriptions that appear in the rubrics.

**For remedial purposes:**

You can use the 4-Point Writing Rubric to identify specific areas of deficiency (organization, word choice, sentence fluency). Do not, however, assign separate scores for each of the writing traits.

# Writing Links: 4-Point Writing Rubric

7-Trait Writing in Science

**Writing About Yourself**

| 4 Excellent | 3 Good | 2 Fair | 1 Unsatisfactory |
|---|---|---|---|
| **Ideas & Content** Demonstrates originality in developing ideas or a story drawn from a personal experience. | **Ideas & Content** Develops reasonably clear ideas that develop a true story about the writer. | **Ideas & Content** Displays difficulty in developing content and fails to show a strong sense of purpose. | **Ideas & Content** Makes no attempt to develop ideas or tell about a true event. |
| **Organization** Crafts a well-organized personal narrative that flows smoothly and moves the reader through the beginning, middle, and end of the text. | **Organization** Crafts a personal narrative that moves the reader through the text without confusion. | **Organization** Crafts a personal narrative that may exhibit organizational problems, such as lack of follow-through after a good beginning. | **Organization** Shows an extreme lack of organization that interferes with comprehension of the text. |
| **Voice** Exhibits a personal voice with a sense of the purpose and audience. | **Voice** Expresses a personal voice and demonstrates an adequate sense of the purpose and audience. | **Voice** Attempts to express a personal voice but is not fully engaged with the audience. | **Voice** Makes no attempt to express a personal voice or share personal insights. |
| **Word Choice** Chooses imaginative words that convey images and feelings in a natural way. | **Word Choice** Makes an effort to choose words that convey images and emotions. | **Word Choice** Chooses words that are often dull and unimaginative. | **Word Choice** Shows an inability to choose words that convey clear pictures or are imaginative. |
| **Sentence Fluency** Produces strong, varied, and purposeful sentences that invite expressive oral reading. | **Sentence Fluency** Produces varied sentences that are easy to read aloud with a little practice. | **Sentence Fluency** Produces sentences with some variation but that may lack an easy flow. | **Sentence Fluency** Produces awkward or incomplete sentences that do not invite oral reading. |
| **Conventions** Exhibits a good grasp of standard writing conventions, including spelling, capitalization, punctuation, and grammar. | **Conventions** Shows a grasp of most standard writing conventions. | **Conventions** May have trouble with some standard writing conventions, including spelling, capitalization, punctuation, or grammar. | **Conventions** Shows an inability to grasp basic writing conventions, severe enough to interfere with readability. |
| **Presentation** Uses neat handwriting or fonts to enhance the reader's ability to connect with the message of the text. | **Presentation** Uses readable handwriting or consistent fonts and font sizes that make the text easy to read. | **Presentation** Uses legible handwriting or experimentation with fonts, although the effect is not consistent throughout the text. | **Presentation** Uses inconsistent handwriting or multiple fonts, making the text difficult or impossible to read. |

# Writing Links: 4-Point Writing Rubric

7-Trait Writing in Science

**Writing That Describes**

| 4 Excellent | 3 Good | 2 Fair | 1 Unsatisfactory |
|---|---|---|---|
| **Ideas & Content** Shows imagination and originality in developing specific descriptive content that is clear, vivid, and focused. | **Ideas & Content** Develops descriptive content in a general way, using ideas that are reasonably clear and focused. | **Ideas & Content** Has difficulty developing clear, focused ideas and specific descriptive content. | **Ideas & Content** Makes no attempt to present clear ideas and specific descriptive content. |
| **Organization** Creates a description that flows smoothly and is well organized in its presentation of details. | **Organization** Organizes a description in a way that groups details, moving the reader through the text without confusion. | **Organization** Creates a description that exhibits organizational problems, such as grouping unlike details together. | **Organization** Demonstrates a lack of organization that interferes with readability and comprehension. |
| **Voice** Uses a strong voice that appeals directly to the audience and expresses the writer's personality. | **Voice** Uses a personal voice that connects the writer and audience. | **Voice** Makes an attempt to use an appealing personal voice but has difficulty maintaining it. | **Voice** Makes no attempt to express a personal voice or appeal to the audience. |
| **Word Choice** Chooses vivid sensory words to create a clear mental picture for the reader. | **Word Choice** Makes an effort to choose words that are clear, vivid, and accurate and that may appeal to the audience's senses. | **Word Choice** Often chooses overused words that fail to capture the audience's imagination. | **Word Choice** Shows an inability to choose words that are correct or appropriate to the description. |
| **Sentence Fluency** Crafts varied, well-paced sentences that are a delight to read aloud. | **Sentence Fluency** Crafts sentences that are generally smooth and natural. | **Sentence Fluency** Crafts some sentences that are choppy, rambling, or awkward to read. | **Sentence Fluency** Produces sentences that are incomplete and difficult to read aloud. |
| **Conventions** Displays a highly developed grasp of writing conventions that make the description easy to read. | **Conventions** Displays a general understanding of writing conventions and applies them to the description. | **Conventions** Often has trouble with spelling, capitalization, punctuation, and grammar. | **Conventions** Demonstrates an inability to grasp basic writing conventions. |
| **Presentation** Consistently uses neat handwriting or appropriate fonts, as well as the correct balance of white space and text, to make the work inviting to the audience. | **Presentation** Uses legible handwriting or consistent fonts, as well as uniform spacing, to invite the reader into the text. | **Presentation** Uses legible handwriting or consistent fonts, as well as uniform spacing, although a different choice of spacing (double spacing) may be preferable. | **Presentation** Uses irregularly formed letters or multiple fonts and font sizes, as well as random spacing, making the text difficult to read and understand. |

# Writing Links: 4-Point Writing Rubric

7-Trait Writing in Science

**Writing a Story**

| 4 Excellent | 3 Good | 2 Fair | 1 Unsatisfactory |
|---|---|---|---|
| **Ideas & Content** Shows imagination in developing story ideas, structure, and content. | **Ideas & Content** Shows some imagination in developing story ideas, structure, and content. | **Ideas & Content** Adequately develops story ideas, structure, and content. | **Ideas & Content** Makes no effort to develop interesting or imaginative story ideas and content; no story structure is evident. |
| **Organization** Exhibits strong organizational skills in creating an interesting beginning, middle, and end of the story. | **Organization** Uses organizational skills to create a beginning, a middle, and an end of the story. | **Organization** Exhibits difficulty in organizing the structure of the story. | **Organization** Shows an inability to create a structure for the story. |
| **Voice** Displays a personal voice that echoes the tone of the story and is highly appealing to the audience. | **Voice** Displays a personal voice that is appropriate and appealing to the audience. | **Voice** Displays a personal voice that tries to engage the audience. | **Voice** Makes no attempt to develop a personal voice and shows no awareness of the audience. |
| **Word Choice** Chooses words carefully to develop the setting, characters, and sequence of events. | **Word Choice** Chooses colorful, accurate words that are appropriate to developing the story. | **Word Choice** Does not choose colorful or specific vocabulary to develop the story. | **Word Choice** Uses words that are incorrect or that confuse the reader. |
| **Sentence Fluency** Crafts interesting and varied sentences that enhance the fluency of the story and invite oral reading. | **Sentence Fluency** Crafts interesting and varied sentences that are easy to read aloud. | **Sentence Fluency** Crafts sentences that can be understood but are sometimes hard to follow or read aloud. | **Sentence Fluency** Writes sentences that are incomplete, confusing, and extremely difficult to read aloud. |
| **Conventions** Exhibits firm knowledge of writing conventions, including spelling, capitalization, punctuation, and grammar. | **Conventions** Exhibits a knowledge of standard writing conventions; work needs little editing. | **Conventions** Exhibits a limited grasp of writing conventions; extensive revising and editing are needed. | **Conventions** Exhibits problems with writing conventions that are severe enough to interfere with readability. |
| **Presentation** Uses neat handwriting or appropriate fonts and font sizes to enhance comprehension and readability. | **Presentation** Uses readable handwriting or successful experimentation with fonts. | **Presentation** Uses legible handwriting, although discrepancies may exist in letter shape or slant; experimentation with fonts may not be effective. | **Presentation** Creates an unclear or confusing story because of problems relating to handwriting, use of fonts, or spacing. |

# Writing Links: 4-Point Writing Rubric

7-Trait Writing in Science

**Writing That Explains**

| 4 Excellent | 3 Good | 2 Fair | 1 Unsatisfactory |
|---|---|---|---|
| **Ideas & Content** Develops a purposeful paper that presents a clear explanation of a task or process. | **Ideas & Content** Develops a paper that presents a reasonably clear explanation of a task or process. | **Ideas & Content** Develops a paper that shows a sense of purpose, but may not explain instructions or a process in a clear way. | **Ideas & Content** Makes no attempt to tell the reader how to do or make something; writing demonstrates no clear purpose. |
| **Organization** Organizes the writing in a way that smoothly moves the reader through the text, step by step, while clearly explaining the specific task or process. | **Organization** Presents the steps in a process in a well-planned manner, with clear transitions. | **Organization** Does not present the information clearly; transitions are weak. | **Organization** Shows an inability to organize the writing or provide connected details. |
| **Voice** Uses a personal voice that demonstrates a strong involvement with the purpose and audience. | **Voice** Makes an effort to explain ideas in a manner appropriate to the purpose and audience. | **Voice** Uses a voice that does not always involve the purpose of the writing or the audience. | **Voice** Makes no effort to involve self with the purpose and audience. |
| **Word Choice** Chooses time-order words, such as *first* and *then*, and spatial words, such as *top* and *bottom*, to present a clear understanding of the steps in the process. | **Word Choice** Chooses functional words that convey the purpose of the paper—to explain a task or process. | **Word Choice** Chooses words that fail to convey a complete understanding of the task or process being explained. | **Word Choice** Shows an inability to choose words that are appropriate to the topic, purpose, and audience. |
| **Sentence Fluency** Crafts effective sentences that flow together and support the content and style of the paper; has control of sentence types and lengths. | **Sentence Fluency** Crafts sentences that make sense and flow together; maintains control of simple sentences. | **Sentence Fluency** Crafts sentences that make sense but are short, choppy, or unvaried. | **Sentence Fluency** Uses sentences or sentence fragments that make little sense and are difficult or impossible to follow. |
| **Conventions** Implements writing conventions accurately and effectively; work needs little editing. | **Conventions** Uses a variety of writing conventions accurately, but some editing is needed. | **Conventions** Makes frequent mistakes in writing conventions, such as spelling, capitalization, punctuation, and grammar. | **Conventions** Shows an inability to use or grasp writing conventions. |
| **Presentation** Uses a pleasing form to present content; successfully aligns text and visuals to support and clarify key information. | **Presentation** Produces easy-to-read text and, for the most part, aligns visuals and content to enable the reader to access the information. | **Presentation** Demonstrates discrepancies in letter shape and slant, as well as spacing; connections between text and visuals are not always clear. | **Presentation** Demonstrates an inability to form regularly shaped and slanted letters, or use consistent spacing; fails to use visuals to support or illustrate key ideas in the text. |

# Writing Links: 4-Point Writing Rubric

7-Trait Writing in Science

**Writing That Compares**

| 4 Excellent | 3 Good | 2 Fair | 1 Unsatisfactory |
|---|---|---|---|
| **Ideas & Content** Develops ideas and content that make a comparison in an informative, purposeful way. | **Ideas & Content** Develops ideas and content to show similarities and differences effectively. | **Ideas & Content** Develops ideas and content that present a comparison but may not hold the reader's interest. | **Ideas & Content** Makes no attempt to develop a comparison. |
| **Organization** Organizes details and information into distinct categories of comparison and contrast. | **Organization** Adequately arranges details and information into categories of comparison and contrast. | **Organization** Arranges some details and information into categories. | **Organization** Shows an inability to organize details or information into categories. |
| **Voice** Presents a personal voice that speaks to the audience in an individual and engaging manner. | **Voice** Presents a personal voice that meets the needs of the audience. | **Voice** Lacks an effective personal voice, or presents a voice that is insensitive to the needs of the audience. | **Voice** Makes no attempt to create a personal voice in the writing. |
| **Word Choice** Chooses compare-and-contrast words, such as *alike* and *different*, to highlight the points of comparison and contrast. | **Word Choice** Chooses compare-and-contrast words to show similarities and differences between items or ideas. | **Word Choice** Chooses words that attempt to support the comparison and link ideas. | **Word Choice** Makes no effort to use words that compare and contrast. |
| **Sentence Fluency** Crafts well-built, interesting sentences that invite expressive oral reading. | **Sentence Fluency** Crafts sentences that may be mechanical but are generally easy to read aloud. | **Sentence Fluency** Crafts short or choppy sentences that may be awkward to read aloud. | **Sentence Fluency** Crafts fragmented, confusing sentences that are difficult to read aloud. |
| **Conventions** Exhibits an excellent grasp of writing conventions, including spelling, capitalization, punctuation, grammar, and paragraph indents. | **Conventions** Exhibits an adequate grasp of standard writing conventions. | **Conventions** Exhibits a limited grasp of writing conventions. | **Conventions** Exhibits a severe inability to employ writing conventions. |
| **Presentation** Presents text that is pleasing to the eye and easy to read; text enables the reader to access key points of comparison and contrast. | **Presentation** Presents clear text that guides the reader to focus on the points of comparison and contrast. | **Presentation** Produces text that does not demonstrate an effective format for presenting points of comparison and contrast. | **Presentation** Presents text that is difficult or impossible to read and understand. |

# Writing Links: 4-Point Writing Rubric

7-Trait Writing in Science

**Writing That Gives Information**

| 4 Excellent | 3 Good | 2 Fair | 1 Unsatisfactory |
|---|---|---|---|
| **Ideas & Content** Develops clear content that supports the main idea and is suited to the purpose and audience. | **Ideas & Content** Develops content that is focused on and suited to the purpose and audience. | **Ideas & Content** Develops content that attempts to support the main idea and hold the interest of the audience. | **Ideas & Content** Makes no attempt to develop content that is focused on or suited to the purpose and audience. |
| **Organization** Exhibits solid organizational skills, with an effective introduction, body, and conclusion. | **Organization** Exhibits good organizational skills, including an effective introduction and a conclusion that summarizes the information. | **Organization** Exhibits limited organizational skills; does not draw a conclusion based on the facts presented. | **Organization** Exhibits extreme organizational problems that interfere with comprehension and readability. |
| **Voice** Expresses a personal voice that is well suited to the topic, purpose, and audience. | **Voice** Expresses a personal voice that is appropriate to the topic, purpose, and audience. | **Voice** Expresses a personal voice that may not fit the topic, purpose, or needs of the audience. | **Voice** Makes no attempt to develop a personal voice. |
| **Word Choice** Uses clear, accurate vocabulary that is well suited to the topic, purpose, and audience. | **Word Choice** Uses vocabulary that helps make the topic clear. | **Word Choice** Uses vocabulary that gets the message across in an adequate, yet ordinary way. | **Word Choice** Uses vocabulary that confuses the reader or is inaccurate. |
| **Sentence Fluency** Crafts a variety of sentences that enhance the understanding and fluency of the piece. | **Sentence Fluency** Crafts sentences that make sense and are easy to read aloud. | **Sentence Fluency** Crafts sentences that may be awkward at times. | **Sentence Fluency** Writes sentence fragments or sentences that are extremely difficult to read. |
| **Conventions** Demonstrates accurate use of standard writing conventions, including spelling, capitalization, punctuation, and grammar. | **Conventions** Demonstrates accurate use of most writing conventions; work needs little editing. | **Conventions** Makes frequent mistakes in spelling, capitalization, punctuation, and grammar; work needs extensive editing. | **Conventions** Makes mistakes in writing conventions that compromise readability and comprehension. |
| **Presentation** Presents a pleasing format that integrates text and illustrations, such as charts and maps, to support and enhance key information. | **Presentation** Uses visuals to clarify points in the text, although the visuals do not always support the key information. | **Presentation** Presents a mostly understandable format, but an integration between text and illustrations may be limited. | **Presentation** Presents a confusing format that denies the reader access to the information in the text. |

# Materials

**Materials required to complete activities in the Student Edition are listed below.**

## Consumable Materials

Based on 6 groups

| MATERIALS | QUANTITY NEEDED PER GROUP | KIT QUANTITY | CHAPTER/ LESSON |
|---|---|---|---|
| Apple | 2 | | 1/2; 2/2; 10/1 |
| Bag, paper | 1 | 6 | 7/3; 11/4 |
| Bag, plastic, resealable (6″ x 8″) | 1 | 6 | 1/3; 11/3 |
| Batteries, size D | 2 | 12 | 8/1, 3; 12/3, 4 |
| Bottle, plastic, 1 L with cap | 1 | | 5/3 |
| Bottle, plastic, 2 L | 1 | | 4/1 |
| Brass fastener | 1 | 100 | 8/2 |
| Brine shrimp eggs (live coupon) | 1/2 tsp | 1 coupon | 4/3 |
| Butter | | | 10/2 |
| Calendar | 1 | | 8/3 |
| Card stock | 1 sheet | | 2/1 |
| Chenille stems | 2 or 3 | 100 | 2/1; 7/1 |
| Chocolate piece | 1 piece | | 10/2 |
| Clay, modeling | 1 box | 2 (1-lb) boxes | 3/3; 5/1; 10/1 |
| Cloth, felt, black (12″ x 20″) | small pieces | 1 | 2/1 |
| Container, plastic with lid (take-out food container) | 1 | | 2/2 |
| Crackers, assorted | 6 | | 9/1 |
| Crackers, fish shape | | | 9/1 |
| Crackers, square shape | | | 9/1; 10/1 |
| Craft sticks | 1 | 30 | 3/2; 7/1; 8/1 |
| Crayons | | | 1/2; 2/2; 3/1, 2, 3; 4/1, 5/1, 2, 3; 6/3; 7/2; 8/1, 2, 3, 4; 9/1; 10/1 |
| Cream, chilled | 1/4 c | | 10/1 |
| Cup, plastic | 3 | 100 | 4/2; 6/2; 7/2; 12/1, 2 |
| Cups, foam | 2 | 25 | 6/2 |
| Earthworms (live coupon) | 1 | 1 coupon | 4/1 |
| Foil, aluminum | 1 | 1 roll | 1/1; 2/1; 11/2; 12/3 |
| Food scraps | 1 to 2 c | | 6/2 |
| Glue | small amount | | 2/1; 3/2; 10/2 |
| Glue stick | 1 | | 8/2 |
| Ice cube | 8 | | 12/1 |
| Index cards | 4 | 1 pack | 2/2; 8/1 |
| Juice, lemon | 3 tbsp | | 10/1 |
| Knife, plastic | 1 | 24 | 10/1, 2 |
| Lamp, miniature | 1 | 12 | 12/4 |
| Leaves | | | 3/3 |
| Lightbulbs | | | 12/4 |
| Magazines, nature | | | 3/1; 5/2; 8/2; 10/2 |

## Non-Consumable Materials

Based on 6 groups

| MATERIALS | QUANTITY NEEDED PER GROUP | KIT QUANTITY | CHAPTER/ LESSON |
|---|---|---|---|
| Balance | 1 | 1 | 9/2; 10/1 |
| Ball, foam | 2 | 12 | 8/3 |
| Bins, recycling | 3 | | 6/3 |
| Books | several | | 11/2 |
| Bottle, plastic, 1 L, with cap | 1 | | 4/3; 5/3 |
| Cardboard | 1 | | 11/2; 12/3 |
| Classroom objects | 6 | | 9/1, 2; 11/2, 4; 12/3 |
| Clock | 1 per classroom | | 6/2, 3; 12/3 |
| Cloth, cotton, white (12″ x 20″) | small square | 1 | 12/3, 4 |
| Cloth, felt, black | small square | 1 | 12/3 |
| Cloth, flannel (12″ x 20″) | small square | 1 | 12/4 |
| Container, plastic | 1 | 6 | 4/3; 5/1; 6/2; 9/3 |
| Container, vacuum sealing | | | 4/2 |
| Cubes, wood | 5 | 6 | 3/3; 11/2, 3 |
| Cup, measuring | 2 | 12 | 4/3; 6/2; 9/3; 10/1, 3 |
| Cups, plastic, 9 oz | 2 | 50 | 9/3; 10/3 |
| Film canister | 6 | 18 | 9/3 |
| Flashlight | 1 | 6 | 8/1, 3; 12/3 |
| Hand lens | 1 | 6 | 1/1, 2, 3; 2/2; 3/3; 4/1, 3; 5/3; 6/1, 2 |
| Jar, plastic 16 oz, with lid | 1 | 12 | 5/3; 10/1 |
| Jar, plastic 16 oz, with lid | 2 | 12 | 4/3 |
| Knife (teacher use only) | 1 | | 1/2; 10/1 |
| Knife, plastic | 1 | 24 | 3/3; 5/1 |
| Lamps or other light source | 1 per classroom | | 1/3 |
| Light socket, miniature | 1 | 2 | 12/4 |
| Magnet, bar | 2 | 12 | 11/4 |
| Magnet, donut | 1 | 6 | 11/4 |
| Magnet, horseshoe | 1 | 6 | 11/4 |
| Meter stick | | | 11/1 |
| Metric weight set | 1 | 1 | 10/1 |
| Mineral kit | 1 | 72 specimens | 6/1 |
| Pail | 1 | | 11/3 |
| Pan, aluminum (8″ x 8″) | 1 | 6 | 9/2, 3 |
| Paper clips | 20 | | 11/4; 12/2 |
| Pennies | 15 | | 11/3 |
| Photo Sorting Cards (11–20, 21–23, 27–29, 31, 34–40) | | | 2/1 |

Materials required to complete activities in the Student Edition are listed below.

## Consumable Materials

Based on 6 groups

| MATERIALS | QUANTITY NEEDED PER GROUP | KIT QUANTITY | CHAPTER/ LESSON |
|---|---|---|---|
| Magazines and newspapers | | | 4/3 |
| Markers, colored | | | 3/1; 4/2; 6/3; 7/2; 8/1, 2; 11/3; 12/1 |
| Mealworms (live coupon) | 1 | 1 coupon | 2/2 |
| Oatmeal | | | 2/2 |
| Paper, construction | 1 sheet | | 2/1; 3/2; 7/1; 8/4; 10/2; 12/3 |
| Paper, crepe, red | 1 sheet (12″ x 20″) | 1 package | 7/1 |
| Paper, drawing | | | 2/3; 5/1, 2, 3; 6/1, 2; 7/3; 8/1, 2, 3; 9/1, 2, 3; 10/1; 11/1, 3; 12/1, 3 |
| Paper, drawing (large) | 1 sheet | | 3/3 |
| Paper, patterned gift wrap | 2 sheets | | 2/3 |
| Paper, poster | 1 sheet | | 6/3; 8/4 |
| Paper, shiny | | | 2/1 |
| Paper, tissue | 1 sheet | | 12/3 |
| Paper, waxed | 12-in. piece | 1 roll | 12/3 |
| Paper towels | 2 sheets | | 1/3; 4/2 |
| Pencils, colored | | | 3/3; 12/3 |
| Pill bugs (live coupon) | 1 | 1 coupon | 4/1 |
| Plant | 1 | | 4/1 |
| Plant, potted, with flower | 2 | | 1/1 |
| Plants, potted, with leaves | 2 | | 1/1 |
| Plates, paper | 2 | 50 | 3/2; 6/2; 8/2, 4; 10/1, 2 |
| Rocks, small | 1 to 2 tbsp | | 4/1 |
| Rubber bands | 1 | | 7/2 |
| Salt | 1/4 c | | 4/3; 10/3 |
| Sand | 1/4 c | 1 (5.5-kg) bag | 10/3 |
| Sandpaper | 1 sheet | 6 sheets | 11/2 |
| Seedling, lima bean | 2 | | 1/3 |
| Seeds, bean | 3 | | 1/3 |
| Seeds, lima bean | 2 or 3 | 1 (8-oz) package | 1/2 |
| Shoebox, cardboard | 1 | | 1/3 |

## Non-Consumable Materials

Based on 6 groups

| MATERIALS | QUANTITY NEEDED PER GROUP | KIT QUANTITY | CHAPTER/ LESSON |
|---|---|---|---|
| Photo Sorting Cards (41–44) | | | 5/1 |
| Plate, plastic | 1 | 6 | 10/3 |
| Pot, for soil | 1 | | 8/1 |
| Prism | 1 | 6 | 12/3 |
| Rock, sandstone chips | 1 | 1 bag | 5/3 |
| Rock kit | 8 to 10 | 72 specimens | 6/1 |
| Rolling pin, roller style | 1 | | 11/3 |
| Rope | about 2 ft | | 11/3 |
| Ruler, plastic | | | 2/2; 7/1; 8/4; 11/1, 2, 3; 12/4 |
| Safety goggles | 1 | | 12/2 |
| Scissors | | | 2/1, 3; 3/2; 4/2; 8/2, 4; 10/1, 2; 12/2, 4 |
| Small toys | | | 9/3 |
| Solids | 4 to 5 | | 9/2 |
| Sponge | 1 | 6 | 1/1 |
| Spoon, metal | 1 | 3 | 9/2 |
| Spoon, plastic | 1 | 24 | 4/1 |
| Spoon, wooden | 1 | 3 | 9/2 |
| Spoons, measuring | 1 | 6 | 4/3 |
| Spray bottle, 16 oz | | | 4/2 |
| Stapler | | | 8/1 |
| Stopwatch | 1 | 1 | 2/3; 11/1, 2 |
| Strainer | 1 | 3 | 6/2 |
| Thermometer | 1 | 18 | 7/1, 2 |
| Thermometers | 3 | 18 | 4/2; 12/1, 3 |
| Toy, plastic, animal | 3 to 5 | | 3/3 |
| Toy, windup | 3 | | 11/1 |
| Toy car | 1 | 6 | 11/2 |
| Toy car | 3 to 5 | 6 | 3/3 |
| Tray, plastic | | | 4/1; 10/3 |
| Tub, plastic | 1 | 6 | 4/3; 5/1; 6/2; 9/2 |
| Tuning fork | 1 | 3 | 12/2 |
| Watch | | | 12/1 |
| Wire cutter | 1 for teacher | 1 | 12/4 |

# Materials

**Materials required to complete activities in the Student Edition are listed below.**

## Consumable Materials

Based on 6 groups

| MATERIALS | QUANTITY NEEDED PER GROUP | KIT QUANTITY | CHAPTER / LESSON |
|---|---|---|---|
| Soil, clay | 1 c | 1 (2.5-kg) bag | 6/2 |
| Soil, potting | 4 c | 3 (8-lb) bags | 4/1, 6/2, 8/1, 12/1 |
| Soil, sandy | 1 c | 1 (2.5-kg) bag | 6/2 |
| Spoon, plastic | 2 | 24 | 9/2, 10/3 |
| Spoon, plastic | 1 | 24 | 6/2, 12/1 |
| Spoons, plastic | 1 | 24 | 12/1 |
| Staples | | | 8/1 |
| Sticky notes | | | 11/4 |
| String, cotton | 12- to 18-in. piece | 200 ft | 8/4, 11/3, 4; 12/2 |
| Tape, masking | small amount | | 7/4, 8/4, 11/1, 2, 3; 12/1 |
| Tape, transparent | 5-in. piece | | 1/3, 2/1, 4/2, 7/1 |
| Tube, cardboard | 2 | | 2/3 |
| Tube, cardboard | 2 | | 2/3 |
| Wire, insulated | 2 (8-in.) pieces | 1 spool | 12/4 |
| Wrap, plastic | 12-in. piece | 1 roll | 4/1, 2; 7/2, 10/1, 11/2, 12/3 |
| Yarn | 12- to 24-in. piece | 1 skein | 2/2, 3/2 |
| Yeast, bakers | small amount | | 4/3 |

# Bibliography

## Unit A: Plants and Animals

***Seeds.*** Robbins, Ken. Atheneum, 2005.

### Chapter 1: Plants

***Dig, Plant, Grow: A Kid's Guide to Gardening.*** Rushing, Felder. Cool Springs Press, 2004.

***From Seed to Pumpkin.*** Pfeffer, Wendy. HarperTrophy, 2004.

***A Grand Old Tree.*** DePalma, Mary Newell. Scholastic Inc., 2006.

***I Wonder Why Trees Have Leaves.*** Charman, Andrew. Kingfisher, 2003.

***Jack's Garden.*** Cole, Henry. Harper Trophy, 1997.

***The Life Cycle of a Flower.*** Aloian, Molly and Kalman, Bobbie. Crabtree Publishing, 2005.

***The Oxford Children's Encyclopedia of Plants and Animals.*** Oxford University Press, 2000.

***Plant Plumbing: A Book About Roots and Stems.*** Blackaby, Susan. Picture Window Books, 2003.

***A Seed in Need: A First Look at the Plant Cycle.*** Godwin, Sam. Picture Window Books, 2004.

***Why Do Plants Have Flowers? Spilsbury, Louise and Spilsbury, Richard.*** Heinemann Library, 2005.

### Chapter 2: Animals

***About Mammals: A Guide for Children.*** Sill, Cathryn. Peachtree Publishers, 2000.

***About Reptiles: A Guide for Children.*** Still, Cathryn. Peachtree Publishers, 1999.

***All About Frogs.*** Arnosky, Jim. Scholastic Reference, 2002.

***Animals, Animals.*** Carle, Eric. Philomel Books, 1989.

***Baby Sea Otter.*** Tatham, Betty. Henry Holt, 2005.

***Beaks!*** Collard, Sneed B. Charlesbridge, 2002.

***Birds Build Nests.*** Winer, Yvonne. Charlesbridge Publishing, 2002.

***How Animal Babies Stay Safe.*** Fraser, Mary Ann. HarperCollins, 2002.

***in the swim.*** Florian, Douglas. Voyager Books, 2001.

***O is for Orca.*** Helman, Andrea. Sasquatch Books, 2003.

***They Call Me Woolly: What Animal Names Can Tell Us.*** DuQuette, Keith. Putnam Juvenile, 2002.

***What Do You Do With a Tail Like This? Jenkins, Steve and Page, Robin.*** Houghton Mifflin, 2003.

## Unit B: Habitats

***Butterflies in the Garden.*** Lerner, Carol. HarperCollins, 2002.

### Chapter 3: Looking at Habitats

***Animal Homes.*** Wilkes, Angela. Kingfisher, 2006.

***Earthshake: Poems from the Ground Up.*** Peters, Lisa Westberg and Felstead, Cathie. Greenwillow, 2003.

***Forest Food Chains.*** Kalman, Bobbie. Crabtree Publishing Co., 2004.

***Gonna Sing My Head Off!*** Krull, Kathleen. Scholastic, 1992.

***A Log's Life.*** Pfeffer, Wendy. Simon & Schuster, 1997.

***Nature in the Neighborhood.*** Morrison, Gordon. Houghton Mifflin/Walter Lorraine Books, 2004.

***One Small Place by the Sea.*** Brenner, Barbara. HarperCollins, 2004.

***One Small Place in a Tree.*** Brenner, Barbara. HarperCollins, 2004.

***River of Life.*** Miller, Debbie S. Clarion Books, 2000.

***Staying Alive: The Story of a Food Chain.*** Bailey, Jacqui. Picture Window Books, 2006.

***Where Are The Night Animals?*** Fraser, Mary Ann. HarperCollins, 1999.

***Who Eats What? Food Chains and Food Webs.*** Lauber, Patricia. HarperTrophy, 1995.

### Chapter 4: Kinds of Habitats

***Antarctic Ice.*** Mastro, Jim. Henry Holt, 2003.

***Cactus Hotel.*** Guibertson, Brenda Z. Henry Holt, 1993.

***Chameleon, Chameleon.*** Cowley, Joy. Scholastic, 2005.

***Deserts.*** Stille, Darlene. Children's Press, 2000.

***Dig Wait Listen: A Desert Toad's Tale.*** Sayre, April Pulley. Greenwillow. 2001.

***Forest Explorer: A Life Sized Field Guide.*** Bishop, Nic. Scholastic, 2004.

***The Great Kapok Tree: A Tale of the Amazon Rain Forest.*** Cherry, Lynne. Voyager Books, 2000.

***A Journey into a Wetland.*** Johnson, Rebecca L. Carolrhoda, 2004.

***Lily Pad Pond.*** Lavies, Bianca. Puffin, 1993.

***Nature's Green Umbrella: Tropical Rain Forests.*** Gibbons, Gail. HarperCollins, 1997.

***On the Way to the Beach.*** Cole, Henry. Greenwillow, 2003.

***Pond.*** Morrison, Gordon. Houghton Mifflin, 2002.

***In the Small, Small Pond.*** Fleming, Denise. Henry Holt, 1998.

# Bibliography

## Unit C: Our Earth


*The Lifecycle of an Earthworm.* Kalman, Bobbie. Crabtree Publishing, Co., 2004.


### Chapter 5: Land and Water


*The Big Rock.* Hiscock, Bruce. Aladdin, 1999.

*Caves and Caverns.* Gibbons, Gail. Voyager Books, 1996.

*Earthquakes.* Branley, Franklyn M. HarperTrophy, 2005.

*Earthsteps: A Rock's Journey Through Time.* Spickert, Diane. Fulcrum, 2000.

*How Mountains are Made.* Zoehfeld, Kathleen Weidner. HarperTrophy, 1995.

*Mountains.* Simon, Seymour. HarperTrophy, 1997.

*Mountains and Our Moving Earth: With Easy-To-Make Geography Projects.* Robson, Pam. Stargazer Books, 2004.

*The Ocean Is...* Kranking, Kathleen W. Henry Holt, 2003.

*Oceans.* Simon, Seymour. Collins, 2006.

*Planet Earth/Inside Out.* Gibbons, Gail. HarperTrophy, 1998.

*River Friendly, River Wild.* Kurtz, Jane. Simon & Schuster, 2000.

*Volcanoes.* Simon, Seymour. Collins, 2006.


### Chapter 6: Earth's Resources


*Dirt: The Scoop on Soil.* Rosinsky, Natalie. Picture Window Books, 2002.

*Everybody Needs a Rock.* Baylor, Byrd. Econo-Clad, 1999.

*A Handful of Dirt.* Bial, Raymond. Walker Books, 2000.

*If You Find a Rock.* Christian, Peggy. Harcourt Children's Books, 2000.

*Let's Go Rock Collecting.* Gans, Roma. HarperTrophy, 1997.

*Looking at Rocks.* Dussling, Jennifer. Grosset & Dunlap, 2001.

*The Pebble in My Pocket: A History of Our Earth.* Hooper, Meredith. Viking Juvenile, 1996.

*Rocks and Fossils.* Pellant, Chris. Kingfisher, 2003.

*Soils.* Ditchfield, Christin. Children's Press, 2002.

*Sylvester and the Magic Pebble.* Steig, William. Aladdin, 2006.

*Wiggling Worms at Work.* Pfeffer, Wendy. HarperCollins, 2003.


## Unit D: Weather and Sky


*In the Snow: Who's Been Here?* George, Lindsey Barrett. HarperTrophy, 1999.


### Chapter 7: Observing Weather


*Cloud Dance.* Locker, Thomas. Voyager Books, 2003.

*Down Comes the Rain.* Branley, Franklyn M. Harper Trophy, 1997.

*Flash, Crash, Rumble and Roll.* Branley, Franklyn. HarperTrophy, 1999.

*Follow a Raindrop: The Water Cycle.* Ward, Elsie. Scholastic, 2000.

*On the Same Day in March: A Tour of the World's Weather.* Singer, Marilyn. HarperTrophy, 2002.

*The Snowflake: A Water Cycle Story.* Waldman, Neil. Millbrook Press, 2003.

*Snowflake Bentley.* Martin, Jacqueline Briggs. Houghton Mifflin, 1998.

*Storms and the Earth.* Bundey, Nikki. Carolrhoda, 2001.

*Weather Forecasting.* Gibbons, Gail. Aladdin, 1993.

*What Will the Weather Be? DeWitt, Lynda.* HarperTrophy, 1993.


### Chapter 8: Earth and Space


*Big Dipper.* Branley, Franklyn M. Harper Trophy, 1991.

*Earth, Our Planet in Space.* Simon, Seymour. Simon & Schuster, 2003.

*Mission to Mars.* Branley, Franklyn M. HarperTropy, 2002.

*The Moon.* Simon, Seymour. Simon & Schuster, 2003.

*On Earth.* Karas, G. Brian. Putnam Juvenile, 2005.

*Once Around the Sun.* Katz, Bobbie. Harcourt, 2006.

*Papa, Please Get the Moon for Me.* Carle, Eric. Simon & Schuster, 1991.

*The Planets.* Gibbons, Gail. Holiday House, rev. ed. 2005.

*Stars.* Tomecek, Steve and Yoshikawa, Sachiko. National Geographic Society, 2003.

*The Sun: Our Nearest Star.* Branley, Franklyn M. HarperTrophy, 2002.

*Sunshine Makes the Seasons.* Branley, Franklyn. HarperTrophy, 2005.

# Unit E: Matter

***The Popcorn Book.*** de Paola, Tomi. Holiday House, 1984.

## Chapter 9: Looking at Matter

***Air is All Around You.*** Branley, Franklyn M. HarperTrophy, 2006.

***Be a Friend to a Tree (Let's-Read-and-Find-Out Science).*** Lauber, Patricia. HarperCollins, 1994.

***Everything Is Matter!*** Bauer, David. Yellow Umbrella Books, 2004.

***Experiments with Solids, Liquids, and Gases.*** Tocci, Salvatore. Children's Press, 2002.

***Joseph Had a Little Overcoat.*** Taback, Simms. Viking Juvenile, 1999.

***Matter and Materials.*** Angliss, Sarah. Kingfisher, 2001.

***Matter: See It, Touch It, Taste It, Smell It.*** Stille, Darlene R. Picture Window Books, 2004.

***Pop! A Book About Bubbles.*** Bradley, Kimberly Brubaker. HarperTrophy, 2001.

***Solids and Liquids.*** Glover, David. Kingfisher, 2002.

***Touch It!*** Materials, Matter, and You. Mason, Adrienne. Kids Can Press, 2005.

***What Is the World Made Of? Zoehfeld, Kathleen W.*** HarperTrophy, 1998.

## Chapter 10: Changes in Matter

***Change It! Solids, Liquids, Gases and You.*** Mason, Adrienne. Kids Can Press, 2006.

***Cooling Investigations.*** Whitehouse, Patricia. Heinemann, 2004.

***Freezing and Melting.*** Nelson, Robin. Lerner Publications, 2003.

***From Wax to Crayon.*** Forman, Michael H. Children's Press, 1997.

***Heating.*** Whitehouse, Patricia. Heinemann Library, 2004.

***Hot and Cold.*** Hewitt, Sally. Children's Press, 2000.

***Is It Hot? Is It Not?*** Blevins, Wiley. Compass Point Books, 2003.

***One Hot Summer Day.*** Crews, Nina. Greenwillow, 1995.

***Slippery or Sticky.*** Parker, Victoria. Raintree, 2005.

***Temperature: Heating Up and Cooling Down.*** Stille, Darlene R. Picture Window Books, 2004.

***Why Does Ice Melt? Pipe, Jim.*** Copper Beech, 2002.

# Unit F: Motion and Energy

***Bat Loves the Night.*** Davies, Nicola. Candlewick, 2004.

## Chapter 11: How Things Move

***Forces Make Things Move.*** Bradley, Kimberly Brubaker. HarperTrophy, 2005.

***I Fall Down.*** Cobb, Vicki. HarperCollins, 2004.

***I Spy.*** Marzollow, Jean and Walter Wick. Cartwheel, 2005.

***I Spy: A Balloon.*** Marzollow, Jean and Wick, Walter. Scholastic, 2006.

***Jump, Frog, Jump.*** Kalan, Robert. HarperCollins, 2003.

***Magnets: Pulling Together, Pushing Apart.*** Rosinsky, Natalie. Picture Window Books, 2002.

***Pulleys.*** Walker, Sally and Feldmann, Roseann. Lerner Publishing Group, 2001.

***Science All Around Me: Magnets.*** Bryant-Mole, Karen. Heinemann, 1998.

***Simple Machines.*** Fowler, Allan. Children's Press, 2001.

***What Is Gravity?*** Trumbauer, Lisa. Children's Press, 2004.

***What Magnets Can Do.*** Fowler, Allan. Children's Press, 1995.

***What Makes a Magnet?*** Branley, Franklyn. HarperCollins, 1996.

***Wheels and Axles.*** Walker, Sally and Feldmann, Roseann. Lerner Publishing Group, 2001.

***Work and Simple Machines.*** Richards, Jon. Stargazer Books, 2004.

## Chapter 12: Using Energy

***All About Light.*** Trumbauer, Lisa. Children's Press, 2004.

***All About Sound.*** Trumbauer, Lisa. Children's Press, 2004.

***Electricity.*** Stille, Darlene. Child's World, 2004.

***Energy: Heat, Light, and Fuel.*** Stille, Darlene. Picture Window Books, 2004.

***Flash, Crash, Rumble and Roll.*** Branley, Franklyn M. HarperTropy, 1999.

***Flick a Switch: How Electricity Gets to Your Home.*** Seuling, Barbara. Holiday House, 2003.

***Sing a Song of Popcorn.*** White, Mary Michaels, Ed., Moore, Eva, Ed., De Regniers, Beatrice Schenk, Ed., and Carr, Jan, Ed. Scholastic, 1988.

***Solar Power.*** Petersen, Christine. Children's Press, 2004.

***Sound and Light.*** Angliss, Sarah. Kingfisher, 2001.

***Switch On, Switch Off.*** Berger, Melvin. HarperCollins, 2001.

***Water Power.*** Petersen, Christine. Children's Press, 2004.

# Life Science • Chapter 1

## Plants

### Lesson 1 What Living Things Need

All living things share certain characteristics: organization, growth, reproduction, the need for food, excretion of waste, respiration, and the ability to respond to stimuli. From the simplest microorganism to the most complex plants and animals, the bodies of all living things consist of cells organized in specific ways. Growth is the increase in size of the total organism, not just of particular parts. Reproduction is the creation of new, similar individuals. All living things require food for energy. Animals and some microorganisms ingest food, whereas plants, algae, and other microorganisms use light to produce their own food. Higher animals have specialized organs and systems to handle excretion. Respiration is the exchange of gases with the environment (for plants it is the intake of carbon dioxide and the release of oxygen; for animals it is the intake of oxygen and release of carbon dioxide). Responses to stimuli generally include movement. Phototropism, a plant's growth toward light, is a form of movement. All living things (with a few exceptions, e.g., mules cannot reproduce) display all of these characteristics. Some nonliving things have from one to a few of these characteristics. Crystals, for example, grow even though they are nonliving, and machines move.

Phototropism

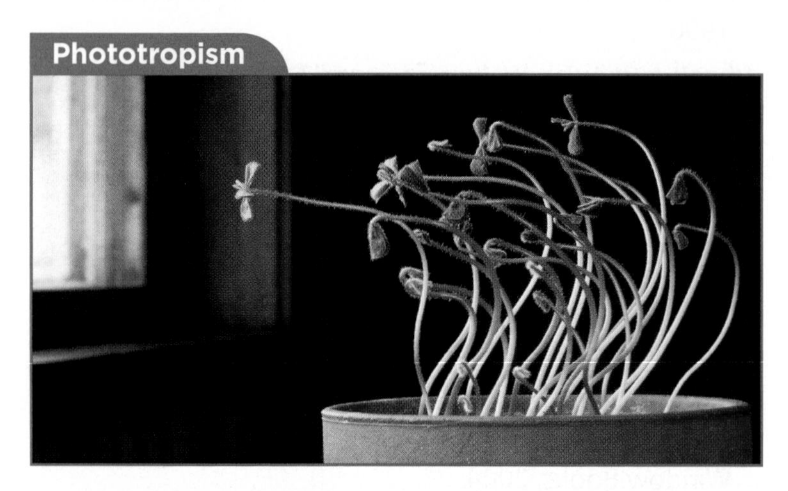

Plants need sunlight, water, carbon dioxide, oxygen, and nutrients that supply elements, such as nitrogen and phosphorus. Animals need air, water, food, and shelter.

Plants are divided into two groups, nonvascular and vascular. The **nonvascular plants** don't have systems that transport water and nutrients. Instead of roots, they have thin growths called rhizoids that help anchor them. Without a system to transport water, they simply soak up water like a sponge. Examples are mosses, liverworts, and hornworts. **Vascular plants**, such as trees, most familiar plants, and grasses, have roots, stems, and leaves.

Roots anchor the plant in the soil; absorb water, oxygen, and minerals; and store organic materials. There are two types of **root** systems. A tap-root system has a main large root, with smaller roots branching off from it. A diffuse root system has numerous slender roots with even smaller roots branching off from them. Roots are covered with many tiny growths called **root hairs**, which increase the surface area, helping the roots to absorb water and minerals. In many plants, the root system is larger than the parts of the plant that are above ground. The slow buildup of pressure exerted by growing roots is enough to split rock.

From the roots, the water and minerals enter the **stem**. Running lengthwise through the roots and stem are vascular bundles, which consist of two types of tissues, **xylem** and **phloem**, separated by a layer called the cambium. Xylem cells, which are on the inner side of the cambium, carry water and mineral nutrients up to the leaves. Phloem cells, which are outside the cambium, carry food made in the leaves to other parts of the plant. Xylem begins as living tissue. As it begins transporting water, however, its cells lose their contents and the dead cells join together to form tiny hollow tubes. The wood of a tree trunk consists mainly of dead xylem tubes that have dried out.

Xylem and phloem carry the water and nutrients throughout the **leaves**. Leaves have small pores called **stomata**. Water evaporates and exits through the stomata in the process called **transpiration**. About 99% of the water that enters a plant through its roots evaporates through the leaves. Leaves are also the site of **photosynthesis**, the process by which plants take energy from sunlight and use it to convert carbon dioxide and water into carbohydrates (also called sugars or starches) and oxygen. Small green structures in the leaves called **chloroplasts** capture the sunlight. Chloroplasts get their color from the green pigment chlorophyll, which absorbs red and blue light and reflects green light. The energy stored as carbohydrates feeds animals, including humans, who eat the plants. It also serves as food for the plant itself. Photosynthesis also supplies all the oxygen required by animals for respiration.

### Lesson 2 Plants Make New Plants

Plants can reproduce by both asexual and sexual reproduction. In some kinds of asexual reproduction, a small piece of plant tissue, such as a cutting, grows into a complete plant. In other kinds, a piece of tissue produces embryos, which then grow into adult plants. In asexual reproduction, the new plants are basically the same as the ones they came from.

In sexual reproduction, the egg of a female plant or plant part is fertilized by the sperm of a male plant or plant part. The new plant has characteristics of both parent plants. Some of the new plants may be less well equipped to survive in the environment, while others may be better equipped than the parent plants. Those that are better able to survive in the environment are more likely to have long life spans and to reproduce.

Most plants are flowering. Botanists classify **flowers** into two groups. Perfect flowers contain both female and male parts. Imperfect flowers contain either the male or female parts. The male part of the flower is the **stamen**, a stalk with an anther on top that produces pollen that contains the sperm. The female part is the **pistil**, which contains the ovary. Inside the ovary are ovules which contain egg cells. When egg and sperm cells join, they form an embryo, which grows inside a **seed**. Perfect flowers can produce seeds without any assistance. Plants with imperfect flowers depend on water, wind, insects, and animals to move pollen to the flowers with eggs to make seeds. Those that need to attract insects tend to have more colorful and fragrant flowers. Some seeds, such as those of some orchids, are as small as a speck of dust. The largest seeds are those of the coconut. The seed contains the plant embryo and nutrients. An outer coat protects the seed. Seeds with very hard coats can remain dormant for a long time; until conditions are advantageous for germination.

The seeds of gymnosperms are naked, and those of angiosperms are enclosed within a fruit. Angiosperms are divided into monocots and dicots, based on the number of **cotyledons**, also called "seed leaves," formed by the embryo to absorb nutrients stored in the seed until the plant is mature enough to produce leaves capable of photosynthesis. **Monocots** have seeds with one cotyledon, and **dicots** have seeds with two cotyledons. The flowers of monocots usually have parts in multiples of three and those of dicots in multiples of five. Some plants, such as the pine tree, form **cones** instead of flowers for reproduction. The same plant usually has male cones that produce pollen and female cones that contain ovaries. Most cone-bearing trees are evergreens;

**Monocot and Dicot Seeds**

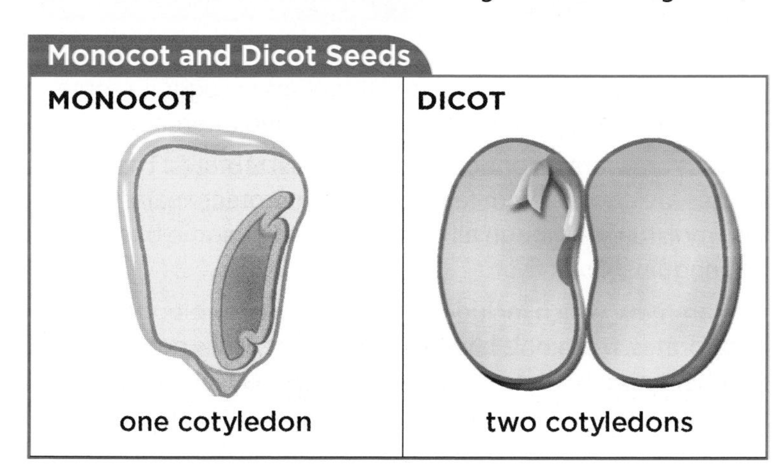


as their name indicates, they have green leaves year round.

Plants use a number of cues that signal when it is time to **germinate**, or sprout. These include the length of daylight, temperature, moisture, and the presence of animal digestive juices (an indication that the seed has been transported). As the seed develops into a seedling and then a mature plant, the roots grow downward and the stem grows upward. Flowers bloom, and when egg cells at the base of the flower are fertilized by sperm contained in pollen, seeds form. The seeds are then dispersed, to await germination and a new cycle.

**Annuals** complete the entire life cycle in one year, dying after seed formation. They include corn, beans, and pansies. A **biennial** completes its life cycle over two years. During the first year it grows roots, a compact stem, and leaves; it then survives the winter on stored food. During the second year it grows a vertical stem, flowers, fruits, and seeds; then it dies. Biennials include carrots, onions, and raspberries. Most plants are **perennials** and can live for many years. Each year the perennial forms seeds that grow into adult plants the following year.

**Female and Male Cones**

## Lesson 3 How Plants Are Alike and Different

Many plant and animal traits are inherited. **Genetics** is the study of heredity, and **genes** are the units of inheritance. The nucleus of every living cell contains **chromosomes**, which consist of **DNA** molecules. Segments of the DNA contain genes, which carry the coded information that is unique to each organism. Every species has a **genome**, which is the sum of genetic information for members of that species. Each unique individual is a variation on the genome of its species. We are all part of the same species, yet we all look slightly different. (Identical twins, who share all the same genes, are an exception.) You will never find a human being with thorns or feathers, as those traits are not part of the human genome.

Plants don't have a nervous system or sensory organs, but they do have ways of responding to their environment. The term **tropism** refers to the involuntary response of an organism to grow toward or away from an external stimulus. The two most important tropisms for a plant are **phototropism**, growth relative to light, and **gravitropism**, growth relative to gravity. (Gravitropism used to be called geotropism.) A plant's roots are negatively phototropic (they grow away from light) and positively gravitropic (they grow toward the force of gravity). The stem is positively phototropic and negatively gravitropic. Stems are also positively hydrotropic, growing toward moisture. Sensitivity to physical contact is called thigmotropism. Carnivorous plants are positively thigmotropic in the presence of prey. The mimosa is negatively thigmotropic, growing away from touch.

# Life Science • Chapter 2
# Animals

## Lesson 1 Animal Groups

The practice of naming and classifying organisms is called **taxonomy**. Scientists place all organisms into one of six kingdoms: plants, animals, fungi, protista, archea, and bacteria. The animal kingdom can be divided into two groups: those with backbones, called **vertebrates**, and those without backbones, called **invertebrates**.

Although many familiar animals are vertebrates, 95% of the more than 1.5 million known animal species are invertebrates. Because of their size and mobility, however, vertebrates tend to dominate their environment. Vertebrates are divided into **warm-blooded** and **cold-blooded**. Warm-blooded animals maintain a relatively constant body temperature. The cold-blooded animals depend on their environment to regulate their body temperature. Only mammals and birds are warm-blooded. Some large fish such as sharks and tuna, however, are able to conserve body heat and maintain body temperatures higher than the water around them.

Vertebrates can also be classified into five classes: fish, amphibians, reptiles, birds, and mammals. Fish are aquatic animals that have fins and internal **gills**. The fins allow them to propel themselves and change direction as they move through the water. The gills, located in slit-like passages between the throat and exterior, enable the fish to breathe in water. Water enters the fish's mouth. As the oxygen-rich water passes over tiny blood-filled filaments in the gills, oxygen crosses from the water into the capillaries of the gills, from which it goes to the body's tissues, and carbon dioxide is released back into the water.

**Fish Gills**

Most fish eggs are fertilized after they are deposited in the water. Because most of the eggs are eaten by other animals, only if the fish lays a tremendous number will any survive. In many species, the female produces as many as 5 million eggs during a single spawning period (as long as several months, depending on the species). As few as 10 out of one million may survive.

Amphibians include frogs and toads, salamanders and newts, and caecilians, which are limbless animals that resemble large worms. Certain species of amphibians can regenerate not only amputated tails and limbs, but also parts of the eye, lower jaw, intestine, and heart. Most amphibians grow from water-dwelling young into an adult that lives on land and in water.

Reptiles, which include turtles, crocodiles, lizards, and snakes, are cold-blooded land-dwelling vertebrates. Although some turtles live in fresh water or in the ocean, all turtles breathe by means of lungs and lay their eggs on land. Some crocodiles and snakes live in the ocean and return to land only to lay eggs. Because they are cold-blooded, they must live in warm climates or bask in the midday sun to bring up their body temperature. Although most reptiles lay eggs, in some species the eggs are incubated and hatched internally.

**Gentoo Penguin with Chicks**

Birds are warm-blooded, egg-laying vertebrates. Their bodies are covered with feathers, which protect against cold and moisture. Although all birds have wings, some birds, such as penguins, don't fly.

Mammals, which include humans, are warm-blooded vertebrates. Mammals have three characteristics not found in other animals: three middle-ear bones, hair, and milk-producing mammary glands. Familiar mammals include cats, dogs, horses, cows, pigs, and rodents. Not all mammals live on land. Aquatic mammals include the sea lion, walrus, whale, porpoise, and dolphin.

The largest group of invertebrates is the **Arthropods**. Of all known animal species, about 75% are Arthropods. Arthropods have a segmented body covered by an exoskeleton (external skeleton) and jointed legs. Arthropods include not only six-legged insects, but also the eight-legged Arachnids, such as spiders, scorpions, and ticks; and Crustaceans, such as shrimp, crabs, and lobsters; millipedes; and centipedes.

Biologists have identified about 800,000 species of insects. There may be as many as 10 million living insect species. Most insects have wings, antennae, and compound eyes (eyes consisting of separate units, each with its own lens).

**Mollusks** make up the second largest group of invertebrates. These include the scallop, clam, oyster, mussel, snail, slug, squid, and octopus. All mollusks have a soft body, and many have a hard external shell. The Cephalopods, a class of mollusks that includes the squid, octopus, cuttlefish, nautilus, and giant squid, can grow very large.

## Lesson 2 Animals Grow and Change

Some animals lay eggs and others give birth to live young. Fish lay eggs, although some, such as guppies and sharks, gestate the eggs inside their body until they hatch. Most frogs and toads lay eggs, but some lizards and snakes bear live young. All birds lay eggs. The only egg-laying mammals are the duckbill platypus and spiny anteater.

The shell of a bird's egg is often colored and marked with streaks or spots that provide camouflage. The shell has tiny pores that allow gases to pass in and out. The yolk supplies nutrients, and the white supplies water and absorbs shock. Birds incubate their eggs and nest in hollow trees.

**Snake Shedding Skin**

The eggs of both fish and amphibians, such as frogs and toads, must be laid in water or they will dry out. A frog spends its early life as a fish-like tadpole, breathing with gills. The *froglet*, a stage between a tadpole and an adult, breathes with lungs before its body absorbs its tail. Reptilian eggs have a leathery shell that prevents them from drying out. In some species, the eggs are incubated and hatched internally.

Some insects go through egg, **larva**, and **pupa** stages on the way to adulthood. Many insect larvae are white and soft. An example is the fly larva, the maggot. Caterpillars are the larvae of moths and butterflies. After sufficient growth, larvae enter the inactive pupa stage, which has a protective case. In the pupa the larva, which does not grow larger, is transformed into an adult, which then sheds its outer covering.

Some animals periodically shed their skin, scales, shell, feathers, or fur. Most birds **molt** every summer. Amphibians and snakes shed their skin every few months. Snakes shed their skin in one piece, crawling out of it over the course of a day. Amphibians shed their skin in one piece and then eat it. In spring, fur-covered mammals shed their heavy winter coat. Insects and crustaceans must shed their exoskeletons in order to grow. Most caterpillars molt several times before pupation.

## Lesson 3 Staying Alive

**Adaptations** are features and behaviors that enable a species to survive in its environment. A popular example is the giraffe's long neck, which enables it to eat leaves high in the trees. The giraffe's neck did not grow long for the purpose of reaching up high. Rather, any giraffes that were born with longer than average necks were better able to survive, especially when food was scarce. If more longer-necked than shorter-necked giraffes survived, they also reproduced more. Eventually, entire giraffe populations had long necks.

A variety of animals have **camouflage** that enables them to blend in with their surroundings and avoid predators. Some animals have camouflage that changes color with the season. The arctic fox has a dark coat in the spring and summer that matches the brown dirt, and a white coat in the winter that matches the snow. Animals often detect the changing seasons by perceiving differences in the length of the day or night. This is a more reliable guide than temperature, which can vary greatly from day to day.

The opposite of camouflage is warning coloration. The wasp's striped belly warns would-be predators of its sting. Some animals, particularly butterflies, mimic the markings of other animals. Many moths have "eye-spots" on their wings, which imitate the eyes of larger animals.

Animals that don't migrate have other ways of dealing with winter. The best known is **hibernation**. A hibernating animal stores enough food in its body fat to enable it to survive the winter. The extra fat also helps it to keep warm. During hibernation, the animal is in a dormant state. The animal does not grow, and it does not eat or drink. All of its bodily processes slow down. A hibernating bear's heart, for example, slows to 20% of its usual rate. Mammals are not the only animals to hibernate. Many fishes, amphibians, and reptiles also hibernate.

**Tau Emperor Moth**

# Life Science • Chapter 3
## Looking at Habitats

### Lesson 1 Places to Live

The word **environment** refers to factors such as climate, soil composition, and native plants and animals that affect an organism. **Habitat** more broadly describes a setting in which a species normally lives. A habitat must meet an organism's basic needs for food, shelter, water, oxygen, and a place to rear young. Some habitats, such as the hot and dry Sahara, the wet rain forest, and the bitterly cold Arctic, are characterized by extreme conditions. But many habitats are less extreme. A city park is a habitat. The **physical factors** of a habitat include elevation, soil type, and water resources. The **biotic factors** include all the other plant and animals species that share that habitat.

All of the habitats in which a plant or animal lives make up its **geographic range**. The geographic range of giraffes, for example, is Africa. Because of our ability to manipulate our surroundings, the geographic range of humans is virtually the entire planet. The geographic range of an inch-long fish called the Devils Hole pupfish, an endangered species, is a single, water-filled cavern (Devils Hole), just east of Death Valley, California. Habitat loss is the main reason species become endangered or extinct. Species with a limited geographic range are more likely to become endangered or extinct.

Animals, and to an extent plants, are categorized as **generalists** or **specialists**. Generalists can live in a variety of conditions and eat a wide range of food. Most omnivores, which eat both plants and animals, are generalists. Some herbivores, which eat only plants, eat a wide enough range of plants that they are considered generalists. Specialists live in a narrow range of habitats and eat a limited diet. The giant panda is a specialist; more than ninety-five percent of its diet consists of bamboo. The majority of species are generalists. One reason is that when habitats degrade or are destroyed, specialist species tend to die out, leaving only generalist species.

Plants and animals have adaptations that enable them to live in particular habitats. An animal with thick fur will survive better in a cold climate than will an animal with sparse fur. The animals with thick fur will be healthier and have more offspring, and the animals with sparse fur will die out. Should the climate suddenly warm up, however, animals with sparse fur will thrive and reproduce more, and the thick-furred animals may die out.

### Lesson 2 Food Chains and Food Webs

Ultimately, most of the energy that supports all life on Earth comes from the Sun. **Food chains** show how energy and organic compounds pass from one living thing to another. Each habitat has its own food chains, composed of producers, consumers, and decomposers. A food chain begins with **producers**, living things that produce their own food, such as plants that produce food by photosynthesis. All of the animals in the chain, called **consumers**, depend on the producers. **Primary consumers** eat the producers (e.g. **herbivores**, such as cows and elephants, eat plants). **Secondary consumers** eat the primary consumers, and **tertiary consumers** eat secondary consumers. Secondary and tertiary consumers that eat meat are called **carnivores**. Consumers that kill and eat other animals are **predators**, and the animals they kill are **prey**. Consumers that find dead animals to eat are called **scavengers**. Each food chain ends with bacteria and fungi that eat the remains of dead animals. These **decomposers** break down the dead material, absorbing some of the products for their own use and returning others to the soil, to be used again by plants in new food chains. There are usually no more than six levels in a food chain.

Changes at a single level can affect an entire food chain. When sea otter populations are reduced by hunting, for example, it causes an increase in the animals eaten by sea otters, such as sea urchins. The sea urchins then overgraze on the kelp, depleting it. This affects not only sea urchins, but also other kelp-eating species.

Interconnected food chains create **food webs**, which more accurately describe the relationships among the populations of a given habitat. A food web is a map of how energy flows, with the food chains as individual routes. Within a web, an animal can belong to more than one consumer group (and chain), depending on what it is eating. A bear is a primary consumer when eating grass, but a secondary or tertiary consumer when eating other animals.

**Energy Pyramid**

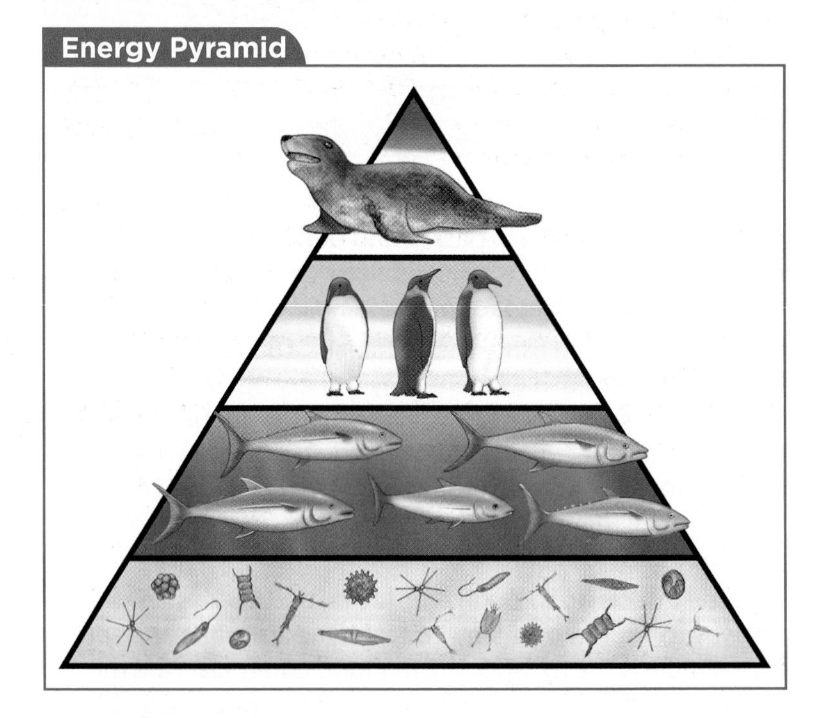

An **energy pyramid** is another way of depicting the flow of energy and matter within a food chain. At each level of the pyramid, the efficiency is about 10% of the efficiency of the layer below. Some plants and animals remain uneaten, and some parts of animals do not get eaten, such as beaks and shells. Also, whenever energy is changed from one form to another, not all of it ends up in a usable form. Some of it is lost as heat. The more levels there are in a food chain, the more energy is lost. Thus, a food chain with only two levels retains a higher percentage of energy from its bottom to its top.

Coral Reef

Lesson 3

## Habitats Change

Although habitats are affected by natural events, such as droughts, earthquakes, and volcanoes, most habitat changes are caused by human activities, including agriculture, urban development, industry, and recreation. Many organisms also alter the structure or chemistry of their surroundings, affecting other species that occupy the same habitat. The best-known example of this is the beaver, which builds dams. Coral reefs protect other species from waves, and many trees provide shade and shelter for animals.

**Adaptations**, which enable plants and animals to survive in their habitats, take place slowly, over generations. As adaptations develop, characteristics of plant and animal populations gradually change. An individual animal with sparse fur cannot suddenly grow thick fur because the local climate suddenly turns much colder. When an individual organism responds to a sudden habitat change, as when an animal migrates during an unusually severe winter, it is called **accommodation**. Species die out, or become extinct, when they are unable to adapt to or accommodate changes in their habitat over the long term. About 99% of the species that have ever lived are extinct.

Part of an organism's adaptation to its habitat is its adaptation to other organisms in it. Each animal's behavior makes life more difficult for some of its neighbors and easier for others. **Invasive species** can have a particularly dramatic effect. These are species that move or are transported from their usual geographic area to invade another. Invasive species have a mostly negative impact. If they are better than the native species at obtaining resources, the number of native species may decrease.

Habitats are constantly changing. When a habitat changes suddenly, the adaptations that helped a species to survive can be ineffectual or even hinder its survival. A plant, for example, may have developed sap that is poisonous to local animals. If humans introduce a new animal species for whom the sap is not dangerous, the plant may not survive. Because adaptations develop over many generations, individual plants would not be able to develop new toxins in time to avoid destruction by the animal. In Alaska, higher temperatures increased the survival rate of a type of beetle called the spruce bark beetle, allowing it to kill about 3 million acres of the trees. The tree had adapted to the beetle by developing a pesticide in its resin, but the pesticide was insufficient to protect it against the enlarged beetle population.

An adaptation that comes about in response to a particular stress may end up increasing a species' survival rate for other reasons. Scientists think that bone tissue originally developed as a way to store inorganic materials such as phosphates. A rigid tissue was best able to do so. Once hard bone tissue had developed, it was also useful as a structural material.

Scientists use a variety of means to track habitat change. Particularly useful are the images obtained from the government's Landsat satellites. The first **Landsat** satellite was launched in 1972 and the most recent, the seventh, in 1999. The satellites remain in the same path, about 700 kilometers (435 miles) above Earth, taking a new image every 16 days. The satellites don't take photographs, but create digital images by detecting the light absorption rate of different objects. The images help scientists to observe changes in such features as land elevation and slope, precipitation, and plants and shrubs.

Because habitats are so complex, a single change can have far-reaching consequences. This makes accurate forecasting very difficult. Still, predictions about how a change will affect a habitat can aid our efforts to slow down habitat destruction and to protect endangered species.

Landsat Images Before and After Tsunami in Thailand

# Life Science • Chapter 4
## Kinds of Habitats

### Lesson 1 Forests

Forests occupy almost 40% of Earth's surface and support a great diversity of plants and animals. Trees are the dominant form of vegetation in forests. There are three basic types of forest, reflecting the different climates in which they occur, tropical, temperate, and boreal (northern).

**Tropical hardwood** forests, which include rain forests, are found in tropical regions (tropical means close to the equator). The trees include ebony, ironwood, mahogany, and teak. They drop their leaves every 2-10 years, at any time of the year rather than at a particular season. **Tropical rain forests** are very warm and humid year round. They have an average yearly rainfall of 50 to 260 inches and a narrow temperature range of 20 degrees Celsius (C) (68 degrees Fahrenheit (F)) to nearly 34 degrees C (about 93 degrees F). Although tropical rain forests cover only 6% of Earth's surface, with their abundant moisture and relatively narrow temperature range, they are hospitable to many kinds of plants and animals.

The lowest layer of a rain forest is the forest floor. The **understory**, the layer above the floor, consists of trees that are about 60 feet tall, as well as trunks of the taller canopy trees, plants, and shrubs. Above the understory is the **canopy**, created by the many treetops growing close together. These trees are 60 to 130 feet tall. Rising above the canopy are the emergent trees, which are spaced far apart and grow to 100 to 240 feet.

Rain Forest Layers

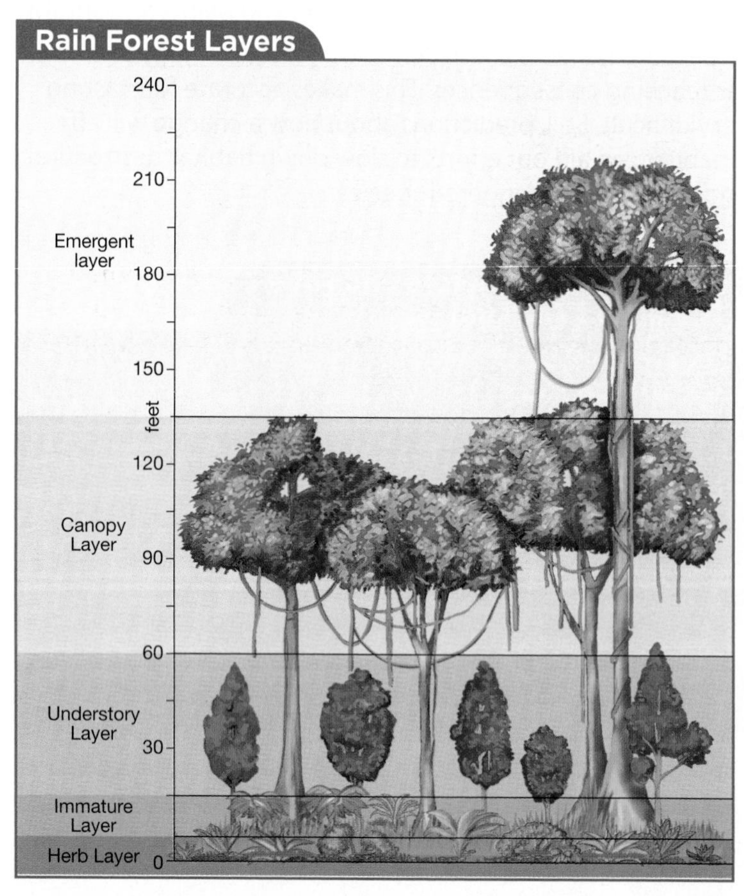


Plants in the dark understory have large leaves that help them absorb as much sunlight as possible. Most rain-forest trees have thin, smooth bark because they don't need to protect the tree from freezing temperatures or water loss. Many trees of the upper canopy have leaves with "spouts" that allow excess water to drip off, which prevents mildew from forming. Vines growing on the trees account for 40% of the canopy leaves.

The forest floor is very dark, as little sunlight reaches it. Insects, frogs, snakes, and mice use leaves and plants that drop from above for food and shelter. A few larger animals, such as wild pigs, also live here. Although the understory gets little light, it is home to many birds and other animals. Large cats, such as the leopard, climb the trees to prey on monkeys and squirrels. Most rain forest animals live in the upper canopy. Sloths, which rarely come down from the trees, sleep for 18 hours at a time, hanging upside down from branches. They move so slowly that algae and moss grow on their fur, giving them a green appearance that helps to camouflage them.

The largest parrots, macaws, nest in holes of the canopy and emergent trees. Their hooked beaks are adapted to eating seeds and fruits and opening nut pods. Unfortunately, they are endangered, due to destruction of the rain forest and the demand for them as pets. The spider monkey has a long tail that it uses like a fifth limb. It gets its name from the long, spindly appearance of its legs. Although the poison-arrow frog is less than an inch long, it holds enough poison to kill about 100 people. An amount of poison smaller than a grain of salt can kill a human. Native hunters put the poison on arrow tips.

Because the soil of tropical rain forests is poor in nutrients, the relationships between plants and animals are crucial to the existence of the rain forest itself. When leaves fall to the forest floor, insects, bacteria, and fungi quickly cause them to rot, releasing the nutrients in them. Termites do the same to fallen twigs and branches. The extensive shallow root systems of the trees, called "root mats," then take up the nutrients.

The **temperate hardwood** forests of Asia, Europe, and North America have seasonal rains. The trees are mainly ash, beech, elm, maple, and oak. They are **deciduous**, losing their leaves at the end of the growing season. In autumn, when the chlorophyll in the leaves decreases, deciduous forests are ablaze with the vivid colors of leaves about to fall off. Unlike tropical rain forests, **temperate rain forests** have seasons. Summer temperatures can reach 27 degrees C (or 80 F), and winter temperatures can drop to 0 degrees C (32 degrees F). Because of the cool winters, temperate rain forests have a less diverse ecology than tropical rain forests.

**Boreal** forests are zones of evergreens that encircle the Northern Hemisphere between the hardwood forests and tundras in the north, and between the hardwood forests and prairies in the south.

## Lesson 2 Hot and Cold Deserts

A **desert** has 10 inches or less annual rainfall. Although many deserts, such as the Sahara, are very hot, tundra are considered cold deserts. Plant and animal life are less abundant in deserts than in other habitats. In moister habitats, most plants are **perennials**, which live for many seasons. In the desert, most are **annuals**, which die after one season. Some annuals produce vast numbers of seeds that can survive for years, providing a "seed bank" until conditions turn favorable.

The ability of succulents, including cacti, to store water is an adaptation to the dryness, as are the very deep roots of some other desert plants. Desert animals, particularly insects, obtain water directly from succulent plants. The abundance of desert insects feeds many birds, bats, and lizards.

Because hot deserts are cooler at night, many animals are **nocturnal**. One nocturnal animal is the kangaroo rat, whose hop resembles that of a kangaroo. Kangaroo rats rarely drink water. They get moisture for their bodies from digested food. Camels do not store water in their bodies; they conserve water. Their body temperature, which is about 34 degrees C (about 93 degrees F) in the morning, can rise to as high as 41 degrees C (about 106 degrees F) during the day before they lose water to perspiration. Camels can go several days without drinking. When water is available, they can drink as much as 15 gallons at a time. A camel's hump contains fat. When food is scarce, the camel uses the fat as an energy source, causing the hump to droop and bend over.

**Bactrian Camel**

**Dromedary Camel**

The arctic **tundra** lies north of the tree-growing areas of North America and Eurasia. About six inches below the soil surface is a layer of permanently frozen ground called **permafrost**. The permafrost can be anywhere from less than a foot to several thousands of feet thick. Because of the permafrost, arctic plants have short roots. During the brief spring and summer, when the snow melts, the permafrost prevents what water there is from draining into the ground, and the soil becomes very soggy. The growing season, during the thaw, is only 50 to 60 days long.

Many arctic plants, such as grasses, mosses, and shrubs, are small. The air is warmer close to the ground, and they are protected from gale winds and blowing snow. Like people huddling together in the cold, arctic plants often grow close together, trapping the warmer air. The plants are often darkly colored, which increases heat absorption. Many arctic plants do not lose their leaves all at one time, enabling them to take advantage of sunlight year round.

Small tundra animals include worms, mites, and spiders. The few large animals in the tundra include the arctic fox, caribou, musk ox, and snowshoe rabbit.

## Lesson 3 Oceans and Ponds

The term **freshwater** refers to bodies of water that, unlike the ocean, are not salty. They include ponds, lakes, rivers, and bogs. As on land, almost all aquatic plant and animal life ultimately depend on photosynthesis. Ponds and lakes have submerged plants, both rooted and rootless; plants such as duckweed, whose leaves float on the surface; and emergent plants, such as the broadleaf arrowhead, which rise above the water surface. Because of its current, a river is less suitable for floating and emergent plants.

**Oceans** cover about 75% of Earth's surface. Because the ocean is so deep, most of it is too dark to support rooted plants. But the ocean contains abundant phytoplankton, which are single-celled algae. (Algae do not belong to the plant kingdom, but because they undergo photosynthesis, they are referred to as plantlike.) Drifting animals called zooplankton feed on the phytoplankton, and larger fish eat the zooplankton.

Some animals at the bottom of the ocean emit light through bioluminescence, a process that transforms chemical energy into light energy. (Fireflies are a familiar example of bioluminescence.) Some fish have light-emitting bacteria living on them, and other fish emit light themselves. The light allows the animals to signal other members of the same species and to find and attract prey.

The fish's fins and streamlined body, which enable it to swim, are its most obvious adaptations to living in water. Most fish also have a swim bladder, an internal gas-filled chamber that increases its volume without increasing its weight. This makes the fish more buoyant. Fish, such as sharks, that don't have a swim bladder must keep moving to avoid sinking.

# Earth and Space Science • Chapter 5
## Land and Water

### Lesson 1 Earth's Land

The term **topography** refers to all of the characteristics of the land, particularly elevation and shape. Topographic maps, also called topo maps, show land features such as mountains, valleys, and plains, as well as bodies of water, forests, roads, urban areas, and distinctive landmarks. Unlike other types of maps, topographic maps depict relief, or three-dimensionality, through contour lines and sometimes cross-hatching. Topographic data have many applications for agriculture, urban development, water resources, weather prediction, conservation, aircraft flight patterns, and military strategy.

The larger a map's scale, the more detail can be shown. A map on which one inch equals 609 meters (about 2,000 feet) can show much more detail than a map on which one inch equals 6 kilometers (3.73 miles).

**Geologic Map of New Jersey**

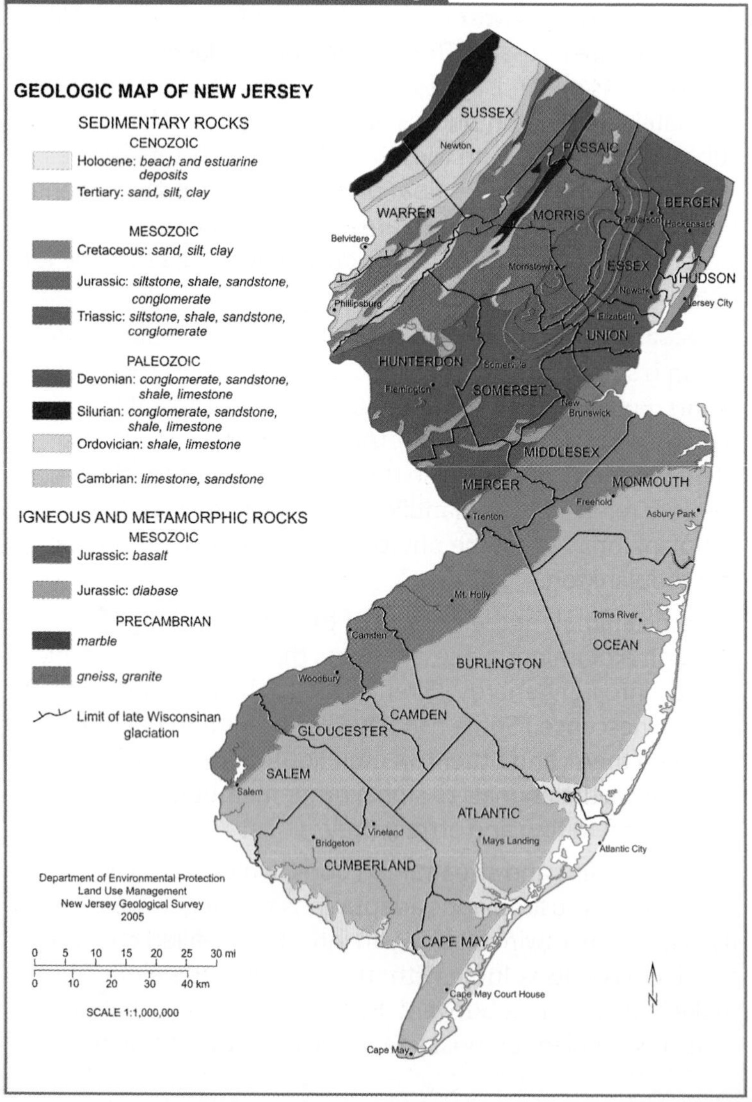
GEOLOGIC MAP OF NEW JERSEY
SEDIMENTARY ROCKS
CENOZOIC
Holocene: beach and estuarine deposits
Tertiary: sand, silt, clay
MESOZOIC
Cretaceous: sand, silt, clay
Jurassic: siltstone, shale, sandstone, conglomerate
Triassic: siltstone, shale, sandstone, conglomerate
PALEOZOIC
Devonian: conglomerate, sandstone, shale, limestone
Silurian: conglomerate, sandstone, shale, limestone
Ordovician: shale, limestone
Cambrian: limestone, sandstone
IGNEOUS AND METAMORPHIC ROCKS
MESOZOIC
Jurassic: basalt
Jurassic: diabase
PRECAMBRIAN
marble
gneiss, granite
Limit of late Wisconsinan glaciation
SUSSEX
PASSAIC
WARREN
MORRIS
BERGEN
ESSEX
HUDSON
UNION
HUNTERDON
SOMERSET
MIDDLESEX
MERCER
MONMOUTH
BURLINGTON
OCEAN
CAMDEN
GLOUCESTER
SALEM
ATLANTIC
CUMBERLAND
CAPE MAY
Newton
Belvidere
Phillipsburg
Morristown
Newark
Jersey City
Elizabeth
Somerville
Flemington
New Brunswick
Trenton
Freehold
Asbury Park
Mt. Holly
Toms River
Camden
Woodbury
Salem
Vineland
Bridgeton
Mays Landing
Atlantic City
Cape May Court House
Cape May
Department of Environmental Protection
Land Use Management
New Jersey Geological Survey
2005
0 5 10 15 20 25 30 mi
0 10 20 30 40 km
SCALE 1:1,000,000
N

Geologic maps show the types and ages of rock that lie beneath the soil; they also provide information about rocks at greater depth. Geologic maps are useful for land- and water-use planning. They also help scientists to understand Earth's composition, structure, and history.

Earth's rigid outer shell, which averages about 100 km (62.14 miles) thick, is called the **lithosphere**. The lithosphere consists of the crust, which ranges from about 4.8 km (3 miles) thick under the oceans and to about 67.59 km (42 miles) thick under some mountains, as well as the uppermost part of the mantle. The lithosphere is divided into plates that form at mid-ocean ridges and include the continents and ocean floors. These moving plates ride on top of the **asthenosphere**, the semi-molten portion of the upper mantle. The mantle, which consists mainly of material that is rich in silicon, iron, and magnesium, is about 2,896 km (about 1,800 miles) thick. The core has a solid center surrounded by an outer liquid core. The liquid core consists of extremely hot, molten metals, mostly iron.

### Lesson 2 Earth's Water

Three-quarters of Earth's surface is covered by interconnected oceans, called the **world ocean**. Most of the remaining surface is covered by land, and a small portion is covered by freshwater bodies. The four main ocean basins are the Pacific, Atlantic, Indian, and Arctic. In 2000, the International Hydrographic Organization recognized a fifth ocean basin, the Southern Ocean, which surrounds Antarctica and extends to 60 degrees latitude. The oceans are subdivided into seas. The largest seas are the South China Sea, Caribbean Sea, and Mediterranean Sea. The average ocean depth is 4.8 km (about 3 miles), and the deepest point, in the western Pacific, is almost 11.3 km (about 7 miles).

Most of the salt in seawater comes from the land. For millions of years, rainwater and rivers have washed over rocks containing salt, dissolving tiny amounts and carrying them to the oceans. **Rivers** are large freshwater streams that flow from headwater areas to their mouths, where they discharge into an ocean or major lake. Smaller tributaries are called brooks, creeks, and branches. **Lakes** are inland bodies of water, and ponds are small, shallow lakes. Although the majority of lakes are freshwater, there are also salt lakes. Salt lakes tend to be in areas with high evaporation.

### Lesson 3 Changes on Earth

**Soil** is a mixture of loose, disintegrated rock and decomposed plant and animal matter. Rocks are broken down and changed by the process of **weathering**. The two main types of weathering are physical and chemical. **Physical weathering** breaks down rocks into smaller pieces. The causes include the freezing and thawing of water, pressure of plant roots, actions of animals, and wind. Because water expands when frozen, water that freezes in the cracks of rocks can exert enough pressure to split rocks apart. Tree roots can exert enough pressure in cracks and joints to break rock.

With **chemical weathering**, rainwater, acids, and oxygen interact with minerals in rocks, changing them into other minerals. Rocks have different susceptibilities to chemical weathering. Limestone and marble react chemically with mildly acidic rainwater. The tops of marble tombstones are often heavily weathered. In limestone caves, dripping water carrying dissolved limestone (calcium carbonate) causes the formation of stalactites that hang from the ceiling and the stalagmites that rise from the ground.

The term **erosion** refers to the loosening and transporting of soil and rock. The main agents of erosion are moving water and wind. All of them combine with gravity to produce erosion. All streams and rivers carry rock fragments and soil acquired from tributaries or rubbed off from their banks. As this material is transported by water, it grinds against the rock beneath the water, creating even more fragments. Ocean waves near the coast loosen and carry away soil and rock. In desert and beach areas, the wind can remove sand from one area and deposit it in another. Erosion is a natural process, but human activities, such as deforestation and poor farming practices, can increase the rate of erosion. When the rate of soil erosion is greater than that of soil-formation, the topsoil gradually disappears.

Landforms shaped by rivers and streams, called **fluvial landforms**, are the most common kind of landforms. Two categories of fluvial landforms are erosional and depositional. The main erosional form is the valley. As water flows, it cuts into the underlying rock through erosion. Landforms caused by deposition include floodplains, deltas, and dunes. Sediment that is deposited by a river during periodic floods produces a broad, flat valley floor called a **floodplain**.

**Glaciers** are large, moving masses of ice that form from compacted, recrystallized snow. Glaciers can alter landforms through both erosion and deposition. Most glacial erosion occurs in valleys, which subsequently deepen and widen. Debris carried in the underside of glaciers acts as an abrasive. Through **plucking**, rock pieces are picked up and incorporated into the glacier. Glaciers can carry sediment many miles before it is deposited.

### Landslides

Most **landslides** are triggered by excessive precipitation or melted snow. Water loosens rock, lubricates the surface between the rock or soil and bedrock below, and adds weight to loose material. Much of the ground in areas prone to landslides consists of fragments of weathered rock that is on its way to becoming soil. This material is less stable than soil, which has already been compacted into horizontal layers. The risk of landslides has risen in recent years, partly due to extensive deforestation and an increase in severe weather caused by global climate change.

### Earthquakes and Volcanoes

In an earthquake, sudden movements within the lithosphere release energy that travels through the surrounding area as **seismic waves**. The largest earthquake in a sequence of earthquakes is called the **main shock**. Smaller shocks called **foreshocks** occur days to weeks before the main shock. Most main shocks are followed by gradually diminishing **aftershocks**, which can continue for days afterward. Scientists use **seismographs** to detect seismic disturbances. By measuring the amplitude of the waves recorded by seismographs, they can determine the magnitude of an earthquake according to the **Richter Scale**. The scale is logarithmic, which means that each increase by one unit corresponds to a tenfold increase in the amplitude of the seismic waves. The largest earthquake has a magnitude of about 9. Very small earthquakes may even have negative magnitudes. The waves of a magnitude 7 earthquake have a 10 times larger amplitude than those of a magnitude 6 earthquake, and 100 times larger than those of a magnitude 5 earthquake.

A volcano is a steep-sided hill or mountain formed by the accumulation of magma (molten lava) that erupted through vents in Earth's crust. In 2000, scientists predicted an eruption of Popocatépetl Volcano, about 40 miles south of Mexico City, just hours before its biggest eruption in 1,200 years. Volcano prediction is based on observing **patterns** of past behavior.

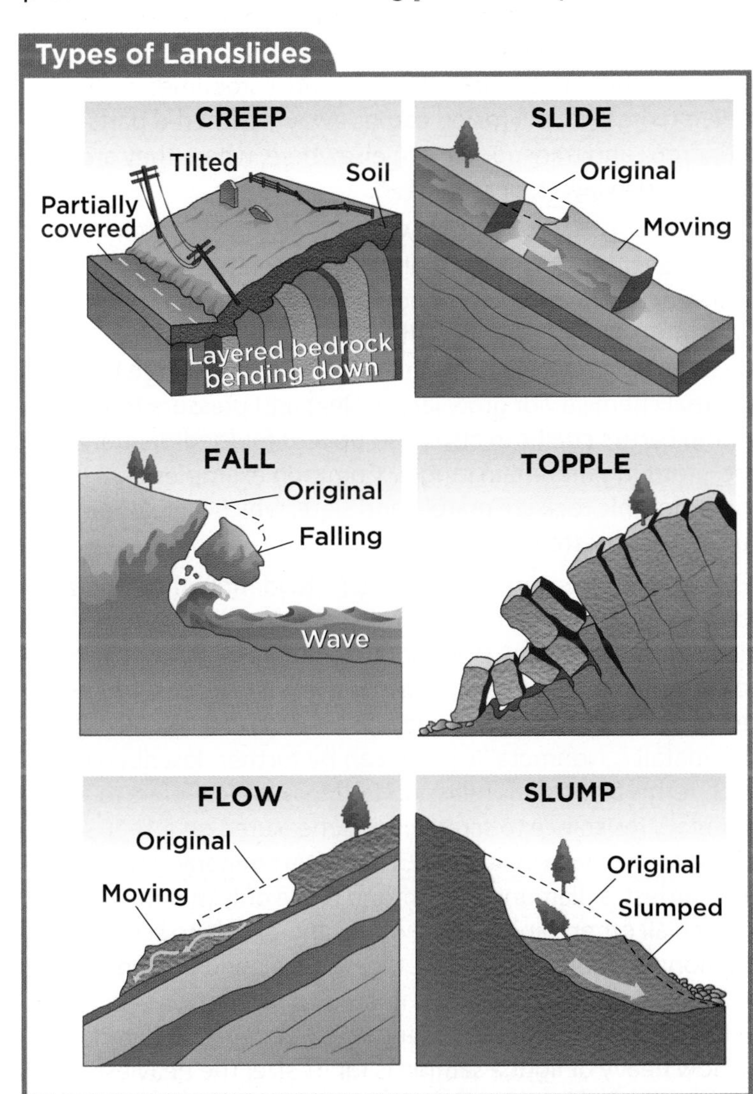

# Earth and Space Science • Chapter 6
## Earth's Resources

### Lesson 1 Rocks and Minerals

A rock is composed of one or more of the **minerals** that make up Earth's crust. A mineral is a naturally-occurring material that has a unique chemical formula and, usually, a crystalline structure. (The molecules of a crystalline structure have a specific arrangement.) Rocks are classified according to their origin. Their properties reflect the processes that formed them.

The three major rock classifications are igneous, sedimentary, and metamorphic. **Igneous** rock forms from the cooling of molten material from Earth's interior. The most common mineral in igneous rocks is feldspar, followed by quartz. Quartz is usually transparent and colorless. Other forms of quartz contain impurities that give them color, creating such semiprecious gemstones as amethyst, citrine quartz, and rose quartz.

Seventy percent of all rocks at Earth's surface are **sedimentary**. Sedimentary rocks form from older igneous, metamorphic, or sedimentary rocks and, sometimes, sediment of living organisms. Erosion carries away weathered parts of older rock and transports them elsewhere, where they are deposited in layers, or **stratified**.

**Metamorphic** rock forms when intense heat and/or pressure within Earth's crust alter the mineral composition and structure of pre-existing igneous, sedimentary, or metamorphic rock. When molten rock intrudes into sedimentary rock, the heat causes minerals to change their internal chemistry or grow larger. Heat and pressure from within Earth's crust can also push up and fold sedimentary rock, forming mountain ranges. Common examples of metamorphic rock are marble and slate, which were originally formed as limestone and shale.

Geologists use various **properties** to identify the minerals that compose rocks. These include color, luster, hardness, cleavage, and density. The color of a mineral is as it appears to the unaided eye. Luster, which is the character of light as reflected from the mineral, may be referred to as metallic or nonmetallic. Nonmetallic luster can be further described by such terms as vitreous (glassy). Hardness, which refers to a mineral's resistance to scratching, is measured on a 1–10 scale called **Mohs Hardness Scale**. The softest mineral is talc, and the hardest is diamond. A diamond is the only mineral able to scratch all other minerals. The only mineral able to scratch a diamond is another diamond. Cleavage, a mineral's tendency to break along flat surfaces, is related to planes of weaknesses in the crystal structure. The density of a mineral is related to how heavy or light a sample is for its size. The heaviest minerals include gold and galena (an ore of lead).

**Mohs Hardness Scale**

| Rating | Reference Mineral | Reference Objects (approximate value) |
|---|---|---|
| 1 | talc | |
| 2 | gypsum | fingernail (2.5) |
| 3 | calcite | copper penny (3.5) |
| 4 | fluorite | |
| 5 | apatite | glass plate (5.5) |
| 6 | potassium feldspar | steel file (6.5) |
| 7 | quartz | |
| 8 | topaz | |
| 9 | corundum | |
| 10 | diamond | |

### Lesson 2 Soil

Soil, the uppermost layer of Earth's surface, generally extends down from about one to six feet. Caused by weathering, it consists of loose, disintegrated rock and decomposed plant and animal matter. **Topsoil**, the uppermost layer of soil, extends down from several inches to several feet. It consists of rock fragments, sand and clay particles, microorganisms, and decayed plant and animal material called **humus**. Humus gives topsoil its dark color. Animals such as beetles, termites, and earthworms help keep the topsoil in good condition. Besides providing organic material, they move through the soil, helping to keep it well mixed and porous. Soil erosion occurs when the topsoil is blown away by wind or washed away by water.

**Soil Erosion**

**Subsoil**, the layer below the topsoil, usually extends down about one and a half to two feet below the topsoil. It contains more clay particles than the topsoil and very little, if any, humus. The absence of humus makes it lighter in color than topsoil. Subsoil is important for drainage. Below the subsoil is the **parent material**, which consists of the type of rock from which the upper two layers formed.

One way to enrich soil, as well as to recycle waste, is through **composting**. In nature, the remains of once-living organisms decompose and the nutrients return to the soil. In composting, kitchen and yard waste are placed in a container, where they are broken down by biological and chemical means. In a typical composter, layers of waste material are alternated with layers of soil, which contain bacteria and other microorganisms that increase the rate of decomposition. One composting method, which can be done indoors, uses worms. Redworms are added to a container filled with moistened waste. The same worms continue to live in the container as composted material is removed and fresh garbage is added.

## Lesson 3 Using Earth's Resources

One of the most effective ways to conserve natural resources is through the use of **renewable** energy sources. Renewable energy refers to any energy source that is replenished as quickly as it is used. These include solar, wind, and hydroelectric energy. Energy sources such as wood, which can be replenished within a reasonable period of time, are considered renewable. **Solar energy** is used to generate electricity and to heat homes and provide hot water. The most common way is through the **photovoltaic**, or solar, cell. The cell absorbs sunlight and converts it to electric current. Because the amount of current is small, and the voltage is low, multiple cells are connected in large **solar panels**.

**Wind energy** is actually converted solar energy. The Sun heats different parts of Earth at different rates. As hot air rises, cool air moves in beneath, and the resulting movement is the wind. The motion of the wind is a form of energy. A windmill or wind turbine is basically the opposite of a fan. A fan uses electricity to move the air. A turbine takes the energy of the moving air and converts it to electricity.

Wind Energy

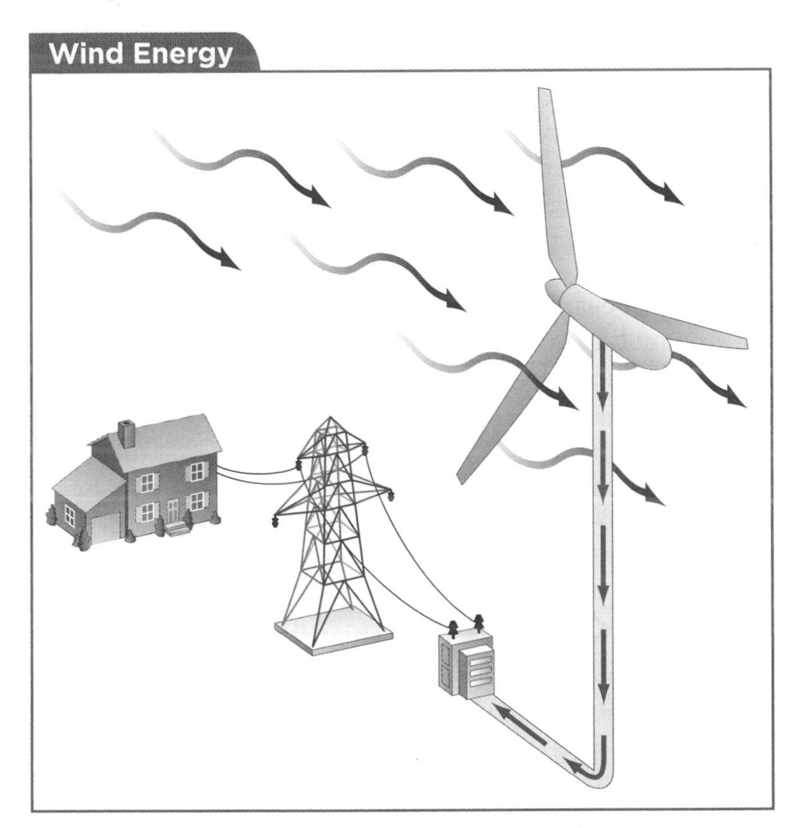

**Hydroelectric power** takes the energy of flowing water and converts it to electricity. In most hydroelectric plants, a dam on a river stores water in a reservoir. When water is released from the reservoir, it flows through a turbine, which activates a generator that produces electricity. **Nuclear energy** harvests the energy in nuclear reactions. It is both renewable, using breeder reactors, and nonrenewable, as it is based on mining uranium.

Glen Canyon Hydroelectric Dam

Reducing the amount of waste we produce is another important way to conserve Earth's resources. Reducing plastic waste, for example, helps to reduce the need for the large underground disposal sites called **landfills**. The fact that plastic is not biodegradable is not significant, as landfills are designed to minimize the biodegrading of waste. This is to protect against groundwater contamination and air pollution. Not all plastic waste ends up in landfills. Plastic bags, in particular, often end up as litter and are transported by the wind. They pose a serious threat to marine animals, which eat them. If they block an animal's stomach, it can starve to death.

Although **recycling** decreases the amount of plastic that ends up in landfills, it is not without its costs. If plastic products undergo chemical reactions to form the materials for new plastic, pollutants may be emitted into the air and workers in the plants may be exposed to harmful compounds.

# Earth and Space Science • Chapter 7
## Observing Weather

### Lesson 1 Weather

The **atmosphere** is the layer of gases surrounding Earth. Most weather takes place in the lowest level, the **troposphere**, which extends upward about seven miles. The term **weather** refers to the atmospheric conditions at a specific time and place. It is caused by the interaction of several factors including temperature, air pressure, humidity, and winds. Weather describes short-term conditions; **climate** refers to the average conditions in an area over an extended period of at least several decades. **Meteorology** is the branch of science that deals with the atmosphere. In fact, "meteor-" = atmosphere; "-ology" = study of. The most practical applications of this field are those related to weather prediction.

Weather is driven by energy from the Sun. The amount of solar energy that reaches any given area on Earth's surface depends on the angle at which the Sun's rays reach Earth and for how long a period each day. At the equator, the Sun is directly overhead, and its light hits this part of Earth at a 90° angle. This area receives the most radiant energy and therefore has the highest temperatures. As you move north or south of the equator, the Sun's rays hit Earth at an angle which is less than 90°, making the radiant energy received less direct than at the equator. Less energy results in lower temperatures.

**Air pressure** is the weight of the air above any point on Earth's surface. Air pressure, also called **barometric pressure**, is measured with a **barometer**. Air pressure is affected by temperature, water vapor, and elevation. Warmer air is less dense than colder air. This is because heat energy causes molecules to move faster and spread farther apart. Molecules in cold air are in slower motion and closer together; therefore, cold air is denser than warm air. Air at higher elevations is also less dense than air at sea level.

**Humidity** is the amount of moisture, or water vapor, in the air. **Relative humidity** is that amount compared to the amount the air can hold. The cliché that "it's not the heat, it's the humidity" is true. High humidity makes warm weather more uncomfortable, because air that contains a lot of water has less room for the evaporation of water given off as perspiration. Humidity is measured with a hygrometer or a sling psychrometer.

**Wind** is the flow of air relative to Earth. The Sun heats various parts of Earth's surface at different rates, causing various parts of the atmosphere to heat up differently. Cool air, which is denser than hot air, exerts greater pressure. The air moves from areas of higher pressure to areas of lower pressure. This movement causes **convection currents**, which we feel as wind.

**Global Wind Patterns**

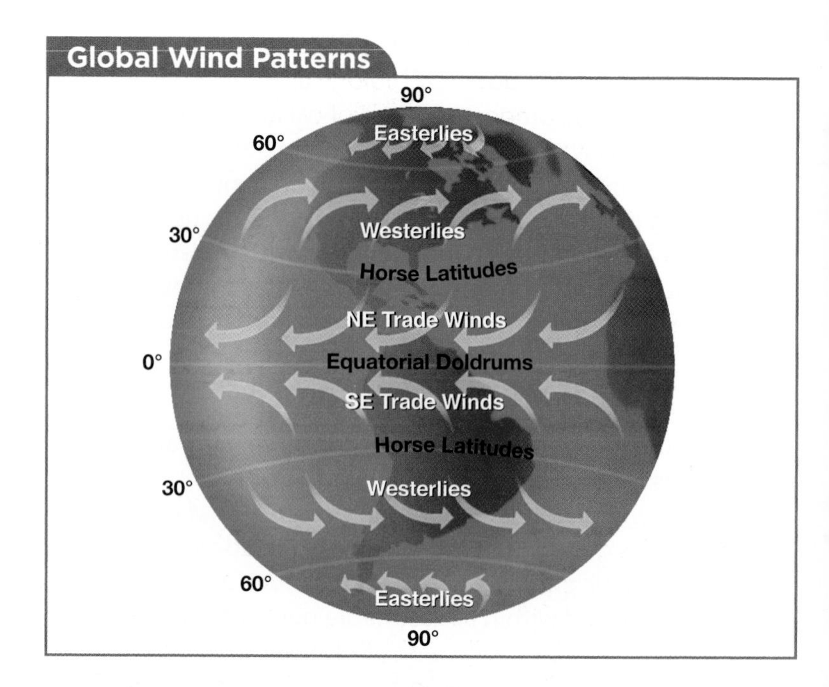


### Lesson 2 The Water Cycle

The water on Earth moves constantly through the **hydrologic cycle,** or **water cycle**. When the Sun heats surface water on Earth, some of the water evaporates as water **vapor**, which air currents carry higher into the atmosphere, where cooler temperatures cause it to condense into water droplets or ice crystals that make up clouds. Air currents within the clouds cause droplets to collide and grow larger. They eventually may fall as rain. In clouds that contain both water droplets and ice crystals, the ice crystals grow larger and the water droplets become smaller. When the ice crystals are large enough, they fall. If the air temperature is above freezing, the ice crystals melt and fall as rain. If the air temperature is below freezing, they fall as snow.

Rain falls into the oceans or onto land, where it soaks into the soil or flows across the ground as **surface runoff**, or **surface water**. Plants take up some of the water, and some is trapped in pores and cracks of rock as **groundwater**. Some rainwater collects in lakes, and some flows into rivers, which then flow into oceans. Over Earth's entire surface, the amount of water that evaporates from the oceans is about the same as the amount that returns as precipitation.

Plants and animals are also part of the water cycle. Plants take in water that has soaked into the ground. The water moves through the plant's vascular system and, eventually, through the leaves in the process of transpiration. Animals take in water by drinking or by eating food, and release it into the environment by exhaling and through wastes.

### Lesson 3 Changes in Weather

Weather describes the day-to-day conditions of the atmosphere. Many factors interact to cause weather.

**Air masses**, large bodies of air with similar temperature and humidity throughout, cause changes in weather. Air masses can cover thousands of square miles and cause the weather conditions below them. When two air masses of different temperature and humidity collide, they form a **front**, which can cause a rapid change in weather.

Air masses are classified by where they form. The most influential air masses in the United States are maritime tropical, maritime polar, continental tropical, and continental polar. Maritime tropical air masses bring hot, humid weather. Maritime polar air masses bring cool temperatures and moist air. Continental tropical air masses bring hot, dry air. Continental polar air masses bring cold temperatures and little moisture.

When two different air masses meet, the boundary between them forms a front. A **cold front** forms when a cold air mass pushes under a warm air mass, forcing warmer moist air upwards, causing condensation and vertically developed clouds. A cold front passing through is characterized by cumulonimbus clouds and heavy rains, sometimes with hail, thunder, or lightning. After it passes, cumulus clouds are present with their associated fair weather. A **warm front** forms when a warm air mass moves over a cold front. As the warm, moist air rises over the cold air, stratus clouds form, accompanied by rain showers.

An **occluded front** occurs when a cold front overtakes and raises a warm front, producing nimbostratus or cumulonimbus clouds along with continuous showers. Afterward there is generally clearing skies and nimbostratus or scattered cumulus clouds. When a warm air mass meets a cold air mass and there is a stand-off where no movement occurs it is called a **stationary front**. Fronts can bring heavy amounts of precipitation.

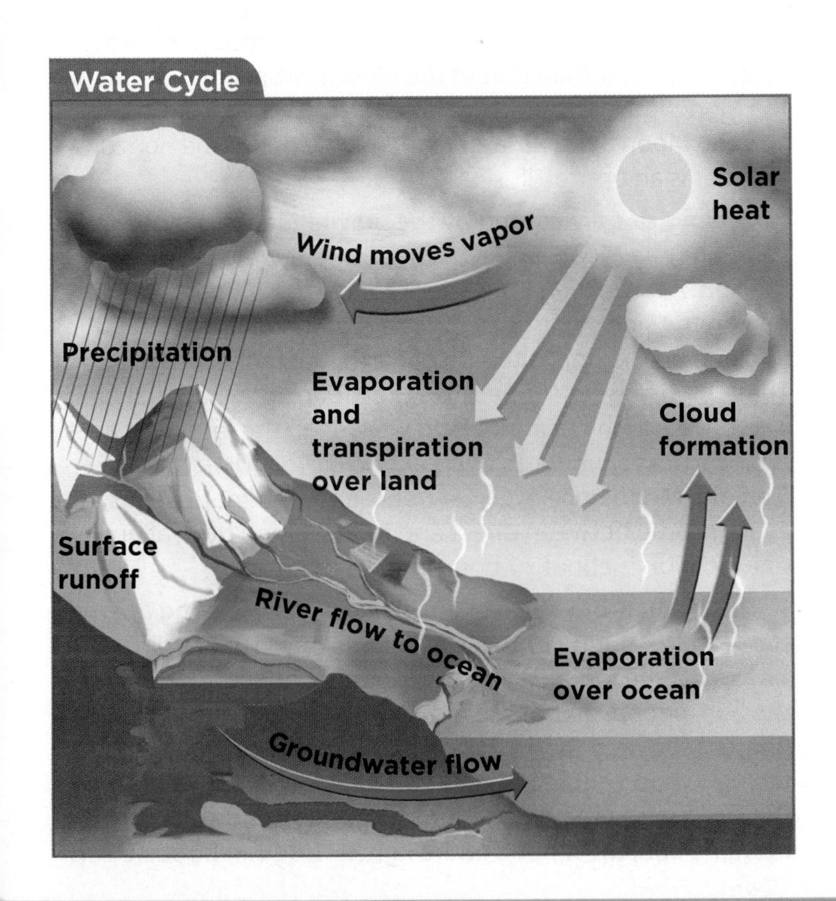


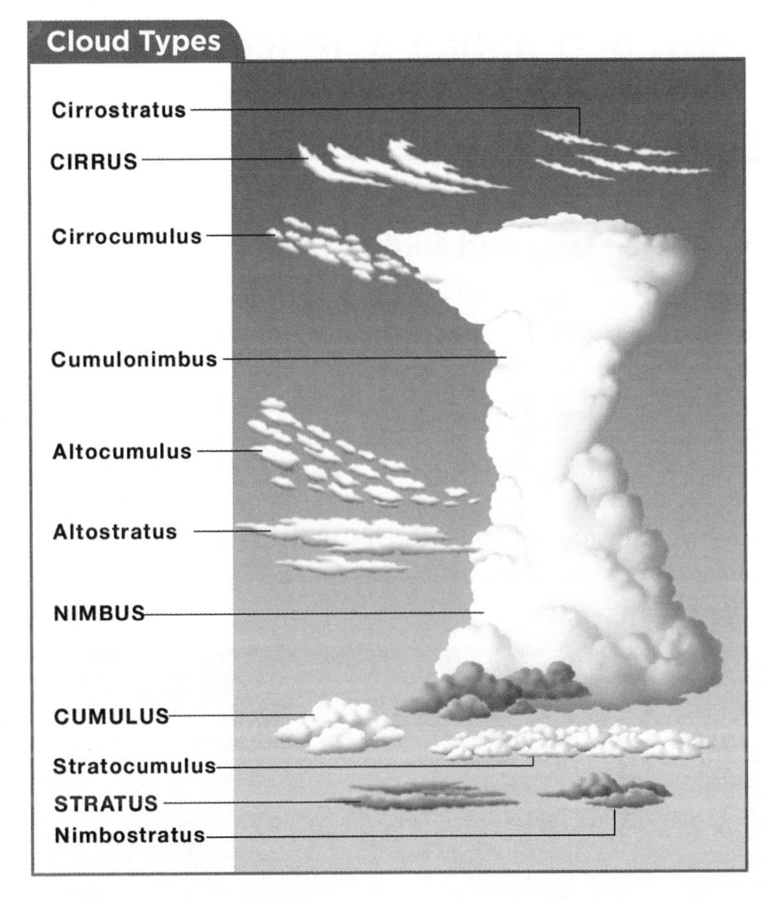


## Clouds

Different **cloud** types are associated with different weather conditions. Clouds form when water droplets form around microscopic solid particles called condensation nuclei found in the atmosphere, including dust, smoke, and salt crystals. These droplets are so small that it may take a million of them to make one raindrop. A smaller portion of particles, called ice nuclei, cause water vapor to solidify into ice crystals without going through the liquid phase.

Clouds are mixtures in which water droplets or ice crystals are suspended in a gas. As the droplets or crystals increase in size they become too heavy to remain suspended and they fall to the ground as rain, or if the ice crystals do not melt on the way down, snow. The accumulation of droplets or crystals depends on the temperature, size, amount, and type of particles available in the atmosphere.

As with air masses, you can tell a lot about a cloud from its name. In Latin, the word "cumulus" means "heap." A cumulus cloud looks like heaps of cotton and is associated with fair weather. "Stratus" means layer. These sheets of clouds generally forecast rain, drizzle, or snow. The word "cirrus," meaning "wisp of hair," and indeed these clouds look wispy as they form at very high altitudes and are made of ice crystals. They often indicate rain or snow. "Nimbus" means rain. Cloud names are usually a combination of these roots. For example, cumulonimbus clouds are like large heaps of cotton and can tower to heights of 39,000 feet. They are associated with powerful thunderstorms. Nimbostratus clouds are dark, low-level clouds that generally bring precipitation, including snow if it is cold enough.

# Earth and Space Science • Chapter 8
## Earth and Space

### Lesson 1 Day and Night

As Earth rotates around its **axis**, which runs between the geographic North and South Poles, the side facing the Sun receives light and the side away from the Sun is dark. One complete rotation takes 24 hours, or one day. Earth rotates from west to east, so the Sun appears to rise in the east and set in the west. Earth's axis is tilted 23.5 degrees from the vertical. This causes the times of sunrise and sunset to vary with the seasons. If it were not for Earth's tilt, both day and night would always be 12 hours long everywhere on Earth.

Earth's Path

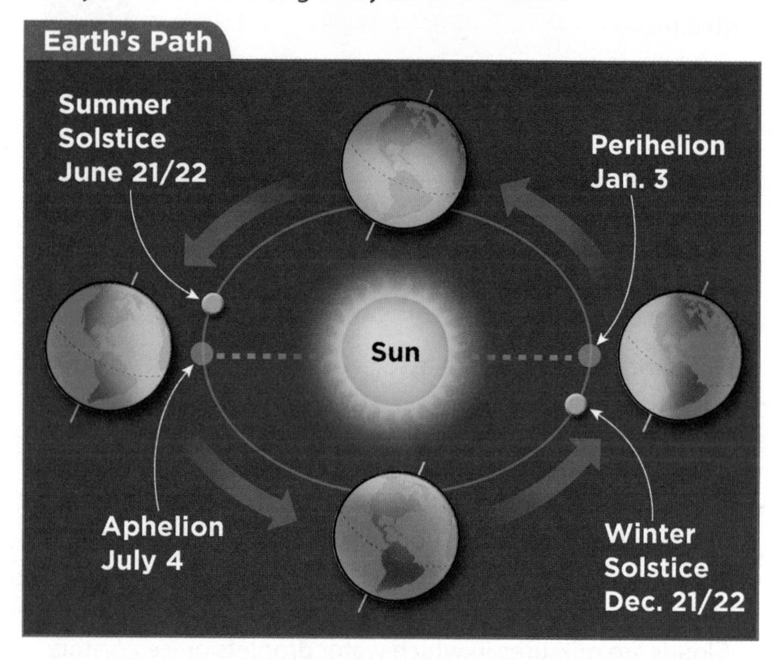


In the Northern Hemisphere, the days grow longer and the nights shorter as the summer solstice approaches, and the days grow shorter and the nights longer as the winter solstice approaches. The longest day and shortest night occur at the summer **solstice**, on June 20 or 21, the beginning of summer. At noon on this day, the Sun is directly overhead at the Tropic of Cancer. The shortest day and longest night occur at the winter solstice, on December 21 or 22. At noon on this day, the Sun is directly overhead at the Tropic of Capricorn. Halfway between the solstices are the **equinoxes**, when the night is the same length everywhere on Earth. The vernal (spring) equinox is on March 20 or 21, and the autumn equinox is on September 22 or 23.

### Lesson 2 Why Seasons Happen

The amount of solar energy that reaches any given point on Earth's surface depends on the angle at which the Sun's rays hit it and the amount of sunlight it receives each day. The areas of Earth with the most direct overhead sunlight and the most hours of daylight receive the most solar energy. If Earth's axis were not tilted, the intensity and duration of sunlight would never change and there would be no seasonal changes. When the Northern Hemisphere is tilted toward the Sun, it is summer. The Sun appears higher in the sky, and its rays hit Earth at a large angle (about 90 degrees) relative to the horizon and are very direct. Temperatures become higher, and the days are longer. When the Northern Hemisphere is tilted away from the Sun, it is winter. The Sun's rays hit Earth at a narrow angle relative to the horizon and are less direct, so the temperatures are lower and the days are shorter. The immediate regions of the North and South poles have six months of daylight followed by six months of darkness. North of the Arctic Circle and south of the Antarctic Circle, there is at least one day a year when the Sun does not rise. Even during the long periods of sunlight, however, the poles remain cold because the sunlight is of low intensity, due to the narrow angle of the Sun's rays relative to the horizon.

At the equator, where the amount and angle of sunlight are about the same all year round, temperatures are relatively constant. The rainfall, however, varies. The Equatorial Convergence Zone, a belt of low pressure formed by the convergence of warm, moist air, produces heavy rains. The Zone shifts back and forth from 15 degrees above the equator to 15 degrees below it, giving tropical areas a dry season and a wet season.

### Lesson 3 The Moon and Stars

Both Earth and the Moon shine by reflecting light from the Sun. The Moon is held in orbit by the mutual gravitational attraction between it and Earth. The Moon appears to rise in the east and set in the west because the motion of Earth's rotation is faster than that of the Moon's orbit. From day to day, the Moon moves from west to east against the background sky. The Moon appears larger when it is close to the horizon, but that is an optical illusion.

Ocean **tides** are caused by the gravitational force between Earth and the Moon, and to a lesser extent, between Earth and the Sun. At any given time, there are two high tides on Earth: one on the side of Earth facing the Moon (the direct high tide) and one on the opposite side of Earth (the indirect high tide). These high tides occur because the Moon's gravitational pull is stronger on the near side and weaker on the far side than it is on Earth as a whole. Every point on Earth is directly across from the Moon once every 24 hours and 50 minutes. Thus, the average time between direct and indirect high tides at any one location is about 12 hours and 25 minutes.

The Moon orbits Earth every 29.5 days, completing one **lunar cycle**. During each cycle, people on Earth see different **phases** of the Moon. When the Moon is between Earth and the Sun, people do not see the Moon's lighted side at all. This phase is the **new moon**. For the few days before and after the new moon, there is a **crescent moon**. Halfway through the cycle, when the Moon is on the opposite side of Earth from the Sun, the Moon's entire lighted side can be seen. This phase is the **full moon**. When the Moon is halfway between a new moon and a full moon, in either half of the cycle, half of its lighted side can be seen. This phase is a **half moon**. A quarter moon is a quarter of the way around its orbit. Halfway between the new moon and full moon is the **first quarter moon**, and halfway between the full moon and new moon is the **last quarter moon**. During the first half of the Moon's cycle, it is waxing (getting bigger), and during the second half, it is waning (getting smaller). More than half of the lighted

**Lunar Tides**

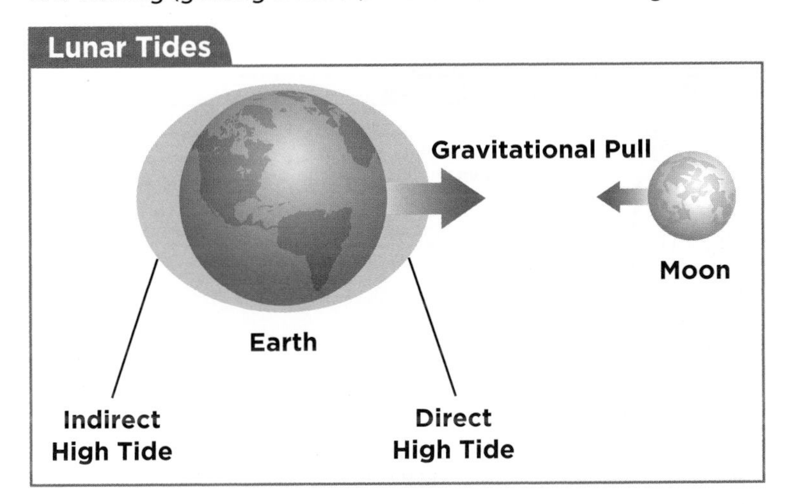


side of the Moon is seen between the first quarter moon and the full moon (**waxing gibbous moon**) and between the full moon and the last quarter (**waning gibbous moon**). The same side of the Moon always faces Earth because the Moon rotates once on its axis in each revolution around Earth.

From Earth, about 6,000 stars can be seen with the unaided eye. **Stars** are celestial objects that glow because of the nuclear energy they produce in their core. They are described and categorized by properties such as mass, size, luminosity (brightness), and temperature. The Sun is a medium-sized star. Stars are in huge groups called galaxies; the Sun is in the Milky Way Galaxy. There are billions of galaxies in the universe, each containing billions of stars. Some stars appear to form patterns, which are called **constellations**. These figures help people to keep track of the stars. Some of the best known of the 88 constellations are Ursa (which contains the Big Dipper), Orion, and Pisces. It is important to remember that the constellations are just arbitrary figures that people devised to locate and identify some of the most prominent stars.

## Lesson 4 The Solar System

The Sun, which is a star, is at the center of our solar system ("solar" means of the Sun). Objects called **planets** revolve around the Sun. The gravitational attraction between the Sun and each planet keeps the planets in orbit. Each planet also rotates on its axis. The planets do not shine with their own light; they can be seen because they reflect the Sun's light.

The planets, in order from closest to the Sun, are: Mercury, Venus, Earth, Mars, Jupiter, Saturn, Uranus, and Neptune. The planets are grouped according to their physical characteristics. The inner planets, Mercury, Venus, Earth, and Mars, are called the **terrestrial**, or rocky, planets. They are similar to Earth, with rocky and metallic composition. Mercury and Venus are the only planets that do not have moons. Venus is about the same size as Earth. Approximately three quarters of Earth, the "Water Planet", is covered with water. Venus was probably once covered with water, too, but because the planet is closer to the Sun, the water was vaporized. As far as scientists know, Earth is the only planet with life on it. Mars was most likely warmer and wetter in the distant past, so it may at one time have supported life. This question is the focus of current research.

The outer planets, Jupiter, Saturn, Uranus, and Neptune, are called the **gas giants**. These large, low-density planets are made up mostly of hydrogen, helium, and ice. The gas giants have rings and numerous moons, and they spin more rapidly than the terrestrial planets.

Beyond the outer gas giant planets is the Kuiper belt, a region of smaller, icy objects. One of the largest of these objects, Pluto, was once known as the ninth planet. But because its size, composition, and orbit do not resemble those of the other planets, in August 2006, the International Astronomical Union (IAU) developed a new definition for Pluto, which became known as a "dwarf planet."

In addition to the planets, the solar system contains comets, asteroids, and meteors. **Comets** are small objects of ice and dust that travel around the Sun in very elongated orbits. Their tails can grow up to 100 million miles long. Perhaps the most famous comet is Halley's comet. It has a period (how long it takes to complete one orbit) of 76 years. Its last visit was in 1986. **Asteroids**, also called minor planets, are small rocky objects. Most of them orbit the Sun between the orbits of Mars and Jupiter. **Meteoroids** are tiny bits of dust (from comets) or rock (from asteroids). Those that enter Earth's atmosphere appear as streaks of light (called "shooting stars"). Those that burn up are called **meteors**, and those that survive their fall to Earth's surface are called **meteorites** ("-ite" is a suffix used for rocks).

**Meteor Crater in Arizona**

# Physical Science • Chapter 9
## Looking at Matter

### Lesson 1 Describing Matter

Matter occupies space, and it cannot share space with other matter. We describe matter in terms of its **properties**, which we perceive with our senses. These include properties that we can perceive with senses other than vision. With our eyes, we perceive size, shape, color, and texture. With our ears, we perceive information about such properties as size, texture, and hardness, often by the sound a material makes when coming in contact with another material. Our sense of touch enables us to perceive size, shape, texture, weight, and hardness. With our senses of smell and taste we perceive properties such as bitterness, saltiness, sweetness, and sourness. The sense of smell also enables us to detect the presence of matter, including some dangerous gases, of which we would otherwise be unaware. People who have lost the use of one of their senses use other senses to determine the properties of matter. A person who is sight impaired, for example, will use the sense of touch to learn an object's size, shape, hardness, and texture.

The molecules that make up matter are in constant motion. They rotate, vibrate, and move around. The three common **states of matter** reflect differences in the rate of the three types of motion of the molecules, which reflects their energy level. The molecules of solid matter have the least energy and move the most slowly. They rotate and vibrate but do not easily move through the solid. The molecules of liquid matter have more energy and move more rapidly. They rotate, vibrate, and move from place to place. Finally, the molecules of gaseous matter have the most energy and move the most rapidly. They have the same types of motion as the molecules of a liquid. The energy level and motion of molecules depend on temperature. The lower the temperature, the lower the energy level and the slower the motion.

**Three States of Matter**

Gas

Liquid

Solid

### Lesson 2 Solids

A **solid** material has a definite shape and volume (how much space it takes up). External forces may be able to alter its shape, but its new shape will then be another definite shape. A good example is clay. Whether you stretch it out into a long tube or roll it into a ball, it is still solid. It still has the same mass and volume. One property of matter is **density**, which is expressed as mass per unit volume, such as grams per cubic centimeter. Forces among the atoms or molecules of a solid keep them very close together, in essentially fixed positions. The closer the molecules are, the denser the material is. Another property of a solid is **hardness**. Hardness is related to the arrangement of the molecules and to how strongly they are held together. If you drop a hard object, the molecules remain fixed relative to each other. The object does not compress or stretch easily. A hard object such as a plate might break, but most of the molecules stay fixed except where the break occurred. In a soft object, some motion is allowed between molecules. If you squeeze and roll a slice of soft bread, you end up with a little ball of bread. The density of the bread has increased because the mass is the same but the volume has decreased. The pieces of a broken plate, however, have the same density as an intact plate. Other properties of a solid include elasticity, ductility (ability to bend), and brittleness (tendency to shatter or break).

**Soft Materials Compress**

Hard materials, such as a wooden chair, do not compress.
Soft materials, such as an upholstered chair, compress.

Two properties of all matter are gravitation and inertia. **Gravity** is the force of attraction that any two objects exert on each other. Because Earth is so massive and so close to us, the only gravitational pull we feel on Earth is that exerted by Earth itself. Just as Earth pulls our bodies toward its center, we pull back on Earth with the same amount of force, but because Earth is so massive, it doesn't budge. **Weight** is a measure of the force of gravity on an object. The farther from Earth you travel, the weaker the pull of Earth's gravity. Although an object weighs less on a mountaintop than in a valley, the difference is so slight that, again, we don't notice it. **Inertia** is the resistance to change in motion. More specifically, it is the tendency of a body that is not moving to resist moving and of a body that is moving to continue moving at the same speed and in the same direction. Scientifically speaking, any change in a body's state of motion, including causing the body to slow down or stop, is considered a form of acceleration. Inertia is a body's resistance to acceleration. **Mass**, the amount of matter in a body, is also a measure of the body's resistance to acceleration. Even in space, where a body is weightless, it will resist acceleration. Therefore, mass is the same everywhere in the universe, whereas weight varies.

First Unattached Space Walk 1984

An object floats if it is less dense than the liquid in which it is immersed. The upward force on the object is called **buoyancy**. The measure of the buoyant force on an object is the weight of the liquid the object displaces. A cork ball will displace an amount of water equal to the part of the cork that is submerged. Because the cork ball is less dense than water, the buoyant force will be sufficient to keep the cork ball afloat. A lead ball the same volume as the submerged portion of the cork ball will displace the same volume, and therefore the same weight, of water. Because the lead ball is more dense than water, however, that same buoyant force will not keep it afloat, and it will sink. The lead ball is more than 11 times heavier than an equal volume of water.

There is always a degree of **error** when we measure any property of an object. If we measure the length of a metal rod to be 1 meter, for example, we are saying that the rod is very close to 1 meter in length. But we can never be certain that it is exactly 1 meter in length. Measuring is different from **counting**, which is saying how many there are of something. In the case of the rods, for example, we can say exactly how many rods there are.

## Lesson 3 Liquids and Gases

The liquid state is partway between solid and gas. The molecules of a liquid are slightly farther apart than those of a solid. That is why matter in a solid state will sink when placed in the same matter in a liquid state. A piece of solid lead, for example, will sink when placed in liquid lead. (Water is the exception. Ice cubes float in liquid because water expands slightly as it freezes. That is, water molecules are slightly farther apart in the solid state than in the liquid state.) Unlike the molecules of a solid, the molecules of a liquid are not in a fixed position. They are free to move about. The molecules of some liquids will be attracted to those of a solid. They tend to coat the surface of a solid, making it "wet". These liquids are said to be **adhesive**. The molecules of other liquids, however, are more attracted to each other. These liquids are said to be **cohesive**. Water is actually a cohesive liquid, its tiny drops held together by the cohesive forces between the molecules. Detergents help to make water less cohesive and therefore "wetter". Another cohesive liquid is mercury. Unlike a soft solid, it cannot be compressed. A liquid's properties include color, viscosity, and transparency or opaqueness.

A **gas** has neither fixed shape nor fixed volume. If it is enclosed, it will not only take on the shape of its container but will also expand to fill the entire container. If it is not enclosed, its molecules will disperse into the surrounding space. There is a lot of space between the molecules of a gas because the molecules are in rapid motion. Because there are relatively few molecules within a given volume of gas, all gases have low density. A gas's properties include odor. Most, but not all, gases are colorless. Chlorine gas is a greenish-yellow. The element iodine got its name from the Greek word *Ioeides*, which means violet, because of its color in gas form.

Sea Otter Floating on Its Back

# Physical Science • Chapter 10
## Changes in Matter

### Lesson 1 Matter Changes

Matter can undergo physical or chemical changes. In a **physical change**, the molecules (or other particles, such as ions) may be rearranged, but the chemical structure (how the atoms are bonded to each other) stays the same. Physical changes include changing from one state to another, such as freezing or melting, and combining to form a mixture.

A **chemical change** results in the formation of new substances, the components of which cannot be separated again by physical means. A mixture usually shares the properties of its individual components, but a chemical change can result in substances with entirely new properties. A common example is baking. Bread dough is a mixture of flour and other ingredients. Once that mixture is baked, the result is a new material, bread.

### Lesson 2 Changes of State

The three common **states of matter** are solid, liquid, and gas. Some substances, such as water, are able to exist in all three states. Changes of matter from one state to another are called **phase changes**.

The molecules that make up all forms of matter are in constant motion. The three common states of matter reflect two factors: the energy of the molecules, which determines how fast they can move, and the attractive forces between molecules, which draw them together. The energy and motion of molecules depend on temperature. Changing the temperature of a type of matter can cause it to change state. Lower temperatures result in lower energy levels and slower motion. Higher temperatures result in higher energy levels and faster motion of the molecules.

A **solid** substance has a definite shape and volume (how much space it takes up). The molecules that make up a solid are arranged tightly together in a well-aligned pattern. Like those in a solid, molecules in a **liquid** state are close together, but they are not in fixed positions. They are free to move about. A liquid has no definite shape of its own, but takes on the shape of its container. Unlike a gas, however, it maintains a constant volume. A **gas** will take on the shape of its container. It has no definite shape or volume. A gas can expand to fill its entire container. A gas's molecules will disperse throughout all available space and are free to move.

#### Changes to Solids

The molecules of a solid move the most slowly; they have insufficient energy to overcome the attractive forces that hold them together. Adding energy to solid matter (through heat) causes the molecules to move faster and overcome some of the attractive forces, causing the solid matter to change into a liquid state. This change of matter from a solid to a liquid is called **melting**. Every solid substance melts at a unique temperature, called its melting point. Thus, the melting point can be used to identify a substance.

Usually, matter changes from solid to liquid, then liquid to gas. **Sublimation** is the direct change from a solid state to a gas. The most familiar example is dry ice. The smoke you see is the water fog that condenses when solid carbon dioxide sublimates. Another example of sublimation is the escape of water vapor from ice cubes in the freezer. This is why ice cubes sometimes appear to shrink or melt if they are not used shortly after freezing.

#### Changes to Liquids

The molecules of a liquid move more rapidly than a solid; they have more energy than those of a solid, but not enough to overcome fully the attractive forces. Molecules of liquid, like all molecules regardless of state, have a range of energy values. Some molecules of the liquid may have sufficient energy to overcome the attractive forces.

Hot Air Balloon

As the temperature of a liquid rises, the molecules move faster and faster. Eventually, individual molecules have enough energy to overcome the forces that bind them together, and they begin to escape from the liquid's surface. These molecules escape from the liquid and become molecules of gas. This process is called **vaporization**. Each substance has a unique boiling, or vaporization, point. The term "gas" refers to a material that normally occurs in the gaseous state. The term "vapor" refers to the gas form of a material that normally exists in a liquid or solid state at room temperature.

When a material evaporates, it changes from a liquid to gaseous state at a temperature that may be below its boiling point. **Evaporation** occurs because some of the molecules near a liquid's surface have enough energy (they are moving fast enough) to escape the forces that bind them to nearby molecules.

When matter changes from a liquid to solid state, we say that it **freezes**. As a liquid is cooled, the molecules move more slowly and with less energy, and they move closer together. (Water is an exception. As it reaches the freezing point, it expands slightly.)

### Changes to Gases

The molecules of a gas move the most rapidly; they have sufficient energy to overcome the attractive forces. However, removing energy, as through cooling, takes away the ability of a gas's molecules to overcome attractive forces.

As a gas is cooled, the molecules move more and more slowly, until they can no longer resist the forces that bind molecules to each other. The molecules then cohere into droplets of liquid. The change of matter from a gas to a liquid is called **condensation**. Condensation is the reverse of vaporization.

## Lesson 3 Mixtures

A mixture is anything that consists of two or more substances physically combined. In a chemical compound the components must be in a specific proportion. In a mixture, however, they can be in any proportion. Because the molecules of the mixture's components do not bind to each other, the components usually can be separated again.

A **heterogeneous** mixture is not the same throughout. An example is gravel. Gravel has different kinds of rocks in different sizes. In the type of heterogeneous mixture called a **suspension**, particles large enough to be seen are mixed into a liquid. An example is muddy water. If left standing long enough, the particles (in this case, the soil) will settle to the bottom.

A **homogeneous** mixture is the same throughout. The most common type of homogeneous mixture is called a **solution**. An example is saltwater made from salt and water. The **solute** (often a solid) is dissolved in the **solvent** (often a liquid). The proportion of solute to solvent is called the **concentration**. If there is a relatively large amount of solute, we say the solution is concentrated. If there is a small amount of solute, we say that it is dilute. The solubility of a solute is the maximum amount of solute that can be dissolved in a given amount of solvent. The solubility of most solids in liquids increases with temperature.

Mixtures can be formed from endless combinations of solids, liquids, and gases. Sand is an example of a mixture of a solid in a solid. Sand in water is a mixture of a solid in a liquid. Smoke consists of solid particles mixed with air, which is a gas. The bubbles observed in soda are a gas in a liquid. The important thing about all of these mixtures is that they can usually be separated into their original components and still have their original properties. This is one thing that differentiates mixtures from compounds, in which the components are chemically combined.

**Sediment Settles to Bottom of Suspension**

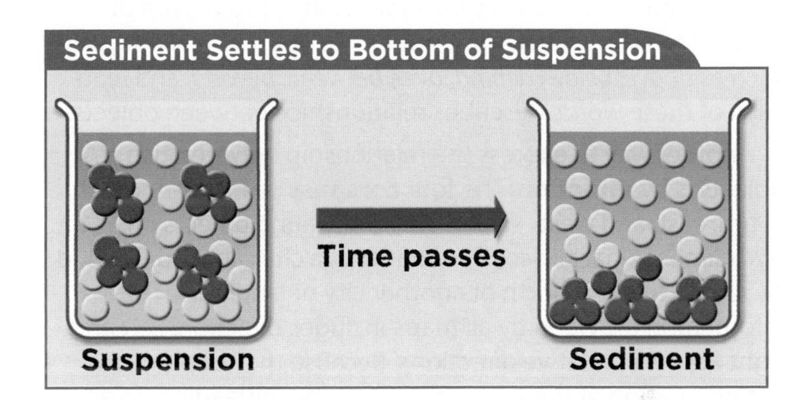


There are a number of ways to separate the components of a mixture. In distillation, a liquid mixture is heated. Those components with lower boiling points vaporize before those with higher boiling points. The vapor travels through a cool tube where it is condensed. Finally the substance is collected in a cooler container in a liquid state. Components of a mixture can also be separated by boiling or evaporating away the liquid until only the solute is left. Some mixtures can be separated by filtering or even plucking larger pieces out, such as rocks from sand.

**Magnet Attracting Iron Away from Sulfur and Iron Mixture**

# Physical Science • Chapter 11
## How Things Move

### Lesson 1 Position and Motion

#### Position

An object's position is described by comparing it to the position of one or more other objects. Even the vague statement that an object is far away is a qualitative comparison of the object's position relative to a person or object. Many qualitative descriptions of position rely on words, such as *above, below, behind, over, between, beyond,* and *in*. Each of these words describes relationships between objects.

Another way to express the relationship between two objects is by **direction**. The four **compass points**—north, south, east, and west—are useful for locating objects on a two-dimensional surface. For example, a city can be described by saying that it is north of another city or landmark. The information provided by all maps includes direction. Left and right are other relative directions. Because they depend on the orientation of the speaker they can be misleading. To a person standing in front of a house, a tree may be on the right; however, to someone facing the back of the house, the tree is on the left.

Directions

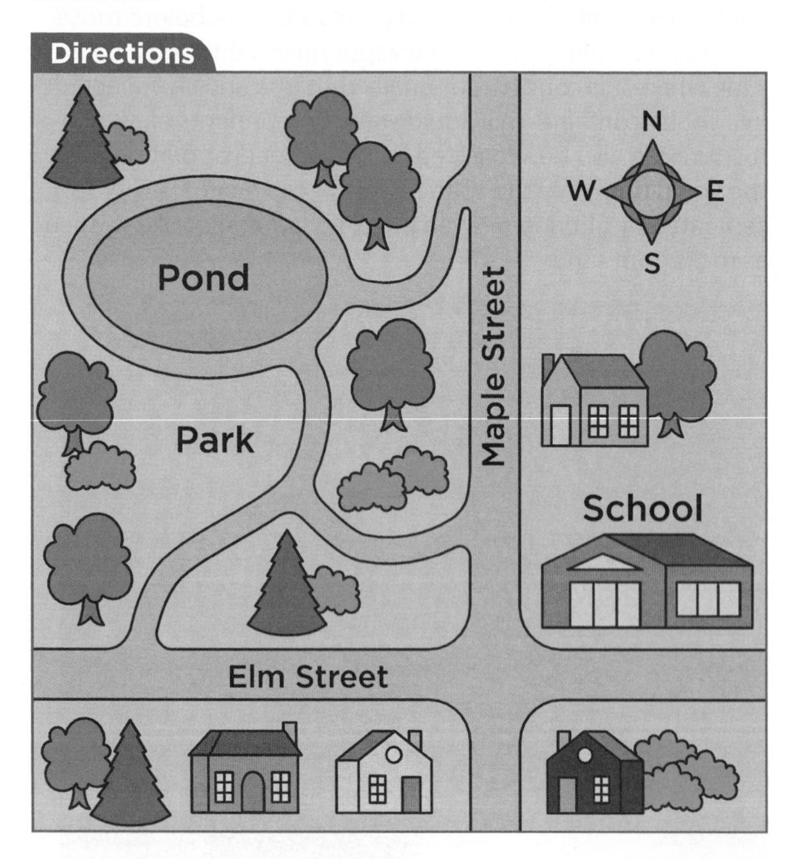


The relationship among objects is also described in terms of **distance**. Distance is how far away one object is from another. By itself, distance does not have direction. The distance from Lexington, Kentucky to Cincinnati, Ohio is 90 miles whether you measure from Lexington to Cincinnati or from Cincinnati to Lexington. Distance is expressed in units of measurement, such as feet, yards, or miles in customary units or in metric units.

The most complete way to indicate position is through combinations of direction, relationship words, and distance. For example: *She hung the picture three feet above the couch, six inches to the right of the mirror.* Or: *The library is two miles east of City Hall.*

#### Motion

When an object is changing its position it is said to be in motion, or moving. The average **speed** is a measure of how far something moves in a set amount of time. Common measurements of speed are miles per hour, feet per second, kilometers per hour, and meters per second. Speed is independent of direction. It does not express where something is moving, only how fast. **Velocity** is the speed of an object in a specified direction. A car's speed might be 50 miles per hour. Its velocity, however, could be 50 miles per hour northeast.

Physical objects are not the only things that move. Sound travels in **waves**. When an object vibrates, it causes a wavelike disturbance in the medium, such as the air, water, or ground surrounding it. The waves travel through the medium until they are absorbed, reflected, or reach a receiver such as the human ear. Light also travels in waves. It travels much faster than sound at a consistent speed of 186,000 miles per second, or nearly 700 million miles per hour.

### Lesson 2 Forces

All materials resist changes to their state of motion. This resistance is called **inertia**. Inertia means that an object that is at rest tends to stay at rest, and an object in motion tends to stay in motion at the same speed and in the same direction. To set an object in motion or to change the course of an object already in motion, it is necessary to apply an external force. Informally, we think of forces as pushes or pulls.

Forces acting on an object from different directions can create a state of **equilibrium**. If there are just two forces they need to be opposite and equal. Equilibrium could also be reached with more than two forces. If one person pushes on a large block, the block may move away from the person. If two people push with the same force on a large block from opposite sides, the block will not move.

### Lesson 3 Using Simple Machines

Machines are devices that help people do tasks. They can increase the amount of force exerted, increase the speed of an action, or change the direction of a force. Common tools like the hammer are **simple machines**. There are six basic types of simple machines: lever, inclined plane, wedge, screw, wheel and axle, and pulley. More complex machines are made from combinations of these six basic ones.

### Types of Simple Machines

When a simple machine increases the amount of force used, there is a trade-off. The force must be applied over a greater distance. A **lever** has a fulcrum (or pivot), a load, and a place where force is applied. A seesaw is a first-class lever. This means the fulcrum is between the two children, the load is the child who is in the air, and the force is the weight of the child at the bottom. A wheelbarrow is an example of a second-class lever, with the fulcrum at the wheel and the force applied at the end of the handles. The load is the material being lifted between the wheel and handles. On a third-class lever, like a broom, the fulcrum is at one end, the effort force is applied close to the fulcrum, and the load is applied at the other end. In a third-class lever the force exerted on the load is smaller than the effort force.

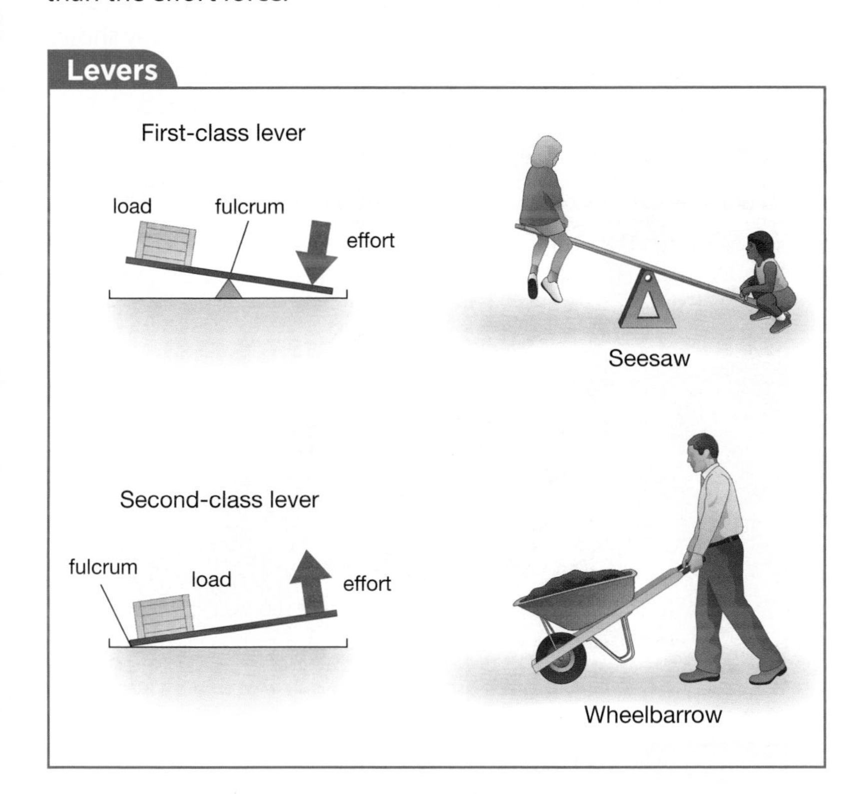


An **inclined plane** moves an object to a higher or lower position without lifting. A ramp, such as one used to raise a wheelchair, is an inclined plane. The longer the ramp, the less force is required to move up it. A shorter, steeper ramp requires more force. A **wedge** is made of two inclined planes. A wedge changes the direction of an applied force. The hull of a boat, which separates the water in front of it, is a wedge.

A **screw** is a essentially a nail with an inclined plane going around it. It takes less force to insert a screw than a nail because the force travels over the greater distance of the thread as the screw turns.

The **wheel and axle** is another type of simple machine. When a wheel makes one complete turn, it covers a greater distance than the axle to which it is connected. Therefore, the force exerted on the rim of a wheel is multiplied on the axle. An example of this is a doorknob. When the knob is turned, it turns the shaft that opens the door. If the knob falls off, it is more difficult to turn the shaft with your fingers because it requires more force. The steering wheel of a large vehicle, like a bus or truck, has a greater diameter than the steering wheel of a car.

A simple **pulley** consists of a wheel and a rope that goes around it. A fixed pulley changes the direction of the force. The person applying the force on one side of the pulley pulls down on the rope and the force on the other side raises the load. Moveable pulleys increase the amount of force applied. An example is a block and tackle used to hoist a piano to an upper-story window.

## Lesson 4 Exploring Magnets

Magnetic force is the attraction between magnets and materials that contain certain elements, such as iron, nickel, and cobalt. Every magnet, including Earth, has two **poles**, called north and south. A magnet's north pole, or north-seeking pole, is the end that points toward Earth's magnetic north pole. Each of the magnet's atoms behaves like a tiny magnet, sensing Earth's magnetic poles and lining up accordingly.

All magnets have two poles. There is no such thing as half a magnet. If you cut a magnet in two, each piece becomes a complete magnet, each with two poles. Permanent magnets maintain their magnetic properties over a long period of time. Metals such as iron and steel can be temporarily magnetized. If you rub a magnet in one direction along a piece of steel, which contains iron, the steel becomes magnetized.

When two magnetic objects are close enough, the opposite poles of the magnets are attracted to each other. Like poles repel each other. A compass needle is a small magnet that rotates on a pivot, its north-seeking pole pointing toward Earth's magnetic north pole. It rotates to align with Earth's magnetic field. The needle of a compass, however, points to Earth's magnetic north pole only when there is no other magnet nearby exerting a stronger magnetic force.

# Physical Science • Chapter 12
## Using Energy

### Lesson 1 Heat

Heat is the flow of energy from one substance to another. The energy always flows from a substance of higher temperature (with higher thermal energy) to a substance of lower temperature until an **equilibrium** is reached. **Temperature** is a measure of the average kinetic energy of a substance's molecules. It can be measured with a **thermometer**. Most thermometers are filled with a liquid, such as alcohol or mercury, that expands when heated and contracts when cooled. The **scale** is determined by choosing two reference points, usually the melting point of ice and the boiling point of water, and dividing the temperatures between them into units called **degrees**. The two most common scales are Fahrenheit (F), on which water has a freezing temperature of 32° and a boiling temperature of 212°; and Celsius (C), on which water has a melting temperature of 0° and a boiling temperature of 100°. When a thermometer is placed in a sample, heat flows from the hotter material to the cooler one until the temperature of the thermometer is the same as that of the sample.

### Lesson 2 Sound

Sound travels by waves. When an object vibrates, it creates **sound waves** that move through the air or other gas (or through a liquid or a solid), eventually reaching the ear of the listener. They cause the eardrum, a membrane, to vibrate. The **ossicles**, three bones in the ear, pass the vibrations to the inner ear. A nerve then sends an impulse to the brain, and the sound is perceived.

**Pitch** refers to how high or low a sound is. The frequency of a sound determines its pitch. **Frequency** is the number of wave crests per unit of time. The greater the frequency of the sound waves, the higher the pitch. Frequency is measured in hertz (Hz), which means cycles, or number of wave crests, per second. When a string is tightened on a violin, or other stringed instrument, it raises the frequency. The thicker strings on a violin have a lower frequency. Also, longer strings have a lower frequency than shorter ones (compare the sounds of a violin to a cello). Different animals are sensitive to different frequencies. For example, dogs can hear higher frequencies than humans, and elephants can hear lower frequencies.

### Lesson 3 Light

**Light** is a form of **electromagnetic** energy. Moving electric charges create fields that in turn create magnetic fields. Many objects in the universe give off light at different wavelengths and frequencies. A **wavelength** is the distance from crest to crest (or trough to trough). A wave's **frequency** is the number of waves that pass through any point in a second. These waves vary from very long wavelengths (low frequencies) to very short wavelengths (high frequencies). Together, all these waves make up the **electromagnetic spectrum**. The different kinds of electromagnetic radiation, from longest to shortest, are radio waves, microwaves, infrared radiation, visible light, ultraviolet radiation, X-rays, and gamma rays.

A point light source, or a light that is distant from an object, causes a sharp, distinctly outlined shadow. A large, diffuse light causes a fuzzier shadow that has a dark central area called an **umbra**, surrounded by a partially lighted area called a **penumbra**. (These are the same as the umbra and penumbra of the shadow cast by the Moon during a solar eclipse and by Earth during a lunar eclipse.)

Light rays are not the only kind of electromagnetic radiation that causes objects to cast shadows. An X-ray machine creates shadows of the objects being observed. In fact, X-ray photographs are also called **shadowgrams**. A dental X-ray shows the shadow cast by the teeth on the film. Metal fillings, which absorb more of the X-rays than the teeth, show up even darker.

**Dental X-Ray with Metal Filling**

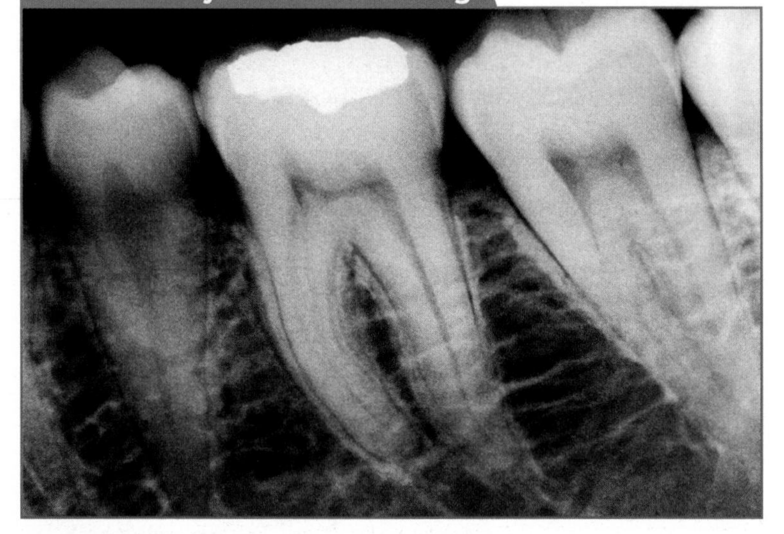

### Lesson 4 Exploring Electricity

All matter is made up of **atoms**. The nucleus, at the center of the atom, contains **protons** and **neutrons**. Surrounding the nucleus are **electrons**. Neutrons, as their name suggests, are neutral. Protons carry a positive electric **charge**, and electrons carry a negative charge. An atom that has the same number of protons and electrons is neutral. If an atom has lost one or more electrons, it has a positive charge, and if it has gained electrons, it has a negative charge. Charged atoms are called **ions**. Electricity is a general term that encompasses all the effects caused by these charges.

The flow of an electrical charge is called a **current**. Because the protons are locked into the nucleus, it is almost always the electrons that travel. Opposite charges attract, so the electrons of a negatively charged object are attracted to positively charged objects.

A material that permits charges to flow through it is a **conductor**. Most metals make good conductors since they have outer electrons that are relatively loosely held by the atom and can move from atom to atom. **Resistance** is the ability of a material to impede the flow of charges. Poor conductors, which have high resistance, are called **insulators**. They are usually nonmetals.

Electric **potential** is the potential per unit of electric charge. It is measured in volts. A difference in electric potential between the ends of a conductor creates a force on charges. The charges flow from the end of a conductor with a high electric potential to the end with a low electric potential. Electric potential is analogous to the water pressure in a pipe.

An electrical **circuit** is an uninterrupted loop of conducting material, through which an electric current moves from a high electrical potential to a low electrical potential. Complicated circuits are also called **networks**. A basic circuit consists of a **source** of electrical energy; a **load**, which changes the electrical energy into a more useful form of energy such as heat, motion, or light; and a **switch**, which interrupts the current to control the circuit. An example of a basic electric circuit is a flashlight. The battery is the source; the bulb, which converts the electrical energy to light and heat, is the load; and the switch turns the flashlight on and off.

All electrical appliances require electric **energy** to maintain the potential difference that drives the current. Although some electricity is naturally produced, people have been unable to make practical use of it. Lightning, for example, is unpredictable and too powerful to harness. Most of the electric energy people use is produced by power plants, which convert the energy stored in coal, oil, natural gas, or radioactive nuclei to electric energy. Power plants heat water to produce steam. The energy in the steam is converted into mechanical energy by a turbine that has blades mounted on a shaft. When the steam strikes the blades the shaft turns, which rotates a generator that produces electric energy. The generator can also be turned by the energy of moving water. A generator consists of a large coil of wire on the rotating shaft surrounded by fixed electromagnets. The fixed electromagnet consists of long wires wound into coils through which there is an electric current. As the shaft rotates, current is induced in the rotating coil. Once generated, the electricity must be transported to the consumers. From the power plants, electricity travels along wires to a **transformer**, which changes the potential from a low to a high voltage. The higher voltage gives the current the pressure it needs to travel long distances with low energy losses. The high-voltage electricity then travels along **transmission lines** to a **substation**. At the substation, transformers change the high-voltage electricity back to low-voltage electricity. Finally, **distribution lines** take the electricity to the homes and businesses where it is needed. The rate at which electric current carries energy is called **power**. Power is measured in **watts** or **kilowatts** (thousands of watts). In the home or in businesses various machines and appliances convert the electricity to heat, light, and motion. Some of these contain **resistors**, which convert electric energy to thermal energy, which is then dissipated as heat.

**Path of Electricity from Power Plant to Residence**

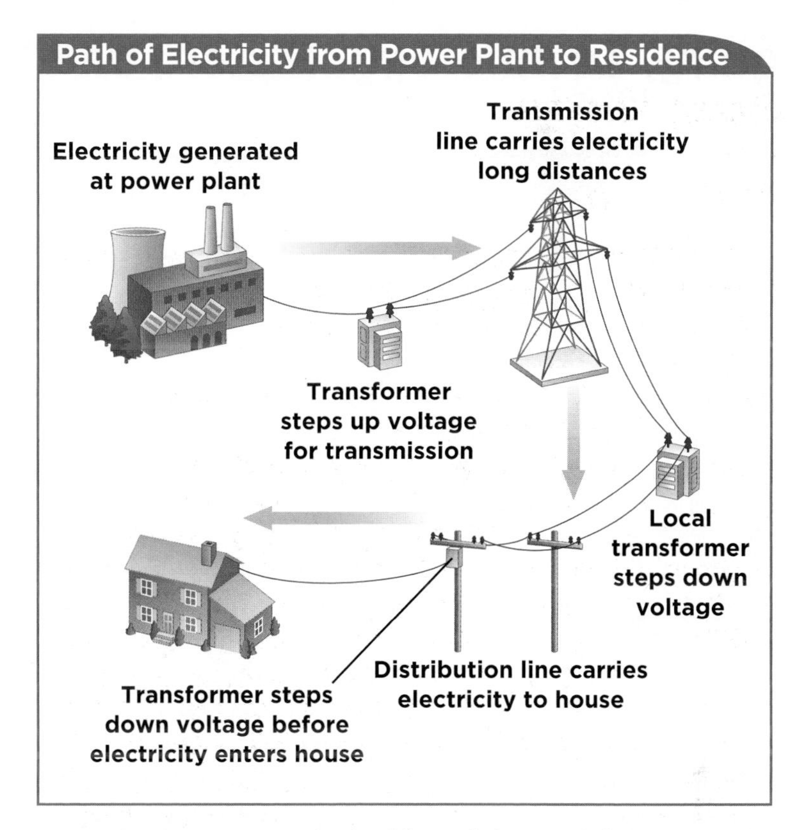
Electricity generated at power plant
Transmission line carries electricity long distances
Transformer steps up voltage for transmission
Local transformer steps down voltage
Distribution line carries electricity to house
Transformer steps down voltage before electricity enters house

**Static electricity** is the buildup of electrical charge in an object. It is often produced by friction. When certain materials are rubbed together, electrons from one material jump over to the other material. This leaves one material with a positive charge and the other with a negative charge. A neutral object can be charged by placing it in contact with a charged object. If the charged object has a positive charge, it will attract electrons from the neutral object, leaving it with a positive charge. If the charged object has a negative charge, some of its electrons will be attracted to the protons of the neutral object, giving it a negative charge. The shock experienced when someone touches an object is the flow of electrons jumping between the object and the person's hand.

The top of a thundercloud has a positive charge and the bottom has a negative charge. Scientists do not yet fully understand how the charges become separated in thunderclouds. They know that the electrons at the bottom of the cloud repel those on the ground, pushing them below the surface. This causes a positive charge on the ground. Eventually, the positive charges on the highest points, or objects, on the ground attract negative charges in the cloud, resulting in an **electrical discharge**. An electrical discharge is a sudden flow of current through a material that is usually an insulator. In the case of lightning, the potential difference between the thundercloud and the ground is so great that it makes the air a conductor. Electric current flows back and forth between the ground and the cloud, causing a lightning bolt. Although cloud-to-ground lightning is the most dangerous to humans, the most common type of discharge occurs between differently charged areas of the same cloud. A discharge can also take place between clouds.

# Scope and Sequence

## Inquiry, Technology, History and Nature of Science

### Inquiry

| Inquiry Skills | K | 1 | 2 | 3 | 4 | 5 | 6 |
|---|---|---|---|---|---|---|---|
| Observe | I | D | D | D | D | D | D |
| Infer | I | D | D | D | D | D | D |
| Predict | I | D | D | D | D | D | D |
| Communicate | I | D | D | D | D | D | D |
| Measure | I | D | D | D | D | D | D |
| Put things in order | I | D | D | | | | |
| Compare | I | D | D | | | | |
| Classify | I | D | D | D | D | D | D |
| Investigate | | I | I | D | D | D | D |
| Make models | | D | D | D | D | D | D |
| Draw conclusions | I | D | D | | | | |
| Use numbers | | | | D | D | D | D |
| Interpret data | | | | D | D | D | D |
| Form a hypothesis | | | | D | D | D | D |
| Experiment | | | | D | D | D | D |

| Special Abilities for Inquiry | K | 1 | 2 | 3 | 4 | 5 | 6 |
|---|---|---|---|---|---|---|---|
| Critical thinking | I | I | I | D | D | D | D |
| Use of mathematics | I | I | I | D | D | D | D |
| Use of technology | I | I | I | D | D | D | D |

| Understandings About Inquiry | K | 1 | 2 | 3 | 4 | 5 | 6 |
|---|---|---|---|---|---|---|---|
| Different questions suggest different kinds of investigations. | I | I | I | D | D | D | D |
| Current scientific knowledge guides investigations. | | | | D | D | D | D |
| Investigations may lead to new ideas. | | | | D | D | D | D |
| Explanations are based on evidence. | I | I | I | D | D | D | D |

### Science and Technology

| Abilities of Technological Design | K | 1 | 2 | 3 | 4 | 5 | 6 |
|---|---|---|---|---|---|---|---|
| Identify appropriate problems. | I | I | D | D | D | D | D |
| Design a solution. | I | I | D | D | D | D | D |
| Implement the solution. | I | I | D | D | D | D | D |
| Evaluate/communicate the solution. | I | I | D | D | D | D | D |

| KEY | | |
|---|---|---|
| **I** = Introduced | **D** = Developed | **R** = Reinforced |

## Science and Technology *cont.*

| Understandings about Science and Technology | K | 1 | 2 | 3 | 4 | 5 | 6 |
|---|---|---|---|---|---|---|---|
| Contributions made by people in different cultures | I | I | I | D | D | D | D |
| Trade-offs: benefits and risks, consequences | I | I | I | D | D | D | D |
| Reciprocity of science and technology | | | I | D | D | D | D |
| Distinguishing natural from human-made objects | I | D | D | R | | | |

## Personal and Social Perspectives

| Personal Health | K | 1 | 2 | 3 | 4 | 5 | 6 |
|---|---|---|---|---|---|---|---|
| Practice of healthful behaviors | I | I | I | D | D | D | D |
| Injury preventions | I | I | I | D | D | D | D |

| Populations, Resources, Environments | K | 1 | 2 | 3 | 4 | 5 | 6 |
|---|---|---|---|---|---|---|---|
| Effect of population density on resources and environment | I | I | I | D | D | D | D |
| Limited supply of many resources | I | I | D | D | D | D | D |
| Natural and human causes of environmental change | I | I | D | D | D | D | R |
| Rapid and gradual environmental change | I | I | D | D | R | R | R |

| Natural Hazards | K | 1 | 2 | 3 | 4 | 5 | 6 |
|---|---|---|---|---|---|---|---|
| Resulting from Earth processes | I | I | D | D | D | D | D |
| Induced by human activities | | | I | D | D | D | D |
| Challenges to society | | | I | D | D | D | D |
| Risks of natural hazards | I | I | D | D | D | D | D |

| Science, Technology, and Society | K | 1 | 2 | 3 | 4 | 5 | 6 |
|---|---|---|---|---|---|---|---|
| Effects of science and technology on society | I | I | I | D | D | D | D |
| Effect of society on science investigation | | | | | I | D | D |
| Many settings for carrying out science and technology | | | | | I | D | D |
| Inability of science to answer all questions | | | | | I | I | D |

## History and Nature of Science

| Science as Human Endeavor | K | 1 | 2 | 3 | 4 | 5 | 6 |
|---|---|---|---|---|---|---|---|
| Carried out by men and women of various backgrounds | | I | I | D | D | D | D |
| The need for different abilities | | I | I | I | D | D | D |
| Varied careers in science | I | I | I | D | D | D | D |

| Nature of Science | K | 1 | 2 | 3 | 4 | 5 | 6 |
|---|---|---|---|---|---|---|---|
| Explanations based on observation and experiments | I | I | D | D | D | D | D |
| Conflicting scientific interpretations | | | | | I | D | D |
| Evaluation of scientific ideas | | | | I | D | D | D |

| History of Science | K | 1 | 2 | 3 | 4 | 5 | 6 |
|---|---|---|---|---|---|---|---|
| The change of accepted ideas because of scientific explanations | | | | | I | D | D |
| Contribution of individuals throughout history | | | | I | D | D | D |

# Scope and Sequence

## Life Science

### Plants

| Needs | K | 1 | 2 | 3 | 4 | 5 | 6 |
|---|---|---|---|---|---|---|---|
| Identified | I | I | I | I | D | D | R |
| Related to habitat | I | D | D | D | R | R | R |
| Related to structures | I | I | I | I | D | D | R |

| Photosynthesis | K | 1 | 2 | 3 | 4 | 5 | 6 |
|---|---|---|---|---|---|---|---|
| Raw materials/products | | I | I | I | D | D | R |
| Chemical formula | | | | | I | D | D |
| Producers | | | | | | | |

| Structures and Functions | K | 1 | 2 | 3 | 4 | 5 | 6 |
|---|---|---|---|---|---|---|---|
| Roots, stems, leaves | I | I | I | I | D | D | R |
| Flowering plants | I | I | I | D | D | D | R |
| ▸ Parts of a flower | I | I | I | I | D | D | R |
| ▸ Monocots/dicots | | | | | I | D | D |
| Nonflowering plants | | | | I | D | D | R |
| ▸ Mosses (parts) | | | | | D | D | R |
| ▸ Ferns (parts) | | | | I | D | D | R |
| ▸ Gymnosperms (parts) | | | | I | D | D | R |

| Reproduction (Life Cycles) | K | 1 | 2 | 3 | 4 | 5 | 6 |
|---|---|---|---|---|---|---|---|
| Flowering plants | I | I | I | I | D | D | R |
| Gymnosperms | | | | I | I | D | R |
| Mosses/ferns | | | | I | I | D | R |
| Bulbs/cuttings/tubers (asexual means) | | I | | I | D | R | |

| Regulation and Responses | K | 1 | 2 | 3 | 4 | 5 | 6 |
|---|---|---|---|---|---|---|---|
| Tropisms | | | | I | I | D | D |
| Other responses to environment | I | I | I | D | D | R | R |

### Animals

| Needs | K | 1 | 2 | 3 | 4 | 5 | 6 |
|---|---|---|---|---|---|---|---|
| Identified | I | I | I | I | D | D | R |
| Related to habitat | I | D | D | D | R | R | R |
| Related to structures | I | I | I | I | D | D | R |

| Structures and Functions | K | 1 | 2 | 3 | 4 | 5 | 6 |
|---|---|---|---|---|---|---|---|
| External characteristics | I | I | I | I | D | D | R |
| Organ systems | | | | | D | D | D |

| KEY | | |
|---|---|---|
| **I** = Introduced | **D** = Developed | **R** = Reinforced |

## Animals *cont.*

| Life Cycles | K | 1 | 2 | 3 | 4 | 5 | 6 |
|---|---|---|---|---|---|---|---|
| Basic examples (e.g., birds, mammals) | I | I | I | I | D | D | R |
| Complete metamorphosis (insect, amphibian) | I | I | I | D | D | D | R |
| Incomplete metamorphosis | | | | I | D | D | R |
| Asexual reproduction (regeneration) | | | | | I | D | R |

| Regulation and Behavior | K | 1 | 2 | 3 | 4 | 5 | 6 |
|---|---|---|---|---|---|---|---|
| Response to environment | I | I | D | D | D | R | R |
| Internal stimuli | | | | I | D | R | R |

## Six Kingdom Classification

| Binomial Classification | K | 1 | 2 | 3 | 4 | 5 | 6 |
|---|---|---|---|---|---|---|---|
| 6 kingdoms defined | | | | | I | D | R |
| Kingdom to species delineations | | | | | I | D | R |
| Species names | | | | | I | D | R |

| Plant Kingdom | K | 1 | 2 | 3 | 4 | 5 | 6 |
|---|---|---|---|---|---|---|---|
| Divisions identified | | | | | I | D | R |
| Nonvascular vs. vascular | | | | | D | D | R |
| Seedless vs. seed | | | | I | D | D | R |
| ▶ Nonflowering vs. flowering | | | | I | D | D | R |
| ▶ Monocots/dicots | | | | | I | D | R |

| Animal Kingdom | K | 1 | 2 | 3 | 4 | 5 | 6 |
|---|---|---|---|---|---|---|---|
| Basic groupings (mammals, insects, reptiles, birds) | I | I | I | I | D | D | R |
| Invertebrates | | | | I | D | D | R |
| ▶ Basic phyla | | | | I | D | D | R |
| Vertebrates | | | | I | D | D | R |
| ▶ Classes | | | | I | D | D | R |

| Microorganisms | K | 1 | 2 | 3 | 4 | 5 | 6 |
|---|---|---|---|---|---|---|---|
| Bacteria Kingdom | | | | I | I | D | D |
| Archaea Kingdom | | | | | I | D | D |
| Protist Kingdom | | | | I | I | D | D |
| Fungus Kingdom | | | | I | I | D | D |
| Reproduction | | | | | I | D | D |
| ▶ Asexual vs. sexual | | | | | | D | D |

# Scope and Sequence

## Cells (Cell Theory of Life)

| Cell Structure | K | 1 | 2 | 3 | 4 | 5 | 6 |
|---|---|---|---|---|---|---|---|
| Cell as unit of life | | | | I | I | D | D |
| Plant cells vs. animal cells | | | | | I | D | R |
| Multicellular vs. single-celled organism | | | | | D | D | R |
| Levels of organization (cell-tissue-organ-system) | | | | | D | D | R |

| Cell Activities | K | 1 | 2 | 3 | 4 | 5 | 6 |
|---|---|---|---|---|---|---|---|
| Photosynthesis/respiration | | I | I | D | D | D | D |
| Cell division | | | | | | D | D |
| ▸ Mitosis/meiosis | | | | | | I | D |
| Diffusion/osmosis | | | | | | | D |

## Heredity and Genetics

| Concepts of Heredity | K | 1 | 2 | 3 | 4 | 5 | 6 |
|---|---|---|---|---|---|---|---|
| Inherited traits | I | I | I | D | D | D | D |
| ▸ In animals | I | I | I | D | D | D | D |
| ▸ In plants | I | I | I | D | D | D | D |
| Learned behavior/acquired traits | I | I | I | D | D | D | D |
| Variation | I | I | I | I | D | D | D |
| Effect of environment | I | I | I | D | D | D | D |
| Natural selection | | | | | | | D |

| Mechanisms of Heredity | K | 1 | 2 | 3 | 4 | 5 | 6 |
|---|---|---|---|---|---|---|---|
| Mendel's patterns | | | | | | I | D |
| Chromosomes | | | | | | I | D |
| Genes | | | | | | I | D |
| DNA | | | | | | | D |

| Genetics | K | 1 | 2 | 3 | 4 | 5 | 6 |
|---|---|---|---|---|---|---|---|
| Pedigree | | | | | | I | D |
| Sex-linked traits | | | | | | | D |
| Genetic engineering | | | | | | I | D |

| KEY | | |
|---|---|---|
| **I** = Introduced | **D** = Developed | **R** = Reinforced |

## Ecology/Ecosystems

| Organization of Ecosystems | K | 1 | 2 | 3 | 4 | 5 | 6 |
|---|---|---|---|---|---|---|---|
| Habitats | I | I | I | D | D | R | R |
| Populations and communities | | | I | D | D | R | R |
| Niches | | | | I | D | D | R |
| Land vs. water | I | I | D | D | D | D | R |
| Biomes | | | | | I | D | D |
| ▶ Latitude | | | | | I | D | D |
| ▶ Comparisons | | | | | I | D | D |
| ▶ Adaptations | | | | | D | D | R |

| Dynamics of Ecosystems | K | 1 | 2 | 3 | 4 | 5 | 6 |
|---|---|---|---|---|---|---|---|
| Energy transfer | | I | I | D | D | D | D |
| ▶ Food chains | | I | I | D | D | D | D |
| ▶ Food webs | | | I | I | D | D | D |
| ▶ Energy pyramids | | | | | I | D | D |
| Cycles of matter | | | | I | I | D | R |
| ▶ Carbon | | | | I | I | D | R |
| ▶ Nitrogen | | | | | | D | R |
| Competition | | | | I | D | D | D |
| Symbiotic relationships | | | | | I | D | D |

| Adaptation and Survival | K | 1 | 2 | 3 | 4 | 5 | 6 |
|---|---|---|---|---|---|---|---|
| Adaptations | I | I | I | D | D | D | D |
| ▶ Animal behaviors | I | I | I | D | D | D | D |
| ▶ Animal structures | I | I | I | D | D | D | D |
| ▶ Plant responses | | | | I | D | D | R |
| ▶ Plant structures | | I | I | D | D | D | R |
| Changes in ecosystems | | | I | D | D | D | D |
| ▶ Destructive change | | | I | D | D | D | D |
| ▶ Effect on populations | | | I | D | D | D | D |
| ▶ Regrowth/succession | | | | | | D | D |
| Extinction | | | I | D | D | D | D |
| ▶ Endangered species | | | I | D | D | D | D |
| ▶ Extinct species | | | I | D | D | D | D |
| ▶ Comparison of present and past species | | | I | D | R | R | D |

# Earth and Space Science

## Earth Processes and Materials

| Earth Structure | K | 1 | 2 | 3 | 4 | 5 | 6 |
|---|---|---|---|---|---|---|---|
| Landforms | I | I | I | D | D | R | R |
| ▶ Comparisons | I | I | I | D | D | R | R |
| ▶ Classified by Earth processes | | | | I | I | D | D |
| Features of ocean bottom | | | | I | I | D | D |
| ▶ Spreading zone | | | | I | I | D | D |
| ▶ Subduction | | | | | | I | D |
| Crust, mantle, core | | | I | I | D | D | R |
| ▶ Lithosphere | | | | | I | I | D |
| Hydrosphere | I | I | I | D | D | D | D |
| ▶ Sources of fresh water | I | I | I | D | D | D | D |
| Glaciers | | | | I | D | D | D |

| Processes of Change | K | 1 | 2 | 3 | 4 | 5 | 6 |
|---|---|---|---|---|---|---|---|
| Weathering | | I | I | D | D | D | D |
| ▶ Causes | | I | I | D | D | D | D |
| ▶ Chemical vs. physical | | | | | I | D | D |
| ▶ Soil as product | | | I | I | I | D | D |
| Erosion/deposition | | I | I | D | D | D | D |
| ▶ Basic agents | | I | I | D | D | D | D |
| ▶ River systems | | | | | I | D | D |
| ▶ Glacier systems | | | | I | D | R | R |
| Plate tectonics | | | | | I | D | D |
| ▶ Evidence | | | | | | D | D |
| ▶ Plate model | | | | | | D | D |
| ▶ Earthquakes | | | | I | I | D | D |
| ▶ Volcanoes | | | | I | I | D | D |
| ▶ Mountain building | | | | | I | D | D |

| Earth History | K | 1 | 2 | 3 | 4 | 5 | 6 |
|---|---|---|---|---|---|---|---|
| Fossil evidence | | | I | D | D | D | D |
| Relative dating | | | | | I | D | D |
| Absolute dating | | | | | | I | D |
| Eras of time | | | | | | I | D |
| Change of environments over time | | | I | D | D | D | D |

| Minerals | K | 1 | 2 | 3 | 4 | 5 | 6 |
|---|---|---|---|---|---|---|---|
| Examples | | | I | I | I | D | D |
| Comparison/observed properties | | | I | I | I | D | D |
| Identification via key | | | | | I | D | D |

| KEY | | |
|---|---|---|
| **I** = Introduced | **D** = Developed | **R** = Reinforced |

## Earth Processes and Materials *cont.*

| Rocks | K | 1 | 2 | 3 | 4 | 5 | 6 |
|---|---|---|---|---|---|---|---|
| Examples | I | I | I | D | D | R | R |
| Comparison/observed properties | I | I | I | D | D | R | R |
| Classification (igneous, sedimentary, metamorphic) | | | | I | D | D | D |
| Rock cycle | | | | | I | D | D |

| Soil | K | 1 | 2 | 3 | 4 | 5 | 6 |
|---|---|---|---|---|---|---|---|
| Formation | | | I | I | D | D | D |
| Components | I | I | I | D | D | R | R |
| Horizons | | | | I | D | D | R |
| Properties | I | I | I | D | D | R | R |
| ▸ Water capacity | | I | I | I | D | D | R |
| ▸ Permeability | | | I | | I | D | R |
| Pollution/conservation | | | | I | D | D | D |

| Renewable and Nonrenewable | K | 1 | 2 | 3 | 4 | 5 | 6 |
|---|---|---|---|---|---|---|---|
| Air | | I | I | D | D | D | D |
| ▸ Pollution/conservation | | I | I | D | D | D | D |
| ▸ Natural recycling | | | | I | D | D | R |
| Water | I | I | I | D | D | R | R |
| ▸ Pollution/conservation | I | I | I | D | D | D | D |
| ▸ Water cycle | | I | I | D | D | R | R |
| Fossil fuels | | | | I | D | D | R |
| ▸ Formation | | | | I | D | D | D |
| ▸ Alternative energy sources | | | | I | D | D | D |

## Weather and Climate

| The Water Cycle | K | 1 | 2 | 3 | 4 | 5 | 6 |
|---|---|---|---|---|---|---|---|
| Processes | | I | I | D | D | R | R |
| ▸ Evaporation/condensation/precipitation | | I | I | D | D | R | R |
| ▸ Runoff/infiltration/transpiration | | | | I | D | D | R |
| Clouds | I | I | D | D | D | D | R |
| ▸ Formation | I | I | D | D | D | D | R |
| ▸ Classification | I | I | D | D | D | D | R |
| ▸ ▸ Shapes | I | I | D | D | R | R | R |
| ▸ Cloud cover (sky conditions) | | | I | D | D | R | R |
| Precipitation | I | I | I | I | D | D | D |
| ▸ Forms | I | I | I | I | D | D | D |
| ▸ Formation | | | | I | D | D | D |
| ▸ Relation to clouds | | | I | I | D | D | R |

Scope & Sequence

# Scope and Sequence

## Weather and Climate *cont.*

| Weather Variables & Measurement | K | 1 | 2 | 3 | 4 | 5 | 6 |
|---|---|---|---|---|---|---|---|
| Air temperature | I | I | I | D | D | R | R |
| ▸ Conceptual model | | | I | I | D | D | R |
| ▸ Thermometer—Celsius/Fahrenheit | I | I | D | D | D | R | R |
| Air pressure | | | | I | D | D | D |
| ▸ Conceptual model | | | | | I | D | D |
| ▸ Barometer | | | | I | D | D | R |
| Wind | I | I | I | D | D | R | R |
| ▸ Relation to air pressure | | | | I | I | D | D |
| ▸ Direction—vanes | I | I | D | D | D | R | R |
| ▸ Speed—anemometer | | I | I | D | D | R | R |
| ▸ Global patterns | | | | | I | D | D |
| ▸ Coriolis effect | | | | | | D | R |

| Weather Prediction | K | 1 | 2 | 3 | 4 | 5 | 6 |
|---|---|---|---|---|---|---|---|
| Air masses/fronts | | | | | I | D | D |
| Station models | | | | | I | D | R |
| Weather maps | | | | I | D | D | D |
| Barometer readings, predictions | | | | | I | D | R |
| Cloud types, trends | | | I | | D | R | R |
| Severe weather | I | I | D | D | D | D | R |
| ▸ Thunderstorms | I | I | D | D | D | D | D |
| ▸ Hurricanes | I | I | D | D | D | D | D |
| ▸ Tornadoes | | | I | I | D | D | D |
| ▸ Hailstorms | | | I | I | D | D | D |

| Climate | | | | | | | |
|---|---|---|---|---|---|---|---|
| Components of | | | | I | D | D | D |
| Comparison/classification | | | | I | D | D | D |
| Seasons | I | D | D | R | R | R | R |
| Factors of | | | | I | D | D | D |
| ▸ Latitude | | | | I | D | D | D |
| ▸ Altitude | | | | I | D | D | D |
| ▸ Coastal/inland | | | | I | D | D | D |
| ▸ Mountain ranges | | | | I | D | D | D |
| ▸ Ocean currents | | | | | I | D | R |
| ▸ Global winds | | | | | I | D | R |

| KEY | | |
|---|---|---|
| **I** = Introduced | **D** = Developed | **R** = Reinforced |

## Astronomy

| Earth-Sun-Moon System | K | 1 | 2 | 3 | 4 | 5 | 6 |
|---|---|---|---|---|---|---|---|
| Day and night | I | I | D | D | D | R | R |
| ▶ Daily observations | I | I | D | D | D | R | R |
| ▶ Rotational model |  | I | I | D | D | D | R |
| Year/seasons | I | I | D | D | D | R | R |
| ▶ Orbital model |  | I | I | D | D | R | R |
| ▶ Solar radiation variations |  |  |  | I | D | R | R |
| Earth's Moon | I | I | I | D | D | D | R |
| ▶ Basic observations/phases | I | I | I | D | D | D | R |
| ▶ Earth-Sun-Moon model |  |  | I | I | D | D | R |
| ▶ Eclipses |  |  |  |  | I | D | D |
| ▶ Tides |  |  |  |  |  | D | D |

| Solar System | K | 1 | 2 | 3 | 4 | 5 | 6 |
|---|---|---|---|---|---|---|---|
| Planets |  | I | I | I | D | D | D |
| ▶ Inner vs. outer |  |  | I |  | I | D | D |
| ▶ Asteroids |  |  |  |  | I | D | D |
| Effects of gravity/inertia |  |  |  |  |  | D | D |
| ▶ Shape of orbital path |  |  |  |  | I | D | R |
| ▶ Speed of planets |  |  |  |  |  | D | R |
| Comets |  |  |  |  | I | D | R |
| ▶ Explanation |  |  |  |  | I | D | R |
| ▶ Historical sightings |  |  |  |  |  | D | R |
| Meteors/meteorites |  |  |  |  | I | D | D |
| ▶ Explanation |  |  |  |  | I | D | D |
| ▶ Impact on Earth |  |  |  |  | I | D | D |
| Formation of solar system |  |  |  |  |  | D | D |

| Stars/Universe | K | 1 | 2 | 3 | 4 | 5 | 6 |
|---|---|---|---|---|---|---|---|
| Star properties/comparisons/constellations | I |  | I | D | D | D | D |
| ▶ Life cycle |  |  |  |  |  | D | D |
| ▶ H-R diagram |  |  |  |  |  | I | D |
| Galaxies |  |  |  |  |  | I | D |
| ▶ Examples/classification |  |  |  |  |  | I | D |
| ▶ Motion/red- vs. blueshift |  |  |  |  |  |  | D |
| Universe |  |  |  |  |  | D | D |
| ▶ Structure |  |  |  |  |  | I | D |
| ▶ Change |  |  |  |  |  | I | D |
| ▶ Formation |  |  |  |  |  | I | D |

# Scope and Sequence

## Physical Science

### Matter

| Properties of Objects/Materials | K | 1 | 2 | 3 | 4 | 5 | 6 |
|---|---|---|---|---|---|---|---|
| Observations | I | I | D | D | D | R | R |
| Classification (physical vs. chemical) | | | | | I | D | D |
| Measurements | I | I | I | D | D | R | R |
| ▸ Length (ruler) | I | I | I | D | D | R | R |
| ▸ Mass (balance) | I | I | I | D | D | R | R |
| ▸ Volume/capacity (containers) | I | I | I | D | D | R | R |
| ▸ Temperature (thermometer) | I | I | I | D | D | R | R |

| Properties of Matter | K | 1 | 2 | 3 | 4 | 5 | 6 |
|---|---|---|---|---|---|---|---|
| Density | | | | | I | D | R |
| Conductivity | | | | I | D | D | R |
| Magnetism | I | D | D | R | R | R | R |
| ▸ Poles | I | D | D | R | R | R | R |
| ▸ Fields | | | I | I | D | D | R |
| ▸ Interactions | I | D | D | R | R | R | R |

| States of Matter | K | 1 | 2 | 3 | 4 | 5 | 6 |
|---|---|---|---|---|---|---|---|
| Descriptions/properties | | I | I | D | D | R | R |
| Particulate models | | | | | I | D | D |
| Melting/freezing point | | | | I | I | D | R |
| Boiling/condensing point | | | | | I | D | R |

| Classification of Matter | K | 1 | 2 | 3 | 4 | 5 | 6 |
|---|---|---|---|---|---|---|---|
| Elements | | | | I | I | D | D |
| ▸ Metals | | | | I | I | D | D |
| ▸ Nonmetals | | | | | I | D | D |
| ▸ Metalloids | | | | | | I | D |
| ▸ Periodic table | | | | | I | D | R |
| Mixtures | | I | I | I | D | D | R |
| ▸ Solid/liquid/gas combinations | | I | I | I | D | D | R |
| ▸ Suspensions | | | I | I | D | D | R |
| ▸ Solutions | | | | | I | D | R |
| Compounds | | | | I | D | D | R |
| ▸ Compared with mixtures | | | | I | D | D | R |
| ▸ Names for | | | | | | D | R |
| ▸ Carbon compounds compared | | | | | | | D |

**KEY**

**I** = Introduced  **D** = Developed  **R** = Reinforced

## Matter *cont.*

| Changes of Matter | K | 1 | 2 | 3 | 4 | 5 | 6 |
|---|---|---|---|---|---|---|---|
| Physical changes | I | I | D | D | D | R | R |
| ▶ Tearing/cutting/molding | I | I | D | D | D | R | R |
| ▶ Change of state | I | I | D | D | D | R | R |
| ▶ Particle/heat model | | | | | I | D | R |
| ▶ Temperature/time graph | | | | | | D | R |
| ▶ Formation of mixtures | | I | I | I | D | D | R |
| ▶ Separation of mixtures | | | I | I | D | D | R |
| ▶ Solubility (means of increasing) | | | | | | D | D |
| Chemical Changes | | | I | I | D | D | D |
| ▶ Chemical reactions | | | | | I | D | D |
| ▶ Signs of chemical change | | | | | I | D | D |
| ▶ Formation of compounds | | | | I | D | D | D |
| ▶ Acid/base/salts | | | | | I | D | D |
| Nuclear changes | | | | | | | D |
| ▶ Radioactivity | | | | | | | D |
| ▶ Fission vs. fusion | | | | | | | D |

| Atomic Model of Matter | K | 1 | 2 | 3 | 4 | 5 | 6 |
|---|---|---|---|---|---|---|---|
| Atoms | | | | | I | D | D |
| ▶ Parts of atoms | | | | | | D | D |
| ▶ Models of atoms of elements | | | | | | D | D |
| ▶ Model through history | | | | | | D | D |
| Molecules | | | | | | D | D |
| ▶ Bonding within | | | | | | I | D |
| ▶ Models of compounds | | | | | | D | D |
| ▶ Models of diatomic elements | | | | | | D | D |

# Scope and Sequence

## Position, Motion, Forces

| Position | K | 1 | 2 | 3 | 4 | 5 | 6 |
|---|---|---|---|---|---|---|---|
| Position words (*above*, *below*, etc.) | I | D | D | R | | | |
| Relative position, fixed/moving | | | | I | D | D | R |

| Motion | K | 1 | 2 | 3 | 4 | 5 | 6 |
|---|---|---|---|---|---|---|---|
| Distance | | | I | D | R | R | R |
| Speed | I | I | I | I | D | D | R |
| ▶ Comparing fast and slow | I | I | D | R | | | |
| ▶ Using mathematical formula | | | | | | I | D |
| ▶ Compared with velocity | | | | | I | D | R |
| Velocity/acceleration | | | | | I | D | R |

| Forces | K | 1 | 2 | 3 | 4 | 5 | 6 |
|---|---|---|---|---|---|---|---|
| Pushes and pulls | I | I | D | D | R | R | R |
| Gravity | | | I | I | D | D | D |
| ▶ Weight change with distance | | | | I | D | D | D |
| ▶ Universal gravitation | | | | | | D | D |
| ▶ Weight on other planets | | | | | I | D | D |
| Friction | | I | I | D | D | D | R |
| ▶ Reduction of, lubrication | | | D | D | D | R | R |
| ▶ Source of heat | | I | I | D | D | R | R |
| ▶ Kinds (rolling, standing, sliding) | | | | | | I | D |
| Magnetism | I | I | D | R | R | R | R |
| ▶ Poles | I | I | D | R | R | R | R |
| ▶ ▶ Attraction vs. repulsion | I | I | D | R | R | R | R |
| ▶ ▶ Magnetic vs. geographic | | | | | I | D | R |
| ▶ Magnetic fields | | | I | I | D | D | R |
| ▶ Magnetic domains | | | | | | D | R |
| Effect on position and motion | | I | I | D | D | D | D |
| ▶ Changes in speed and direction (acceleration) | | I | I | D | D | D | D |
| ▶ Balanced vs. unbalanced forces | | | | I | D | D | D |
| ▶ Inertia | | | | | I | D | D |
| ▶ Circular motion | | | | | | D | D |
| ▶ Newton's First Law | | | I | I | I | D | D |
| Newton's Second Law | | | I | I | I | D | D |
| ▶ Effect of changing mass on constant force | | | I | I | I | D | D |
| ▶ Effect of changing force on constant mass | | | I | I | I | D | D |
| Newton's Third Law (action-reaction) | | | | | | D | R |
| Momentum | | | | | | I | D |

| KEY | | |
|---|---|---|
| **I** = Introduced | **D** = Developed | **R** = Reinforced |

## Work and Energy

| Work | K | 1 | 2 | 3 | 4 | 5 | 6 |
|---|---|---|---|---|---|---|---|
| Examples of | | | | I | D | D | R |
| Mathematical formula | I | I | | | I | D | D |
| Calculations in joules | | | | | | I | D |

| Energy | K | 1 | 2 | 3 | 4 | 5 | 6 |
|---|---|---|---|---|---|---|---|
| Potential | | | | I | D | D | R |
| ▶ Examples of | | | | I | D | D | R |
| ▶ Gravitational potential | | | | I | D | D | R |
| Kinetic | | | | I | D | D | R |
| ▶ Examples of | | | | I | D | D | R |
| ▶ Relation to temperature | | | | | I | D | D |
| Transformation of potential and kinetic | | | | I | I | D | R |
| Law of Conservation of Energy | | | | | I | D | R |

| Simple Machines | K | 1 | 2 | 3 | 4 | 5 | 6 |
|---|---|---|---|---|---|---|---|
| Relation to needs, found in nature | | I | I | D | D | R | R |
| Examples of | | I | I | D | D | R | R |
| ▶ Levers | | I | I | D | D | D | R |
| ▶ Fulcrum | | | I | I | D | D | R |
| ▶ Classification | | | I | I | D | D | R |
| ▶ Pulleys | I | I | I | I | D | D | R |
| ▶ Fixed vs. movable | | | | I | D | D | R |
| ▶ Wheel-and-axles | I | I | I | I | D | D | R |
| ▶ Inclined planes | | I | I | D | D | D | R |
| ▶ Compared with lifting | | I | I | D | D | D | R |
| ▶ Effects of texture, angle | | | | D | D | D | R |
| ▶ Wedges | | I | I | I | D | D | R |
| ▶ Screws | | I | I | I | D | D | R |
| Mechanical advantage | | | | | | I | D |
| Friction/efficiency | | | | | | I | D |
| Relation to compound machines | | | | D | D | D | R |

# Scope and Sequence

## Transfer of Energy

| Heat | K | 1 | 2 | 3 | 4 | 5 | 6 |
|---|---|---|---|---|---|---|---|
| Sources | | I | I | D | D | R | R |
| Transfer | | | | I | D | D | D |
| ▶ Conduction | | | | I | D | D | D |
| ▶ Convection | | | | I | D | D | D |
| ▶ Radiation | | | | I | D | D | D |
| Related to temperature | | I | I | I | D | D | D |
| Kinetic energy model | | | | | I | D | D |
| Heating of different surfaces, trends | | | | | I | D | D |

| Sound | K | 1 | 2 | 3 | 4 | 5 | 6 |
|---|---|---|---|---|---|---|---|
| Sources (vibrations) | I | I | I | D | D | D | R |
| Transfer of sound | | | I | D | D | D | R |
| ▶ Through solids, liquids, gases | | | I | D | D | D | R |
| ▶ Compression wave model | | | | | I | D | D |
| Pitch variations | I | I | I | D | D | D | R |
| ▶ Frequency | | | | | I | D | R |
| ▶ Hertz measurements | | | | | | I | D |
| ▶ Doppler effect | | | | | | | D |
| Volume (loudness) variations | I | I | I | D | D | D | R |
| ▶ Amplitude | | | | | I | D | D |
| ▶ Decibel measurements | | | | | | I | D |
| Absorption (insulation) | | | | | | D | D |
| Reflection (echoes) | | | | | I | D | R |

| Light | K | 1 | 2 | 3 | 4 | 5 | 6 |
|---|---|---|---|---|---|---|---|
| Sources | | I | I | D | D | D | R |
| ▶ Natural vs. artificial | | I | I | I | D | D | R |
| Straight-line path | | I | I | D | D | D | R |
| ▶ Opaque, transparent, translucent materials | | | | I | D | D | R |
| ▶ Shadows | I | I | I | D | D | D | R |
| Reflection | | | I | I | D | D | D |
| ▶ Law of reflection | | | | | I | D | D |
| ▶ Mirrors | | | I | I | D | D | D |
| ▶ Convex vs. concave | | | | | I | D | D |
| Refraction | | | I | I | D | D | D |
| Lenses | | | I | I | D | D | D |
| ▶ Convex vs. concave | | | | | I | D | D |

**KEY**
**I** = Introduced **D** = Developed **R** = Reinforced

## Transfer of Energy *cont.*

| **Light** *cont.* | K | 1 | 2 | 3 | 4 | 5 | 6 |
|---|---|---|---|---|---|---|---|
| Prism | | | I | I | D | D | D |
| ▶ Visible spectrum (colors) | | | I | I | I | D | D |
| ▶ Mixing colors | | | | | | D | R |
| ▶ Mixing pigments | | | | | | D | R |
| Wave model | | | | | | D | D |
| ▶ Parts (crest, trough) | | | | | | D | D |
| ▶ Photons | | | | | | D | D |
| Electromagnetic spectrum | | | | | | I | D |
| ▶ Descriptions, uses | | | | | | I | D |
| ▶ Wavelengths/frequencies compared | | | | | | I | D |

| **Electricity** | K | 1 | 2 | 3 | 4 | 5 | 6 |
|---|---|---|---|---|---|---|---|
| Static | | | I | | D | D | R |
| ▶ Attraction vs. repulsion | | | I | | D | D | R |
| ▶ Induced charge | | | | | I | D | D |
| ▶ Discharge | | | | | I | D | D |
| ▶ Insulators/conductors | | | | I | D | D | R |
| Current/circuits | | | I | I | D | D | R |
| ▶ Closed vs. open | | | | I | D | D | R |
| ▶ Parts of | | I | I | D | D | D | R |
| ▶ Batteries/wet-dry cells | | I | I | I | D | D | R |
| ▶ Transformation of energy in | | I | I | I | D | R | R |
| ▶ Series vs. parallel | | | | | D | D | D |
| Electromagnetism | | | | | D | D | D |
| ▶ Making electromagnets | | | | | D | D | D |
| ▶ Increasing strength of | | | | | D | D | D |
| ▶ Common uses | | | | I | D | D | D |
| ▶ Motors | | | | I | D | D | R |
| ▶ Generating current | | | | | D | D | R |
| ▶ AC vs. DC | | | | | D | D | R |
| ▶ Power transmission | | | | | D | D | R |
| Household uses | I | I | I | D | D | R | R |
| ▶ Safety | | | | | D | D | D |

# Index



Page references followed by an asterisk * indicate activities.

## C



## D



Page references followed by an asterisk * indicate activities.

# Index



Page references followed by an asterisk * indicate activities.

## F



## G



## H



Page references followed by an asterisk * indicate activities.

# Index



## I



## J



## K



## L



Page references followed by an asterisk * indicate activities.

## M



Page references followed by an asterisk * indicate activities.

Page references followed by an asterisk * indicate activities.

# P



Page references followed by an asterisk * indicate activities.

# Index



Page references followed by an asterisk * indicate activities.

Page references followed by an asterisk * indicate activities.

## T



Page references followed by an asterisk * indicate activities.



## U



## V

## W



## Y



Page references followed by an asterisk * indicate activities.

# Correlation to National Science Education Standards

## Grades K-4

The Science content standards are presented in the *National Science Education Standards*. These standards offer a structured guide to what scientifically literate students should know, understand, and be able to do at different grade levels. The standards are clustered for Grades K–4, 5–8, and 9–12.

The following table shows where the content standards are met in Science: *A Closer Look*, Grades K–4. Grade K page numbers refer to the Teacher's Edition. Grades 1–2 page numbers refer to the Student Edition or, when preceded by AF, to the Activity Flipchart. Grades 3–4 page numbers refer to the Student Edition.

| Science As Inquiry | |
|---|---|
| **Abilities Necessary to Do Scientific Inquiry** | |
| *Ask a question about objects, organisms, and events in the environment.* This aspect of the standard emphasizes students asking questions that they can answer with scientific knowledge, combined with their own observations. Students should answer their questions by seeking information from reliable sources of scientific information and from their own observations and investigations. | **Grade K:** 12, 52, 65, 82, 89, 96, 97, 102, 116, 127, 130, 131, 136, 144, 150, 158, 161, 164, 202, 218, 237, 254, 255, 257, 260, 266, 269, 272, 273<br>**Grade 1:** 3, 11, 23, 29, 37, 53, 59, 67, 87, 95, 103, 109, 127, 133, 141, 163, 171, 179, 195, 201, 209, 229, 235, 241, 249, 265, 271, 279, 299, 307, 315, 329, 333, 341, 361, 367, 373, 381, 397, 403, 411, 419, AF4, AF11, AF16, AF27, AF30, AF34, AF45, AF50, AF54<br>**Grade 2:** 3, 11, 23, 29, 39, 55, 61, 69, 89, 95, 103, 121, 129, 135, 155, 163, 171, 187, 195, 201, 223, 231, 237, 253, 259, 267, 275, 295, 301, 309, 325, 331, 339, 361, 367, 377, 385, 399, 405, 415, 421, AF4, AF10, AF16, AF19, AF22, AF26, AF28, AF34, AF38, AF41, AF46, AF50, AF54, AF56, AF59<br>**Grade 3:** 3, 21, 31, 43, 53, 69, 81, 91, 107, 119, 133, 151, 161, 173, 191, 203, 213, 227, 239, 249, 259, 279, 289, 303, 317, 327, 337, 347, 363, 373, 383, 397, 407, 417, 433, 443, 453, 463, 479, 489, 499, 511<br>**Grade 4:** 3, 21, 33, 45, 59, 77, 89, 99, 109, 129, 137, 149, 165, 175, 183, 203, 213, 225, 237, 251, 263, 273, 285, 295, 313, 323, 335, 345, 359, 369, 379, 393, 411, 421, 431, 445, 457, 467, 483, 493, 503, 513, 529, 539, 551, 563, 575 |
| *Plan and conduct a simple investigation.* In the earliest years, investigations are largely based on systematic observations. As students develop, they may design and conduct simple experiments to answer questions. The idea of a fair test is possible for many students to consider by fourth grade. | **Grade K:** 30, 38, 44, 52, 58, 68, 76, 82, 88, 96, 102, 110, 116, 130, 136, 144, 150, 158, 164, 176, 182, 188, 196, 202, 216, 222, 228, 236, 246, 254, 260, 266, 272<br>**Grade 1:** 59, 99, 195, 201, 209, 271, 279, 329, 331, 333, 335, 341, 361, 367, 370, 373, 381, 397, 403, 419, AF4, AF11, AF27, AF30, AF34, AF50, AF54<br>**Grade 2:** 39, 73, 132, 137, 163, 325, 339, 344, 380, 388, AF22, AF28, AF34, AF46, AF50, AF54, AF56<br>**Grade 3:** 40–41, 144–145, 268–269, 334–335, 422–423, 450–451, 496–497<br>**Grade 4:** 56–57, 106–107, 244–245, 270–271, 282–283, 352–353, 400–401, 472–473, 522–523, 560–561, 572–573 |

| Abilities Necessary To Do Scientific Inquiry continued | |
|---|---|
| *Employ simple equipment and tools to gather data and extend the senses.* In early years, students develop simple skills, such as how to observe, measure, cut, connect, switch, turn on and off, pour, hold, tie, and hook. Beginning with simple instruments, students can use rulers to measure the length, height, and depth of objects and materials; thermometers to measure temperature; watches to measure time; beam balances and spring scales to measure weight and force; magnifiers to observe objects and organisms; and microscopes to observe the finer details of plants, animals, rocks, and other materials. Children also develop skills in the use of computers and calculators for conducting investigations. | **Grade K:** 52, 58, 68, 76, 82, 102, 110, 117, 127, 130, 131, 133, 136, 139, 144, 150, 151, 153, 164, 165, 168F, 176, 177, 197, 199, 202, 203, 222, 223, 229, 231, 243, 246, 255, 261, 267, 268, 269, 272, 273, TR7, TR8, TR9, TR10, TR11<br>**Grade 1:** 29, 53, 87, 95, 127, 171, 229, 235, 249, 271, 307, 311, 315, 329, 341, 343, 376, 381, 384, 397, 403, 411, 419, R3, R4, R5, R6, AF4, AF27, AF34, AF45, AF50, AF54<br>**Grade 2:** 3, 29, 39, 69, 89, 95, 103, 129, 135, 171, 187, 195, 223, 231, 253, 259, 267, 275, 301, 309, 325, 339, 367, 377, 385, 399, 405, 415, 421, R3, R4, R5, R6, AF4, AF10, AF16, AF22, AF28, AF34, AF46, AF50, AF54, AF56, AF59<br>**Grade 3:** 26, 31, 40, 41, 53, 69, 81, 107, 119, 133, 144, 145, 161, 173, 200–201, 213, 231, 239, 243, 259, 268, 269, 307, 317, 334–335, 339, 347, 349, 363, 373, 380, 381, 383, 397, 407, 417, 422, 423, 438, 450, 460–461, 481, 486–487, 489, 496–497, 499, 511, 516, R2, R4, R5, R6, R7, R8, R9<br>**Grade 4:** 21, 30, 33, 39, 45, 56, 93, 99, 106–107, 109, 129, 134–135, 137, 143, 149, 172–173, 175, 237, 241, 251, 263, 267, 270–271, 285, 289, 295, 332–333, 352–353, 359, 363, 369, 379, 393, 400, 411, 421, 425, 431, 445, 461, 464–465, 467, 483, 490–491, 493, 503, 513, 522–523, 536–537, 539, 551, 560–561, 575, 580, R4, R5, R6, R7, R8, R9 |
| *Use data to construct a reasonable explanation.* This aspect of the standard emphasizes the students' thinking as they use data to formulate explanations. Even at the earliest grade levels, students should learn what constitutes evidence and judge the merits or strength of the data and information that will be used to make explanations. After students propose an explanation, they will appeal to the knowledge and evidence they obtained to support their explanations. Students should check their explanations against scientific knowledge, experiences, and observations of others. | **Grade K:** 7, 30, 58, 76, 79, 99, 113, 158, 176, 203, 228, 243, 246, 260<br>**Grade 1:** 11, 141, 397, 403, 411, 419, AF4, AF8, AF27, AF30, AF34, AF38, AF40, AF45, AF50, AF54<br>**Grade 2:** 11, 129, 171, 201, 237, 259, 267, 275, 309, 325, 361, AF4, AF10, AF22, AF26, AF28, AF34, AF38, AF41, AF46, AF50, AF54, AF56, AF59<br>**Grade 3:** 21, 31, 41, 43, 53, 69, 81, 91, 107, 119, 133, 145, 151, 161, 173, 191, 203, 213, 227, 239, 249, 259, 269, 279, 289, 303, 317, 327, 335, 337, 347, 363, 373, 383, 397, 407, 417, 422, 423, 433, 443, 450, 453, 463, 479, 489, 497, 499, 511<br>**Grade 4:** 21, 33, 45, 57, 59, 77, 89, 99, 107, 109, 129, 137, 149, 165, 175, 183, 203, 213, 225, 237, 245, 251, 263, 271, 273, 283, 285, 295, 313, 323, 335, 345, 352, 359, 369, 379, 393, 401, 411, 421, 431, 445, 457, 467, 473, 483, 493, 503, 513, 523, 529, 539, 551, 561, 563, 573, 575 |

| Abilities Necessary To Do Scientific Inquiry continued | |
|---|---|
| *Communicate investigations and explanations.* Students should begin developing the abilities to communicate, critique, and analyze their work and the work of other students. This communication might be spoken or drawn as well as written. | **Grade K:** 26, 30, 32, 38, 40, 44, 46, 52, 54, 58, 64, 68, 70, 76, 78, 82, 84, 88, 90, 96, 98, 102, 104, 110, 112, 116, 126, 130, 132, 136, 138, 144, 146, 150, 152, 158, 160, 164, 170, 176, 178, 182, 184, 188, 190, 196, 198, 202, 212, 216, 218, 222, 224, 228, 230, 236, 242, 246, 248, 254, 256, 260, 262, 266, 268, 272<br>**Grade 1:** 3, 29, 32, 37, 56, 59, 67, 95, 103, 127, 133, 141, 163, 195, 235, 271, 279, 299, 315, 329, AF4, AF11, AF16, AF27, AF30, AF34, AF50, AF54, AF58<br>**Grade 2:** 11, 29, 55, 59, 63, 89, 93, 95, 97, 106, 163, 187, 195, 201, 261, 331, 361, AF4, AF10, AF16, AF22, AF28, AF34, AF38, AF45, AF46, AF50, AF54, AF56, AF59<br>**Grade 3:** 26, 35, 43, 47, 53, 85, 91, 93, 114, 116–117, 144, 151, 173, 203, 207, 239, 251, 265, 268, 327, 334, 349, 363, 367, 373, 387, 397, 423, 433, 438, 453, 463, 479, 499, 511<br>**Grade 4:** 12, 21, 45, 57, 77, 129, 151, 203, 225, 251, 260, 261, 273, 277, 295, 299, 323, 339, 349, 352, 359, 467, 472, 523, 539, 561, 563, 573, 575 |
| **Understanding About Scientific Inquiry** | |
| Scientific investigations involve asking and answering a question and comparing the answer with what scientists already know about the world. | **Grade K:** 30, 38, 44, 52, 58, 68, 76, 82, 88, 96, 102, 110, 116, 130, 136, 144, 150, 158, 164, 176, 182, 188, 196, 202, 216, 222, 228, 236, 246, 254, 260, 266, 272<br>**Grade 1:** 3, 11, 23, 29, 37, 53, 59, 67, 87, 95, 103, 109, 127, 133, 141, 163, 171, 179, 195, 201, 209, 229, 235, 241, 249, 265, 271, 279, 299, 307, 315, 329, 333, 341, 361, 367, 373, 381, 397, 403, 411, 419, AF4, AF30, AF34, AF50, AF54, AF55<br>**Grade 2:** 3, 11, 23, 29, 39, 55, 61, 69, 89, 95, 103, 121, 129, 135, 155, 163, 171, 187, 195, 201, 223, 231, 237, 253, 259, 267, 275, 295, 301, 309, 325, 331, 339, 361, 367, 377, 385, 399, 405, 415, 421, AF4, AF10, AF26, AF28, AF34, AF46, AF50, AF54, AF56, AF59<br>**Grade 3:** 3, 21, 31, 40–41, 43, 53, 69, 81, 91, 107, 119, 133, 144–145, 151, 161, 173, 191, 203, 213, 227, 239, 249, 259, 268–269, 279, 289, 303, 317, 327, 334–335, 337, 347, 363, 373, 383, 397, 407, 417, 422–423, 433, 443, 450–451, 453, 463, 479, 489, 493, 496–497, 499, 511<br>**Grade 4:** 3, 21, 33, 45, 56–57, 59, 77, 89, 99, 106–107, 109, 129, 137, 149, 165, 175, 183, 203, 213, 225, 237, 244–245, 251, 263, 270–271, 273, 282–283, 285, 295, 313, 323, 335, 345, 352–353, 359, 369, 379, 393, 400–401, 411, 421, 431, 445, 457, 467, 472–473, 483, 493, 503, 513, 522–523, 529, 539, 551, 560–561, 563, 572–573, 575 |

| Understanding About Scientific Inquiry continued | |
|---|---|
| Scientists use different kinds of investigations depending on the questions they are trying to answer. Types of investigations include describing objects, events, and organisms; classifying them; and doing a fair test (experimenting). | **Grade K:** 30, 38, 44, 52, 58, 68, 76, 82, 88, 96, 102, 110, 116, 130, 136, 144, 150, 158, 164, 176, 182, 188, 196, 202, 216, 222, 228, 236, 246, 254, 260, 266, 272<br>**Grade 1:** 3, 11, 23, 25, 29, 32, 37, 40, 53, 56, 59, 63, 67, 70, 87, 91, 95, 99, 103, 107, 109, 113, 127, 130, 133, 135, 141, 144, 163, 166, 171, 173, 179, 183, 195, 198, 201, 204, 209, 211, 229, 232, 235, 238, 241, 243, 249, 251, 265, 267, 271, 273, 279, 281, 299, 301, 307, 311, 315, 319, 329, 331, 333, 335, 341, 343, 361, 364, 367, 370, 373, 376, 381, 384, 397, 399, 403, 405, 411, 414, 419, 421, AF8, AF27, AF30, AF34, AF38, AF50, AF54<br>**Grade 2:** 3, 11, 23, 26, 29, 32, 39, 42, 55, 59, 61, 63, 69, 73, 89, 93, 95, 97, 103, 106, 121, 124, 129, 132, 135, 137, 155, 161, 163, 166, 171, 173, 187, 191, 195, 199, 201, 207, 223, 226, 231, 234, 237, 240, 253, 256, 259, 261, 267, 272, 275, 278, 295, 298, 301, 304, 309, 313, 325, 328, 331, 335, 339, 344, 361, 364, 367, 371, 377, 380, 385, 388, 399, 403, 405, 407, 415, 418, 421, 424, AF4, AF8, AF22, AF26, AF28, AF34, AF38, AF50, AF54, AF56, AF59<br>**Grade 3:** 3, 21, 26, 31, 35, 43, 47, 53, 55, 69, 73, 81, 85, 91, 93, 107, 114, 119, 127, 133, 137, 151, 155, 161, 167, 173, 177, 191, 195, 203, 207, 213, 217, 227, 231, 239, 243, 249, 251, 259, 265, 279, 283, 289, 293, 303, 307, 317, 319, 327, 331, 337, 339, 347, 349, 363, 367, 373, 377, 383, 387, 397, 401, 407, 412, 417, 419, 433, 438, 443, 447, 453, 457, 463, 469, 479, 481, 489, 493, 499, 505, 511, 516<br>**Grade 4:** 3, 21, 27, 33, 39, 45, 53, 59, 64, 77, 80, 89, 93, 99, 102, 109, 116, 129, 131, 137, 143, 149, 151, 165, 169, 175, 177, 183, 186, 203, 207, 213, 219, 225, 231, 237, 241, 251, 255, 263, 267, 273, 277, 285, 289, 295, 299, 313, 317, 323, 328, 335, 339, 345, 349, 359, 363, 369, 373, 379, 384, 393, 397, 411, 415, 421, 425, 431, 435, 445, 449, 457, 461, 467, 470, 483, 487, 493, 498, 503, 507, 513, 515, 529, 533, 539, 544, 551, 557, 563, 569, 575, 580 |

| Understanding About Scientific Inquiry continued | |
|---|---|
| Simple instruments, such as magnifiers, thermometers, and rulers, provide more information than scientists obtain using only their senses. | **Grade K:** 52, 58, 68, 76, 82, 102, 110, 117, 127, 130, 131, 133, 136, 139, 144, 150, 151, 153, 164, 165, 168F, 176, 177, 197, 199, 202, 203, 222, 223, 229, 231, 243, 246, 255, 261, 267, 268, 269, 272, 273, TR7, TR8, TR9, TR10, TR11<br>**Grade 1:** 29, 53, 87, 95, 127, 171, 229, 235, 249, 271, 307, 311, 315, 329, 341, 343, 376, 381, 384, 397, 403, 411, 419, R3, R4, R5, R6, AF4, AF34<br>**Grade 2:** 3, 29, 39, 69, 89, 95, 103, 129, 135, 171, 187, 195, 223, 231, 253, 259, 267, 275, 301, 309, 325, 339, 367, 377, 385, 399, 405, 415, 421, R3, R4, R5, R6, AF28, AF34, AF46, AF50, AF54, AF56, AF59<br>**Grade 3:** 26, 31, 40, 41, 53, 69, 81, 107, 119, 133, 144, 145, 161, 173, 200–201, 213, 231, 239, 243, 259, 268, 269, 307, 317, 334–335, 339, 347, 349, 363, 373, 380, 381, 383, 397, 407, 417, 422, 423, 438, 450, 460–461, 481, 486–487, 489, 496–497, 499, 511, 516, R2, R4, R5, R6, R7, R8, R9<br>**Grade 4:** 21, 30, 33, 39, 45, 56, 93, 99, 106–107, 109, 129, 134–135, 137, 143, 149, 172–173, 175, 237, 241, 251, 263, 267, 270–271, 285, 289, 295, 332–333, 352–353, 359, 363, 369, 379, 393, 400, 411, 421, 425, 431, 445, 461, 464–465, 467, 483, 490–491, 493, 503, 513, 522–523, 536–537, 539, 551, 560–561, 575, 580, R4, R5, R6, R7, R8, R9 |
| Scientists develop explanations using observations (evidence) and what they already know about the world (scientific knowledge). Good explanations are based on evidence from investigations. | **Grade K:** 7, 30, 58, 76, 79, 99, 113, 158, 176, 203, 228, 243, 246, 260<br>**Grade 1:** 11, 141, 397, 403, 411, 419, AF4, AF11, AF16, AF30, AF50, AF54<br>**Grade 2:** 11, 129, 171, 201, 237, 259, 267, 275, 309, 325, 361, AF4, AF10, AF22, AF26, AF28, AF34, AF38, AF46, AF50, AF54, AF56, AF59<br>**Grade 3:** 21, 31, 41, 43, 53, 69, 81, 91, 107, 119, 133, 145, 151, 161, 173, 191, 203, 213, 227, 239, 249, 259, 269, 279, 289, 303, 317, 327, 335, 337, 347, 363, 373, 383, 397, 407, 417, 422, 423, 433, 443, 450, 453, 463, 479, 489, 497, 499, 511<br>**Grade 4:** 21, 33, 45, 57, 59, 77, 89, 99, 107, 109, 129, 137, 149, 165, 175, 183, 203, 213, 225, 237, 245, 251, 263, 271, 273, 283, 285, 295, 313, 323, 335, 345, 352, 359, 369, 379, 393, 401, 411, 421, 431, 445, 457, 467, 473, 483, 493, 503, 513, 523, 529, 539, 551, 561, 563, 573, 575 |

| Understanding About Scientific Inquiry continued | |
|---|---|
| Scientists make the results of their investigations public; they describe the investigations in ways that enable others to repeat the investigations. | **Grade K:** 6, 7, 12, 13, 19, 26, 27, 30, 31, 32, 33, 38, 39, 41, 44, 45, 46, 47, 52, 53, 54, 55, 58, 59, 64, 65, 68, 69, 70, 71, 76, 77, 79, 83, 88, 90, 91, 97, 98, 99, 102, 103, 104, 105, 110, 111, 113, 116, 117, 126, 127, 132, 133, 136, 137, 138, 139, 146, 147, 150, 151, 152, 153, 158, 159, 160, 161, 164, 165, 170, 171, 177, 178, 179, 182, 183, 184, 185, 188, 189, 190, 191, 196, 197, 198, 199, 203, 212, 213, 217, 219, 222, 223, 224, 225, 229, 231, 236, 237, 242, 243, 246, 247, 248, 255, 256, 257, 261, 262, 263, 266, 267, 269, 273<br>**Grade 1:** 3, 29, 32, 37, 56, 59, 67, 95, 103, 127, 133, 141, 163, 195, 235, 271, 279, 299, 315, 329, AF11, AF16, AF30, AF34, AF50<br>**Grade 2:** 11, 29, 55, 59, 63, 89, 93, 95, 97, 106, 163, 187, 195, 201, 261, 331, 361, AF10, AF54<br>**Grade 3:** 7, 26, 35, 43, 47, 53, 85, 91, 93, 114, 116–117, 144, 151, 173, 203, 207, 239, 251, 265, 268, 327, 334, 349, 363, 367, 373, 387, 397, 423, 433, 438, 453, 463, 479, 499, 511<br>**Grade 4:** 12, 21, 45, 57, 77, 129, 151, 203, 225, 251, 260, 261, 273, 277, 295, 299, 323, 339, 349, 352, 359, 467, 472, 523, 539, 561, 563, 573, 575 |
| Scientists review and ask questions about the results of other scientists' work. | **Grade 2:** 280–281, AF16<br>**Grade 3:** 10–11, 180–181<br>**Grade 4:** 106–107, 387 |

## Physical Science

| Properties of Objects and Materials | |
|---|---|
| Objects have many observable properties, including size, weight, shape, color, temperature, and the ability to react with other substances. Those properties can be measured using tools, such as rulers, balances, and thermometers. | **Grade K:** 210E, 211, 212, 213, 214, 215, 218, 223, 224, 226, 227, 228, 230, 231, 232, 233, 234, 235, 236, 237, 238, 239, 240J, 249, 254, 255, 260, 268, 269, 270, 271, 273, 274, 275, 276<br>**Grade 1:** 298, 299, 300, 301, 302, 303, 304, 305, 306, 307, 308–311, 314, 315, 316–319, 320–323, 324, 325, 328, 329, 330–331, 332–337, 338, 339, 341, 342–345, 346, 347, 348–351, 352, 353, AF45, AF54<br>**Grade 2:** 295, 296, 297, 298, 299, 300, 301, 302, 303, 304, 305, 306–307, 308, 309, 310, 311, 312, 313, 314, 315, 316–319, 320, 321, 324, 325, 326–329, 331, 332–335, 336–337, 348–351, 352, 353, AF45<br>**Grade 3:** 358–359, 362, 363, 364, 365, 366, 367, 368, 369, 372, 373, 374, 375, 376, 377, 378, 379, 380–381, 383, 384–389, 390, 391, 392, 393, 406, 407, 408–413, 414–415, 416, 417, 418–421, 422–423, 424, 425, 479, 482, 483, 485<br>**Grade 4:** 411, 412, 413, 414, 415, 416, 417, 418, 419, 421, 422, 423, 424, 425, 426, 427, 428–429, 431, 432, 433, 434, 435, 436, 437, 440, 441, 445, 446, 447, 448, 449, 450, 451, 452, 453, 454–455, 456, 463, 467, 468, 469, 470, 471, 472–473, 474, 475 |

| Properties of Objects and Materials continued | |
|---|---|
| Objects are made of one or more materials, such as paper, wood, and metal. Objects can be described by the properties of the materials from which they are made, and those properties can be used to separate or sort a group of objects or materials. | **Grade K:** 210E, 210J, 210, 212, 214, 215, 216, 217, 218, 219, 220, 221, 222, 225, 226, 229, 239, 271<br>**Grade 1:** 312–313, 325, 346–347<br>**Grade 2:** 306–307, 336–337, 339, 340, 341, 342, 343, 344, 345, 346, 347, 348–351, 352, 353<br>**Grade 3:** 366, 367, 368, 369, 370–371, 409, 410, 411, 412, 413, 414–415, 424, 425<br>**Grade 4:** 54, 406–407, 444, 456, 457, 458, 459, 460, 461, 462, 463, 474, 475 |
| Materials can exist in different states—solid, liquid, and gas. Some common materials, such as water, can be changed from one state to another by heating or cooling. | **Grade K:** 230, 231, 232, 233, 237, 238, 239<br>**Grade 1:** 294, 295, 306, 307, 308, 309, 310, 311, 314, 315, 316, 317, 318, 319, 320–323, 324, 325, 341, 342, 343, 344, 345, 352, 353, AF50, AF54<br>**Grade 2:** 299, 300–305, 308, 309, 310–313, 314, 315, 316–319, 320, 321, 327, 330, 331, 332–335, 336–337, 348–351, 352, 353, 400, AF41, AF46<br>**Grade 3:** 380–381, 382, 383, 384, 385, 386, 387, 388, 389, 392, 393, 396, 397, 398, 399, 400, 401, 402, 403, 404–405, 408, 424, 425, 483, 485<br>**Grade 4:** 323, 324, 325, 331, 410, 414, 415, 417, 446, 447, 448, 449, 453, 534, 535 |
| **Position and Motion of Objects** | |
| The position of an object can be described by locating it relative to another object or the background. | **Grade 1:** 361, 362, 363, 365, 392, 393<br>**Grade 2:** 360, 361, 362, 363, 365, 394, 395<br>**Grade 3:** 432, 433, 434, 435, 439<br>**Grade 4:** 484, 485, 503 |
| An object's motion can be described by tracing and measuring its position over time. | **Grade K:** 240F, 242, 243, 250, 251, 252, 253, 255<br>**Grade 1:** 356–357, 364, 365, 392<br>**Grade 2:** 363, 364, 365, 367, 372, 373, 383, 390–393, 394, 395<br>**Grade 3:** 429, 436, 437, 438, 439, 443, 449<br>**Grade 4:** 482, 483, 484, 485, 486, 487, 488, 489, 490–491, 496, 497, 498, 499, 503, 524, 525 |
| The position and motion of objects can be changed by pushing or pulling. The size of the change is related to the strength of the push or pull. | **Grade K:** 240, 241, 244, 245, 246, 247, 248, 249, 252, 253, 255, 274, 275, 277<br>**Grade 1:** 366, 367, 368, 369, 370, 371, 373, 388–391, 392, 393<br>**Grade 2:** 367, 368, 369, 370, 371, 372, 373, 390–393, 394, 395<br>**Grade 3:** 442, 443, 444, 445, 446, 447, 448, 449, 450–451<br>**Grade 4:** 486, 487, 488, 489, 490–491, 492, 493, 494, 495, 496, 497, 498, 499, 500, 501, 503, 504, 505, 521, 524, 525 |

| Position and Motion of Objects continued | |
|---|---|
| Sound is produced by vibrating objects. The pitch of the sound can be varied by changing the rate of vibration. | **Grade K:** 264, 265, 266, 267, 274, 275<br>**Grade 1:** 402, 403, 404, 405, 406, 407, 428, 429<br>**Grade 2:** 356–357, 405, 406, 407, 409, 410, 411, 412, 429, 432<br>**Grade 3:** 489, 490, 491, 492, 493, 494, 495, 496–497, 520, 521<br>**Grade 4:** 538, 539, 540, 541, 542, 543, 544, 545, 546, 547, 548, 549, 588, 589 |
| **Light, Heat, Electricity, and Magnetism** | |
| Light travels in a straight line until it strikes an object. Light can be reflected by a mirror, refracted by a lens, or absorbed by the object. | **Grade K:** 198, 199, 200, 201, 202, 203<br>**Grade 1:** 411, 412, 413, 414, 415, 416, 417, 426, 429<br>**Grade 2:** 414, 416, 417, 419, 430, 432<br>**Grade 3:** 498, 499, 500, 501, 502, 503, 504, 505, 506, 507, 508–509, 520, 521<br>**Grade 4:** 506, 509, 550, 551, 552, 553, 554, 556, 557, 558, 559, 560–561, 588, 589 |
| Heat can be produced in many ways, such as burning, rubbing, or mixing one substance with another. Heat can move from one object to another by conduction. | **Grade 1:** 396, 397, 400, 401, 428, 429, AF47, AF58<br>**Grade 2:** 399, 400, 401, 403, 428, 433<br>**Grade 3:** 478, 479, 480, 481, 482, 483, 484, 485, 486–487, 520, 521<br>**Grade 4:** 507, 509, 530, 531, 532, 533, 534, 535, 536–537, 588, 589 |
| Electricity in circuits can produce light, heat, sound, and magnetic effects. Electrical circuits require a complete loop through which an electrical current can pass. | **Grade 1:** 418, 419, 420, 421, 422, 423, 427, 428, 429<br>**Grade 2:** 420, 421, 422, 423, 425, 426–427, 431, 432, 433<br>**Grade 3:** 511, 514, 515, 516, 517, 518, 520, 521<br>**Grade 4:** 508, 509, 562, 566, 567, 568, 569, 570, 571, 580, 581, 582, 583, 584, 585, 586–587, 588 |
| Magnets attract and repel each other and certain kinds of other materials. | **Grade K:** 240J, 268, 269, 270, 271, 272, 273, 274, 275<br>**Grade 1:** 380, 381, 382, 383, 384, 385, 386, 387, 392<br>**Grade 2:** 384, 385, 386, 387, 388, 389, 393, AF54<br>**Grade 3:** 367, 446, 450–451<br>**Grade 4:** 574, 575, 576, 577, 578, 579, 580, 581, 582, 583, 585, 588, 589 |

## Life Science

| The Characteristics of Organisms | |
|---|---|
| Organisms have basic needs. For example, animals need air, water, and food; plants require air, water, nutrients, and light. Organisms can survive only in environments in which their needs can be met. The world has many different environments, and distinct environments support the life of different types of organisms. | **Grade K:** 26, 32, 33, 34, 35, 37, 38, 39, 43, 44, 45, 60, 61, 67, 70, 71, 72, 73, 74, 75, 77, 82, 83, 92, 93, 98, 120, TR3, TR4, TR5, TR6<br>**Grade 1:** 24, 26, 27, 30, 42–43, 49, 50, 56, 58, 59, 67, 95, 96, 97, 99, 100, 104, 118, 135, 202, AF11<br>**Grade 2:** 23, 24, 25, 26, 27, 47, 51, 56, 59, 77, 89, 90, 93, 94<br>**Grade 3:** 20, 24, 25, 27, 32, 33, 36, 37, 39, 40–41, 43, 46, 47, 49, 64, 65, 69, 70, 71, 78–79, 82, 119, 120, 121, 122–128<br>**Grade 4:** 22, 46, 48, 49, 50, 51, 55, 59, 64, 72 |

| The Characteristics of Organisms continued | |
|---|---|
| Each plant or animal has different structures that serve different functions in growth, survival, and reproduction. For example, humans have distinct body structures for walking, holding, seeing, and talking. | **Grade K:** 5, 10–11, 26, 27, 28, 29, 30, 31, 42, 43, 60, 61, 78, 80, 81, 86, 87, 88, 90, 91, 92, 93, 94, 95, 97, 99, 100, 101, 103, 105, TR9<br>**Grade 1:** 19, 28, 29, 31, 32, 33, 35, 49, 52, 54, 55, 56, 57, 59, 60, 62, 63, 64–65, 68, 69, 70, 71, 78, 79, 82–83, 89, 90, 92, 93, 95, 98, 99, 102, 103, 105, 106, 107, 114, 115, 123, 129, 130, 131, 133, 136, 137, 138, 139, 154, AF13, AF16, AF52<br>**Grade 2:** 11, 12, 13, 26, 27, 30, 31, 32, 33, 34, 35, 36, 39, 42, 46–49, 50, 57, 58, 59, 69, 70, 71, 72, 73, 74, 77, 78, 80, 89, 93, 123, 125, 129, 130, 131, 132, 133, 137, 139, R8, R9, R10, R11, AF4<br>**Grade 3:** 28–29, 31, 32, 33, 34, 35, 36, 37, 38, 39, 42, 43, 44, 45, 46, 47, 48, 49, 50–51, 53, 54, 55, 56, 57, 58, 59, 60, 61, 62, 64, 65, 70, 72, 73, 75, 76, 77, 102–103, 132, 133, 134, 135, 136, 137, 138, 139, 140, 141, 142, 143, 144–145, 146, 147, 449, 494, 506, R14, R15, R16, R17, R18, R19, R20, R21, R22<br>**Grade 4:** 17, 27, 45, 46, 47, 49, 50, 51, 52, 53, 55, 56–57, 60, 61, 62, 63, 70, 71, 72, 82, 83, 84, 85, 88, 89, 90, 91, 92, 93, 95, 98, 100, 101, 102, 103, 104, 105, 106–107, 110, 120, 121, 124–125, 164, 165, 166, 167, 168, 169, 171, 172–173, 174, 178, 179, 194, 195, 528, 529, R14, R15, R16, R17, R18, R19, R20, R21, R22 |
| The behavior of individual organisms is influenced by internal cues (such as hunger) and by external cues (such as a change in the environment). Humans and other organisms have senses that help them detect internal and external cues. | **Grade 1:** 17, 19<br>**Grade 2:** 42, 58, 72, 73, R9, AF4<br>**Grade 3:** 24, 25, 44, 49, R21<br>**Grade 4:** 99, 100, 101, 105, 120, 478–479 |
| **Life Cycles of Organisms** | |
| Plants and animals have life cycles that include being born, developing into adults, reproducing, and eventually dying. The details of this life cycle are different for different organisms. | **Grade K:** 40, 41, 42, 43, 44, 47, 104, 105, 106, 107, 108, 109, 110, 111, 119<br>**Grade 1:** 54, 57, 60, 61, 62, 63, 78, 79, 91, 110, 111, 112, 113, 114, 115, 116, 117, 123<br>**Grade 2:** 8, 9, 25, 30, 31, 34, 35, 48, 49, 50, 62, 63, 64, 65, 67, 78, 80, 81, AF10, AF13<br>**Grade 3:** 17, 23, 27, 64, 68, 70, 71, 72, 73, 74, 75, 76, 77, 80, 81, 82, 83, 84, 85, 86, 87, 88, 98, 99<br>**Grade 4:** 47, 52, 53, 58, 59, 62, 63, 64, 65, 68, 69, 72, 94, 108, 109, 110, 111, 112, 113, 114, 115, 117, 118–119, 120, 158 |
| Plants and animals closely resemble their parents. | **Grade K:** 104, 106, 107, 108, 109, 115<br>**Grade 1:** 108, 109, 110, 111, 112, 113, 123<br>**Grade 2:** 24, 35, 40, 41, 43, 47, 61<br>**Grade 3:** 22, 84, 85, 86, 88, 90, 91, 92, 93<br>**Grade 4:** 66, 69, 112 |

| Life Cycles of Organisms continued | |
|---|---|
| Many characteristics of an organism are inherited from the parents of the organism, but other characteristics result from an individual's interactions with the environment. Inherited characteristics include the color of flowers and the number of limbs of an animal. Other features, such as the ability to ride a bicycle, are learned through interactions with the environment and cannot be passed on to the next generation. | **Grade K:** 109<br>**Grade 1:** 113<br>**Grade 2:** 42, 43, 47, 75, 78<br>**Grade 3:** 22, 23, 86, 91, 92, 93, 94, 95, 96–97, 98, 99<br>**Grade 4:** 66, 67, 69, 114, 116, 117, 478–479 |
| **Organisms and Their Environments** | |
| All animals depend on plants. Some animals eat plants for food. Other animals eat animals that eat the plants. | **Grade K:** 56, 57, 58, 59<br>**Grade 1:** 12, 13, 14, 104, 105, 142, 144, 145, 154, 155, 198<br>**Grade 2:** 27, 92, 94, 95, 96, 97, 98, 99, 100, 101, 116, 117, 136, 138<br>**Grade 3:** 106, 107, 108, 109, 110, 111, 112, 113, 114, 115, 116–117, 146, 147<br>**Grade 4:** 44, 54, 124–125, 131, 148, 149, 150, 151, 152, 153, 154, 155, 156, 157, 158, 159, 160, 161, 164, 198–199 |
| An organism's patterns of behavior are related to the nature of that organism's environment, including the kinds and numbers of other organisms present, the availability of food and resources, and the physical characteristics of the environment. When the environment changes, some plants and animals survive and reproduce, and others die or move to new locations. | **Grade K:** 89, 102, 103<br>**Grade 1:** 11, 12, 13, 14, 15, 66, 67, 68, 69, 70, 71, 97, 126, 127, 128, 130, 131, 132, 133, 134, 135, 136, 137, 138, 141, 144, 145, 150–153, 154, 155, 241, AF19<br>**Grade 2:** 42, 43, 90, 91, 92, 104, 105, 106, 108, 109, 110–111, 112–115, 116, 117, 123, 130, 131, 132, 133, 135, 136, 137, 138, 139, 142–145, 147, 150–151<br>**Grade 3:** 107, 108, 109, 115, 123, 125, 126, 128, 136, 137, 138, 139, 141, 142, 143, 146, 147, 152, 153, 164, 165, 166, 167, 168, 169, 170, 171, 174, 175, 179, 182, 183, 274–275<br>**Grade 4:** 23, 96, 98, 99, 118–119, 121, 124–125, 141, 142, 143, 144, 155, 158, 168, 170, 171, 175, 176, 177, 179, 180, 181, 185, 188, 189, 194, 195, 238 |
| All organisms cause changes in the environment where they live. Some of these changes are detrimental to the organism or other organisms, whereas others are beneficial. | **Grade K:** 124–125, 154, 155, 160, 161, 162, 163, 164, 165<br>**Grade 1:** 143, 146, 147, 195, AF30<br>**Grade 2:** 105, 106, 107, 109, 126–127<br>**Grade 3:** 28–29, 114, 115, 128, 150, 151, 152, 153, 154, 155, 156, 157, 182, 183<br>**Grade 4:** 42–43, 124–125, 185, 191 |
| Humans depend on their natural and constructed environments. Humans change environments in ways that can be either beneficial or detrimental for themselves and other organisms. | **Grade K:** 124–125, 150, 152, 153, 154, 155, 156, 157, 158, 159, 160, 161, 162, 163, 164, 165<br>**Grade 1:** 64–65, 146, 147, 195, 204, 205, 246–247<br>**Grade 2:** 44, 105, 106, 107, 109, 126–127<br>**Grade 3:** 28–29, 154, 155, 156, 157, 158–159, 170–171, 182, 183, 184, 210–211, 218, 219<br>**Grade 4:** 54, 96, 118–119, 146–147, 159, 186, 187, 190, 191, 232, 233 |

# Correlation to National Standards

| Earth and Space Science | |
|---|---|
| **Properties of Earth Materials** | |
| Earth materials are solid rocks and soils, water, and the gases of the atmosphere. The varied materials have different physical and chemical properties, which make them useful in different ways, for example, as building materials, as sources of fuel, or for growing the plants we use as food. Earth materials provide many of the resources that humans use. | **Grade K:** 124E, 126, 127, 128, 129, 130, 131, 132, 133, 134, 135, 136, 137, 146, 147, 148, 149, 151, 166, 167<br>**Grade 1:** 64–65, 158–159, 162, 163, 164, 165, 166, 167, 168, 169, 170, 171, 172, 173, 174, 175, 176–177, 179, 185, 186–189, 190, 191, 196, 197, 198, 199, 200, 201, 202, 203, 204, 205, 206–207, 312–313, AF23<br>**Grade 2:** 155, 162, 163, 164, 165, 166, 167, 168, 169, 171, 172, 173, 178–181, 182, 183, 186, 187, 188, 189, 190, 191, 192, 193, 194, 195, 196, 197, 198, 199, 200, 201, 202, 203, 210–213, 214, 215, 244–247, 401<br>**Grade 3:** 190, 191, 192, 193, 194, 195, 196, 197, 199, 200–201, 213, 214, 215, 216, 219, 220, 221, 222, 223, 226, 227, 228, 229, 230, 231, 232, 233, 234, 235, 236, 238, 239, 240, 241, 242, 243, 244, 245, 251, 258, 259, 260, 261, 262, 263, 267, 268–269, 270, 271, 278, 279, 280, 281<br>**Grade 4:** 250, 251, 252, 253, 254, 255, 256, 257, 258, 259, 260–261, 262, 263, 264, 265, 266, 267, 268, 269, 270–271, 284, 285, 286, 287, 288, 289, 290, 291, 304, 305, 314, 315, 319, 428–429, 438–439 |
| Soils have properties of color and texture, capacity to retain water, and ability to support the growth of many kinds of plants, including those in our food supply. | **Grade K:** 126, 127, 128, 129, 130, 131<br>**Grade 1:** 174, 175, 190, 198, 199, AF27<br>**Grade 2:** 194, 195, 196, 197, 198, 199, 203, 210, 211, 214, 215, AF28<br>**Grade 3:** 238, 239, 240, 241, 242, 243, 244, 245, 246, 247<br>**Grade 4:** 262, 263, 264, 265, 266, 267, 268, 269, 270–271, 304, 305 |
| Fossils provide evidence about the plants and animals that lived long ago and the nature of the environment at that time. | **Grade 1:** 148–149<br>**Grade 2:** 108, 109, 110, 111, AF16<br>**Grade 3:** 172, 173, 174, 175, 176, 177, 178, 179, 180–181, 182, 183, 248, 249, 250, 251, 255, 270<br>**Grade 4:** 189, 255, 272, 273, 274, 275, 276, 277, 281, 282–283, 304 |

| Objects in the Sky | |
|---|---|
| The sun, moon, stars, clouds, birds, and airplanes all have properties, locations, and movements that can be observed and described. | **Grade K:** 190, 191, 192, 193, 194, 195, 196, 197, 203, 205, 256, 257, 258, 259, 261<br>**Grade 1:** 235, 237, 238, 239, 262, 264, 265, 266, 267, 268, 269, 270, 272, 273, 276, 278, 279, 280, 281, 286–289, 290, 291, AF38, AF40<br>**Grade 2:** 232, 233, 234, 236, 237, 238, 239, 248, 249, 253, 254–257, 262, 263, 265, 266, 267, 268, 269, 270, 271, 272, 273, 280–281, 286, 287, 374–375, AF38<br>**Grade 3:** 273, 290, 291, 292, 293, 316, 317, 318, 319–321, 322, 323, 326, 327, 328, 329, 330, 331, 332, 333, 334–335, 346, 347, 348, 349, 350, 351, 352–353, 354, 355, 370–371<br>**Grade 4:** 326, 328, 329, 331, 343, 358, 361, 363, 364, 365, 368, 369, 370, 371, 372, 373, 374, 375, 390–391, 392, 393, 394, 395, 396, 397, 398, 399, 400–401, 402 |
| The sun provides the light and heat necessary to maintain the temperature of the earth. | **Grade K:** 198, 199, 200, 201, 202, 203<br>**Grade 1:** 243, 245, 267, 268, 269, 271, 273, 286–289, 290, 291, AF40, AF58<br>**Grade 2:** 254–257, 262, 263, 265, 273, 400, 416, AF59<br>**Grade 3:** 294, 305, 316, 319, 320, 321, 322, 323, 480<br>**Grade 4:** 48, 50, 360, 366, 367, 398, 506 |
| **Changes in the Earth and Sky** | |
| The surface of the earth changes. Some changes are due to slow processes, such as erosion and weathering, and some changes are due to rapid processes, such as landslides, volcanic eruptions, and earthquakes. | **Grade K:** 138, 140, 141, 142, 143, 144, 148, 149<br>**Grade 1:** 179, 180, 181, 182, 183, 184, 191<br>**Grade 2:** 170, 171, 172, 173, 174, 175, 176–177, 178–181, 182, 183<br>**Grade 3:** 202, 203, 204, 205, 206, 207, 208, 209, 210–211, 212, 213, 214, 215, 216, 217, 219, 220, 221, 222, 223<br>**Grade 4:** 3–11, 184, 197, 198–199, 202, 204, 206, 210–211, 213, 214, 215, 216, 217, 218, 220, 221, 224, 225, 226, 227, 228, 229, 230, 231, 233, 234, 235, 236, 237, 238, 242, 243, 244–245, 246, 247 |

| Changes in the Earth and Sky continued | |
|---|---|
| Weather changes from day to day and over the seasons. Weather can be described by measurable quantities, such as temperature, wind direction and speed, and precipitation. | **Grade K:** 4–5, 6, 7, 168E, 168F, 168J, 168, 169, 170, 171, 172, 173, 174, 175, 176, 177, 178, 179, 180, 181, 182, 183, 184, 185, 186, 187, 188, 189, 204, 205, 206, 207<br>**Grade 1:** 228, 229, 230, 231, 232, 233, 235, 236, 237, 238, 239, 240, 241, 242, 243, 244, 246–247, 248, 249, 250, 251, 252, 253, 254, 255, 256–259, 260, 261, 275, AF32, AF34<br>**Grade 2:** 176–177, 220, 222, 223, 224, 225, 226, 227, 228, 229, 236, 237, 240, 241, 242–243, 248, 249, 258, 259, 260, 261, 262, 263, 264, 265, AF32<br>**Grade 3:** 274–275, 278, 280, 281, 282, 283, 284, 285, 286–287, 288, 289, 290, 291, 292, 293, 294, 295, 296, 297, 298, 299, 300–301, 303, 304, 305, 306, 307, 308, 309, 312, 313, 321, 324<br>**Grade 4:** 240, 241, 243, 312, 313, 316, 317, 318, 319, 320, 321, 325, 327, 328, 329, 330, 331, 334, 335, 336, 337, 338, 339, 340, 341, 342–343, 345, 346, 350, 351, 352–353, 354, 355 |
| Objects in the sky have patterns of movement. The sun, for example, appears to move across the sky in the same way every day, but its path changes slowly over the seasons. The moon moves across the sky on a daily basis much like the sun. The observable shape of the moon changes from day to day in a cycle that lasts about a month. | **Grade K:** 190, 192, 193, 194, 195, 196, 197, 199, 200, 201, 203, 205<br>**Grade 1:** 272, 273, 274, 275, 276, 279, 280, 281, 282, 283, 284–285, 286–289, 290, 291, AF38<br>**Grade 2:** 253, 254, 255, 256, 257, 262, 263, 268, 269, 270, 271, 273, 275, 276–279, 280–281, 282–285, 286, 287, AF34, AF38<br>**Grade 3:** 316, 317, 318, 319, 320, 321, 323, 324, 327, 328, 329, 330, 331, 332, 333, 334–335, 350, 354, 355<br>**Grade 4:** 358, 359, 360, 361, 362, 363, 364, 365, 372, 373, 374, 375, 376–377, 402, 403 |

## Science and Technology

| Abilities of Technological Design | |
|---|---|
| *Identify a simple problem.* In problem identification, children should develop the ability to explain a problem in their own words and identify a specific task and solution related to the problem. | **Grade 1:** 229, 249, 271, 403, 421<br>**Grade 2:** 377, 380, 405, 421<br>**Grade 3:** 144–145, 289, 443, 460–461, 463, 511<br>**Grade 4:** 106–107, 210–211, 335, 483, 503<br>See also *Technology: A Closer Look*, Grades K–2 and 3–4. |
| *Propose a solution.* Students should make proposals to build something or get something to work better; they should be able to describe and communicate their ideas. Students should recognize that designing a solution might have constraints, such as cost, materials, time, space, or safety. | **Grade K:** 247<br>**Grade 1:** 271, 373, 376, 405<br>**Grade 2:** 234, 367, 380, 405, 411<br>**Grade 3:** 203, 289, 443, 450–451, 463, 475, 486–487, 489, 511<br>**Grade 4:** 89, 102, 335, 483, 493, 503, 525<br>See also *Technology: A Closer Look*, Grades K–2 and 3–4. |

| **Abilities of Technological Design** continued | |
|---|---|
| *Implementing proposed solutions.* Children should develop abilities to work individually and collaboratively and to use suitable tools, techniques, and quantitative measurements when appropriate. Students should demonstrate the ability to balance simple constraints in problem solving. | **Grade K:** 246<br>**Grade 1:** 204, 249, 373, 376, 403, 405<br>**Grade 2:** 171, 267, 380, 421<br>**Grade 3:** 55, 293, 463, 469, 486–487<br>**Grade 4:** 102, 210–211, 483, 503, 522–523<br>See also *Technology: A Closer Look*, Grades K–2 and 3–4. |
| *Evaluate a product or design.* Students should evaluate their own results or solutions to problems, as well as those of other children, by considering how well a product or design met the challenge to solve a problem. When possible, students should use measurements and include constraints and other criteria in their evaluations. They should modify designs based on the results of evaluations. | **Grade 1:** 229, 249, 373, 379, 403<br>**Grade 2:** 129, 367, 411<br>**Grade 3:** 203, 283, 443, 463, 469, 486–487, 495, 511<br>**Grade 4:** 89, 106–107, 210–211, 483, 493, 503, 522–523<br>See also *Technology: A Closer Look*, Grades K–2 and 3–4. |
| *Communicate a problem, design, and solution.* Student abilities should include oral, written, and pictorial communication of the design process and product. The communication might be show and tell, group discussions, short written reports, or pictures, depending on the students' abilities and the design project. | **Grade 1:** 229, 249, 373, 379, 403<br>**Grade K:** 150<br>**Grade 1:** 95, 271<br>**Grade 2:** 95, 421<br>**Grade 3:** 289, 443, 460–461, 463<br>**Grade 4:** 106–107, 490–491, 522–523<br>See also *Technology: A Closer Look*, Grades K–2 and 3–4. |
| **Understanding About Science and Technology** | |
| People have always had questions about their world. Science is one way of answering questions and explaining the natural world. | **Grade K:** 4–5, 6, 7, 10–11, 12–13, 16–17, 18–19<br>**Grade 1:** 206–207, 284–285, 292<br>**Grade 2:** 110–111, 242–243, 276, 277, 280–281<br>**Grade 3:** 4, 130–131, 352–353, 370–371<br>**Grade 4:** 3–11, 38, 276–277, 282–283, 381, 382, 406–407, 438–439, 552 |
| People have always had problems and invented tools and techniques (ways of doing something) to solve problems. Trying to determine the effects of solutions helps people avoid some new problems. | **Grade K:** 150<br>**Grade 1:** 374–377, 378–379, 398, 399, 408–409, 414, 415, 421, 422<br>**Grade 2:** 176–177, 189, 242–243, 280–281, 354, 426–427<br>**Grade 3:** 28–29, 210–211, 254, 262, 263, 300–301, 342, 440–441, 471<br>**Grade 4:** 43, 218, 219, 279, 280, 288, 289, 302–303, 382, 383, 389, 390–391, 397, 477, 510–511, 586–587<br>See also *Technology: A Closer Look*, Grades K–2 and 3–4. |
| Scientists and engineers often work in teams with different individuals doing different things that contribute to the results. This understanding focuses primarily on teams working together and secondarily, on the combination of scientist and engineer teams. | **Grade 2:** 82, 110–111, 288<br>**Grade 3:** 3–11, 180–181, 187<br>**Grade 4:** 3–11, 119<br>See also *Technology: A Closer Look*, Grades K–2 and 3–4. |

| Understanding About Science and Technology continued | |
|---|---|
| Women and men of all ages, backgrounds, and groups engage in a variety of scientific and technological work. | **Grade K:** 116, 252<br>**Grade 1:** 80, 116–117, 148–149, 156, 176–177, 206–207, 222, 284–285, 292, 430<br>**Grade 2:** 66–67, 82, 110–111, 126–127, 148, 216, 288, 354, 374–375, 434<br>**Grade 3:** 96–97, 100, 130–131, 184, 272, 352–353, 356, 370–371, 426, 522<br>**Grade 4:** 118–119, 122, 196, 222–223, 306, 404, 438–439, 476, 590<br>See also *Technology: A Closer Look*, Grades K–2 and 3–4. |
| Tools help scientists make better observations, measurements, and equipment for investigations. They help scientists see, measure, and do things that they could not otherwise see, measure, and do. | **Grade K:** TR7, TR8<br>**Grade 1:** 266, 284–285, 292, R2, R3, R4, R5, R6, R7<br>**Grade 2:** 224, 225, 226, 227, 242–243, 374–375, R3, R4, R5, R6<br>**Grade 3:** 6, 7, 26, 209, 281, 282, 283, 284, 300–301, 342, 344–345, 356, 379, R2, R4, R5, R6, R7, R8, R9<br>**Grade 4:** 28, 218, 219, 276, 318, 319, 342–343, 381, 382, 383, 390–391, 392, 397, 531, R2, R3, R4, R5, R6, R7, R8, R9<br>See also *Technology: A Closer Look*, Grades K–2 and 3–4. |
| **Abilities to Distinguish Between Natural Objects and Objects Made by Humans** | |
| Some objects occur in nature; others have been designed and made by people to solve human problems and enhance the quality of life. | **Grade K:** 276–277<br>**Grade 1:** 195, 312–313, 325, 414<br>**Grade 2:** 188, 189, 190, 202, 203, 208–209, 216, 306–307<br>**Grade 3:** 210–211, 252, 253, 262, 263, 414–415<br>**Grade 4:** 54, 258, 406–407, 459, 463, 474, 476, 506<br>See also *Technology: A Closer Look*, Grades K–2 and 3–4. |
| Objects can be categorized into two groups, natural and designed. | **Grade K:** 276–277<br>**Grade 1:** 312–313, 414<br>**Grade 2:** 190<br>**Grade 3:** 409, 414–415<br>**Grade 4:** 54, 463, 476<br>See also *Technology: A Closer Look*, Grades K–2 and 3–4. |

## Science in Personal and Social Perspectives

| Personal Health | |
|---|---|
| Safety and security are basic needs of humans. Safety involves freedom from danger, risk, or injury. Security involves feelings of confidence and lack of anxiety and fear. Student understandings include following safety rules for home and school, preventing abuse and neglect, avoiding injury, knowing whom to ask for help, and when and how to say no. | **Grade K:** TR13, TR14<br>**Grade 1:** 16, 103, 127, 163, 403, R16, R17<br>**Grade 2:** R16, R17, R18<br>**Grade 3:** 14, 49, 299, 323, 517<br>**Grade 4:** 239, 340, 341, 398, 570, 571 |

| Personal Health continued | |
|---|---|
| Individuals have some responsibility for their own health. Students should engage in personal care—dental hygiene, cleanliness, and exercise—that will maintain and improve health. Understandings include how communicable diseases, such as colds, are transmitted and some of the body's defense mechanisms that prevent or overcome illness. | **Grade K:** TR9, TR10<br>**Grade 1:** R14<br>**Grade 2:** R9, R14, R15<br>**Grade 3:** 323, R23, R26<br>**Grade 4:** 319, 398, R23, R26 |
| Nutrition is essential to health. Students should understand how the body uses food and how various foods contribute to health. Recommendations for good nutrition include eating a variety of foods, eating less sugar, and eating less fat. | **Grade K:** TR11<br>**Grade 1:** R13<br>**Grade 2:** R12, R13<br>**Grade 3:** 27, R23, R24, R25, R26<br>**Grade 4:** R23, R24, R25, R26 |
| Different substances can damage the body and how it functions. Such substances include tobacco, alcohol, over-the-counter medicines, and illicit drugs. Students should understand that some substances, such as prescription drugs, can be beneficial, but that any substance can be harmful if used inappropriately. | **Grade 1:** R14, R15<br>**Grade 2:** R15<br>**Grade 3:** R23, R26<br>**Grade 4:** R26 |
| **Characteristics and Changes in Populations** | |
| Human populations include groups of individuals living in a particular location. One important characteristic of a human population is the population density—the number of individuals of a particular population that live in a given amount of space. | **Grade 4:** 145, 186 |
| The size of a human population can increase or decrease. Populations will increase unless other factors such as disease or famine decrease the population. | This standard is covered in the Macmillan/McGraw-Hill *Health & Wellness* program. |
| **Types of Resources** | |
| Resources are things that we get from the living and nonliving environment to meet the needs and wants of a population. | **Grade K:** 146, 152, 153, 154, 155, 156, 157, 158, 159, 160, 161, 162, 163, 164, 165<br>**Grade 1:** 176–177, 195, 196, 197, 198, 199, 200, 201, 202, 203, 216–219, 220, 221, 222<br>**Grade 2:** 162, 163, 164, 165, 167, 168, 184, 188, 189, 190, 191, 200, 201, 202, 203, 204, 206, 207, 210–213, 214, 215<br>**Grade 3:** 152, 153, 244, 245, 252, 253, 254, 255, 256–257, 258, 259, 260, 261, 262, 263, 264, 265, 267, 268–269, 270, 271, 518<br>**Grade 4:** 248, 258, 259, 278, 279, 280, 281, 290, 291, 292, 293, 304, 305<br>See also *Technology: A Closer Look*, Grades K–2 and 3–4. |
| Some resources are basic materials, such as air, water, and soil; some are produced from basic resources, such as food, fuel, and building materials; and some resources are nonmaterial, such as quiet places, beauty, security, and safety. | **Grade K:** 146, 152, 153, 154, 155, 156, 157, 158, 159, 160, 161, 162, 163, 164, 165, 276, 277<br>**Grade 1:** 64–65, 194, 195, 196, 197, 198, 199, 200, 201, 202, 203, 216–219, 220, 221, 222, 312–313<br>**Grade 2:** 162, 163, 164, 165, 168, 188, 189, 190, 191, 194–199, 200, 201, 202, 203, 204, 208–209, 210–213, 214, 401, 426<br>**Grade 3:** 26, 27, 244, 252, 253, 254, 255, 256–257, 259, 260, 261, 262, 263, 267, 270, 271, 518<br>**Grade 4:** 54, 278, 279, 280, 281, 290, 291, 294, 296, 304, 305<br>See also *Technology: A Closer Look,* Grades K–2 and 3–4. |

# Correlation to National Standards

| Types of Resources continued | |
|---|---|
| The supply of many resources is limited. If used, resources can be extended through recycling and decreased use. | **Grade K:** 154, 155, 158, 159, 160, 161, 162, 163, 164, 165<br>**Grade 1:** 192, 197, 201, 203, 205, 209, 210, 211, 212, 213, 214, 215, 216–219, 220, 221<br>**Grade 2:** 200, 201, 203, 204, 205, 206, 207, 210–213, 215<br>**Grade 3:** 156, 157, 158–159, 183, 253, 256–257, 264, 265, 266, 267, 270, 271, 415, 426, 518, 519<br>**Grade 4:** 187, 292, 293, 298, 299, 300, 301, 302–303, 304, 305, 416<br>See also *Technology: A Closer Look*, Grades K–2 and 3–4. |
| **Changes in Environments** | |
| Environments are the space, conditions, and factors that affect an individual's and a population's ability to survive and their quality of life. | **Grade K:** 87, 89, 124–125<br>**Grade 1:** 11, 66, 68, 69, 70, 71, 126, 128, 129, 130, 131, 132, 133, 134, 135, 136, 137, 141, 150–153<br>**Grade 2:** 90–93, 120, 121, 122–125, 128, 130–133, 134, 135, 136–139, 140, 141, 142–145, 146<br>**Grade 3:** 24, 25, 28–29, 108, 109, 118, 120, 121, 122, 123, 124, 125, 126, 128, 129, 146, 147, 184<br>**Grade 4:** 128, 129, 130, 131, 132, 133, 144, 145, 146–147, 159, 176, 177, 179, 180, 181, 192–193 |
| Changes in environments can be natural or influenced by humans. Some changes are good, some are bad, and some are neither good nor bad. Pollution is a change in the environment that can influence the health, survival, or activities of organisms, including humans. | **Grade K:** 138, 140, 141, 142, 143, 144, 148, 149, 154, 155, 162, 163, 164, 165<br>**Grade 1:** 131, 146, 147, 155, 184, 204, 205, 206–207, 209, 220<br>**Grade 2:** 44, 102, 104, 105, 106, 107, 108, 109, 112–115, 116, 117, 126–127, 176–177, 204, 205, 207, 215<br>**Grade 3:** 28–29, 142, 150, 151, 152, 153, 154, 155, 156, 157, 162, 163, 168, 169, 170, 174, 175, 182, 183, 218, 264, 265, 266, 267, 426<br>**Grade 4:** 42–43, 96, 118–119, 124–125, 146–147, 159, 183, 184, 185, 186, 187, 188, 189, 190, 191, 192–193, 194, 195, 232, 233, 244–245, 294, 295, 296, 297, 298, 301, 302–303 |
| Some environmental changes occur slowly, and others occur rapidly. Students should understand the different consequences of changing environments in small increments over long periods as compared with changing environments in large increments over short periods. | **Grade K:** 138, 140, 141, 142, 143, 144, 148, 149, 154, 155, 162, 163, 164, 165<br>**Grade 1:** 131, 146, 147, 155, 182<br>**Grade 2:** 104, 105, 108, 109, 112–115, 176–177, 178–181<br>**Grade 3:** 28–29, 114, 150, 151, 152, 153, 154, 155, 162, 163, 168, 169, 170, 174, 175, 179, 182, 183, 202–223<br>**Grade 4:** 42–43, 118–119, 124–125, 146–147, 190, 191, 192–193, 232, 233, 244–245 |

| Science and Technology in Local Challenges | |
|---|---|
| People continue inventing new ways of doing things, solving problems, and getting work done. New ideas and inventions often affect other people; sometimes the effects are good and sometimes they are bad. It is helpful to try to determine in advance how ideas and inventions will affect other people. | **Grade K:** 244, 245<br>**Grade 1:** 284–285, 346–347, 378–379, 398, 399, 408–409, 414–415, 422<br>**Grade 2:** 126–127, 148, 176–177, 189, 242–243, 374–375, 426–427<br>**Grade 3:** 210–211, 254, 256–257, 300–301, 342, 440–441, 508–509, 518<br>**Grade 4:** 218, 219, 279, 280, 288, 289, 290, 342–343, 375, 381, 382, 383, 389, 390–391, 397, 501, 510–511, 586–587<br>See also *Technology: A Closer Look*, Grades K–2 and 3–4. |
| Science and technology have greatly improved food quality and quantity, transportation, health, sanitation, and communication. These benefits of science and technology are not available to all of the people in the world. | **Grade K:** 150, 244, 245<br>**Grade 1:** 42–43, 64–65, 222, 312–313, 346–347, 378–379, 408–409, 422<br>**Grade 2:** 242–243, 280–281, 306–307, 336–337, 354, 426–427<br>**Grade 3:** 28–29, 300–301, 440–441, 508–509<br>**Grade 4:** 43, 54, 218, 219, 280, 342–343, 489, 500, 501, 506, 507, 508, 510–511, 513, 584, 585, 586–587<br>See also *Technology: A Closer Look*, Grades K–2 and 3–4. |

## History and Nature of Science

| Science as a Human Endeavor | |
|---|---|
| Science and technology have been practiced by people for a long time. | **Grade 1:** 284–285, 346–347, 378–379, 398, 399, 408–409, 414–415, 422<br>**Grade 2:** 126–127, 148, 176–177, 189, 242–243, 374–375, 426–427<br>**Grade 3:** 210–211, 254, 256–257, 300–301, 342, 440–441, 508–509, 518<br>**Grade 4:** 218, 219, 279, 280, 288, 289, 290, 342–343, 375, 381, 382, 383, 389, 390–391, 397, 501, 510–511, 586–587<br>See also *Technology: A Closer Look*, Grades K–2 and 3–4. |
| Men and women have made a variety of contributions throughout the history of science and technology. | **Grade K:** 252<br>**Grade 1:** 116–117, 148–149, 176–177, 206–207, 284–285, 378–379<br>**Grade 2:** 66–67, 110–111, 126–127, 280–281, 374–375<br>**Grade 3:** 96–97, 130–131, 180–181, 256–257, 333, 352–353, 355, 370–371, 440–441<br>**Grade 4:** 48, 118–119, 222–223, 375, 381, 382, 390–391, 434, 436, 438–439, 552, 586–587 |

| Science as a Human Endeavor continued | |
|---|---|
| Although men and women using scientific inquiry have learned much about the objects, events, and phenomena in nature, much more remains to be understood. Science will never be finished. | **Grade K:** 4–5, 6, 7, 10–11, 12–13, 16–17, 18–19<br>**Grade 1:** 148–149, 284–285<br>**Grade 2:** 4–9, 44–45, 279, 374–375<br>**Grade 3:** 4, 11, 96–97, 172, 177, 180–181, 342, 352–353, 355, 370–371<br>**Grade 4:** 10–11, 34–35, 40, 222–223, 279, 383, 389 |
| Many people choose science as a career and devote their entire lives to studying it. Many people derive great pleasure from doing science. | **Grade K:** 116<br>**Grade 1:** 80, 116–117, 148–149, 156, 176–177, 206–207, 222, 284–285, 292, 430<br>**Grade 2:** 66–67, 82, 110–111, 126–127, 148, 216, 288, 354, 374–375, 434<br>**Grade 3:** 96–97, 100, 130–131, 184, 272, 352–353, 356, 370–371, 426, 522<br>**Grade 4:** 118–119, 122, 196, 222–223, 306, 404, 438–439, 476, 590 |

# Teacher's Edition Credits

## Cover Photos

Natural Selection Stock Photography

## Illustrations

All children's realia are by Macmillan/McGraw-Hill, Jenny Vainisi, and Marta Pernas for MMH except as noted below:

**TR57** Peter Gunther; **TR58** Accurate Art, Inc.; **TR60** Greg Harris; **TR62** XNR Productions; **TR65** Kenneth Batelman; **TR66** Macmillan/The McGraw-Hill Companies; **TR67** Kenneth Batelman; **TR68** Barb Cousins; **TR69** (l) Barb Cousins, (r) Robert Roper; **TR70** Barb Cousins; **TR71** Kenneth Batelman; **TR72** (l) Barb Cousins, (r) Greg Harris; **TR74 TR75** Barb Cousins; **TR76** Jean Wisenbaugh; **TR77 TR79** Kenneth Batelman.

## Photography

**Tiii** (tl) Erik Stenbakken/stenbakken.com, (tc) Scott Stewart, (tr) Scott Schauer Photographs, Scottsdale, AZ, (cl) Dan Donovan, (c) Ken Cavanagh / Macmillan/The McGraw-Hill Companies, (cr) Nancy Brown, (bl) Deborah Attoinese/Overall Productions, (bc) Joshua P. Roberts; **Tvii** Stockbyte/Punchstock; **17A** Macmillan/The McGraw-Hill Companies ; **17B** (tl r) Frank Krahmer/Masterfile, (bl) Dwight Kuhn Photography; **20B** (tl bl) Ken Karp, (cl) Michael Scott/Macmillan/The McGraw-Hill Companies, (br) Ken Cavanagh/Macmillan/The McGraw-Hill Companies; **42** Ken Cavanagh/Macmillan/The McGraw-Hill Companies; **43** (l) Ray Grover/Alamy Images, (r) Alamy Images; **52B** (tl) Kevin Schafer/zefa/CORBIS, (tr) Ken Cavanagh/ Macmillan/The McGraw-Hill Companies, (cl) Nancy Ney/ Digital Vision/Getty Images, (cr) Liquidlibrary/Jupiterimages, (b)Ken Karp; **58** Ken Cavanagh/Macmillan/The McGraw-Hill Companies; **63** Liquidlibrary/Jupiterimages; **65** (l) Image Source/PunchStock, (r) L. Hobbs/PhotoLink/Getty Images; **79** PhotoLink/Getty Images; **80** Jacques Cornell/ Macmillan/The McGraw-Hill Companies; **83B** (tl r) Victoria McCormick/Animals Animals/Earth Scenes, (bl) David Pike/naturepl.com; **86B** (tl) NHPA/Stephen Krasemen, (tr) Russell Illig/Getty Images, **(cl)** Gary Meszaros/Visuals Unlimited, (bl)Ken Karp; **92** Russell Illig/ Getty Images; **116** Jacques Cornell/Macmillan/The McGraw-Hill Companies; **118B** Ken Karp ;**139** Digital Vision/Getty Images; **145** (t) Digital Vision, (b) PhotoLink/Photodisc/Getty Images; **146** Jacques Cornell/Macmillan/The McGraw-Hill Companies; **149A** Macmillan/The McGraw-Hill Companies; **149B** (tl) Ian Grant/Alamy Images, (bl) Digital Vision/Getty Images, (r) Nicholas Pitt/Alamy Images; **152B** Eastcott Momatiuk/The Image Bank/Getty Images; **161** Punchstock; **175** USDOC (U.S. Department of Commerce); **181** Getty Images; **184B** (t) Michael Scott/ Macmillan/The McGraw-Hill Companies; (c b)Ken Karp; **191** Ken Cavanagh/Macmillan/The McGraw-Hill Companies; **217A** Macmillan/The McGraw-Hill Companies; **217B** (tl) CORBIS, (bl) Brand X Pictures/Punchstock, (r)Galen Rowell/ CORBIS; **220B** (tl)Stockbyte/Getty Images, (cl)Ken Karp, (tr b) Macmillan/McGraw-Hill Companies; **235** Russell Illig/Getty Images; **241** CORBIS; **289B** (tl) Andre Jenny/Alamy Images, (bl) David Chasey/Getty Images, (r) Matthias Kulka/CORBIS; **292B** (tl) Macmillan/The McGraw-Hill Companies, (cl bl) Ken Karp, (br) Ken Cavanagh/Macmillan/The McGraw-Hill Companies; **312** Ken Cavanagh/Macmillan/The McGraw-Hill Companies; **322B** (l)Ken Karp , (r) Thinkstock/Alamy Images ; **334** (l) Medioimages/PunchStock, (c) Digital Vision/PunchStock, (r) Thinkstock/Alamy Images; **345** PhotoAlto/PunchStock; **355A** Macmillan/The McGraw-Hill Companies; **355B** (tl) Jean Brooks/Getty Images, (bl) Kelly-Mooney Photography/CORBIS, (r) Lester Leftkowitz/Getty Images; **358B** (tl bl) Michael Scott/ Macmillan/The McGraw-Hill Companies, (cl)Ken Karp, (tcr bcr) Macmillan/McGraw-Hill Companies, (br) Jacques Cornell/ Macmillan/The McGraw-Hill Companies; **396B** (tl bcl bl)Ken Karp/ Macmillan/The McGraw-Hill Companies, (tcl) Michael Scott/Macmillan/The McGraw-Hill Companies; **411** Photodisc Collection/Getty Images; **TR1** Ingram Publishing/Alamy Images; **TR2** CORBIS; **TR52** Ken Cavanagh/Macmillan/The McGraw-Hill Companies; **TR56** Cathlyn Melloan/Stone/Getty Images; **TR57** (tl) BOB GIBBONS/SCIENCE PHOTO LIBRARY, (tr) Andrew Park/Oxford Scientific/Jupiter Images; **TR58** Digital Vision/PunchStock; **TR59** (l) LESZCZYNSKI, ZIGMUND/Animals Animals-Earth Scenes, (r) Stockbyte; **TR61** (cl) Digital Vision/ Getty Images, (bl br) GeoEye/Photo Researchers, Inc.; **TR63** (t) blickwinkel/Alamy Images, (b) Getty Images; **TR64** The New Jersey Geological Survey has generously given permission to use the following map: Geologic Map of New Jersey (New Jersey Geological Survey, scale 1 to 1,000,000 2005).; **TR66** (b) PhotoLink/Getty Images, (1) Dr. Parvinder Sethi/Macmillan/The McGraw-Hill Companies, (2) Scientifica/Visuals Unlimited, (3) Albert J. Copley/Visuals Unlimited, (4 6) Mark A. Schneider/ Visuals Unlimited, (5) Harry Taylor/Dorling Kindersley, (7) Colin Keates © Dorling Kindersley, courtesy of the Natural History Museum, London, (8) Marli Miller/Visuals Unlimited, (9) Charles D. Winters/Photo Researchers, Inc, (10) PHOTOTAKE Inc./Alamy Images; **TR67** Dr. Parvinder Sethi/Macmillan/The McGraw-Hill Companies; **TR71** U.S. Geological Survey; **TR73** (l) StockTrek/ Getty Images, (r) Photodisc/Getty Images; **TR74** PhotoLink/ Getty Images; **TR75** Stephen Frisch/Macmillan/The McGraw-Hill Companies; **TR78** Kul Bhatia/CORBIS.

# Teacher's Notes

## Teacher's Notes

# Teacher's Notes

# Teacher's Notes

# Teacher's Notes